The Student Edition of MATLAB®

The Ultimate Computing Environment for Technical Education

User's Guide

The MATLAB® Curriculum Series

PRENTICE HALL, Englewood Cliffs, NJ 07632

The MATH WORKS Inc.

Library of Congress Cataloging-in-Publication Data

The student edition of MATLAB : version 4 : user's guide / the MathWorks Inc. ; with tutorial by Duane Hanselman and Bruce Littlefield.
p. cm. -- (The MATLAB curriculum series)
Includes index.
ISBN 0-13-184979-4 (pbk.)
1. MATLAB. 2. Numerical analysis--Data processing.
I. MathWorks, Inc. II. Series.
QA297.S8427 1995
519.4'0285'5369--dc20 94-43322
CIP

Acquisitions Editors: Don Fowley, Marcia Horton
Production Editor: Joe Scordato
Cover Designer: The MathWorks, Inc.
Buyer: Dave Dickey
Editorial Assistant : Naomi Goldman

Published by Prentice-Hall, Inc.
A Simon & Schuster Company
Englewood Cliffs, New Jersey 07632

Printed in the United States of America

10 9 8 7 6 5 4 3 2

ISBN 0-13-184979-4

Prentice-Hall International (UK) Limited, London
Prentice-Hall of Australia Pty. Limited, Sydney
Prentice-Hall Canada Inc., Toronto
Prentice-Hall Hispanoamericana, S.A., Mexico
Prentice-Hall of India Private Limited, New Delhi
Prentice-Hall of Japan, Inc., Tokyo
Simon & Schuster Asia Pte. Ltd., Singapore
Editora Prentice-Hall do Brasil, Ltda., Rio de Janeiro

The MathWorks, Inc.
24 Prime Park Way
Natick, Massachusetts 01760

Contents

Preface to the Instructor

For hundreds of thousands of industrial, government, and academic users spanning a broad range of engineering, scientific, and other applications, MATLAB® has become the premier technical computing environment. Now students can affordably use this powerful numeric computation, data analysis, and visualization software in their undergraduate and graduate studies, while getting acquainted with a tool that will prove invaluable throughout their careers.

As computers have become indispensable for creative work in science and engineering, academic institutions are increasingly aware of the importance of computer use and promoting software "literacy." However, the high cost of commercial-quality software and the challenge of integrating it into the curriculum have made it difficult to turn that awareness into positive results. In response, we created *The Student Edition of MATLAB* and this *User's Guide* so students can be introduced to this powerful tool early in their academic careers. *The Student Edition of MATLAB* encapsulates algorithmic mathematics in a form that can be easily applied to a wide range of disciplines, in courses such as Digital Signal Processing, Control Theory, Linear Algebra, Signals and Systems, Numerical Methods, Applied Mathematics, and Advanced Engineering Mathematics. By itself or when coupled with the other texts, MATLAB can be incorporated effectively into the curriculum to enhance the understanding of both fundamental and advanced topics, while enabling the student actively to put theory into practice.

We are pleased and somewhat surprised to see how quickly this movement is already happening. As we visit many colleges and universities from Stanford to MIT and travel on to the leading institutions throughout Europe and the Pacific Rim, we are consistently rewarded by the sight of students using MATLAB, not simply to get the answers, but rather to understand how to get the answers. It is happening across departments at schools, large and small, around the world. We sincerely hope that you enjoy taking part.

Technical Support for Instructors

The MathWorks provides technical support to registered instructors who have adopted *The Student Edition of MATLAB* for use in their courses. For technical support questions, instructors may direct inquiries:

- via e-mail: `support@mathworks.com`
- via telephone: (508) 653-2452 ext. 300
- via fax: (508) 653-2997

New Tutorial Sections

To make this *User's Guide* a "user friendly" vehicle for the beginning-level student's first exposure to MATLAB, we have added new tutorial sections, Chapters 5, 6, and 7, authored by Duane Hanselman and Bruce Littlefield of University of Maine. These sections have been carefully developed to reach the first-time user of MATLAB who has had limited exposure to advanced mathematics.

Other Information Sources for Instructors and Students

- Use the MATLAB online help facility by typing `help <command>` at the MATLAB prompt (where `<command>` is the name of the function about which you would like information). Even better, access help from the menu.
- Students and instructors with access to Usenet newsgroups can participate in the MATLAB newsgroup, `comp.soft-sys.matlab`. Here, an active community of MATLAB users—spanning industries, countries, applications, and schools—exchanges ideas, helps with each other's questions and problems, shares user-written functions and tools, and generally "talks" MATLAB. Members of the MathWorks staff also participate, and the newsgroup has become a stimulating, open, and free-flowing forum that embodies the spirit of MATLAB.

- On the World Wide Web (WWW), use Mosaic or another browser to reach the MathWorks Home Page using the URL `http://www.mathworks.com`.
- The MathWorks maintains an electronic archive of user-contributed routines, product information, and other useful things. It can be reached using anonymous ftp to `ftp.mathworks.com`, or from the MATLAB Forum in the MathWorks Home Page on the WWW.
- The quarterly MathWorks newsletter *MATLAB News & Notes* provides information on new products, technical notes and tips, application articles, a calendar of trade shows and conferences, and other useful information. *MATLAB News & Notes* is free for registered users of *The Student Edition of MATLAB.*

MATLAB-Based Books

A wide selection of texts may be used with *The Student Edition of MATLAB*, many featuring MATLAB-based exercises, problem sets, and supplemental M-files. These include standard texts or supplemental workbooks in a broad range of courses, such as Digital Signal Processing, Control Theory, Linear Algebra, Signals and Systems, Linear Systems, Numerical Methods, Applied Mathematics, Advanced Engineering Mathematics, Probability, and Statistics, and Calculus.

For a current list of MATLAB-based books, consult the MathWorks Home Page on the WWW at `http://www.mathworks.com` in the MATLAB Forum or the MathWorks anonymous ftp server at `ftp.mathworks.com` in `pub/books/booklist/`. Or contact your MathWorks educational account representative at (508) 653-1415 (e-mail: `info@mathworks.com`).

Acknowledgments

The Student Edition of MATLAB is the product of a collaborative effort between The MathWorks and Prentice Hall. Many people contributed to the development of *The Student Edition of MATLAB.* At The MathWorks, we especially want to acknowledge Cleve Moler, Julie Orofino, Liz Callanan, Jim Tung, Jim Boyles, Jason Kinchen, Josh Tillotson, Helen Paret, Bryan Shumway, Dan Checkoway, Carmela Storm, Donna Sullivan, and Steve Weingart. At Prentice

Hall, there have been contributions from Marcia Horton, Don Fowley, Joe Scordato, Kathleen Schiaparelli, Dave Dickey, Gary June, Meghan Dacey, Tom McElwee, Jennifer Klein, Amy Folsom, Linda Ratts Engelman, Frank Nicolazzo, and Dave Ostrow.

We would also like to thank the countless individuals who have provided feedback on the first release of *The Student Edition of MATLAB*. The shape of Version 4 has, in many respects, been influenced by such feedback.

Part One

Getting Started

1
To the Student

The Student Edition of MATLAB brings us back to our roots. The very first version of MATLAB, written at the University of New Mexico and Stanford University in the late 1970s, was intended for use in courses in matrix theory, linear algebra, and numerical analysis. We had been involved in the development of LINPACK and EISPACK, which were FORTRAN subroutine packages for matrix manipulation, and we wanted our students to be able to use these packages without writing FORTRAN programs.

Today, MATLAB's capabilities extend far beyond the original "Matrix Laboratory." MATLAB is an interactive system and programming language for general scientific and technical computation. Its basic data element is a matrix that does not require dimensioning. This allows solution of many numeric problems in a fraction of the time it would take to write a program in a language such as FORTRAN, Basic, or C. Furthermore, problem solutions are expressed in MATLAB almost exactly as they are written mathematically.

We've continued to maintain close ties to the academic community and have offered academic discounts and classroom licensing arrangements. We have been pleased with the popularity of MATLAB in the computer labs on campus. *The Student Edition of MATLAB* makes it practical for students to use MATLAB on their own personal computers in their homes, dorms, or wherever they study.

Mathematics is the common language of much of science and engineering. Matrices, differential equations, arrays of data, plots, and graphs are the basic building blocks of both applied mathematics and MATLAB. It is the underlying mathematical base that makes MATLAB accessible and powerful. One professor, who is a big MATLAB fan, told us, "The reason why MATLAB is so useful for signal processing is that it wasn't designed specifically for signal processing, but for mathematics." MATLAB has been used in many different fields:

- A physics grad student analyzing and visualizing data from her experiments with magnetic fields of superconductors.
- An internationally known amusement park modeling the control systems for its water rides.
- A large food company analyzing how microwave ovens cook pizzas.
- A cable television company investigating encoding and compression schemes for digital TV.
- A sports equipment manufacturer modeling golf swings.
- A third grader learning her multiplication tables.

In all these cases, and thousands more, MATLAB's mathematical foundation made it useful in places and applications far beyond those we contemplated originally.

1.1 About the Cover

The cover of this guide depicts a solution to a problem that has played a small, but interesting, role in the history of numerical methods during the last 30 years. The problem involves finding the modes of vibration of a membrane supported by an L-shaped domain consisting of three unit squares. The nonconvex corner in the domain generates singularities in the solutions, thereby providing challenges for both the underlying mathematical theory and the computational algorithms. There are important applications, including wave guides, structures, and semiconductors.

Two of the founders of modern numerical analysis, George Forsythe and J. H. Wilkinson, worked on the problem in the 1950s. (See G. E. Forsythe and W. R. Wasow, *Finite-Difference Methods for Partial Differential Equations,* Wiley, 1960.) One of the authors of this guide (Moler) used finite difference techniques to compute solutions in 1965. Typical computer runs took up to half an hour of dedicated computer time on what were then Stanford University's primary computers, an IBM 7090 and a Burroughs B5000. The first version of the approach we now use was published in 1967 by L. Fox, P. Henrici, and C. Moler (*SIAM J. Numer. Anal.* 4, 1967, pp. 89–102). It replaced finite differences by combinations of distinguished fundamental solutions to the underlying differential equation formed from Bessel and trigonometric functions. The idea is a generalization of the fact that the real and imaginary parts of complex analytic functions are solutions to Laplace's equation. In the early 1970s, new matrix algorithms, particularly Gene Golub's orthogonalization techniques for least squares problems, provided further algorithmic improvements. Today, MATLAB allows you to express the entire algorithm in a few dozen lines, to compute the solution with great accuracy in a few minutes on a computer at home, and to manipulate readily color three-dimensional displays of the results. We have included our MATLAB program, `membrane.m`, with the M-files supplied with this *Student Edition of MATLAB*.

1.2 The Student Edition of MATLAB vs. Professional MATLAB

The *Student Edition* is available for Microsoft Windows compatible personal computers and Macintoshes. It is identical to the professional version 4.2 of MATLAB except for the following:

- A math coprocessor is strongly recommended but not required.
- Matrix size capability is limited to 8192 elements, with either the number of rows or columns limited to 32.
- Prints to Windows, Macintosh, and PostScript printing devices only.
- Available in single-user licenses only (no networking).
- Cannot dynamically link C or FORTRAN subroutines (MEX-files).
- The only toolboxes it can be used with are the included *Symbolic Math Toolbox* and *Signals and Systems Toolbox*.
- Provides a Student User Upgrade Discount for purchase of the professional versions.

1.3 How to Upgrade to Professional MATLAB

You can obtain a professional version of MATLAB for Microsoft Windows and Macintosh personal computers; UNIX workstations from Sun, Hewlett-Packard, IBM, Silicon Graphics, and Digital; and VMS computers. For product information or to place an order, call or write to your educational account representative at The MathWorks at:

The MathWorks, Inc.
University Sales Department
24 Prime Park Way

Natick, Massachusetts 01760-1500
Phone: (508) 653-1415
Fax: (508) 653-2997
E-mail: `info@mathworks.com`

1.4 Support, Registration, and Warranty

1.4.1 Student Support Policy

Neither Prentice Hall, Inc. nor The MathWorks, Inc. provides technical support to student users of *The Student Edition of MATLAB.* If you encounter difficulty while using the Student Edition software:

1. Read the relevant section of this *User's Guide* containing tutorial and/or reference information on the commands or procedures you are trying to execute.
2. Use the software's online help facility by typing `help <command>` at the MATLAB prompt where `<command>` is the name of the function you are executing.
3. Write down the sequence of procedures you were executing so that you can explain to your instructor the nature of the problem. Be certain to note the exact error message you encountered.
4. If you have consulted this *User's Guide* and the online help and are still stymied, you may post your question to the `comp.soft-sys.matlab` newsgroup, if you have access to Usenet newsgroups. Many active MATLAB users participate in the newsgroup, and they are a good resource for answers or tips about MATLAB usage.

1.4.2 Student User Registration

Students who have purchased the software package will find a card in the package for registering as a user of *The Student Edition of MATLAB.* Take a moment now to complete and return this card to us. Registered student users:

- Are entitled to replace defective disks at no charge.
- Qualify for a discount on upgrades to professional versions of MATLAB.
- Become active members of the worldwide MATLAB user community.

It is VERY important that you return this card. Otherwise you will not be on our mailing list and you will not qualify for student upgrades and other promotions.

1.4.3 Defective Disk Replacement

Contact Prentice Hall at (201) 592-3096 for disk replacement. You must send us your damaged or defective disk, and we will provide you with a new one.

1.4.4 Limited Warranty

No warranties, express or implied, are made by The MathWorks, Inc. that the program or documentation is free of error. Further, The MathWorks, Inc. does not warrant the program for correctness, accuracy, or fitness for a task. *You rely on the results of the program solely at your own risk.* The program should not be relied on as the sole basis to solve a problem whose incorrect solution could result in injury to person or property. If the program is employed in such a manner, it is at the user's own risk, and The MathWorks, Inc. disclaims all liability for such misuse. Neither The MathWorks, Inc. nor anyone else who has been involved in the creation, production, or delivery of this program shall be liable for any direct or indirect damages.

1.5 Toolboxes

MATLAB is both an environment and a programming language, and one of its great strengths is the fact that the MATLAB language allows you to build your own reusable tools. You can easily create your own special functions and programs (known as M-files) in MATLAB code. As you write more and more MATLAB functions to deal with certain problems, you might be tempted to group related functions together into special directories for convenience. This leads directly to the concept of a *Toolbox*: a specialized collection of M-files for working on particular classes of problems.

Toolboxes are more than just collections of useful functions; they represent the efforts of some of the world's top researchers in fields such as controls, signal processing, system identification, and others. Because of this, the MATLAB Application Toolboxes let you "stand on the shoulders" of world class scientists.

The Student Edition of MATLAB contains two toolboxes, bundled free with the software: the *Signals and Systems Toolbox* and the *Symbolic Math Toolbox*. These toolboxes are educational versions of the commercially available *Signal Processing Toolbox, Control System Toolbox*, and *Symbolic Math Toolbox*.

All toolboxes are built directly on top of MATLAB. This has some very important implications for you:

- Every toolbox builds on the robust numerics, rock-solid accuracy, and years of experience in MATLAB.
- You get seamless and immediate integration with SIMULINK® and any other toolboxes you may own.
- Since all toolboxes are written in MATLAB code, you can take advantage of MATLAB's open-system approach. You can inspect M-files, add to them, or use them for templates when you're creating your own functions.
- Every toolbox is available on any computer platform that runs MATLAB.

Here is a list of professional toolboxes currently available from The MathWorks. This list is by no means static—more are being created every year.

Control System Toolbox The *Control System Toolbox*, the foundation of the MATLAB control design toolbox family, contains functions for modeling, analyzing, and designing automatic control systems. The application of automatic control grows with each year as sensors and computers get cheaper. As a result, automatic controllers are used not only in highly technical settings for automotive and aerospace systems, computer peripherals, and process control, but also in less obvious applications such as washing machines and cameras.

Frequency-Domain System Identification Toolbox The *Frequency-Domain System Identification Toolbox* by István Kollár in cooperation with Johan Schoukens and researchers at the Vrije Universiteit in Brussels, is a set of M-files for modeling linear systems based on measurements of the system's frequency response.

Fuzzy Logic Toolbox The *Fuzzy Logic Toolbox* provides a complete set of GUI-based tools for designing, simulating, and analyzing fuzzy inference systems. Fuzzy logic provides an easily understandable, yet powerful way to map an input space to an output space with arbitrary complexity, with rules and relationships specified in natural language. Systems can be simulated in MATLAB or incorporated into a SIMULINK block diagram, with the ability to generate code for stand-alone execution.

Higher-Order Spectral Analysis Toolbox The *Higher-Order Spectral Analysis Toolbox*, by Jerry M. Mendel, C. L. (Max) Nikias, and Ananthram Swami, provides tools for signal processing using higher-order spectra. These methods are particularly useful for analyzing signals originating from a nonlinear process or corrupted by non-Gaussian noise.

Image Processing Toolbox The *Image Processing Toolbox* contains tools for image processing and algorithm development. It includes tools for filter design and image restoration; image enhancement; analysis and statistics; color, geometric, and morphological operations; and 2-D transforms.

Model Predictive Control Toolbox The *Model Predictive Control Toolbox* was written by Manfred Morari and N. Lawrence Ricker. Model predictive control is especially useful for control applications with many input and output variables, many of which have constraints. As a result, it has become particularly popular in chemical engineering and other process control applications.

Mu-Analysis and Synthesis Toolbox The *Mu-Analysis and Synthesis Toolbox*, by Gary Balas, Andy Packard, John Doyle, Keith Glover, and Roy Smith, contains specialized tools for H_∞ optimal control and μ-analysis and synthesis, an approach to advanced robust control design of multivariable linear systems.

NAG Foundation Toolbox The *NAG Foundation Toolbox* includes over 200 numeric computation functions from the well-regarded NAG Fortran subroutine libraries. It adds to MATLAB specialized tools for boundary-value problems, optimization, adaptive quadrature, surface and curve-fitting, and other applications.

Neural Network Toolbox The *Neural Network Toolbox* by Howard Demuth and Mark Beale is a collection of MATLAB functions for designing and simulating neural networks. Neural networks are computing architectures, inspired by biological nervous systems, that are useful in applications where formal analysis is extremely difficult or impossible, such as pattern recognition and nonlinear system identification and control.

Nonlinear Control Design Toolbox The *Nonlinear Control Design Toolbox* is an add-on to SIMULINK. It allows the design of linear and nonlinear control systems, using a time-domain based optimization technique. It includes a graphical user interface for drawing performance constraints and tuning controller parameters.

Optimization Toolbox The *Optimization Toolbox* contains commands for the optimization of general linear and nonlinear functions, including those with constraints. An optimization problem can be visualized as trying to find the lowest (or highest) point in a complex, highly contoured landscape. An optimization algorithm can thus be likened to an explorer wandering through valleys and across plains in search of the topological extremes.

Quantitative Feedback Theory Toolbox The *Quantitative Feedback Theory Toolbox* by Yossi Chait, Craig Borghesani, and Oded Yaniv implements

QFT, a frequency-domain approach to controller design for uncertain systems, that provides direct insight into the trade-offs between controller complexity (hence the ability to implement it) and meeting specifications.

Robust Control Toolbox The *Robust Control Toolbox* provides a specialized set of tools for the analysis and synthesis of control systems that are "robust" with respect to uncertainties that can arise in the real world. The *Robust Control Toolbox* was created by controls theorists Richard Y. Chiang and Michael G. Safonov.

Signal Processing Toolbox The *Signal Processing Toolbox* contains tools for signal processing. Applications include audio (e.g., compact disc and digital audio tape), video (digital HDTV, image processing, and compression), telecommunications (fax and voice telephone), medicine (CAT scan, magnetic resonance imaging), geophysics, and econometrics.

SIMULINK *SIMULINK* is an extension to MATLAB that adds a graphical environment for modeling, simulating, and analyzing dynamic linear and nonlinear systems. It supports continuous, discrete-time, multirate, and hybrid systems. *The Student Edition of SIMULINK* is now available for use with *The Student Edition of MATLAB.* See your local college bookstore for availability.

SIMULINK Real-Time Workshop The *SIMULINK Real-Time Workshop* is an add-on to SIMULINK for rapid prototyping, real-time simulation, and implementing control strategies on real-time hardware. It provides code generation of SIMULINK block diagrams and templates for real-time operating systems and various hardware environments.

Spline Toolbox The *Spline Toolbox* by Carl de Boor, a pioneer in the field of splines, provides a set of M-files for constructing and using splines, which are piecewise polynomial approximations. Splines are useful because they can approximate other functions without the nasty side-effects that result from other kinds of approximations, such as piecewise linear curves.

Statistics Toolbox The *Statistics Toolbox* provides a set of M-files for statistical data analysis, modeling, and Monte Carlo simulation, with GUI-based tools for exploring fundamental concepts in statistics and probability.

Symbolic Math Toolbox The *Symbolic Math Toolbox* gives MATLAB an integrated set of tools for symbolic computation and variable-precision arithmetic, based on Maple V®. The *Extended Symbolic Math Toolbox* adds support for Maple programming plus additional specialized functions.

System Identification Toolbox The *System Identification Toolbox*, written by Lennart Ljung, is a collection of tools for estimation and identification. System identification is a way to find a mathematical model for a physical system (like an electric motor, or even a financial market) based only on a record of the system's inputs and outputs.

2 MATLAB for Microsoft Windows

2.1 System Requirements

- An IBM, Compaq, or 100 percent compatible system with an Intel (or compatible) 386, 486, or Pentium processor
- Microsoft Windows 3.1
- High density 3.5" floppy disk drive
- Microsoft Windows-supported mouse and monitor
- 15 MB of free space on your hard drive
- 8 MB of extended memory

The following items are strongly recommended:

- Additional memory
- 8-bit graphics adapter and display (for 256 simultaneous colors)
- Microsoft Windows-supported graphics accelerator card
- Microsoft Windows-supported printer
- Microsoft Windows-supported sound card

2.2 The MATLAB Notebook for Windows

The *MATLAB Notebook* is a dynamic integration of MATLAB and Microsoft Word 6.0, which allows you to construct interactive word processing documents containing live MATLAB commands and graphics.

Using the Notebook, you can create an *M-book*, a standard Microsoft Word document that contains not only text, but also MATLAB commands and the output from those commands. You can think of an M-book as MATLAB running within Word, with a full set of powerful word processing capabilities available to annotate and document your MATLAB work. If you are a Microsoft Word user, you will find the Notebook to be a natural environment that combines a familiar and powerful word processor with the computational and graphing capabilities of MATLAB. The MATLAB Notebook benefits any MATLAB user who needs to create interactive, technical documents, such as:

- Electronic textbooks
- Electronic handbooks
- Interactive MATLAB sessions
- Technical reports and project design logbooks
- Interactive product specifications
- Electronic class notes and homework

The enabling technology of the MATLAB Notebook is *Dynamic Data Exchange (DDE)*, which provides the communication between MATLAB and Word. With a single keystroke, MATLAB commands entered in the Notebook are formatted and sent via DDE to MATLAB for evaluation. The result of the command is returned to the Notebook and inserted in the Word document. The Notebook handles both text and graphics output from MATLAB.

To use the Notebook, first make sure that you have installed Word for Windows version 6.0, then follow the installation instructions below. After you have installed the Notebook, double-click on the M-book titled `MATLAB Notebook README` in the Student Edition Program Group for more information. Other example M-books are included with *The Student Edition of MATLAB*; they are described in the README M-book.

2.3 Installation

Please be sure that Windows is installed and running properly in 386-enhanced mode before attempting to install MATLAB. You can check this by selecting **About Program Manager...** from the **Help** menu of the Windows Program Manager. It should say "386-Enhanced Mode." If it does not, please refer to your Windows documentation for instructions on configuring 386-enhanced mode.

To install MATLAB, follow these steps:

1. Insert the MATLAB distribution disk into your floppy disk drive.
2. Start Windows.
3. When the Windows Program Manager is displayed, select the **Run...** option from the **File** menu.
4. At the **Command** line prompt, enter `a:setup` (or `b:setup`), and click on the **OK** button.
5. At the **Directory** prompt, enter the name of the directory into which you want to install MATLAB, and click on the **OK** button. The default is `C:\MATLAB`, but you can specify any other drive and/or directory.
6. At the prompt, type in your name and your school and click on the **Continue** button.
7. The **Insert Next Disk** sign will prompt you for each disk in succession. When you see the prompt, remove the floppy disk in the drive, insert the indicated disk, and click the **OK** button. Continue until all four disks are installed. To finish the installation procedure, click on the **OK** button. The setup program will create a new Program Group in the Windows Program Manager containing the MATLAB application and MATLAB help icons.
8. When the MATLAB installation is finished, you may choose to install the MATLAB Notebook for Word. *Proceed with these steps only if you have already installed Microsoft Word version 6.0.* Insert the MATLAB Notebook diskette into your floppy disk drive and repeat steps 3, 4, and 5.
9. A dialog box titled **Installation Location** will appear. Use it to indicate where Microsoft Word is installed on your hard disk; the default is `C:\WINWORD`. Click on the **OK** button to exit the dialog box.
10. You will be asked whether you would like to view the M-book. If you want to see the features of the MATLAB Notebook, click on the **Yes** button; otherwise, click on **No** to finish the Notebook installation.

11. Configure your swap file and SMARTdrive as described in the section "Optimizing Performance" presented later in this chapter.

2.4 Disk Contents

After installation, your MATLAB directory will hold the following files and subdirectories containing the various components of the MATLAB system:

\BIN	MATLAB binary and associated files
\TOOLBOX	The MATLAB toolboxes
MATLABRC.M	MATLAB's system-wide startup file
PRINTOPT.M	User-configurable print options

The subdirectories and their contents are listed below.

MATLAB\BIN

MATLAB.EXE	The MATLAB binary
MATLAB.BAT	Batch file for starting MATLAB at DOS prompt
MATLAB.HLP	MATLAB help file
ML_BANG.PIF	PIF file for configuring bang (!) command behavior
ML_DOS.PIF	PIF file for configuring DOS command behavior
*.EXE	Miscellaneous utility files

MATLAB\TOOLBOX\MATLAB

\COLOR	Color controls and lighting model functions
\DATAFUN	Data analysis and Fourier transform functions
\ELFUN	Elementary math functions
\ELMAT	Elementary matrices and matrix manipulation
\FUNFUN	Function functions—nonlinear numerical methods
\GENERAL	General purpose commands
\GRAPHICS	General purpose graphics commands
\IOFUN	Low-level file I/O functions
\LANG	Language constructs and debugging

\MATFUN	Matrix functions—numerical linear algebra
\OPS	Operators and special characters
\PLOTXY	Two-dimensional graphics
\PLOTXYZ	Three-dimensional graphics
\POLYFUN	Polynomial and interpolation functions
\SOUNDS	Sound functions
\SPARFUN	Sparse matrix functions
\SPECFUN	Specialized math functions
\SPECMAT	Specialized matrices
\STRFUN	Character string functions
\DEMOS	Demonstrations and samples
MATLAB\TOOLBOX	
\SIGSYS	*Signals and Systems Toolbox*
\SYMBOLIC	*Symbolic Math Toolbox*

2.5 Optimizing Performance

2.5.1 Resource Considerations

MATLAB is a rich and powerful environment designed to take full advantage of the latest hardware capabilities. As a result, the performance you see when running MATLAB will depend on the configuration of the hardware you are using. The following factors can play a significant role in determining the speed at which MATLAB runs:

- Processor type and clock rate
- Amount of physical memory available
- Disk access speed and throughput
- Graphics accelerator card

In particular, it is important that your computer has enough memory to tackle the class of problems for which you will be using *The Student Edition of MATLAB*. At a minimum, MATLAB for Windows requires 8 MB of extended memory. Typically, the more memory you have, the more you can do with multiple color maps, movies, interpolated shading, and lighting models.

In addition to adding more physical memory to your computer, you can also increase the amount of memory available to MATLAB by taking advantage of Windows' virtual memory capabilities as described in the following section.

2.5.2 Maximizing Available Memory

We recommend maximizing the amount of memory available to MATLAB by making use of Windows' virtual memory capabilities. Virtual memory is a technology supported in 386-enhanced mode that enables you to effectively increase your available system memory through the use of a swap file. Windows automatically "swaps" unused pages of memory out to disk, thus freeing them up for re-use. A complete discussion on installing and using swap files can be found in the "Optimizing Windows" chapter of your *Microsoft Windows User's Guide*. To obtain the best performance, use a "permanent" swap file.

When using a swap file, it is also beneficial to use SMARTdrive. SMARTdrive is a disk caching utility provided with Windows that can dramatically improve the performance of your hard drive. Normally, when you install Windows, the setup program adds the `smartdrv` command to your `autoexec.bat` file. Information about using and installing the SMARTdrive utility can be found in the "Optimizing Windows" chapter of your *Microsoft Windows User's Guide*. Note that although the default size of the SMARTdrive disk cache for a system with 6 MB of extended memory or more is 2 MB, our experience has shown that a cache size of 512 KB is usually sufficient. In conjunction with the above suggestions, we recommend that you follow the general guidelines described in Chapter 14 of the *Microsoft Windows User's Guide* to optimize the performance of your system.

2.5.3 Color Considerations

When operating in 256-color mode, Microsoft Windows uses a sophisticated color palette manager that optimizes colors in the current or foreground window at the expense of background windows. This means that if you display several figure windows, each of which uses a different color map or color map with many colors, you may see some color changes when you shift the focus between these windows. These color shifts result from the palette manager resetting the system color table for the foreground, or active, window. Although the shifts may result in background windows displaying what appear to be the incorrect colors, this behavior is normal and cannot be avoided.

3 MATLAB for Macintosh Computers

3.1 Macintosh System Requirements

- Macintosh equipped with a 68030 or 68040 microprocessor
- 15 MB of free space on your hard drive
- 4 MB memory partition for MATLAB, which translates into 8 MB of total system memory. A memory partition of 8 MB or more (at least 12 MB of total system memory) is recommended to use MATLAB's 3-D color graphics.
- A SuperDrive (1.4 MB) floppy disk drive
- Color QuickDraw
- System 6.0.5 with 32-bit QuickDraw installed or System 6.0.7 or later
- A 12″ or larger monitor

The following items are strongly recommended:

- System 7 or later
- Additional memory to bring the total system memory to at least 12 MB
- 8-bit graphics capability and display (for 256 simultaneous colors)
- Apple LaserWriter or other PostScript printer
- Suggested total system memory assumes a reasonably large System Folder. Your memory requirements may vary depending upon the exact size of your System Folder.

3.2 Installation

The Student Edition of MATLAB is shipped in compressed form and must be decompressed before it can be used. The installation process automatically decompresses the files and installs them onto your hard disk.

Before you begin the installation process, you should make a backup copy of your set of disks. In addition, you should lock the disks by sliding the write-protect tab so that the square hole is open.

Note: Some system extensions (INITs), such as virus checkers, may interfere with the installation process. If the installation fails, you should reboot your machine with extensions off (hold down the **Shift** key as you reboot) and repeat the installation.

To install MATLAB, perform the following steps:

1. Insert Disk 1 into your 1.4 MB drive and double-click on it to open it.
2. Double-click on the file, **MATLAB Installer**. A splash screen appears. Click on the **Continue** button.
3. A save dialog box with the prompt **Install software as:** will appear with a suggested folder name, which you may change. Navigate to the place where you want to install MATLAB, and then click on the **Save** button.
4. A dialog box will appear, showing the decompression progress. You will be prompted to insert the remaining disks as needed by the installer.

3.3 Disk Contents

MATLAB is distributed in compressed format on floppy disks. The installation procedure moves the files on these disks to your hard disk, decompresses them, and installs them in the Macintosh environment. The main MATLAB system includes:

`MATLAB`	The MATLAB application
`Toolbox`	A folder containing the MATLAB Toolbox M-files
`README`	A file containing the latest release notes

The subdirectories and their contents are listed below.

The subdirectories and their contents are listed below.

MATLAB:Toolbox:matlab

`matlabcolor`	Color controls and lighting model functions
`datafun`	Data analysis and Fourier transform functions
`demos`	Demonstrations and samples
`elfun`	Elementary math functions
`elmat`	Elementary matrices and matrix manipulation
`funfun`	Function functions—nonlinear numerical methods
`general`	General purpose commands
`graphics`	General purpose graphics commands
`iofun`	Low-level file I/O functions
`lang`	Language constructs and debugging
`matfun`	Matrix functions—numerical linear algebra
`ops`	Operators and special characters
`plotxy`	Two dimensional graphics
`plotxyz`	Three dimensional graphics
`polyfun`	Polynomial and interpolation functions
`sounds`	Sound functions
`sparfun`	Sparse matrix functions
`specfun`	Specialized math functions
`specmat`	Specialized matrices
`strfun`	Character string functions

Matlab:Toolbox:local

`matlabrc.m`	MATLAB's system-wide startup file
`printopt.m`	User-configurable print options

Matlab:Toolbox:

`sigsys`	*Signals and Systems Toolbox*
`symbolic`	*Symbolic Math Toolbox*

3.4 Optimizing Performance

3.4.1 Resource Considerations

MATLAB is designed to take full advantage of the latest hardware capabilities. As a result, the performance you see when running MATLAB will depend on the configuration of the hardware you are using. The following factors can play a significant role in determining the speed at which MATLAB runs:

- Processor type and clock rate
- Amount of physical memory available
- Disk access speed and throughput
- Graphics accelerator card

It is important that your computer has enough memory to tackle the class of problems for which you will be using *The Student Edition of MATLAB.* At a minimum, MATLAB requires a 4 MB memory partition, which translates into 8 MB of total system memory. Typically, the more memory you have, the more you can do with multiple color maps, movies, interpolated shading, and lighting models.

In addition to adding more physical memory to your computer, you can also increase the amount of memory available to MATLAB by taking advantage of System 7's virtual memory capabilities as described in the following section.

3.4.2 Maximizing Available Memory

You may want to maximize the amount of memory available to MATLAB by making use of System 7's virtual memory capabilities via the Memory control panel. Virtual memory is a technology that enables you to increase your available system memory through the use of your computer's hard disk. System 7 automatically swaps unused memory out to disk, thus freeing memory for reuse. A complete discussion on configuring virtual memory can be found in your Macintosh system documentation. While the use of virtual memory increases the amount of memory available to MATLAB, it does degrade performance. For maximum performance, you may want to consider adding physical memory (RAM) to your computer rather than using the virtual memory option.

3.4.3 MATLAB's Memory Management

MATLAB uses (i.e., dynamically allocates) memory from the memory partition made available to it by the Finder. If you are running under System 7 or MultiFinder, you can change the amount of memory the Finder makes available to MATLAB. To do so, quit MATLAB if you are currently running it. Then at the Finder, click on MATLAB's icon to select it. Choose **Get Info** from the **File** menu. The box labeled **Current size** is the amount of memory (in kilobytes) that the Finder attempts to make available to MATLAB when it is run. For example, if you want to give MATLAB 4 MB of memory, enter `4096` in the box.

3.4.4 Resolving Memory Errors

If you do not have a large enough contiguous chunk of memory to allocate a matrix, an out of memory error may occur even though you seem to have enough available memory. To consolidate the fragmented memory, you can either use the MATLAB `pack` command, allocate larger matrices earlier in the MATLAB session, or give MATLAB more memory as described in the previous section.

3.4.5 Color Considerations

When operating in 256-color mode, the Macintosh uses a sophisticated color palette manager that optimizes colors in the current or foreground window at the expense of background windows. This means that if you display several figure windows, each of which uses a different color map, you may see some color changes when you shift the focus between these windows. These color shifts result from the palette manager resetting the system color table for the foreground, or active, window. Although the shifts may result in background windows displaying what appear to be the incorrect colors, this behavior is normal and cannot be avoided.

This situation is a result of trying to display more colors than your monitor can handle (256). The Macintosh palette manager does its best to arbitrate which colors should be displayed. If your monitor displays thousands or millions of colors, you will not see the shift, since your monitor is capable of displaying all of the desired colors.

All of MATLAB's graphic features are designed specifically for 256-color mode. If your monitor displays 2 colors (black and white), 4 colors or grays, or 16 colors or grays, MATLAB will do its best to provide a reasonable representation of your graphics by dithering images and polygons. *Dithering* is a technique that blends colors such that they can be displayed with however many colors you have.

If your Macintosh has multiple monitors and a figure window spans multiple displays, MATLAB draws colors for the device with the most colors.

3.4.6 Figure Window Default Background Color

The default background color of figure windows is black. If you want to make the default background color white, you should consider calling `whitebg` from your `startup.m` file. When you start MATLAB, it automatically executes the M-file `matlabrc.m`, which invokes the file `startup.m` if it exists on MATLAB's search path. For more information about `startup.m`, see the `startup` command in Chapter 8, *Matlab Reference*.

The command `whitebg` changes the default properties so that plots in Figure windows use a white background. For more information about `whitebg`, type `help whitebg`.

4 Upgrading from MATLAB 3.5 to MATLAB 4.0

This section contains important information if you use the previous version of *The Student Edition of MATLAB* or professional MATLAB 3.5. Otherwise you may disregard this section.

MATLAB 4.0 is a major upgrade to MATLAB. Although The MathWorks endeavors to maintain full upwards compatibility between subsequent releases of MATLAB, inevitably there are situations where this is not possible. In the case of MATLAB 4.0, there are a number of changes that you need to know about in order to migrate your code from MATLAB 3.5 to MATLAB 4.0.

It is useful to introduce two terms in discussing this migration. The first step in converting your code to MATLAB 4.0 is to make it Version 4.0 *compatible*. This involves a rather short list of possible changes. After you have made these changes, your M-files will run under MATLAB 4.0. The second step is to make it Version 4.0 *compliant*. This means making further changes so that your M-file is not using obsolete, but grandfathered, features of MATLAB. It also can mean taking advantage of MATLAB 4.0 features like the new graphics, user-interface controls, etc.

There are a relatively small number of things that are likely to be in your code that you will have to change to make your M-files MATLAB 4.0 compatible. Most of these are in the graphics area and have to do with the fact that MATLAB 4.0's graphics is now a fully object-oriented WYSIWYG (what-you-see-is-what-you-get) system.

There is a somewhat larger number of things you can do (but don't have to) to make your M-files fully MATLAB 4.0 compliant. To help you gradually make your code compliant, MATLAB 4.0 displays warning messages when you use functions that are obsolete, even though they still work correctly.

Major Incompatibilities

Nine major commands in MATLAB 4.0 are enhanced in a manner that is not fully backward-compatible with MATLAB 3.5. They are:

```
axis
clg
contour
global
hold
mesh
meshdom
meta
subplot
```

Some of the differences are minor and you probably won't notice them. Others are more major and require changes to your existing M-files to make them run under MATLAB 4.0.

Most of the affected commands are in the graphics area. The changes we made to these commands make MATLAB's graphics more comprehensive and powerful while still maintaining consistency. In general, those of you who tried hardest to make MATLAB's graphics achieve the most complicated effects may find that you have to make the most changes, but you will also benefit the most from the increased power and consistency.

The following sections provide detailed conversion instructions for each of these commands.

axis

- `axis` now takes effect immediately (immediate mode). In version 3.5, axis limits were set before `plot` statements. With version 4.0, the limits are set after `plot` statements.
- `axis`, by itself, no longer toggles the state of the axis limits – it just returns the axis limits.
- `axis('auto')` is now used to set the axis limits back to autoranging.
- If the last plot was a 3-D plot, `axis` now returns a six-element vector, with the additional elements being the z-axis limits.

To convert MATLAB 3.5 code to MATLAB 4.0 code:

1. Move `axis` statements from before `plot` statements to after `plot` statements:

MATLAB 3.5	MATLAB 4.0
`axis([0 10 -5 5])`	`plot (x, y)`
`plot(x,y)`	`axis([0 10 -5 5])`

2. Change instances of `axis` with no arguments that were used to freeze the axis limits at the current values to `axis(axis)`.

3. Change instances of `axis` with no arguments that were used to set the axis limits back to autoranging to `axis('auto')`.

Under MATLAB 4.0, note that code sequences like

```
plot(1:10)
axis([0 20 0 20])
pause
```

will not "flash" at the `axis` command because the screen is not updated until the `pause` command, a `drawnow` command, or a return to the MATLAB prompt is encountered.

clg

- `clg` has been superseded by `clf` in MATLAB 4.0, which behaves similarly in that it deletes all axes (subplots) from the screen. Because `hold` is now a property of the individual axes, `clf` has the effect of resetting the hold state since the next `plot` command will result in the creation of a new set of axes with default properties. The old command name `clg` is grandfathered but obsolete.

To convert MATLAB 3.5 code to MATLAB 4.0 code, in situations where `clg` was used in conjunction with `hold` to draw graphs without axes being displayed, add `axis('off')` commands.

MATLAB 3.5	MATLAB 4.0
`plot(0:50)`	`plot(0:50)`
`hold on`	`hold on`
`clg`	`axis(axis)`
`plot(2:30)`	`axis('off')`
	`cla`
	`plot(2:30)`

contour

- The appearance of the graph is flipped in the up/down vertical direction. Version 3.5 `contour` is equivalent to `contour(flipud(z))` using version 4.0. However, if you used `meshgrid` to generate the data for the plot instead of `meshdom`, the graph will still appear the same because `meshgrid` is `flipud` of `meshdom`.
- The order of the arguments to `contour` has been switched to make it consistent with the new `mesh` and `surf`:

MATLAB 3.5	MATLAB 4.0
`n/a`	`contour(x,y,z)`
`contour(z,n,x,y)`	`contour(x,y,z,n)`
`contour(z,v,x,y)`	`contour(x,y,z,v)`

The older order still works, but we recommend that you switch to the new one.

global

Global variable scoping rules have been modified substantially for MATLAB 4.0, thereby greatly improving the functionality of global variables. Why the change? In MATLAB 3.5 global variables were visible everywhere and could not be shadowed. It was not possible for two functions to share a global variable without throwing it into the scope of all functions. This often led to unexpected name clashes with existing variables. Additionally you were not able to declare a variable to be global from within an M-file function.

You now declare specific variables global only in those workspaces that need access to them. These include local workspaces of functions as well as MATLAB's base workspace. Thus you can think of there being a separate "global workspace."

To convert MATLAB 3.5 code to MATLAB 4.0 code, you must put global declarations inside every function M-file that requires access to it. You must also issue a `global` command in the base workspace (or in an M-file script) for any global variables that you wish to manipulate from the MATLAB prompt.

Additional changes affecting the use of global variables include:

- `who global` now shows a list of global variables in this workspace.
- `clear`, by itself, no longer clears global variables.
- `clear a`, if a is a global variable, clears the link to the global variable, not the global variable itself.

- `clear global` clears all variables in the global workspace.
- `clear global a` clears just the global variable a.

hold

`hold` has changed slightly. In MATLAB 4.0, to put `hold` on means "add the next plot to the current axes" rather than "freeze the current plot up on the window." It holds the data, title, labels, and any manually specified axes limits on the current axes. It now also holds all additional states associated with an axes.

`hold` is now a property of the individual axes, rather than being global in nature. Therefore the `clf` command (formerly `clg`) has the effect of resetting the hold state, since the next `plot` command will result in the creation of a new set of axes with default properties.

mesh

`mesh` plots now include 3-D axes. (This was the most frequently requested enhancement that we received on our technical support lines.)

The appearance of the graph is flipped in the up/down (vertical) direction. Version 3.5 `mesh(z)` is equivalent to `mesh(flipud(z))` using Version 4.0. However, if you used `meshgrid` to generate the data for the plot instead of `meshdom`, the graph will still appear the same because `meshgrid` is `flipud` of `meshdom`.

Viewpoint and scale factors in the `mesh` command—`mesh(z,m)`, `mesh(z,s)`, `mesh(z,m,s)`—are no longer supported (although they are grandfathered). Use `axis` to set the mesh scale factor and `view` to set the viewpoint.

To convert MATLAB 3.5 code to MATLAB 4.0 code, where you used `m = [az el]` to set the viewpoint,

```
m = [az el]
mesh(z,m)
```

change it to

```
mesh(z)
view(az,el)
```

meshdom

- `meshdom` has been superseded by `meshgrid`. `meshgrid` is the same as the old `meshdom` except that it returns the `Y` matrix as `flipud(Y)` of `meshdom`. `meshdom` has been grandfathered.

Why the change? Data analysis with `mesh`, `contour`, and `surf` is better done in a Cartesian coordinate system, where the row dimension of the matrix increases with increasing *y*-axis values, rather than the reverse, as was done in version 3.5.

To convert MATLAB 3.5 code to MATLAB 4.0 code, change instances of `meshdom` to `meshgrid`, when you used `meshdom` in conjunction with `mesh` and `contour`. If your code used `meshdom` for some other purpose, add a `flipud` call to the `Y` output of `meshgrid`:

```
[X,Y]=meshgrid(x,y);
Y = flipud(Y);
```

meta

- Metafiles and GPP no longer exist. The MATLAB 3.5 metafile was a closed, binary representation of a graph or a series of graphs. A new command, `print`, has been introduced to save graphics to files. `print` generates PostScript code (both regular PostScript and Encapsulated PostScript (EPSF)) as well as output for other commonly available printers.
- The `meta` command has been grandfathered so that it operates; however, it now directly generates PostScript output files.

subplot

- We now call subplots "axes." Where you used to say, "there are several *subplots* in my graph window," you would now say, "there are several *axes* in my figure."
- A new command, `axes`, can be used to place axes anywhere in the graph window. The `subplot` command, however, is still the preferred way to conveniently put several axes on the graph window in a tiled fashion.
- `subplot(m,n,p)` is now preferred over `subplot(mnp)`, where `mnp` is a three digit decimal integer. The old form is grandfathered.
- Subplots no longer autocycle after `plot` commands. You must issue an explicit `subplot` command to move to the next set of axes.

- Each set of axes now has its own properties (state variables), rather than there being a single set of global properties. Version 3.5 state variables include axes limits and `hold`. Version 4.0 adds a number of additional state variables for each axis. Modifying these properties only affects the current axis.

To convert MATLAB 3.5 code to MATLAB 4.0 code:

1. Change instances of `subplot(mnp)` to `subplot(m,n,p)`.
2. Insert explicit `subplot(m,n,p)` statements in places where you relied on the undocumented subplot autocycling feature.
3. Beware of the *absence* of side effects as a result of each subplot now having its own state variables.

Additional Changes and Enhancements

This section goes through the various categories of MATLAB functions and describes changes and enhancements to existing features that are available in MATLAB 4.0. It also discusses changes to those commands that behave differently from their MATLAB 3.5 counterparts.

General

- The help facility has been completely reworked. `help`, by itself, displays MATLAB's search path. `help` with a directory name as its argument displays the `Contents.m` file of the corresponding directory.
- You can now type `what dirname` to get a list of M-files in a given directory.
- Smart command-line recall has been added. Typing one or several characters of a previous command and pressing the up-arrow key recalls the most recent command starting with those letters. Pressing the up-arrow key again recalls the next most recent command that also matches the specified set of letters. The down-arrow key works similarly.
- The command `cd` now works to change your current directory. You can use `cd ..` and `cd ../foo` now, too.
- The master startup M-file `matlab.m` has been renamed to `matlabrc.m`.
- `format rat` now causes MATLAB to display numbers in rational approximation format.

- To scroll output one screen at a time, use the new function `more on`. Disable it with `more off` and override the default number of lines with `more n`.
- New commands:

`lookfor`	Keyword search through the help entries.
`ver`	Display MATLAB and toolbox version numbers.
`whatsnew`	Display readme files.
`which`	Locate functions and files.
`path`	Control MATLAB's search path.
`unix`	Execute operating system command and return result.

Elementary Matrix Functions

- The syntax for `zeros`, `ones`, `eye`, and `rand` has been changed to eliminate the ambiguity that arose when the input argument was a scalar. In MATLAB 3.5, there is an ambiguity in the syntax for the four statements

```
Y = zeros(X)
Y = ones(X)
Y = eye(X)
Y = rand(X)
```

The matrix `Y` returned by these statements is the same size as the input argument, `X`, except when `X` is a scalar. If `X` is a scalar, its value is used instead and the resulting `Y` is `X`-by-`X`.

These statements still work in MATLAB 4.0, but a warning message is printed if `X` is a matrix whose dimensions are not 1-by-1, 1-by-2, or 2-by-1. The message indicates that this use is obsolete and *will be eliminated* in future versions of MATLAB.

If you want to create a matrix the same size as `X`, the preferred way is now

```
Y = zeros(size(X))
Y = ones(size(X))
Y = eye(size(X))
Y = rand(size(X))
```

If you want to create a matrix of a specified size, the following statements produce the desired result in all versions of MATLAB.

```
Y = zeros(n)                    Y = zeros(m,n)
Y = ones(n)                     Y = ones(m,n)
Y = eye(n)                      Y = eye(m,n)
Y = rand(n)                     Y = rand(m,n)
```

- `rand('normal')` now produces a warning message. Use the new function `randn` to generate normally distributed numbers.
- New functions:

`tic, toc`	Stopwatch timer functions
`str2mat`	Form text matrix from individual strings
`upper`	Convert string to upper case
`lower`	Convert string to lower case
`hadamard`	Hadamard matrix
`rosser`	Classic symmetric eigenvalue test problem
`wilkinson`	Wilkinson's eigenvalue test matrix

The MATLAB Language

- Line continuation used to work with two consecutive periods; three are now required (`...`).
- In the expression `1./x`, where `x` is an array, `x` the interpreter now associates the `.` with the `/` operator instead of with the numeric value 1.
- Matrices local to a function are no longer initialized to the empty matrix. You must now explicitly initialize matrices used inside functions.
- Scripts called from within functions now act on the function's workspace rather than the base workspace.
- In MATLAB 3.5, there is a distinction between *commands*, like

```
clear
load X
echo on
```

and *functions* like

```
sin(X)
plot(Z)
```

If commands have any arguments, then they must be given on the command line, separated by blanks, with no parentheses. Commands alter the environment, but do not return results. New commands cannot be added with M-files.

In MATLAB 4.0, commands and functions are now "duals." Commands are considered to be functions that take string arguments. The two constructions

```
command argument
```

and

```
command('argument')
```

are equivalent. Therefore

```
load X.dat
```

can also be accomplished with

```
load('X.dat')
```

or

```
filename = 'X.dat'
load(filename)
```

Other examples are `hold on` and `hold('on')`; `disp finished` and `disp('finished')`; and `axis square` and `axis('square')`. This enhancement makes it possible to generate command arguments with string manipulations. It also makes it possible to create new commands with M-files.

- You can now enter imaginary numbers as `4i` or `4j` instead of having to use `4*i`. So `a = 4i` and `b = 3 + 2j` are now legal statements.
- `NaN` and Infinity handling has been improved.
- `NaN == NaN` now follows the IEEE standard and returns false (`0`), rather than true (`1`) or `NaN` as it does in previous versions of MATLAB. Use `isnan(x)` to detect NaNs.
- `ieee` has been renamed to `isieee`.
- New functions:

`xor`	Logical exclusive OR
`isinf`	True for infinite elements
`realmax`	Largest floating point number
`realmin`	Smallest positive floating point number

Elementary and Specialized Math Functions

- `inverf` has been changed to `erfinv`. The old command name has been grandfathered.
- `gammac` has been renamed `gamma` and computes only the complete gamma function. Use `gammainc` to compute the incomplete gamma function.
- `erf(x,'high')` no longer works; use `erf(x)` instead. The `erf` algorithm has been improved and calculates answers as accurately as possible.
- `ellipk` is now `ellipke` and returns elliptic integrals of both the first and second kinds.
- The algorithms for hyperbolic sine, cosine, and tangent have been improved.
- The `bessel` function algorithm has been improved.
- New functions:

`log2`	Dissect floating point numbers
`pow2`	Scale floating point numbers
`trapz`	Numerical integration using trapezoidal method
`del2`	Five-point discrete Laplacian

Matrix Functions – Numerical Linear Algebra

- `norm(Z,1)` now computes the actual norm, `max(sum(abs(Z)))`, instead of the "fast and dirty" norm

  ```
  max(sum(abs(real(Z))+abs(imag(Z))))
  ```
- `norm(Z,inf)` now computes `max(sum(abs(Z')))` instead of the "fast and dirty" norm

  ```
  max(sum(abs(real(Z'))+abs(imag(Z'))))
  ```
- `lscov` is a new function for least-squares problems in the presence of known covariance.
- `chol` now allows detection of non-positive definite input matrices under program control.
- `det` handles NaNs better (doesn't go into an infinite loop).
- `\` and `/` now detect systems that are already in triangular form, in which case they only perform the backsubstitution step.

Polynomial, Signal Processing, and Interpolation Functions

- `roots1` has been removed due to instabilities in the algorithm. Use `roots` instead.
- `residue`'s treatment of repeated roots has been improved.
- `spline` has been modified to use sparse matrices. This allows much larger data sets to be interpolated.
- Interpolation functions in MATLAB 3.5 (`interp1`, `interp2`, `interp3`, `interp4`, `table1`, `table2`, `spline`) have been reorganized into two main routines, `interp1`, for 1-D interpolation and table lookup, and `interp2`, for 2-D interpolation and table lookup. An optional string controls the choice of algorithms.
- `interpft` now implements interpolation using the FFT method (the old `interp1`).
- `griddata` now does gridding (used to be overloaded onto the old `interp2`).
- `filter2` is a new function for 2-D FIR filtering.
- `fsolve` is now in the *Optimization Toolbox*.
- `fmin` is now much faster.

Graphics

- `polar` is now an M-file, rather than being built-in. It no longer takes multiple input-pairs. Use `hold` if you need to add multiple lines.
- `text` no longer uses screen coordinates. Where you previously used the string `'sc'` to indicate screen coordinates, you might want to open a new axes and set its `Position` property to `[0 0 1 1]`, its `XLim` property to `[0 1]`, and its `YLim` property to `[0 1]` to achieve a similar effect.
- `quiver`'s arguments have been reordered to be consistent with the other new 3-D graphics commands – `quiver(x,y,px,py)`.
- Grid lines can now be removed from graphs after they have been added.
- `prtsc` no longer exists.
- Use of `'c1','c2',...` is discouraged. Instead use explicit colors or Handle Graphics™ commands to specify colors.
- Invisible `'i'` is no longer a valid line color. It was used for a variety of effects in MATLAB 3.5, including deleting lines, and animation. If you want to delete a line in MATLAB 4.0, use the `delete` function on the line's

handle. Animation can be performed using the movie command, or with clever use of the EraseMode property of line objects. Although the 'i' line color is grandfathered, its use is strongly discouraged because it works by creating a new line in the background color. This can fill up memory very quickly if it is called many times, because each call creates a new line object.

- When called a second time, title, xlabel, and ylabel now replace their text, rather than overwriting it.
- inquire is now obsolete, but grandfathered. Use the Handle Graphics get command to inquire about the state of object properties.
- The meaning of normalized coordinates has changed, especially for the purposes of ginput and text. In MATLAB 3.5, (0,0) and (1,1) were outside the actual axes and were the corners of the region surrounding titles, labels, etc. In 4.0, these same coordinates refer to the axes limits.
- For the two undocumented commands, sc2dc and dc2sc, the disparity between normalized screen coordinates, which used to span an imaginary area larger than the axes, and the plot axes box has been removed—the two are now the same. For upward compatibility, these functions are grandfathered; however, they are now M-files that simply end up returning the axis limits.
- Axis limits for log axes are now given as the actual values rather than their logs.
- The autocycling sequence for colors when using the plot command is now different. The new cycle is yellow, magenta, cyan, red, green, and blue. The sequence order can now be changed by modifying the ColorOrder property of axes objects.
- The color abbreviations are now y, m, c, r, g, b, w, k (k is black).

Part Two

Learning MATLAB and the Toolboxes

5 MATLAB Tutorial

5.1 Introduction

Now that you've installed MATLAB it's time to see what it can do. In this tutorial you will be shown some of its capabilities. To show all of what MATLAB can do would simply take too much time. Hopefully as you follow this tutorial you will begin to see the power of MATLAB to solve a wide variety of problems important to you. You may find it beneficial to go through this tutorial while running MATLAB. In doing so you will be able to enter the MATLAB statements as described, confirm the results presented, and develop a hands-on understanding of MATLAB.

Perhaps the easiest way to visualize MATLAB is to think of it as a full-featured calculator. Like a basic calculator, it does simple math such as addition, subtraction, multiplication, and division. Like a scientific calculator, it handles complex numbers, square roots and powers, logarithms, and trigonometric operations such as sine, cosine, and tangent. Like a programmable calculator, you can store and retrieve data; you can create, execute and save sequences of commands to automate the computation of important equations; you can make logical comparisons and control the order in which commands are executed. Like the most powerful calculators available, it allows you to plot data in a wide variety of ways, perform matrix algebra, manipulate polynomials, integrate functions, manipulate equations symbolically, etc.

In reality, MATLAB offers many more features and is more multifaceted than any calculator. MATLAB is a tool for making mathematical calculations. MATLAB is a user-friendly programming language with features more advanced and much easier to use than computer languages such as BASIC, Pascal, or C. MATLAB provides a rich environment for data visualization through its powerful graphics capabilities. MATLAB is an application development platform, where sets of intelligent problem-solving tools for specific application areas, often called *Toolboxes*, can be developed with relative ease. For example, this *Student Edition of MATLAB* includes the *Symbolic Math Toolbox* and the *Signals and Systems Toolbox*.

Because of the vast power of MATLAB, it is important to start with the basics. That is, rather than throw everything at you and hope that you understand some of it, in the beginning it is helpful to think of MATLAB as a calculator. First as a basic calculator. Next as a scientific calculator. Then as a programmable calculator. And finally as a top-of-the-line calculator. By using this calculator analogy, you will see the ease with which MATLAB solves everyday problems, and begin to see how MATLAB can be used to solve complex problems in a flexible, straightforward manner.

Depending on your background, you may find parts of this tutorial boring or some of it may be over your head. In either case, find a point in the tutorial where you're comfortable, start up MATLAB, and begin.

5.2 Basic Features

Running MATLAB creates one or more windows on your monitor. Of these, the command window is the primary place where you interact with MATLAB. This window has an appearance as shown below. The character string `EDU»` is the MATLAB prompt in the *Student Edition of MATLAB*. In other versions of MATLAB, the prompt is simply ». In this text » will be used to denote the MATLAB prompt. When the command window is active, a cursor (most likely blinking) should appear to the right of the prompt as shown in the figure. This cursor and the MATLAB prompt signify that MATLAB is waiting to answer a mathematical question.

Here is the command window as it appears on the Macintosh.

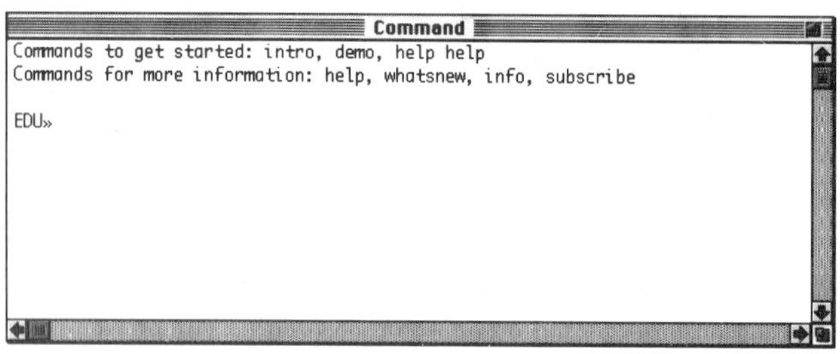

Here is the command window as it appears on MS-Windows computers.

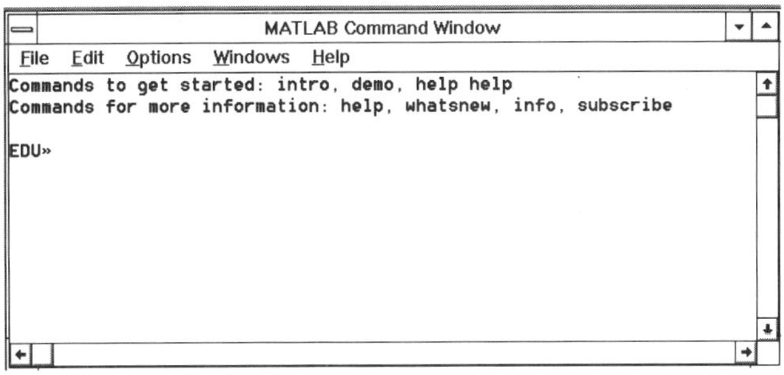

5.2.1 Simple Math

Just like your calculator, MATLAB does simple math. Consider the following simple example:

> *Homer buys 4 apples for 25 cents each, 6 bananas for 22 cents each and 2 cantaloupes for 99 cents each. When he gets home, Marge asks him how many pieces of fruit he bought and how much did he spend?*

To solve this using your calculator you enter:

$$4 + 6 + 2 = 12 \text{ pieces of fruit}$$
$$(4 \times 25) + (6 \times 22) + (2 \times 99) = 430 \text{ cents}$$

In MATLAB this can be solved in a number of different ways. First, the above calculator approach can be taken:

```
» 4+6+2
ans =
    12

» 4*25 + 6*22 + 2*99
ans =
   430
```

(For convenience the general MATLAB prompt » is shown in this text rather than the Student Edition prompt `EDU»`.)

Note that MATLAB doesn't care about spaces for the most part and that multiplication takes precedence over addition. Note also that MATLAB calls the result `ans` (short for answer) for both computations.

As an alternative, the above can be solved by storing information in MATLAB variables:

```
» apples=4
apples =
     4

» bananas=6
bananas =
     6

» cantaloupes=2;
» fruit=apples+bananas+cantaloupes
fruit =
    12

» cost=apples*25+bananas*22+cantaloupes*99
cost =
   430
```

Here we created three MATLAB variables `apples`, `bananas`, and `cantaloupes` to store the number of each fruit. After entering each statement, MATLAB displayed the results except in the case of `cantaloupes`. The semicolon at the end of the line

```
» cantaloupes=2;
```

tells MATLAB to evaluate the line but not tell us the answer. Finally, rather than calling the results `ans`, we told MATLAB to call the number of fruit purchased `fruit` and the total price paid `cost`.

If the statement is too long to fit on one line, an ellipsis consisting of three periods (. . .) followed by **Enter** indicates that the statement continues on the next line.

```
» cost=apples*25+bananas*22+...
    cantaloupes*99
cost =
   430
```

At each step MATLAB remembered past information. Because MATLAB remembers things, let's suppose that:

Bart asks, "What was the average cost per fruit?"

```
» average_cost=cost/fruit
average_cost =
   35.8333
```

Because average cost is two words and MATLAB variable names must be one word, the underscore was used to create the single MATLAB variable `average_cost`.

In addition to addition and multiplication, MATLAB offers the following basic arithmetic operations:

Operation	Symbol	Example
addition, $a + b$	+	5+3
subtraction, $a - b$	–	23–12
multiplication, $a \times b$	*	3.14*0.85
division, $a \div b$	/ or \	56/8 = 8\56
power, a^b	^	5^2

The order in which these operations are evaluated in a given expression is given by the usual rules of precedence that can be summarized as follows:

> *Expressions are evaluated from left to right with the power operation having the highest order of precedence, followed by both multiplication and division having equal precedence, followed by both addition and subtraction having equal precedence.*

Parentheses can be used to alter this usual ordering, in which case evaluation initiates within the innermost parentheses and proceeds outward. To illustrate these concepts, convince yourself that the following statements are true:

```
3^2-5-6/3*2 = 0
3^2-5-6/(3*2) = 3
4*3^2+1 = 37
(4*3)^2+1 = 145
```

Before we illustrate other computational features, let's take some time to discuss some fundamental operational features of MATLAB.

5.2.2 The MATLAB Workspace

As you work in the command window, MATLAB remembers the commands you enter as well as the values of any variables you create. These commands and variables are said to reside in the *MATLAB Workspace* and can be recalled whenever you wish. For example, to check the value of `cantaloupes` all you have to do is ask MATLAB for it by entering its name at the prompt:

```
» cantaloupes
cantaloupes =
      2
```

If you can't remember the name of a variable, you can ask MATLAB for a list of the variables it knows by using the MATLAB command who:

```
» who

Your variables are:

ans            average_cost   cantaloupes    fruit
apples         bananas        cost

leaving 3041192 bytes of memory free.
```

Note that MATLAB doesn't tell you the value of all the variables; it just gives you their names. To find their values you must enter their names at the MATLAB prompt. Just like a calculator, there's only so much room to store variables. The last line of the who command response says that there is lots of room left! Each piece of data takes 8 bytes, so there is room to create 380,149 more pieces of data. How much workspace room you have depends on the computer you are using.

To recall previous commands, MATLAB uses the cursor keys, ←, →, ↑, ↓, on your keyboard. For example, pressing the ↑ key once recalls the most recent command to the MATLAB prompt. Repeated pressing scrolls back through prior commands one at a time. In a similar manner pressing the ↓ key scrolls forward through commands. At any time the ← and → keys can be used to move the cursor within the command at the MATLAB prompt. In this manner the command can be edited. One convenient use of command recall and editing is fixing errors in the most recent command. Continuing with the apples, bananas, and cantaloupes example, suppose we entered cantaloupes as:

```
» canteloupes=2;
```

Here we misspelled cantaloupes by changing the second a to an e. To fix the error, press the ↑ key once to recall the command. Then move to the left using the ← key and change the e to an a. On an MS-Windows PC this is accomplished by moving the cursor over the e, pressing **Delete**, then entering a. On a Macintosh this is accomplished by moving the cursor between the e and l, pressing **Delete**, then entering a. On both computers pressing **Return** or **Enter**, with the cursor at the corrected a, tells MATLAB to evaluate the entire expression again.

5.2.3 Saving and Retrieving Data

In addition to remembering variables, MATLAB can save and load data from files on your computer. The **Save Workspace as...** menu item in the **File** menu opens a standard file dialog box for saving all current variables. Similarly, the **Load Workspace...** menu item in the **File** menu opens a dialog box for loading variables from a previously saved workspace. Saving variables does not delete them from the MATLAB workspace. Loading variables of the same name as those found in the MATLAB workspace changes the variable values to those loaded from the file.

If the **File** menu approach does not meet your needs, MATLAB provides two commands, `save` and `load`, which offer more flexibility. For further information on these commands see Section 8.2, *MATLAB Commands and Functions.*

5.2.4 Number Display Formats

When MATLAB displays numerical results, it follows several rules. By default, if a result is an integer, MATLAB displays it as an integer. Likewise, when a result is a real number, MATLAB displays it with approximately four digits to the right of the decimal point. If the significant digits in the result are outside of this range, MATLAB displays the result in scientific notation similar to scientific calculators. You can override this default behavior by specifying a different numerical format using the **Numerical Format** menu item in the **Options** menu or by typing the appropriate MATLAB command at the prompt. Using the variable `average_cost` from the above example, these numerical formats are:

MATLAB Command	average_cost	Comments
`format long`	`35.83333333333334`	16 digits
`format short e`	`3.5833e+01`	5 digits plus exponent
`format long e`	`3.583333333333334e+01`	16 digits plus exponent
`format hex`	`4041eaaaaaaaaaab`	hexadecimal
`format bank`	`35.83`	2 decimal digits
`format +`	`+`	positive, negative, or zero
`format rat`	`215/6`	rational approximation
`format short`	`35.8333`	default display

It is important to note that MATLAB does not change the internal representation of a number when different display formats are chosen; only the display of the number changes.

5.2.5 About Variables

Like any other computer language, MATLAB has rules about variable names. Earlier it was noted that variable names must be a single word containing no spaces. More specifically, MATLAB variable naming rules are:

Rule	Comments
Variables are case sensitive.	`fruit`, `Fruit`, `FrUiT`, and `FRUIT` are all different MATLAB variables.
Variables can contain up to 19 characters.	Characters beyond the 19th are ignored.
Variables must start with a letter, followed by any number of letters, digits, or underscores.	Punctuation characters are not allowed since many have special meaning to MATLAB.

In addition to these naming rules, MATLAB has several special variables. They are:

Variable	Value
`ans`	Default variable name used for results
`pi`	Ratio of the circumference of a circle to its diameter
`eps`	Smallest number such that when added to 1 creates a floating-point number greater than 1 on the computer
`inf`	Infinity, e.g., 1/0
`NaN`	Not-a-Number, e.g., 0/0
`i` and `j`	`i = j =` $\sqrt{-1}$
`realmin`	The smallest usable positive real number
`realmax`	The largest usable positive real number

As you create variables in MATLAB, there may be instances where you wish to redefine one or more variables. For example:

```
» apples=4;
» bananas=6;
» cantaloupes=2;
» fruit = apples+bananas+cantaloupes
fruit =
    12

» apples = 6
apples =
     6

» fruit
fruit =
    12
```

Here, using the first example again, we found the number of fruit Homer purchased. Afterward we changed the number of apples to 6, overwriting its prior value of 4. In doing so, the value of `fruit` has not changed. MATLAB does not recalculate the number of fruit based on the new value of `apples`. When MATLAB performs a calculation, it does so using the values it knows at the time the requested command is evaluated. In the above example, if you wish to recalculate the number of fruit, the total cost, and the average cost, it is necessary to recall the appropriate MATLAB commands and ask MATLAB to evaluate them again. Later, with the introduction of M-files, you will be shown a simple and powerful way to reevaluate a group of MATLAB commands.

Note: The special variables given above follow this guideline also. When you start MATLAB, they have the values given above. If you change their values, the original special values are lost until you clear the variables or you restart MATLAB. With this in mind, avoid redefining special variables, unless absolutely necessary.

Variables in the MATLAB workspace can be unconditionally deleted by using the command `clear`. For example:

`» clear apples`	deletes just the variable `apples`.
`» clear cost fruit`	deletes both `cost` and `fruit`.
`» clear`	deletes all variables in the workspace! You are not asked to confirm this command. All variables are cleared and cannot be retrieved!

Needless to say the `clear` command is dangerous and should be used with caution. Thankfully, there is seldom a need to clear variables from the workspace.

5.2.6 Other Basic Features

- All text after a percent sign (%) is taken as a comment statement:

```
» apples=4  % Number of apples.
apples =
     4
```

  The variable apples is given the value 4 and MATLAB simply ignores the percent sign and all text following it. This feature makes it easy to document what you are doing.

- Multiple commands can be placed on one line if they are separated by commas or semicolons:

```
» apples=4, bananas=6; cantaloupes=2
apples =
     4
cantaloupes =
     2
```

  Commas tell MATLAB to display the results; semicolons suppress printing.

- You can interrupt MATLAB at any time by pressing **Ctrl-C** (pressing the **Ctrl** and **C** keys simultaneously) on a PC. Pressing **⌘-.** (pressing the ⌘ and **.** keys simultaneously) on a Macintosh does the same thing.
- Typing the command quit terminates MATLAB.

5.2.7 Summary

- MATLAB knows addition (+), subtraction (–), multiplication (*), division (/ or \), and powers (^).
- MATLAB evaluates an expression from left to right giving precedence to powers over multiplication and division and these over addition and subtraction.
- A semicolon (;) at the end of a MATLAB statement suppresses printing of results.
- If a statement is too long, type three periods (...) followed by **Enter** to continue a MATLAB statement on the next line.
- As a default, MATLAB stores results in the variable ans.
- Only the first 19 characters of a variable name are remembered.
- Variable names must begin with a letter. After that, any ordering of letters, numbers, and the underscore is valid.

- MATLAB is case sensitive. `fruit`, `Fruit`, `FrUiT`, and `FRUIT` are all different variables.

5.3 Scientific Features

Like most scientific calculators, MATLAB offers many common functions important to mathematics, engineering, and the sciences. Moreover, MATLAB handles complex numbers with ease.

5.3.1 Common Mathematical Functions

A partial list of the common functions that MATLAB supports is shown in the table below. Most of these functions are used the same way you would write them mathematically:

```
» x=sqrt(2)/2
x =
    0.7071

» y=asin(x)
y =
    0.7854

» y_deg=y*180/pi
y_deg =
   45.0000
```

These commands find the angle where the sine function has a value of $\sqrt{2}/2$. While your calculator may work in degrees or radians, MATLAB only works in radians, where 2π radians is equal to 360 degrees. Other examples include:

```
» y=sqrt(3^2 + 4^2)  % show 3-4-5 are sides of a right triangle
y =
     5

» y=rem(23,4)
y =
     3

» x=2.6, y1=fix(x), y2=floor(x), y3=ceil(x), y4=round(x)
x =
    2.6000
y1 =
     2
y2 =
     2
y3 =
     3
y4 =
     3
```

See Section 8.2, *MATLAB Commands and Functions* for more specific information on these functions. Further illustration of these functions will appear throughout this tutorial.

Elementary Math Functions	
`abs(x)`	Absolute value or magnitude of complex number
`acos(x)`	Inverse cosine
`acosh(x)`	Inverse hyperbolic cosine
`angle(x)`	Angle of complex
`asin(x)`	Inverse sine
`asinh(x)`	Inverse hyperbolic sine
`atan(x)`	Inverse tangent
`atan2(x,y)`	Four quadrant inverse tangent
`atanh(x)`	Inverse hyperbolic tangent
`ceil(x)`	Round towards plus infinity
`conj(x)`	Complex conjugate
`cos(x)`	Cosine

Elementary Math Functions (Continued)	
`cosh(x)`	Hyperbolic cosine
`exp(x)`	Exponential: e^x
`fix(x)`	Round towards zero
`floor(x)`	Round towards minus infinity
`imag(x)`	Complex imaginary part
`log(x)`	Natural logarithm
`log10(x)`	Common logarithm
`real(x)`	Complex real part
`rem(x,y)`	Remainder after division: `rem(x,y)` gives the remainder of `x/y`
`round(x)`	Round toward nearest integer
`sign(x)`	Signum function: return sign of argument, e.g., `sign(1.2)=1`, `sign(–23.4)=–1`, `sign(0)=0`
`sin(x)`	Sine
`sinh(x)`	Hyperbolic sine
`sqrt(x)`	Square root
`tan(x)`	Tangent
`tanh(x)`	Hyperbolic tangent

5.3.2 Complex Numbers

To illustrate complex numbers, consider the quadratic equation $ax^2 + bx + c = 0$. The values of x where this equation is true, i.e., the roots of this equation, are given by

$$x_1, x_2 = \frac{-b \pm \sqrt{b^2 - 4ac}}{2a}$$

If $a = 1$, $b = 5$, and $c = 6$, the solution can be found using MATLAB as:

```
» a=1; b=5; c=6;
» x1=(-b+sqrt(b^2-4*a*c))/(2*a)
x1 =
    -3

» x2=(-b-sqrt(b^2-4*a*c))/(2*a)
x2 =
    -2

» a*x1^2+b*x1+c  % substitute x1 to check answer
ans =
     0

» a*x2^2+b*x2+c  % substitute x2 to check answer
ans =
     0
```

The last two statements were used to confirm the results. In this case, the term inside the square root is positive and the two roots are real numbers. However, if $a = 1$, $b = 4$, and $c = 13$ the solutions are

$$x_1, x_2 = \frac{-4 \pm \sqrt{4^2 - 4 \times 1 \times 13}}{2 \times 1} = \begin{cases} x_1 = -2 + 3\sqrt{-1} \\ x_2 = -2 - 3\sqrt{-1} \end{cases}$$

The $\sqrt{-1}$ term cannot be simplified further and as a result the solutions are said to be *complex*. The terms –2 in x_1 and x_2 are the *real* part of the solution. The terms 3 and –3 are the *imaginary* parts of the solutions. MATLAB computes the solutions in this case as:

```
» a=1; b=4; c=13;
» x1=(-b+sqrt(b^2-4*a*c))/(2*a)
x1 =
  -2.0000 + 3.0000i

» x2=(-b-sqrt(b^2-4*a*c))/(2*a)
x2 =
  -2.0000 - 3.0000i

» a*x1^2+b*x1+c  % substitute x1 to check answer
ans =
     0

» a*x2^2+b*x2+c  % substitute x2 to check answer
ans =
     0
```

x1 and x2 above display the complex results following the common form where a complex number is written as $a + bi$ in which a is the real part, b is the imaginary part, and $i = \sqrt{-1}$.

Entering complex numbers follows a similar approach. However, since the letter *j* rather than *i* is commonly associated with $\sqrt{-1}$ in engineering, it too can be used to denote the imaginary part. Examples of complex numbers are:

```
» c1=1-2i
c1 =
   1.0000 - 2.0000i

» c1=1-2j   % j also works
c1 =
   1.0000 - 2.0000i

» c2=3*(2-sqrt(-1)*3)
c2 =
   6.0000 - 9.0000i

» c3=sqrt(-2)
c3 =
        0 + 1.4142i

» c4=6+sin(.5)*i
c4 =
   6.0000 + 0.4794i

» c5=6+sin(.5)*j
c5 =
   6.0000 + 0.4794i
```

In the last two examples, the MATLAB default values of `i = j =` $\sqrt{-1}$ are used to form the imaginary part. Multiplication by `i` or `j` is required in these cases since `sin(.5)i` and `sin(.5)j` have no meaning to MATLAB. Termination with the characters `i` and `j`, as shown in the first two examples above, only works with simple numbers, not expressions.

Some programming languages require special handling for complex numbers wherever they appear. In MATLAB no special handling is required. Mathematical operations on complex numbers are written the same as those for real numbers:

```
» c6=(c1+c2)/c3  % from the above data
c6 =
  -7.7782 - 4.9497i

» check_it_out=i^2  % sqrt(-1) squared must be -1!
check_it_out =
  -1.0000 + 0.0000i

» check_it_out=real(check_it_out)  % show the real part
check_it_out =
    -1
```

In general, operations on complex numbers lead to complex numbers. Thus, even though $i^2 = -1$ is strictly real, MATLAB keeps the zero imaginary part. As shown, the MATLAB function `real` extracts the real part of a complex number.

As a final example of complex arithmetic, consider the Euler (sounds like *Oiler*) identity which relates the polar form of a complex number to its rectangular form:

$$M\angle\theta \equiv M \cdot e^{j\theta} = a + bi$$

where the polar form is given by a magnitude M and an angle θ, and the rectangular form is given by $a + bi$. The relationships among these forms are:

$$M = \sqrt{a^2 + b^2}$$
$$\theta = \tan^{-1}(b/a)$$
$$a = M\cos\theta$$
$$b = M\sin\theta$$

In MATLAB the conversion between polar and rectangular forms makes use of the functions `real`, `imag`, `abs`, and `angle`:

```
» c1
c1 =
   1.0000 - 2.0000i

» mag_c1=abs(c1)
mag_c1 =
    2.2361

» angle_c1=angle(c1)
angle_c1 =
   -1.1071

» deg_c1=angle_c1*180/pi
deg_c1 =
  -63.4349

» real_c1=real(c1)
real_c1 =
     1

» imag_c1=imag(c1)
imag_c1 =
    -2
```

The MATLAB function `abs` computes the magnitude of complex numbers or the absolute value of real numbers, depending upon which one you give it. Likewise, the MATLAB function `angle` computes the angle of a complex number in radians.

5.3.3 Summary

- MATLAB has many common mathematical functions.
- Complex numbers require no special treatment in MATLAB. You use them in rectangular form just as you would real numbers.
- The functions `real`, `imag`, `abs`, and `angle` are useful for converting between complex polar and rectangular forms.
- The default value of `i` and `j` is $\sqrt{-1}$. If you reassign them in some MATLAB command, they can no longer be used to create complex values. However, one can always assign them back or clear them.
- Appending `i` or `j` to the end of a number tells MATLAB to make the number the imaginary part of a complex number, e.g., `x=3i` produces `0 + 3i`.

5.4 Online Help

You probably have the sense that MATLAB has many more commands than you could possibly remember. To help you find commands, MATLAB provides assistance through its extensive *online help* capabilities. These capabilities are available in three forms: the MATLAB command `help`, the MATLAB command `lookfor`, and interactively using help from the menu bar.

5.4.1 The `help` Command

The MATLAB `help` command is the simplest way to get help if you know the topic you want help on. Typing `help <topic>` displays help about that topic if it exists, e.g.,

```
» help sqrt

 SQRT  Square root.
       SQRT(X) is the square root of the elements of X. Complex
       results are produced if X is not positive.

       See also SQRTM.
```

Here we received help on MATLAB's square root function. On the other hand,

```
» help cows

cows not found.
```

simply says that MATLAB knows nothing about cows.

The `help` command works well if you know the exact topic you want help on. Since many times this isn't true, `help` provides guidance to direct you to the exact topic you want by simply typing `help` with no `topic`:

```
» help

HELP topics:

matlab:general     - General purpose commands.
matlab:ops         - Operators and special characters.
matlab:lang        - Language constructs and debugging.
matlab:elmat       - Elementary matrices and matrix
                     manipulation.
matlab:specmat     - Specialized matrices.
matlab:elfun       - Elementary math functions.
matlab:specfun     - Specialized math functions.
matlab:matfun      - Matrix functions-numerical linear algebra.
matlab:datafun     - Data analysis and Fourier transform
                     functions.
matlab:polyfun     - Polynomial and interpolation functions.
matlab:funfun      - Function functions-nonlinear numerical
                     methods.
matlab:sparfun     - Sparse matrix functions.
matlab:plotxy      - Two dimensional graphics.
matlab:plotxyz     - Three dimensional graphics.
matlab:graphics    - General purpose graphics functions.
matlab:color       - Color control and lighting model functions.
matlab:strfun      - Character string functions.
matlab:iofun       - Low-level file I/O functions.
Toolbox:local      - Local function library.
Toolbox:sigsys     - Signals and Systems Toolbox.
Toolbox:symbolic   - Symbolic Math Toolbox.

For more help on directory/topic, type "help topic".
```

Your display may differ slightly from the above. In any case, this display describes categories from which you can ask for help. For example, `help general` returns a list (too long to show here) of general MATLAB topics that you can use the `help topic` command to get help on.

While the `help` command allows you to access help, it is not the most convenient way to do so unless you know the exact topic you are seeking help on. When you are uncertain about the spelling or existence of a topic, the other two approaches to obtaining help are often more productive.

5.4.2 The `lookfor` Command

The `lookfor` command provides help by searching through all the first lines of MATLAB help topics and returning those that contain a key word you specify. *Most important is that the key word need not be a MATLAB command.* For example:

```
» lookfor complex

CONJ       Complex conjugate.
IMAG       Complex imaginary part.
REAL       Complex real part.
CDF2RDF    Complex diagonal form to real block diagonal form.
RSF2CSF    Real block diagonal form to complex diagonal form.
CPLXPAIR   Sort numbers into complex conjugate pairs.
```

The key word `complex` is not a MATLAB command, but was found in the help descriptions of six MATLAB commands. You may find more or fewer. Given this information, the `help` command can be used to display help about a specific command, e.g.,

```
» help conj

 CONJ        Complex conjugate.
             CONJ(X) is the complex conjugate of X.
```

In summary, the `lookfor` command provides a way to find MATLAB commands and help topics given a general key word.

5.4.3 Menu-Driven Help

As an alternative to getting help in the command window, menu-driven help is available from the menu bar. On a Macintosh, menu-driven help is available by selecting **About MATLAB...** from the **Apple (🍎)** menu or by selecting

MATLAB Help from the **Balloon Help** menu. Selecting **About MATLAB...** produces the dialog box:

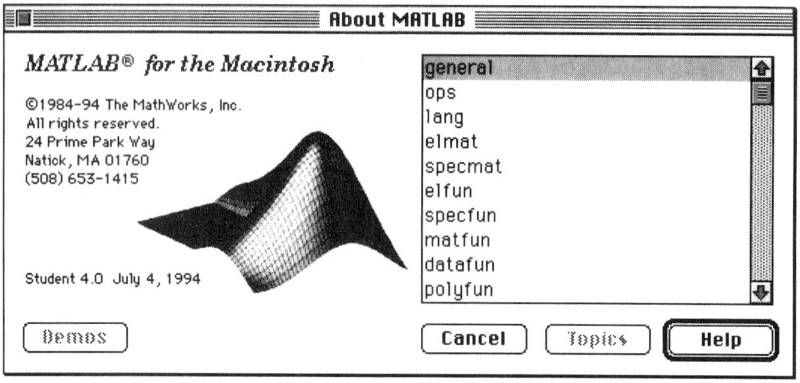

The list on the right is the categories within which you can ask for help. Following common Macintosh usage, double-clicking on or selecting **Help** on a category displays a list of topics for which help can be obtained. Double-clicking on or selecting **Help** on the topic of your choice brings up the help text for that topic. Selecting **Topics** returns you to the previous help listing. Selecting **Cancel** or the window close box closes the dialog box and returns you to the **Command** window. In case you're wondering, selecting **Demos** shows you some of the features of MATLAB.

On a PC, menu-driven help is available by selecting **Table of Contents...** or **Index...** from the **Help** menu. Doing so creates a **MATLAB Help** window that lets you double-click to select any topic or function from the displayed list. The **MATLAB Help** window uses the standard MS Windows help format that allows you to search for topics, set bookmarks, annotate topics, and print help screens.

5.5 Array Operations

All of the computations considered to this point have involved single numbers called *scalars*. Operations involving scalars are the basis of mathematics. At the same time, when one wishes to perform the same operation on more than one number at a time, repeated scalar operations are time-consuming and cumbersome.

5.5.1 Simple Arrays

Consider the problem of computing values of the sine function over one half of its period, namely: $y = \sin(x)$ over $0 \le x \le \pi$. Since it is impossible to compute $\sin(x)$ at all points over this range (there are an infinite number of them!), we must choose a finite number of points. In doing so we are *sampling* the function. To pick some number, let's evaluate $\sin(x)$ every 0.1π in this range, i.e., let $x = 0, 0.1\pi, 0.2\pi, \ldots, 1.0\pi$. If you were using your scientific calculator to compute these values, you would start by making a list or *array* of the values of x as shown below. Then you would enter each value of x into your calculator, find its sine, and write down the result as the second array shown below.

$$x = [0,\ 0.1\pi,\ 0.2\pi,\ 0.3\pi,\ 0.4\pi,\ 0.5\pi,\ 0.6\pi,\ 0.7\pi,\ 0.8\pi,\ 0.9\pi,\ \pi]$$

$$y = [0,\ 0.31,\ 0.59,\ 0.81,\ 0.95,\ 1.00,\ 0.95,\ 0.81,\ 0.59,\ 0.31,\ 0]$$

x and y are ordered lists of numbers, i.e., the first value or element in y is associated with the first value or element in x, the second element in y is associated with the second element in x, and so on. Because of this ordering, it is common to refer to individual values or elements in x and y with subscripts, e.g., x_1 is the first element in x, y_5 is the fifth element in y, x_n is the nth element in x.

Given the above array for x, what would be the easiest way for MATLAB to compute the associated values of y? It would be cumbersome to mimic the above hand calculations by entering each value of x as a separate variable, x_1, $x_2, x_3, \ldots, x_{11}$, then computing the sine of each term, $y_1 = \sin(x_1), y_2 = \sin(x_2), \ldots, y_{11} = \sin(x_{11})$. MATLAB must have a better way. As it turns out, MATLAB knows about arrays and handles them intuitively. Creating arrays is easy—just follow the visual organization given above:

```
» x=[0 .1*pi .2*pi .3*pi .4*pi .5*pi .6*pi .7*pi .8*pi .9*pi pi]
x =
  Columns 1 through 7
         0    0.3142    0.6283    0.9425    1.2566    1.5708    1.8850
  Columns 8 through 11
    2.1991    2.5133    2.8274    3.1416

» y=sin(x)
y =
  Columns 1 through 7
         0    0.3090    0.5878    0.8090    0.9511    1.0000    0.9511
  Columns 8 through 11
    0.8090    0.5878    0.3090    0.0000
```

To create an array in MATLAB all you have to do is start with a left bracket, enter the desired values separated by spaces (or commas), then close the array with a right bracket. Notice that finding the sine of the values in x follows naturally. MATLAB understands that you want to find the sine of each element in x and place the results in an associated array called y. This fundamental capability makes MATLAB different than other computer languages.

Since spaces separate array values, complex numbers entered as array values cannot have embedded spaces unless expressions are enclosed in parentheses. For example, `[1 -2i 3 4 5+6i]` contains five elements whereas the identical arrays `[(1 -2i) 3 4 5+6i]` and `[1-2i 3 4 5+6i]` contain four.

5.5.2 Array Addressing

Now since x in the above example has more than one element, namely it has 11 values separated into columns, MATLAB gives you the result back with the columns identified. As shown above, x is an array having one row and eleven columns; in other words it is a one-by-eleven array or simply an array of length 11.

In MATLAB, individual array elements are accessed using *subscripts*, e.g., `x(1)` is the first element in x, `x(2)` is the second element in x, and so on. For example:

```
» x(3)  % The third element of x
ans =
    0.6283

» y(5)  % The fifth element of y
ans =
    0.9511
```

To access a block of elements at one time MATLAB provides *colon notation*:

```
» x(1:5)
ans =
         0    0.3142    0.6283    0.9425    1.2566
```

This is the first through fifth elements in x. `1:5` says “start with 1 and count up to 5.”

```
» y(3:-1:1)
ans =
    0.5878    0.3090         0
```

This is the third, second, and first elements in reverse order. `3:-1:1` says “start with 3, count down by 1, and stop at 1.”

```
» x(2:2:7)
ans =
    0.3142    0.9425    1.5708
```

This is the second, fourth, and sixth elements in x. 2:2:7 says "start with 2, count up by 2, and stop when you get to 7." In this case adding 2 to 6 gives 8, which is greater than 7 so the eighth element is not included.

```
» y([8 2 9 1])
ans =
    0.8090    0.3090    0.5878         0
```

Here we used another array [8 2 9 1] to extract the elements of the array y in the order we wanted them! The first element taken is the eighth, the second is the second, the third is the ninth, and the fourth is the first. Try further examples by yourself. For example, try to pick out the fifteenth element (it doesn't exist, so see what MATLAB does).

5.5.3 Array Construction

Earlier we entered the values of x by typing each individual element in x. While this is fine when there are only 11 values in x, what if there are 111 values? Using the colon notation, two other ways of entering x are:

```
» x=(0:0.1:1)*pi
x =
  Columns 1 through 7
        0    0.3142    0.6283    0.9425    1.2566    1.5708    1.8850
  Columns 8 through 11
   2.1991    2.5133    2.8274    3.1416

» x=linspace(0,pi,11)
x =
  Columns 1 through 7
        0    0.3142    0.6283    0.9425    1.2566    1.5708    1.8850
  Columns 8 through 11
   2.1991    2.5133    2.8274    3.1416
```

In the first case above, the colon notation (0:0.1:1) creates an array that starts at 0, increments or counts by 0.1, and ends at 1. Each element in this array is then multiplied by π to create the desired values in x. In the second case, the MATLAB function linspace is used to create x. This function's arguments are described by

```
linspace(first_value,last_value,number_of_values)
```

Both of these array creation forms are common in MATLAB. The colon notation form allows you to directly specify the increment between data points, but not the number of data points. linspace, on the other hand, allows you to directly specify the number of data points, but not the increment between the data points.

Both of the above array creation forms create arrays where the individual elements are linearly spaced with respect to each other. For the special case where a logarithmically spaced array is desired, MATLAB provides the `logspace` function:

```
» logspace(0,2,11)
ans =
  Columns 1 through 7
   1.0000    1.5849    2.5119    3.9811    6.3096   10.0000   15.8489
  Columns 8 through 11
  25.1189   39.8107   63.0957  100.0000
```

Here, we created an array starting at 10^0, ending at 10^2, containing 11 values. The function arguments are described by

```
logspace(first_exponent,last_exponent,number_of_values)
```

Though it is common to begin and end at integer powers of ten, `logspace` works equally well with nonintegers.

Sometimes an array is required that is not conveniently described by a linearly or logarithmically spaced element relationship. There is no uniform way to create these arrays. However, array addressing and the ability to combine expressions can help eliminate the need to enter individual elements one at a time:

```
» a=1:5,b=1:2:9
a =
     1     2     3     4     5
b =
     1     3     5     7     9
```

creates two arrays. Remember that multiple statements can appear on a single line if they are separated by commas or semicolons.

```
» c=[b a]
c =
     1     3     5     7     9     1     2     3     4     5
```

creates an array `c` composed of the elements of `b` followed by those of `a`.

```
» d=[a(1:2:5) 1 0 1]
d =
     1     3     5     1     0     1
```

creates an array `d` composed of the first, third, and fifth elements of `a` followed by three additional elements.

5.5.4 Scalar-Array Mathematics

In the first example above, the array x is multiplied by the scalar π. Other simple mathematical operations between scalars and arrays follow the same natural interpretation. Addition, subtraction, multiplication, and division by a scalar simply applies the operation to all elements of the array:

```
» a–2
ans =
    –1    0    1    2    3
```

subtracts 2 from each element in a.

```
» 2*a–1
ans =
     1    3    5    7    9
```

multiplies each element in a by 2 and subtracts 1 from each element of the result. Note that scalar-array mathematics uses the same order of precedence used in scalar expressions to determine the order of evaluation.

5.5.5 Array-Array Mathematics

Mathematical operations between arrays are not quite as simple as those between scalars and arrays. Clearly, array operations between arrays of different lengths are difficult to define and of even more dubious value. *However, when two arrays have the same length, addition, subtraction, multiplication, and division apply on an element-by-element basis.* For example:

```
» a,b
a =
     1    2    3    4    5
b =
     1    3    5    7    9
```

recalls the arrays used earlier.

```
» a+b
ans =
     2    5    8   11   14
```

adds the two arrays element by element and places the result in the default variable ans.

```
» ans–b
ans =
     1    2    3    4    5
```

subtracts b from the most recent result, giving us back the values in a.

```
» 2*a–b
ans =
      1     1     1     1     1
```

multiplies all elements of a by 2, subtracts b from them, and places the result in ans. Note that array-array mathematics also uses the same order of precedence used in scalar expressions to determine the order of evaluation.

Element-by-element multiplication and division work similarly but use slightly unconventional notation:

```
» a.*b
ans =
      1     6     15     28     45
```

Here we multiplied the arrays a and b element by element using the *dot multiplication* symbol .*. The dot preceding the standard asterisk multiplication symbol tells MATLAB to perform element-by-element array multiplication. Multiplication without the dot signifies *matrix multiplication*, which will be discussed in Section 5.10 of this tutorial. For this particular example, matrix multiplication is not defined:

```
»a*b
??? Error using ==> *
Inner matrix dimensions must agree.
```

Array division, or *dot division*, also requires use of the dot symbol:

```
» a./b
ans =
    1.0000    0.6667    0.6000    0.5714    0.5556

»b.\a
ans =
    1.0000    0.6667    0.6000    0.5714    0.5556
```

As with scalars, division is defined using both the forward and backward slashes. In both cases, the array below the slash is divided into the array above the slash.

Division without the dot is the *matrix division* operation, which is an entirely different operation:

```
»a/b
ans =
     0.5758

»a\b
ans =
           0          0          0          0          0
           0          0          0          0          0
           0          0          0          0          0
           0          0          0          0          0
      0.2000     0.6000     1.0000     1.4000     1.8000
```

Matrix division gives results that are not the same size as a and b. Matrix operations are discussed in Section 2.10 of this tutorial.

Array powers are defined in several ways. As with multiplication and division, ^ is reserved for matrix powers and .^ is used to denote element-by-element powers:

```
» a,b
a =
      1     2     3     4     5
b =
      1     3     5     7     9
```

recalls the arrays used earlier.

```
» a.^2
ans =
      1     4     9    16    25
```

squares the individual elements of a.

```
» 2.^a
ans =
      2     4     8    16    32
```

raises 2 to the power of each element in the array a.

```
» b.^a
ans =
              1              9            125           2401          59049
```

raises the elements of b to the corresponding elements in a.

```
» b.^(a-3)
ans =
     1.0000     0.3333     1.0000     7.0000    81.0000
```

shows that scalar and array operations can be combined.

5.5.6 Array Orientation

In the above examples, arrays contained one row and multiple columns. As a result of this row orientation they are commonly called *row vectors*. It is also possible for an array to be a *column vector* having one column and multiple rows. In this case, all of the above array manipulation and mathematics apply without change. The only difference is that results are displayed as columns rather than rows.

Since the array creation functions illustrated above all create row vectors, there must be some way to create column vectors. The most straightforward way to create a column vector is to specify it element by element and separating values with semicolons:

```
» c=[1;2;3;4;5]
c =
     1
     2
     3
     4
     5
```

Based on this example, *separating elements by spaces or commas specifies elements in different columns; separating elements by semicolons specifies elements in a different rows.*

To create a column vector using the colon notation `start:increment:end`, or the functions `linspace` and `logspace`, one must *transpose* the resulting row into a column using the MATLAB transpose operator ('):

```
» a=1:5
a =
     1     2     3     4     5
```

creates a row vector using the colon notation format.

```
» b=a'
b =
     1
     2
     3
     4
     5
```

uses the transpose operator to change the row vector a into the column vector b.

```
» c=b'
c =
     1     2     3     4     5
```

applies the transpose again and changes the column back to a row.

In addition to the simple transpose above, MATLAB also offers a transpose operator with a preceding dot. In this case the *dot-transpose operator* is interpreted as the non-complex conjugate transpose. When an array is complex, the transpose (`'`) gives the complex conjugate transpose, i.e., the sign on the imaginary part is changed as part of the transpose operation. On the other hand, the dot-transpose (`.'`) transposes the array but does not conjugate it.

```
» c=a.'
c =
     1
     2
     3
     4
     5
```

shows that `.'` and `'` are identical for real data.

```
» d=a+i*a
d =
  Columns 1 through 4
  1.0000+1.0000i 2.0000+2.0000i 3.0000+3.0000i 4.0000+4.0000i
  Column 5
  5.0000 + 5.0000i
```

creates a simple complex row vector from the array a using the default value `i=sqrt(-1)`.

```
» e=d'
e =
   1.0000 - 1.0000i
   2.0000 - 2.0000i
   3.0000 - 3.0000i
   4.0000 - 4.0000i
   5.0000 - 5.0000i
```

creates a column vector `e` that is the *complex conjugate* transpose of `d`.

```
» f=d.'
f =
   1.0000 + 1.0000i
   2.0000 + 2.0000i
   3.0000 + 3.0000i
   4.0000 + 4.0000i
   5.0000 + 5.0000i
```

creates a column vector `f` that is the transpose of `d`.

Earlier in the array-array mathematics section it was stated "when two arrays have the same length, addition, subtraction, multiplication, and division apply on an element by element basis." Now that it is known that arrays or vectors can have row or column orientation, this statement must be restated to include orientation: When two vectors have the same length and *orientation*, addition, subtraction, multiplication, and division apply on an element by element basis. In other words, operations between arrays of different orientations are either not defined or are not defined on an element-by-element basis.

If an array can have one row and multiple columns (a row vector) or one column and multiple rows (a column vector), it makes intuitive sense that arrays could just as well have both multiple rows and multiple columns. In particular, this *rectangular* orientation is useful when all rows have the same number of columns. Arrays having multiple rows and columns are called *matrices*. Creation of matrices follows that of row and column vectors. Commas or spaces are used to separate elements in a specific row, and semicolons are used to separate individual rows:

```
» g=[1 2 3 4;5 6 7 8]
g =
        1        2        3        4
        5        6        7        8
```

Here g is an array or matrix having 2 rows and 4 columns, i.e., it is a 2-by-4 matrix. The semicolon tells MATLAB to start a new row between the 4 and 5.

```
» g=[1 2 3 4
5 6 7 8
9 10 11 12]
g =
        1        2        3        4
        5        6        7        8
        9       10       11       12
```

In addition to semicolons, pressing the **Return** or **Enter** key while entering a matrix also tells MATLAB to start a new row.

```
» h=[1 2 3;4 5 6 7]
???  All rows in the bracketed expression must have the same
number of columns.
```

MATLAB strictly enforces the fact that all rows must contain the same number of columns.

Given this rectangular array format, the array-array mathematical concepts discussed earlier apply as long as the size, i.e., the number of rows and columns, of the arrays being operated on are identical.

5.5.7 Other Features

Earlier in the discussion of scalars, the who command was illustrated as a command that displays the names of all user-created variables. In the case of arrays, it is also important to know their size. In MATLAB the command `whos` provides this additional information:

```
» whos
  Name      Size      Elements     Bytes    Density   Complex

  a        1 by 5         5          40       Full       No
  b        5 by 1         5          40       Full       No
  c        1 by 5         5          40       Full       No
  d        1 by 5         5          80       Full       Yes
  e        5 by 1         5          80       Full       Yes
  f        5 by 1         5          80       Full       Yes
  g        3 by 4        12          96       Full       No

Grand total is 42 elements using 456 bytes

leaving 2316084 bytes of memory free.
```

In addition to the names and sizes of the variables, `whos` identifies the total number of elements in each variable, the total number of bytes occupied, whether the variable is full or sparse, and whether the data contains complex data. Of these pieces of information, `Density` is beyond the scope of this text and can be ignored.

5.5.8 Summary

- Arbitrary arrays are entered as: variable = [(a list of numbers separated by spaces or commas)], e.g., `x=[4 1 sqrt(2) sin(.5) 23.2]`.
- Functions of arrays apply the function to individual elements of the array.
- Arrays containing linearly spaced data elements can be formed using the colon notation `first:increment:last` or `first:last`. In the latter form, the increment is assumed to be one.
- Arrays containing linearly spaced data elements can be formed using the MATLAB function `linspace(first, last, number_of_points)`.
- Arrays containing logarithmically spaced data elements can be formed using the MATLAB function `logspace(first_exp, last_exp, number_of_points)`.
- Values within a data array can be addressed by using subscripts, e.g., `x(2:6)` is the second through sixth elements of `x`.
- Basic mathematical operations between scalars and arrays apply the operation to all elements in the array, e.g., `x–2` subtracts two from all elements in `x`.

- Basic mathematical operations between arrays are valid if the arrays are the same size. Again, the result is given by the element-by-element application of the operation. Multiplication and division are symbolized by `.*` and `./` or `.\`, respectively.

- Array powers have multiple definitions depending on the form of the expression. For example, if `a` and `b` are arrays, `a.^(0.5)` finds the square root of the elements of `a`, `2.^b` raises `2` to the powers given in `b`, and `a.^b` raises each element in `a` by the corresponding element in `b`. As with multiplication and division, the dot notation is necessary.

- Arrays can be built from subsets of other arrays. That is, if `x` and `y` are row oriented arrays, `z=[x y]` produces an array composed of `x` followed by `y` and `z=[x(1:5) -x(6:10) 5.6 2 10 y(5:23)]` produces an array composed of subarrays of `x`, `y` and several additional values.

- MATLAB arrays can have any rectangular orientation. Arrays containing one row are commonly called row vectors. Arrays containing one column are commonly called column vectors. Arrays containing multiple rows and columns are commonly called matrices.

- Row vectors can be converted to column vectors and vice-versa by using the dot-apostrophe transpose operation, (`.'`), e.g., if `x` is a row, `x.'` is a column.

- Dot-apostrophe denotes transpose. A simple apostrophe denotes complex conjugate transpose.

- The MATLAB command `whos` (as opposed to `who`) displays MATLAB variable names as well as the row and column sizes of the variables.

The following table illustrates basic array operations.

MATLAB Array Operations	
Illustrative data: `a = [a1 a2...an], b = [b1 b2...bn], c= {a scalar}`	
Scalar addition	`a+c = [a1+c a2+c...an+c]`
Scalar multiplication	`a*c = [a1*c a2*c...an*c]`
Array addition	`a+b = [a1+b1 a2+b2...an+bn]`
Array multiplication	`a.*b = [a1*b1 a2*b2...an*bn]`
Array right division	`a./b = [a1/b1 a2/b2...an/bn]`
Array left division	`a.\b = [a1\b1 a2\b2...an\bn]`

Array powers	`a.^c = [a1^c a2^c...an^c]`
	`c.^a = [c^a1 c^a2...c^an]`
	`a.^b = [a1^b1 a2^b2...an^bn]`

5.6 Simple Plots

Before rigorously considering MATLAB's extensive graphics capabilities, let's generate some simple plots. Plots are a powerful visual way to interpret data. Consider the process you would follow to plot a sine function over one period by hand, i.e., $y = \sin(x)$ for $0 \le x \le 2\pi$. As discussed earlier, you first choose data points for the independent variable x. This data forms the horizontal axis of the plot. Then, the sine of each data point is found, which provides the vertical axis data. Each pair of data points $\{x_n, y_n\}$ is then marked on a suitable set of axes. Finally, to guide the eye, it is common to draw a straight line between each marked point.

By using arrays, MATLAB follows this same approach to plotting. Consider the task of plotting a sine wave as described above. First, create 30 points between 0 and 2π.

```
» x=linspace(0,2*pi,30);
```

Then find the sine of the points in x.

```
» y=sin(x);
```

The plot command generates a plot:

```
» plot(x,y)
```

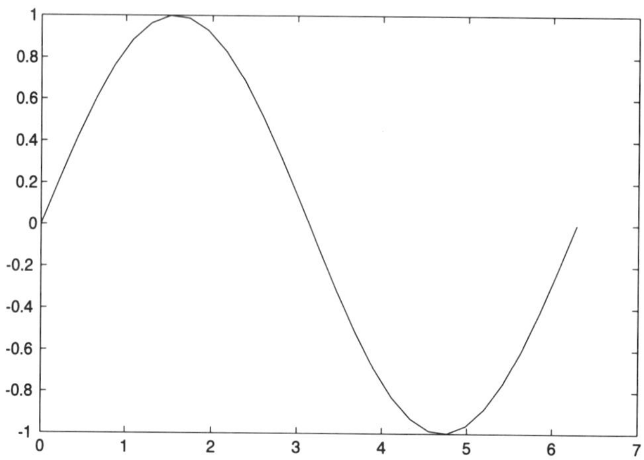

The MATLAB function plot is extremely powerful. It automatically chooses axis limits, marks the individual data points, and draws straight lines between them. Options in the plot command allow you to plot multiple data sets on the same axes, use different line types such as dotted and dashed, mark just the data points without drawing lines between them, use different colors for curves. In addition you can place labels on the axes, a title at the top, draw a grid at the tick marks, and so one. To illustrate some of these features consider the following examples.

A plot of both sine and cosine can be made on the same axes.

```
» z=cos(x);
» plot(x,y,x,z)
```

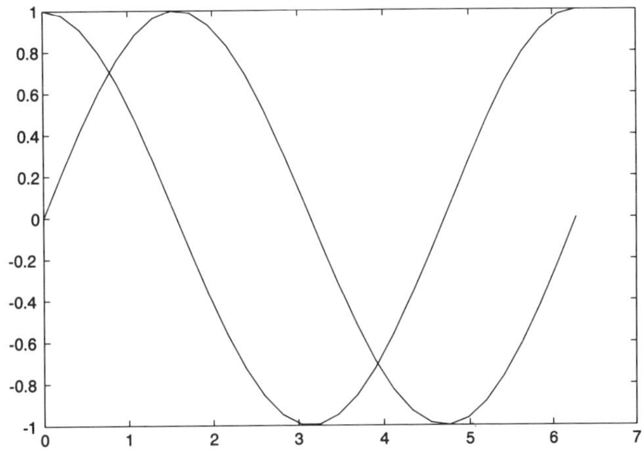

The following code plots sine twice, once with lines connecting the data points, the second with the data points marked with the symbol +.

```
» plot(x,y,x,y,'+')
```

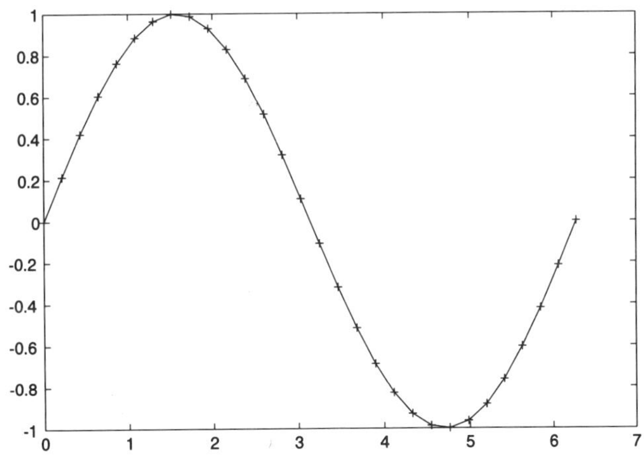

This code plots sine versus cosine.

```
» plot(y,z)
```

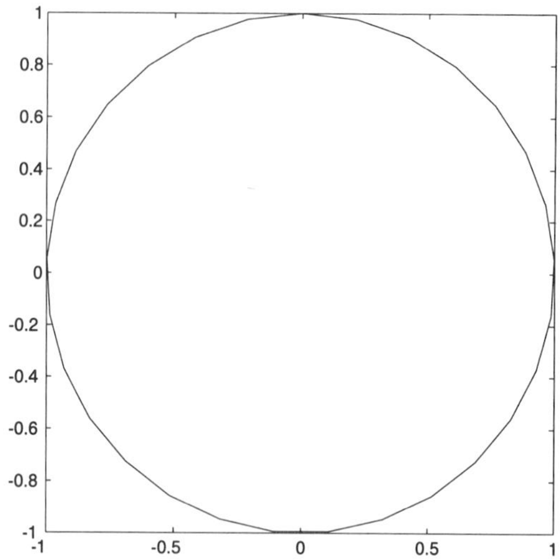

The following example illustrates the trig identity 2sinθcosθ = sin2θ. The plot of 2sinθcosθ is plotted using dashed lines.

```
» plot(x,y,x,2*y.*z,'--')
```

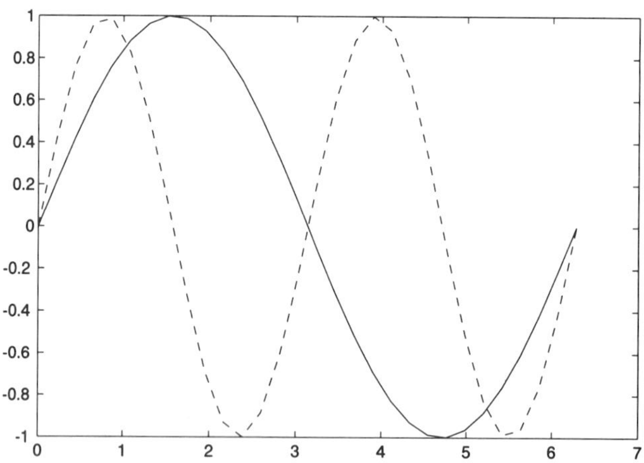

To place a grid at the tick marks of the current plot, use

```
» grid
```

To place an x-axis label on the current plot, use

```
» xlabel('independent variable X')
```

To place a y-axis label on the current plot, use

```
» ylabel('dependent variables')
```

To place a title on the current plot, use

```
» title('2sin(x)cos(x) = sin(2x)')
```

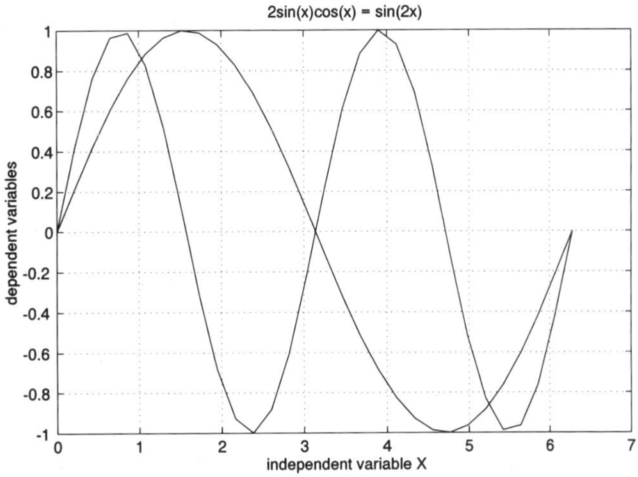

The following example generates a 3-D plot with a grid.

```
» plot3(y,z,x),grid
```

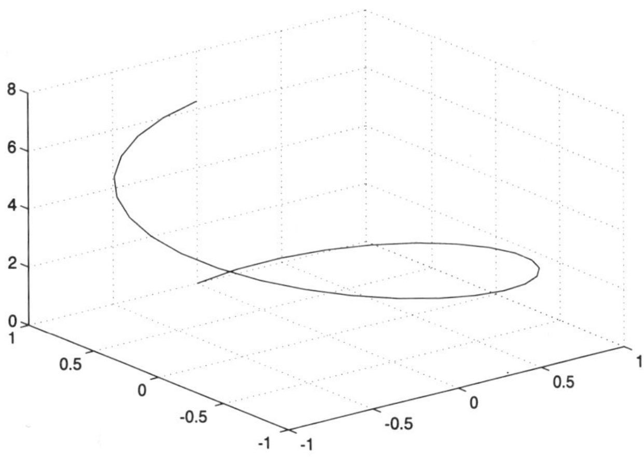

Hopefully this introduction to plotting gives you enough information to make simple plots. More comprehensive information can be found in Sections 5.19 and 5.20 of this tutorial as well as in Section 8.2, *MATLAB Commands and Functions*. Online help can also be used, e.g., `help plot`.

5.7 Script Files

By now it should be apparent that MATLAB answers whatever you ask of it in the order in which you ask it. For simple problems, entering your requests at the MATLAB prompt is fast and efficient. However, as the number of commands increases or in the case when you wish to change the value of one or more variables and reevaluate a number of commands, typing at the MATLAB prompt becomes tedious. MATLAB provides a logical solution to this problem. It allows you to place MATLAB commands in a simple text file and then tell MATLAB to open the file and evaluate commands exactly as it would if you had typed them at the MATLAB prompt. These files are called script files or M-files. The term *script* symbolizes the fact that MATLAB simply follows the script found in the file. The term *M-file* recognizes the fact that script file names must end with the extension 'm', e.g., `test.m`.

To create a script M-file choose **New** from the **File** menu and select **M-file**. This procedure brings up a text editor window where you enter MATLAB commands.

After this file is saved as the M-file `example.m` on your disk, MATLAB will execute the commands in `example.m` by simply typing `example` at the MATLAB prompt:

```
» example
```

When MATLAB interprets this statement, it prioritizes current MATLAB variables and built-in MATLAB commands ahead of M-file names. Thus, if `example` is not a current MATLAB variable or a built-in MATLAB command (it isn't) MATLAB opens the file `example.m` (if it can find it) and evaluates the commands found there just as if they were entered directly at the **Command** window prompt. As a result, commands within the M-file have access to all variables in the MATLAB workspace and all variables created in the M-file become part of the workspace. Normally, the commands read in from the M-file are not displayed as they are evaluated. The `echo on` command tells MATLAB to display or echo commands to the **Command** window as they are read and evaluated. You can probably guess what the `echo off` command does.

With this M-file feature it is simple to answer *what if?* questions. For example, one can repeatedly open the `example.m` M-file using the **Open M-file** item in the **File** menu, change the number of apples, bananas, or cantaloupes, save the file, then ask MATLAB to reevaluate the commands in the file. The power of this capability cannot be overstated. Moreover, by creating M-files, your commands are saved on disk for future MATLAB sessions.

Script files are also convenient for entering large arrays that may, for example, come from laboratory measurements. By using a text editor to enter one or more arrays, the editing capabilities of the editor make it is easy to correct mistakes without having to type the whole array in again. As above, this approach also saves the data on disk for future use.

The utility of MATLAB comments is readily apparent when using script files as shown in `example.m`. Comments allow you to document the commands found in a script file so that they are not forgotten when viewed in the future. In addition, the use of semicolons at the end of lines to suppress the display of results allows you to control script file output so that only important results are shown.

MATLAB provides several file management commands that allow you to list file names, view and delete M-files, show and change the current directory or folder. A summary of these commands is given in the following table.

Command	Description
`what`	Return a listing of all M-files in the current directory or folder.
`dir`	List all files in the current directory or folder.
`ls`	Same as `dir`.
`type test`	Display the M-file `test.m` in the command window.
`delete test`	Delete the M-file `test.m`.
`cd path`	Change to directory or folder given by `path`.
`chdir path`	Same as `cd path`.
`cd`	Show present working directory or folder.
`chdir`	Same as `cd`.
`pwd`	Same as `cd`.
`which test`	Display the directory path to `test.m`.

5.8 Text

MATLAB's true power is in its ability to crunch numbers. However, there are times when it is desirable to manipulate text, such as when putting labels and titles on plots. In MATLAB, text is called *character strings* or simply *strings*. Character strings are handled just like row vectors:

```
» t='How about this character string?'
t =
How about this character string?
```

A character string is simply text surrounded by single quotes.

```
» u=t(16:24)
u =
character
```

Strings are addressed just like arrays. Here elements 16 through 24 are the word `character`.

```
» u=t(24:-1:16)
u =
retcarahc
```

Array manipulation also works. This is the word `character` spelled backward.

```
» v=['Character strings having more than'
     'one row must have the same number '
     'of columns just like matrices!    ']
v =
Character strings having more than
one row must have the same number
of columns just like matrices!
```

As with matrices, character strings can have multiple rows, but each row must have an equal number of columns. Therefore blanks are required above to make all rows the same length. Since character strings are just arrays with quotes, all the previous material on array manipulation applies. Mathematical operations on strings are possible too! However, once you perform some mathematical operation on a string, the string no longer displays as a string. Instead it is displayed as an array of numbers in the *ASCII* standard. This standard associates the integers 0 to 255 with 256 different characters that includes all the character keys on your keyboard. To see the ASCII representation of a string, take its absolute value or add zero to it:

```
» s='ABCDEFG'
s =
ABCDEFG

» m=abs(s)
m =
    65    66    67    68    69    70    71

» m=s+0
m =
    65    66    67    68    69    70    71
```

As you can see, the letters of the alphabet are in numerical order. It is now clear that strings and arrays are really the same. After a string has been converted to its ASCII representation, the string can be restored so that it displays as a character string by using the MATLAB function `setstr`:

```
» setstr(m)
ans =
ABCDEFG
```

How about adding 5 to s then converting back to a string:

```
» n=s+5
n =
     70    71    72    73    74    75    76

» setstr(n)
ans =
FGHIJKL
```

Finally, change to lower case characters by adding the difference between a and A:

```
» n=s+'a'-'A'
n =
     97    98    99   100   101   102   103

» setstr(n)
ans =
abcdefg
```

5.9 Relational and Logical Operations

In addition to traditional mathematical operations, MATLAB supports relational and logical operations. You may be familiar with these if you've had some experience with other programming languages. The purpose of these operators and functions is to provide answers to True/False questions. One important use of this capability is to control the flow or order of execution of a series of MATLAB commands (usually in a M-file) based on the results of True/False questions.

As inputs to all relational and logical expressions, MATLAB considers any nonzero number to be True and zero to be False. The output of all relational and logical expressions produces 1 for True and 0 for False.

5.9.1 Relational Operators

MATLAB relational operators include all common comparisons:

Operator	Description
<	Less than
<=	Less than or equal to
>	Greater than
>=	Greater than or equal to
==	Equal to
~=	Not equal to

MATLAB relational operators can be used to compare two arrays of the same size, or to compare an array to a scalar. In the latter case, the scalar is compared with all elements of the array and the result has the same size as the array. Some examples include:

```
» A=1:9,B=9-A
A =
     1     2     3     4     5     6     7     8     9
B =
     8     7     6     5     4     3     2     1     0

» tf=A>4
tf =
     0     0     0     0     1     1     1     1     1
```

finds elements of `A` that are greater than 4. Zeros appear in the result where `A` is not greater than 4 and ones appear where `A > 4`.

```
» tf=A==B
tf =
     0     0     0     0     0     0     0     0     0
```

finds elements of `A` that are equal to those in `B`. Notice that = and == mean two different things: == compares two variables and returns ones where they are equal and zeros where they are not; = on the other hand is used to assign the output of an operation to a variable.

```
» tf=B-(A>2)
tf =
      8     7     5     4     3     2     1     0    -1
```

finds where A > 2 and subtracts the result from B. This example shows that since the output of logical operations are numerical arrays of ones and zeros, they can be used in mathematical operations too.

```
» B=B+(B==0)*eps
B =
  Columns 1 through 7
    8.0000 7.0000 6.0000 5.0000 4.0000 3.0000 2.0000
  Columns 8 through 9
    1.0000 0.0000
```

is a demonstration of how to replace zero elements in an array with the special MATLAB number eps, which is approximately 2.2e–16 (or 2.2×10^{-16}). This particular expression is sometimes useful to avoid dividing by zero as in:

```
» x=(-3:3)/3
x =
   -1.0000 -0.6667 -0.3333       0 0.3333 0.6667 1.0000

» sin(x)./x

Warning: Divide by zero
ans =
    0.8415 0.9276 0.9816      NaN 0.9816 0.9276 0.8415
```

computing the function $\sin(x)/x$ gives a warning because the fifth data point is zero. Since sin(0)/0 is undefined, MATLAB returns NaN (meaning Not-a-Number) at that location in the result. Try again, after replacing the zero with eps:

```
» x=x+(x==0)*eps;

» sin(x)./x
ans =
    0.8415 0.9276 0.9816 1.0000 0.9816 0.9276 0.8415
```

Now $\sin(x)/x$ for $x = 0$ gives the correct limiting answer.

5.9.2 Logical Operators

Logical operators provide a way to combine or negate relational expressions. MATLAB logical operators include:

Operator	Description
&	AND
|	OR
~	NOT

Some examples of the use of logical operators are:

```
» A=1:9;B=9–A;
» tf=A>4
tf =
     0     0     0     0     1     1     1     1     1
```

finds where A is greater than 4.

```
» tf=~(A>4)
tf =
     1     1     1     1     0     0     0     0     0
```

negates the above result, i.e., swaps where the ones and zeros appear.

```
» tf=(A>2)&(A<6)
tf =
     0     0     1     1     1     0     0     0     0
```

returns ones where A is greater than 2 AND less than 6.

Finally, the above capabilities make it easy to generate arrays representing signals with discontinuities or signals that are composed of segments of other signals. The basic idea is to multiply those values in an array that you wish to keep with ones, and multiply all other values with zeros. For example:

```
» x=linspace(0,10,100); % create data
» y=sin(x);             % compute sine
» z=(y>=0).*y;          % set negative values of sin(x) to zero
» z=z + 0.5*(y<0);      % where sin(x) is negative add 1/2
» z=(x<=8).*z;          % set values past x=8 to zero
» plot(x,z)
» xlabel('x'),ylabel('z=f(x)'),
» title('A Discontinuous Signal')
```

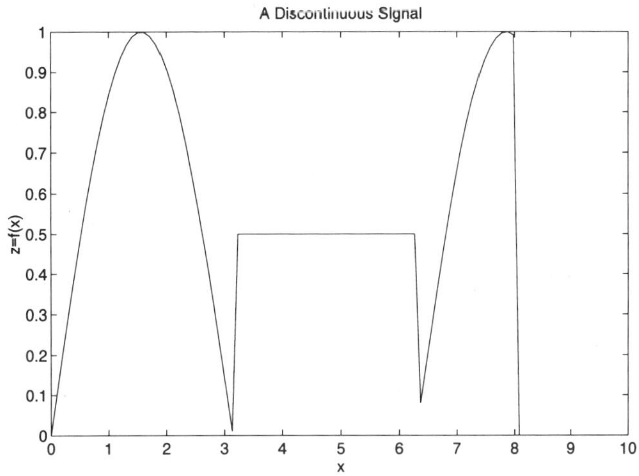

In addition to the above basic relational and logical operators, MATLAB provides a number of additional relational and logical functions including:

Function	Description
`xor(x,y)`	Exclusive OR operation. Return ones where either x or y is nonzero (True). Return zeros where both x and y are zero (False) or both are nonzero (True).
`any(x)`	Return one if any element in a vector x is nonzero. Return one for each column in a matrix x that has nonzero elements.
`all(x)`	Return one if all elements in a vector x are nonzero. Return one for each column in a matrix x that has all nonzero elements.
`isnan(x)`	Return ones at NaNs in x.
`isinf(x)`	Return ones at Infs in x.
`finite(x)`	Return ones at finite values in x.

The following chart shows the order of precedence for arithmetic, logical, and relational orders, with the top row taking precedence.

MATLAB Operator Preference Table
^ .^ ' .'
* / \ .* ./ .\
+ – ~ +(unary) –(unary)
: > < >= <= == ~=
| &

5.10 Linear Algebra and Matrices

5.10.1 Major Features

The material in this section assumes that you have at least some familiarity with linear algebra. If you do not, the material presented here may be confusing. If you would like to know more about linear algebra, please refer to a calculus text or perhaps a text specifically on linear algebra.

Historically, MATLAB was written to simplify the linear algebra computations that appear in many applications. At the heart of many applications is the problem of solving a set of linear equations. To illustrate this particular problem, consider the following example:

> *Homer buys one apple, two bananas, and three cantaloupes and pays \$3.66. Marge buys four apples, five bananas, and six cantaloupes and pays \$8.04. Bart doesn't like cantaloupe so he buys seven apples, eight bananas, and no cantaloupes and pays \$3.51. What was the individual price of each fruit?*

To solve this problem, write each statement as a mathematical statement. If we let x_1 be the price per apple, x_2 the price per banana, and x_3 the price per cantaloupe, then

$$\begin{aligned} 1x_1 + 2x_2 + 3x_3 &= 366 \\ 4x_1 + 5x_2 + 6x_3 &= 804 \\ 7x_1 + 8x_2 + 0x_3 &= 351 \end{aligned}$$

describes the relationships given in the problem statement where cost is expressed in cents for convenience. These equations can be organized into a *matrix* equation by the progression of statements:

$$\begin{bmatrix} 1x_1 + 2x_2 + 3x_3 = 366 \\ 4x_1 + 5x_2 + 6x_3 = 804 \\ 7x_1 + 8x_2 + 0x_3 = 351 \end{bmatrix}$$

$$\begin{bmatrix} 1x_1 + 2x_2 + 3x_3 \\ 4x_1 + 5x_2 + 6x_3 \\ 7x_1 + 8x_2 + 0x_3 \end{bmatrix} = \begin{bmatrix} 366 \\ 804 \\ 351 \end{bmatrix}$$

$$\begin{bmatrix} 1 & 2 & 3 \\ 4 & 5 & 6 \\ 7 & 8 & 0 \end{bmatrix} \times \begin{bmatrix} x_1 \\ x_2 \\ x_3 \end{bmatrix} = \begin{bmatrix} 366 \\ 804 \\ 351 \end{bmatrix}$$

$$Ax = b$$

where the multiplication is now defined in the matrix sense as opposed to the array sense discussed earlier. In MATLAB, this matrix multiplication is denoted with the asterisk notation `*`. The above equations define the matrix product of the matrix A and the vector x as being equal to the vector b. Recall that array operations between arrays of different sizes and shapes are undefined. However, in linear or matrix algebra there are strict rules beyond the scope of this tutorial that govern mathematical operations between matrices. If you are interested in learning more, linear algebra can be found in most calculus texts.

The existence of solutions to the above equation is a fundamental question in linear algebra. Moreover, when a solution does exist there are numerous approaches to finding the solution such as Gaussian elimination, LU factorization, or direct use of A^{-1}. Perhaps you recall solving problems like this in a math course. Analytically the solution is written as $x = A^{-1}b$. It is beyond the scope of this tutorial to discuss the many analytical and numerical issues of linear or matrix algebra. We only wish to demonstrate how MATLAB can be used to solve problems like the one above.

To solve the above problem it is necessary to enter A and b:

```
» A=[1 2 3;4 5 6
7 8 0]
A =
        1    2    3
        4    5    6
        7    8    0

» b=[366;804;351]
b =
    366
    804
    351
```

As discussed earlier, the entry of the matrix A shows the two ways that MATLAB distinguishes between rows. The semicolon between the 3 and 4 signifies the start of a new row as does the new line between the 6 and 7. The vector b is a column because each semicolon signifies the start of a new row.

Without explanation, this problem has a unique answer whenever the *determinant* of the matrix A is nonzero:

```
» det(A)
ans =
     27
```

Since this is true, MATLAB can find the solution of $A\ x = b$ in two ways, one of which is preferred. The less favorable but more straightforward method is to take $x = A^{-1}b$ literally:

```
» x=inv(A)*b
x =
    25.0000
    22.0000
    99.0000
```

Here inv(A) is a MATLAB function that computes A^{-1} and the matrix multiplication operator *, without a preceding dot, is matrix multiplication. The preferable solution is found using the matrix left division operator:

```
» x=A\b
x =
    25.0000
    22.0000
    99.0000
```

This equation utilizes the LU factorization approach, which is a modification of Gaussian elimination, and expresses the answer as the left division of A into b. The division operator \ has no preceding dot as this is a matrix operation, not an element-by-element array operation. There are many reasons why this second solution is preferable. Of these, the simplest is that the latter method requires fewer internal multiplications and divisions and as a result is faster.

In addition, this solution is generally more accurate for larger problems. In either case, if MATLAB cannot find a solution or cannot find it accurately, it displays an error message.

If you've studied linear algebra rigorously, you know that when the number of equations and number of unknowns differ, a single unique solution usually does not exist. However, with further constraints a practical solution usually can be found.

In MATLAB when there are more equations than unknowns, i.e., the overdetermined case, use of a division operator / or \ automatically finds the solution that minimizes the squared error in $Ax - b = 0$. This solution is of great practical value and is called the *least squares solution.*

On the other hand, when there are fewer equations than unknowns, i.e., the underdetermined case, an infinite number of solutions exist. Of these solutions, MATLAB computes two in a straightforward way. Use of the division operator gives a solution that has zeros for some of the elements of x. Alternatively, computing `x=pinv(A)*b` gives a solution where the Euclidean length or norm of `x` is smaller than all other possible solutions. This solution, based on the pseudoinverse, also has great practical value and is called the *minimum norm solution.*

5.10.2 Other Features

- The *Symbolic Math Toolbox* discussed in *Symbolic Math Toolbox Tutorial* offers many capabilities for solving linear algebra problems.
- `A.'` is the transpose of the matrix `A`. The complex conjugate transpose of the matrix `A` is written as `A'` (with no dot).
- `d=eig(A)` returns the eigenvalues associated with the square matrix `A` as a column vector.
- `[V,D]=eig(A)` returns the eigenvectors in the matrix `V` and the eigenvalues as the diagonal elements in the matrix `D`.

  Note: `[a,b]` is the standard MATLAB form for returning two variables `a` and `b` from a function.
- `[L,U]=lu(A)` computes the LU factorization of the square matrix `A`.
- `[Q,R]=qr(A)` computes the QR factorization of the matrix `A`.
- `[U,S,V]=svd(A)` computes the singular value decomposition of the matrix `A`.
- `rank(A)` returns the rank of the matrix `A`.

- `cond(A)` returns the condition number of the matrix `A`.
- `norm(A)` computes the norm of the matrix `A`. 1-norm, 2-norm, F-norm, and ∞-norm are supported.
- `poly(A)` finds the characteristic polynomial associated with the square matrix `A`.
- `polyvalm(v,A)` evaluates the matrix polynomial `v` using the square matrix `A`.

MATLAB supports many linear algebra operations. See the quick reference table entitled *Matrix Functions—Numerical Linear Algebra* in Chapter 8, *MATLAB Reference,* and online help for more detailed information. Information regarding matrices and linear algebra can be found in any number of references on the subject.

5.10.3 Summary

- `*` and `\` or `/` define multiplication and division respectively in the matrix sense rather than in the element-by-element array sense.
- `det(A)` returns the determinant of the square matrix `A`.
- `inv(A)` finds the inverse of a square matrix A. If `det(A)=0`, or is close to zero, a warning message is given and the results may have no value.
- `pinv(A)` finds the pseudoinverse of a rectangular matrix A. `pinv(A)` is useful for finding the minimum norm solution of an underdetermined set of linear equations.
- The solution of a set of linear equations using a division operator `/` or `\` gives different results depending on the properties of the coefficient matrix. If the set of equations is square and `det(A) ≠ 0`, the exact solution is computed. If the set of equations is overdetermined, the least squares solution is computed.

5.11 Matrix Manipulation

Since matrices are fundamental to MATLAB, there are many ways to manipulate them in MATLAB. Once matrices are formed, MATLAB provides powerful ways to insert, extract, and rearrange subsets of them by identifying subscripts of interest. Knowledge of these features is a key to using MATLAB efficiently. Some of these features, like the powerful colon notation, were introduced earlier in the *Array Operations* section. To illustrate the matrix and array manipulation features of MATLAB, consider the following examples:

```
» A=[1 2 3;4 5 6;7 8 9]
A =
        1     2     3
        4     5     6
        7     8     9

» A(3,3)=0
A =
        1     2     3
        4     5     6
        7     8     0
```

changes the element in the third row and third column to zero.

```
» A(2,6)=1
A =
        1     2     3     0     0     0
        4     5     6     0     0     1
        7     8     0     0     0     0
```

places one in the second row, sixth column. Since A does not have six columns, the size of A is increased as necessary and filled with zeros so that the matrix remains rectangular.

```
» A=[1 2 3;4 5 6;7 8 9];

» B=A(3:-1:1,1:3)
B =
        7     8     9
        4     5     6
        1     2     3
```

creates a matrix B by taking the rows of A in reverse order.

```
» B=A(3:-1:1,:)
B =
        7     8     9
        4     5     6
        1     2     3
```

does the same as the above example. Here the final single colon means take all columns. That is, : is short for 1:3 in this example because A has three columns.

```
» C=[A B(:,[1 3])]
C =
        1     2     3     7     9
        4     5     6     4     6
        7     8     9     1     3
```

creates C by appending all rows in the first and third columns of B to the right of A.

```
» B=A(1:2,2:3)
B =
       2     3
       5     6
```

creates B by extracting the first two rows and last two columns of A.

```
» C=[1 3]
C =
       1     3

» B=A(C,C)
B =
       1     3
       7     9
```

uses the array C to index the matrix A rather than specifying them directly using the colon notation `start:increment:end` or `start:end`. In this example, B is formed from the first and third rows and first and third columns of A.

```
» B=A(:)
B =
       1
       4
       7
       2
       5
       8
       3
       6
       9
```

builds B by stretching A into a column vector taking its columns one at a time.

```
» B=B.'
B =
     1     4     7     2     5     8     3     6     9
```

illustrates the dot-transpose operation introduced earlier.

```
» B=A
B =
       1     2     3
       4     5     6
       7     8     9

» B(:,2)=[]
B =
       1     3
       4     6
       7     9
```

redefines B by throwing away all rows in the second column of original B. *When you set something equal to the empty matrix [ ], it gets deleted, causing the matrix to collapse to what remains.*

```
» B=B.'
B =
       1     4     7
       3     6     9
```

illustrates the transpose of a matrix. In general, the *i*th row becomes the *i*th column of the result, so the original 3-by-2 matrix becomes a 2-by-3 matrix.

```
» B(2,:)=[]
B =
       1     4     7
```

throws out the second row of B.

```
» A(2,:)=B
A =
       1     2     3
       1     4     7
       7     8     9
```

replaces the second row of A with B.

```
» B=A(:,[2 2 2 2])
B =
       2     2     2     2
       4     4     4     4
       8     8     8     8
```

creates B by duplicating all rows in the second column of A four times.

```
» A(2,2)=[]
???  In an assignment  A(matrix,matrix) = B, the number of
rows in B and the number of elements in the A row index matrix
must be the same.
```

shows that you can only throw out entire rows or columns. MATLAB does not know how to collapse a matrix when partial rows or columns are thrown out.

```
» B=A(4,:)
???  Index exceeds matrix dimensions.
```

since A is does not have a fourth row, MATLAB doesn't know what to do and says so.

```
» B(1:2,:)=A
???  In an assignment A(matrix,:) = B, the number of columns
in A and B must be the same.
```

shows that you can't squeeze one matrix into one having a different size.

```
» B(3:4,:)=A(2:3,:)
B =
     0     0     0
     0     0     0
     1     4     7
     7     8     9
```

But you can place the second and third columns of A into same size area of B. Since B did not exist, its undefined first and second rows are filled with zeros.

```
» C(1:6)=A(:,2:3)
C =
     2     4     8     3     7     9
```

creates a row vector C by extracting all rows in the second and third columns of A.

In addition to addressing matrices based on their subscripts, relational operations or arrays containing zeros and ones, *0-1 logical arrays*, can also be used if the size of the array is equal to that of the array it is addressing. In this case, True (1) elements are retained and False (0) elements are discarded.

```
» x=-3:3  % Create data
x =
    -3    -2    -1     0     1     2     3

» abs(x)>1
ans =
     1     1     0     0     0     1     1
```

gives ones where the absolute value of x is greater than 1.

```
» y=x(abs(x)>1)
y =
    -3    -2     2     3
```

creates y by taking those values of x where its absolute value is greater than 1.

```
» y=x([1 1 1 1 0 0 0])
y =
    -3    -2    -1     0
```

creates y by selecting the first four values only, discarding others. Compare this result with

```
» y=x([1 1 1 1])
y =
    -3    -3    -3    -3
```

Here y is created by taking the first element of x four times. In the former example, the array [1 1 1 1 0 0 0] is the same size as x, thus y is created using the 0-1 logical array approach described here. In the latter example, the array [1 1 1 1] is smaller than x, so y is created using [1 1 1 1] as a numerical index as described earlier in this section. If this difference in usage is too subtle and confusing, try several examples on your own.

```
» y=x([1 0 1 0])
???  Index into matrix is negative or zero.
```

produces an error because `[1 0 1 0]` is not the same size as `x` and 0 is not a valid subscript.

```
» x(abs(x)>1)=[]
x =
    -1     0     1
```

throws out values of `x` where `abs(x)>1`. This example illustrates that relational operations can appear on the left-hand side of an assignment statement also.

Relational operations work on matrices as well as vectors also:

```
» b=[5 -3;2 -4]
b =
     5    -3
     2    -4

» x=abs(b)>2
x =
     1     1
     0     1
```

Likewise, 0-1 logical array extraction works for matrices as well.

```
» y=b(abs(b)>2)
y =
     5
    -3
     4
```

However, the results are converted to a column vector, since there is no way to define a matrix having only three elements.

As an alternative to the above approaches, MATLAB includes the function `find` that returns the subscripts where a relational expression is True:

```
» x=-3:3
x =
    -3    -2    -1     0     1     2     3

» k=find(abs(x)>1)
k =
     1     2     6     7
```

finds those subscripts where `abs(x)>1`

```
» y=x(k)
y =
      -3      -2       2       3
```

creates `y` using the indexes in `k`.

The `find` function also works for matrices:

```
» A=[1 2 3;4 5 6;7 8 9]
A =
      1       2       3
      4       5       6
      7       8       9

» [i,j]=find(A>5)
i =
      3
      3
      2
      3
j =
      1
      2
      3
      3
```

Here the indices stored in `i` and `j` are the associated row and column indices respectively where the relational expression is True. That is, `A(i(1),j(1))` is the first element of `A` where `A>5`, and so on.

Note that when a MATLAB function returns two or more variables, they are enclosed by square brackets on the left hand side of the equal sign. This syntax is different than the matrix manipulation syntax discussed above where `[i,j]` on the right hand side of the equal sign builds a new array where `j` appended to the right of `i`.

Finally, in those cases where the size of a matrix or vector is unknown, MATLAB provides two utility functions `size` and `length`:

```
» A=[1 2 3 4;5 6 7 8]
A =
      1       2       3       4
      5       6       7       8

» B=pi:0.01:2*pi;

» s=size(A)
s =
      2       4
```

With one output argument, the size function returns a row vector whose first element is the number of rows and whose second element is the number of columns.

```
» [r,c]=size(A)
r =
      2
c =
      4
```

With two output arguments, size returns the number of rows in the first variable and the number of columns in the second variable.

```
» length(A)
ans =
      4
```

returns the number of rows or the number of columns, whichever is larger.

```
» size(B)
ans =
      1    315
```

shows that B is a row vector.

```
» length(B)
ans =
    315
```

Since vectors have one dimension equal to one, length returns the length of the vector.

```
» size([])
ans =
      0      0
```

shows that the empty matrix has zero size, as it should.

5.11.1 Other Features

- flipud(A) flips a matrix upside down.
- fliplr(A) flips a matrix left to right.
- rot90(A) rotates a matrix counterclockwise.
- reshape(A,m,n) returns an m-by-n matrix whose elements are taken columnwise from A. A must contain m*n elements.
- diag(v) creates a diagonal matrix with the vector v on the diagonal.
- diag(A) extracts the diagonal of the matrix A as a column vector.

Further information regarding these functions can be found in *MATLAB Reference*.

5.11.2 Summary

- Matrix elements are addressed in row,column format: `A(rows,columns)`.
- Values internal to a matrix are accessed by identifying the subscripts of the desired elements.
- Using the colon symbol as the rows or columns designation implies all the rows or columns respectively, e.g., `A(:,1)` is all the rows in column one, `A(2,:)` is all the columns in row 2.
- Placing data outside the current range of a matrix fills unspecified areas with zeros to maintain a rectangular matrix form.
- Setting rows or columns of a matrix equal to the empty matrix `[]` discards those rows or columns.
- Using the colon alone, e.g., `A(:)`, rearranges a matrix into a column vector, taking the columns one at a time.
- 0-1 logical vectors can also be used to address parts of a vector. In this case, the 0-1 logical vector must be the same size as the vector it addresses. False (0) elements are thrown out, True (1) elements are retained.
- The `find` function returns the subscripts or indices where a relational expression is True.
- The function `size` returns the row and column size of a matrix. The function `length` returns the length of a vector or the maximum dimension of a matrix.

5.12 Special Matrices

MATLAB offers a number of special matrices; some of them are general utilities, while others are matrices of interest to specialized disciplines. The general utility matrices include:

```
» zeros(3)
ans =
     0     0     0
     0     0     0
     0     0     0
```

a 3-by-3 matrix of zeros.

```
» ones(2,4)
ans =
     1     1     1     1
     1     1     1     1
```

a 2-by-4 matrix of ones.

```
» ones(3)*pi
ans =
    3.1416    3.1416    3.1416
    3.1416    3.1416    3.1416
    3.1416    3.1416    3.1416
```

an example of creating a 3-by-3 matrix with all elements equal to π.

```
» rand(3,1)
ans =
    0.2190
    0.0470
    0.6789
```

a 3-by-1 matrix of uniformly distributed random numbers between zero and one.

```
» randn(2)
ans =
    1.1650    0.0751
    0.6268    0.3516
```

a 2-by-2 matrix of normally distributed random numbers with zero mean and unit variance.

```
» eye(3)
ans =
     1     0     0
     0     1     0
     0     0     1
```

a 3-by-3 identity matrix. (Or maybe it should be spelled *eye*dentity!)

```
» eye(3,2)
ans =
     1     0
     0     1
     0     0
```

a 3-by-2 identity matrix.

In addition to specifying the size of a matrix explicitly, you can also use the function `size` to create a special matrix the same size as another:

```
» A=[1 2 3;4 5 6];

» ones(size(A))
ans =
     1     1     1
     1     1     1
```

a matrix of ones the same size as `A`.

5.13 Decision Making: Control Flow

Computer programming languages and programmable calculators offer features that allow you to control the flow of command execution based on decision making structures. If you have used these features before, this section will be very familiar to you. On the other hand, if control flow is new to you, this material may seem complicated the first time through. If this is the case, take it slow.

Control flow is extremely powerful since it lets past computations influence future operations. MATLAB offers three decision making or control flow structures. They are: `for` loops, `while` loops, and `if-else-end` structures. Because these structures often encompass numerous MATLAB commands, they often appear in M-files, rather than being typed directly at the MATLAB prompt.

5.13.1 For Loops

`for` loops allow a group of commands to be repeated a fixed, predetermined number of times. The general form of a `for` loop is

```
for x = array
        commands
end
```

The commands between the for and end statements are executed once for every column in array. At each iteration, x is assigned to the next column of array, i.e., during the *i*th time through the loop, x=array(:,i). For example:

```
» for n=1:10
   x(n)=sin(n*pi/10);
  end
» x
x =
  Columns 1 through 7
   0.3090   0.5878   0.8090   0.9511   1.0000   0.9511   0.8090
  Columns 8 through 10
   0.5878   0.3090   0.0000
```

In words, the first statement says: *for* i *equals one to ten evaluate all statements until the next* end *statement*. The first time through the for loop n = 1, the second time n = 2, and so on until the n = 10 case. After the n = 10 case, the for loop ends and any commands after the end statement are evaluated, which in this case is to display the computed elements of x. Other important aspects of for loops are:

- A for loop cannot be terminated by reassigning the loop variable n within the for loop:

```
» for n=1:10
   x(n)=sin(n*pi/10);
   n=10;
  end
» x
x =
  Columns 1 through 7
    0.3090 0.5878 0.8090 0.9511 1.0000 0.9511 0.8090
  Columns 8 through 10
    0.5878 0.3090 0.0000
```

- The statement `1:10` is a standard MATLAB array creation statement. Any valid MATLAB array is acceptable in the `for` loop:

```
» data=[3 9 45 6; 7 16 -1 5]
data =
     3     9    45     6
     7    16    -1     5

» for n=data
    x=n(1)-n(2)
  end
x =
    -4
x =
    -7
x =
    46
x =
     1
```

- `for` loops can be nested as desired:

```
» for n=1:5
    for m=5:-1:1
      A(n,m)=n^2+m^2;
    end
    disp(n)
  end
     1
     2
     3
     4
     5

» A
A =
     2     5    10    17    26
     5     8    13    20    29
    10    13    18    25    34
    17    20    25    32    41
    26    29    34    41    50
```

- `for` loops should be avoided whenever there is an equivalent array or matrix approach to solving a given problem. For example, the first example above can be rewritten as:

```
» n=1:10;
» x=sin(n*pi/10)
x =
  Columns 1 through 7
     0.3090   0.5878   0.8090   0.9511   1.0000   0.9511   0.8090
  Columns 8 through 10
    0.5878 0.3090 0.0000
```

  While both approaches lead to identical results, the latter approach executes faster, is more intuitive, and requires less typing.

5.13.2 While Loops

As opposed to a `for` loop that evaluates a group of commands a fixed number of times, a `while` loop evaluates a group of statements an indefinite number of times.

The general form of a `while` loop is

```
while expression
      commands
end
```

The `commands` between the `while` and `end` statements are executed as long as all elements in `expression` are True. For example:

```
» num=0;EPS=1;
» while (1+EPS)>1
    EPS=EPS/2;
    num=num+1;
  end

» num
num =
     53

» EPS=2*EPS
EPS =
    2.2204e-16
```

This example shows one way of computing the special MATLAB value `eps`. Here we used upper case `EPS` so that the MATLAB value is not overwritten. In this example `EPS` starts at 1. As long as `(1+EPS)>1` is True (nonzero), the commands inside the `while` loop are evaluated. Since `EPS` is continually divided in two, `EPS` eventually gets so small that adding `EPS` to 1 is no longer greater than 1. (Recall that this happens because a computer uses a fixed number of digits to represent numbers. MATLAB uses 16 digits so one would expect eps to be near 10^{-16}.) At this point, (1+EPS)>1 is False (zero) and the while loop terminates. Finally, EPS is multiplied by 2 because the last division by 2 made it too small by a factor of 2.

5.13.3 If-Else-End Structures

Many times, sequences of commands must be conditionally evaluated based on a relational test. In programming languages this logic is provided by some variation of an if-else-end structure. The simplest if-else-end structure is

```
if expression
   commands
end
```

The commands between the if and end statements are evaluated if all elements in expression are True (nonzero). For example:

```
» apples=10;        % number of apples
» cost=apples*25    % cost of apples
cost =
   250

» if apples>5 % apply 20% quantity discount
   cost=(1-20/100)*cost;
  end
» cost
cost =
   200
```

In cases where there are two alternatives the if-else-end structure is

```
if expression
     commands evaluated if True
else
     commands evaluated if False
end
```

Here the first set of commands are evaluated if expression is True; the second set are evaluated if expression is False.

When there are three or more alternatives, the `if-else-end` structure takes the form

```
if expression1
     commands evaluated if expression1 is True
elseif expression2
     commands evaluated if expression2 is True
elseif expression3
     commands evaluated if expression3 is True
elseif expression4
     commands evaluated if expression4 is True
elseif ...
       .
       .
       .
else
    commands evaluated if no other expression is True
end
```

In this last form only the commands associated with the first True expression encountered are evaluated; ensuing relational expressions are not tested and the rest of the `if-else-end` structure is skipped. Furthermore, the final `else` command may or may not appear.

Now that we know how to make decisions with `if-else-end` structures, it is possible to show a legal way for jumping or breaking out of `for` loops and `while` loops, but not `if-else-end` structures:

```
» EPS=1;
» for num=1:1000
   EPS=EPS/2;
   if (1+EPS)<=1
     EPS=EPS*2
     break
   end
  end
EPS =
   2.2204e-16

» num
num =
    53
```

This example demonstrates another way of estimating `eps`. In this case, the `for` loop is instructed to run some sufficiently large number of times. The `if-else-end` structure tests to see if EPS has gotten small enough. If it has, EPS is multiplied by 2 and the `break` command forces the `for` loop to end prematurely, i.e., at `num = 53` in this case.

5.13.4 Summary

MATLAB control flow features can be summarized as:

<table>
<tr><th>Control Flow Structure</th><th>Description</th></tr>
<tr><td><pre>for x = array
     commands
end</pre></td><td>A <code>for</code> loop that on each iteration assigns <code>x</code> to the ith column of <code>array</code> and executes <code>commands</code>.</td></tr>
<tr><td><pre>while expression
     commands
end</pre></td><td>A <code>while</code> loop that executes <code>commands</code> as long as all elements of <code>expression</code> are True or nonzero.</td></tr>
<tr><td><pre>if expression
     commands
end</pre></td><td>A simple <code>if-else-end</code> structure where <code>commands</code> are executed if all elements in <code>expression</code> are True or nonzero.</td></tr>
<tr><td><pre>if expression
     commands evaluated if True
else
     commands evaluated if False
end</pre></td><td>An <code>if-else-end</code> structure with two paths. One group of commands are executed if <code>expression</code> is True or nonzero. The other set are executed if <code>expression</code> is False or zero.</td></tr>
<tr><td><pre>if expression1
   commands evaluated if
   expression1 is True
elseif expression2
   commands evaluated if
   expression2 is True
elseif expression3
   commands evaluated if
   expression3 is True
elseif expression4
   commands evaluated if
   expression4 is True
elseif _
       .
       .
       .
else
   commands evaluated if no
   other expression is True
end</pre></td><td>The most general <code>if-else-end</code> structure. Only the commands associated with the first True expression are evaluated.</td></tr>
<tr><td><pre>break</pre></td><td>Terminates execution of <code>for</code> loops and <code>while</code> loops.</td></tr>
</table>

5.14 M-File Functions

5.14.1 Major Features

When you use MATLAB functions such as `inv`, `abs`, `angle`, and `sqrt`, MATLAB takes the variables you pass it, computes the required results using your input, then passes those results back to you. The commands evaluated by the function as well as any intermediate variables created by those commands are hidden. All you see is what goes in and what comes out.

These properties make functions very powerful tools for evaluating commands that encapsulate useful mathematical functions or sequences of commands that appear often when solving some larger problem. Because of this power, MATLAB provides a structure for creating functions of your own in the form of text M-files stored on your computer. The MATLAB function `fliplr` is a good example of an M-file function:

```
function y = fliplr(x)
%FLIPLR   Flip matrix in the left/right direction.
%  FLIPLR(X) returns X with row preserved and
%  columns flipped in the left/right direction.
%
%   X = 1 2 3      becomes  3 2 1
%       4 5 6               6 5 4
%
%  See also FLIPUD, ROT90.

%  Copyright (c) 1984–93 by The MathWorks, Inc.

[m,n] = size(x);
y = x(:,n:-1:1);
```

A function M-file is similar to a script file in that it is a text file having a `.m` extension. As with script M-files, function M-files are not entered in the command window, but rather are external text files created with a text editor. A function M-file is different than a script file in that a function communicates with the MATLAB workspace only through the variables passed to it and through the output variables it creates. Intermediate variables within the function do not appear in or interact with the MATLAB workspace. As can be seen in the above example, the first line of a function M-file defines the M-file as a function, specifies its name and its input and output variable names. The next continuous sequence of comment lines are the text displayed in response to the help command: `help fliplr`. Finally, the remainder of the M-file contains MATLAB commands that create the output variables.

5.14.2 Summary and Other Features

- The function name and file name must be identical. For example, the function `fliplr` is stored in a file named `fliplr.m`.
- Comment lines up to the first noncomment line in a function M-file are the help text returned when one requests help, e.g., `help fliplr` returns the first eight comment lines above.
- The very first help line, known as the H1 line, is the line interrogated by the `lookfor` command.
- All variables within a function are isolated from the MATLAB workspace. The only connections between the variables within a function and the MATLAB workspace are the input and output variables. If a function changes the value of any input variable, the changes appear within the function only and do not affect the variables in the MATLAB workspace.
- When a function has more than one output variable, the output variables are enclosed in brackets, e.g. `[V,D] = eig(A)`. This is a somewhat confusing syntax since `[V,D]` on the right hand side of an equal sign is the composition of a matrix from `V` and `D`.
- The first time MATLAB executes an M-file function, it opens the corresponding text file and compiles the commands into an internal representation that speeds their execution for all ensuing calls to the function.
- MATLAB searches for function M-files in the same way it does for script M-files. For example, if you type `cow` at the prompt, MATLAB first considers `cow` to be a variable. If it's not, then it considers it to be a built-in function. If it's not, then it checks the current directory or folder for `cow.m`. If it's not there, it checks all directories or folders on the MATLAB search path for `cow.m`.
- The number of input variables passed to a function is available within the function in the variable `nargin`. The number of output variables requested when a function is called is available within the function in the variable

nargout. The MATLAB function linspace is a good example of the use of nargin:

```
function y = linspace(d1, d2, n)
%LINSPACE Linearly spaced vector.
%   LINSPACE(x1, x2) generates a row vector of 100
%   linearly equally spaced points between x1 and
%   x2. LINSPACE(x1, x2, N) generates N points
%   between x1 and x2.
%
%   See also LOGSPACE, :.

%   Copyright (c) 1984–94 by The MathWorks, Inc.

if nargin == 2
    n = 100;
end
y = [d1+(0:n–2)*(d2–d1)/(n–1) d2];
```

If no value for n is given, e.g., x=linspace(0,pi), linspace supplies a default value of 100 using an if-else-end structure to check for the existence of n. If the number of input arguments used, nargin, is equal to 2, n was not provided and is set equal to 100, otherwise the user-supplied value of n is used.

5.15 Data Analysis

5.15.1 Major Features

Because of its matrix orientation, MATLAB readily performs statistical analyses on data sets. By convention, *data sets are stored in column-oriented matrices.* That is, each column of a matrix represents a different measured variable and each row represents individual samples of the variables. For example, let's assume that the daily high temperature (in Celsius) of three cities over a 31-day month was recorded and assigned to the variable temps in a script M-file.

Running the M-file puts the variable `temps` in the MATLAB workspace. If you do this yourself, the variable `temps` contains:

```
» temps
temps =
    12     8    18
    15     9    22
    12     5    19
    14     8    23
    12     6    22
    11     9    19
    15     9    15
     8    10    20
    19     7    18
    12     7    18
    14    10    19
    11     8    17
     9     7    23
     8     8    19
    15     8    18
     8     9    20
    10     7    17
    12     7    22
     9     8    19
    12     8    21
    12     8    20
    10     9    17
    13    12    18
     9    10    20
    10     6    22
    14     7    21
    12     5    22
    13     7    18
    15    10    23
    13    11    24
    12    12    22
```

Each row contains the high temperatures for a given day. Each column contains the high temperatures for a different city. To visualize the data, plot it:

```
» d=1:31;    % number the days of the month
» plot(d,temps)
» xlabel('Day of Month'),ylabel('Celsius')
» title('Daily High Temperatures in Three Cities')
```

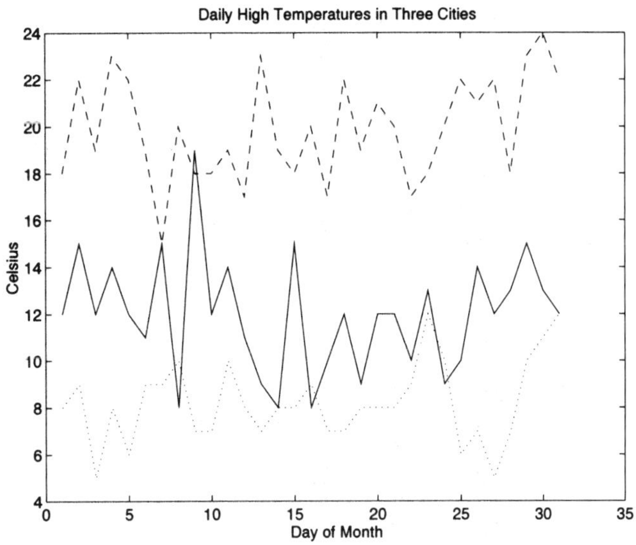

The plot command above illustrates yet another form of plot command usage. The variable d is a vector of length 31; whereas temps is a 31-by-3 matrix. Given this data, the plot command plots each column of temps versus d.

To illustrate some of the data analysis capabilities of MATLAB, consider the following commands based on the above temperature data:

```
» avg_temp=mean(temps)
avg_temp =
   11.9677    8.2258   19.8710
```

shows that the third city has the highest average temperature. Here MATLAB found the average of each column individually. Taking the average again gives:

```
» avg_avg=mean(avg_temp)
avg_avg =
   13.3548
```

which finds the overall average temperature of the three cities. When the input to a data analysis function is a row or column vector, MATLAB simply performs the operation on the vector, returning a scalar result.

```
» max_temp=max(temps)
max_temp =
     19     12     24
```

finds the maximum high temperature of each city over the month.

```
» [max_temp,x]=max(temps)
max_temp =
     19     12     24
x =
      9     23     30
```

finds the maximum high temperature of each city and the row index x where the maximum appears. For this example x identifies the day of the month when the highest temperature occurred.

```
» min_temp=min(temps)
min_temp =
      8      5     15
```

finds the minimum high temperature of each city.

```
» [min_temp,n]=min(temps)
min_temp =
      8      5     15
n =
      8      3      7
```

finds the minimum high temperature of each city and the row index n where the minimum appears. For this example n identifies the day of the month when the lowest high temperature occurred.

```
» s_dev=std(temps)
s_dev =
    2.5098    1.7646    2.2322
```

finds the standard deviation in `temps`.

```
» daily_change=diff(temps)
daily_change =
     3     1     4
    -3    -4    -3
     2     3     4
    -2    -2    -1
    -1     3    -3
     4     0    -4
    -7     1     5
    11    -3    -2
    -7     0     0
     2     3     1
    -3    -2    -2
    -2    -1     6
    -1     1    -4
     7     0    -1
    -7     1     2
     2    -2    -3
     2     0     5
    -3     1    -3
     3     0     2
     0     0    -1
    -2     1    -3
     3     3     1
    -4    -2     2
     1    -4     2
     4     1    -1
    -2    -2     1
     1     2    -4
     2     3     5
    -2     1     1
    -1     1    -2
```

computes the difference between daily high temperatures, which describes how much the daily high temperature varied from day to day.

5.15.2 Summary and Other Features

Data analysis in MATLAB is performed on column-oriented matrices. Different variables are stored in individual columns and each row represents a different observation of each variable. MATLAB statistical functions include:

Statistical Function	Description
`corrcoef(x)`	Correlation coefficients.
`cov(x)`	Covariance matrix.
`cumprod(x)`	Cumulative product of columns.
`cumsum(x)`	Cumulative sum of columns.
`diff(x)`	Compute differences between elements.
`hist(x)`	Histogram or bar chart.
`mean(x)`	Mean or average value of columns.
`median(x)`	Median value of columns.
`prod(x)`	Product of elements in columns.
`rand(x)`	Uniformly distributed random numbers.
`randn(x)`	Normally distributed random numbers.
`sort(x)`	Sort columns in ascending order.
`std(x)`	Standard deviation of columns.
`sum(x)`	Sum of elements in each column.

5.16 Polynomials

5.16.1 Roots

Finding the roots of a polynomial, i.e., the values for which the polynomial is zero, is a problem common to many disciplines. MATLAB solves this problem and provides other polynomial manipulation tools as well. In MATLAB, a polynomial is represented by a row vector of its coefficients in descending order. For example the polynomial $x^4 - 12x^3 + 0x^2 + 25x + 116$ is entered as:

```
» p=[1 -12 0 25 116]
p =
       1   -12     0    25   116
```

Note that terms with zero coefficients must be included. MATLAB has no way of knowing which terms are zero unless you specifically identify them. Given this form, the roots of a polynomial are found by using the function `roots`:

```
» r=roots(p)
r =
  11.7473
   2.7028
  -1.2251 + 1.4672i
  -1.2251 - 1.4672i
```

Since both a polynomial and its roots are vectors in MATLAB, MATLAB adopts the convention that polynomials are *row* vectors and roots are *column* vectors. Given the roots of a polynomial, it is also possible to construct the associated polynomial. In MATLAB the command `poly` performs this task:

```
» pp=poly(r)
pp =
   1.0e+02 *
  Columns 1 through 4
   0.0100        -0.1200          -0.0000            0.2500
  Column 5
   1.1600 + 0.0000i

» pp=real(pp)  % throw away spurious imaginary part
pp =
    1.0000  -12.0000   -0.0000   25.0000  116.0000
```

Since MATLAB deals seamlessly with complex quantities, it is common for the results of `poly` to have some small imaginary part due to roundoff error in recomposing a polynomial from its roots if some of them have imaginary parts. Eliminating the spurious imaginary part is simply a matter of using the function `real` to extract the real part, as shown above.

5.16.2 Multiplication

Polynomial multiplication is supported by the function conv (which performs the convolution of two arrays). Consider the product of the two polynomials $a(x) = x^3 + 2x^2 + 3x + 4$ and $b(x) = x^3 + 4x^2 + 9x + 16$:

```
» a=[1 2 3 4]; b=[1 4 9 16];
» c=conv(a,b)
c =
         1     6    20    50    75    84    64
```

This result is $c(x) = x^6 + 6x^5 + 20x^4 + 50x^3 + 75x^2 + 84x + 64$. Multiplication of more than two polynomials requires repeated use of conv.

5.16.3 Addition

MATLAB does not provide a direct function for adding polynomials. Standard array addition works if both polynomial vectors are the same size. Add the polynomials $a(x)$ and $b(x)$ given above:

```
» d=a+b
d =
         2     6    12    20
```

which is $d(x) = 2x^3 + 6x^2 + 12x + 20$. When two polynomials are of different orders, the one having lower order must be padded with leading zeros to make it have the same effective order as the higher order polynomial. Consider the addition of polynomials c and d above:

```
» e=c+[0 0 0 d]
e =
         1     6    20    52    81    96    84
```

which is $e(x) = x^6 + 6x^5 + 20x^4 + 52x^3 + 81x^2 + 96x + 84$. Leading zeros are required rather than trailing zeros because coefficients associated with like powers of x must line up.

If desired, you can create a function M-file using a text editor to perform general polynomial addition:

```
function p=polyadd(a,b)
%POLYADD Polynomial addition.
% POLYADD(A,B) adds the polynomials A and B

if nargin<2,
     error('Not enough input arguments'),
     end  % error checking
a=a(:).';  % make sure inputs are row vectors
b=b(:).';
na=length(a);  % find lengths of a and b
nb=length(b);
p=[zeros(1,nb-na) a]+[zeros(1,na-nb) b];  % zero pad
```

Now, to illustrate the use of `polyadd`, reconsider the above example:

```
» f=polyadd(c,d)
f =
     1     6    20    52    81    96    84
```

which is the same as `e` above. Of course, `polyadd` can also be used for subtraction:

```
» g=polyadd(c,-d)
g =
     1     6    20    48    69    72    44
```

which is $g(x) = x^6 + 6x^5 + 20x^4 + 48x^3 + 69x^2 + 72x + 44$.

5.16.4 Division

In some special cases it is necessary to divide one polynomial into another. In MATLAB, this is accomplished with the function `deconv`. Using the polynomials `b` and `c` from above:

```
» [q,r]=deconv(c,b)
q =
     1     2     3     4
r =
     0     0     0     0     0     0     0
```

This result says that `b` divided into `c` gives the quotient polynomial `q` and the remainder `r`, which is zero in this case since the product of `b` and `q` is exactly `c` (`c` was formed as the product of `b` and `q = a` earlier).

5.16.5 Derivatives

Because differentiation of a polynomial is simple to express, MATLAB offers the function `polyder` for polynomial differentiation:

```
» g
g =
       1       6      20      48      69      72      44

» h=polyder(g)
h =
       6      30      80     144     138      72
```

5.16.6 Evaluation

Given that you can add, subtract, multiply, divide, and differentiate polynomials based on row vectors of their coefficients, you should be able to evaluate them also. In MATLAB this is accomplished with the function `polyval`:

```
» x=linspace(-1,3);
```

chooses 100 data points between –1 and 3.

```
» p=[1 4 -7 -10];
```

uses the polynomial $p(x) = x^3 + 4x^2 - 7x - 10$

```
» v=polyval(p,x);
```

evaluates $p(x)$ at the values in x and stores the result in v. The result is then plotted using

```
» plot(x,v),title('x^3 + 4x^2 – 7x –10'),xlabel('x')
```

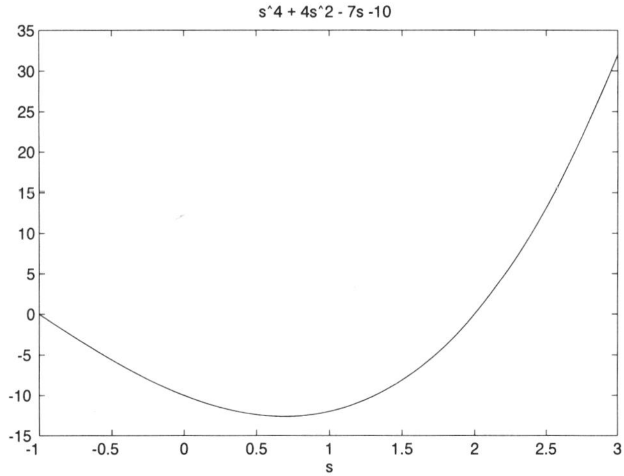

5.16.7 Summary and Other Features

- The *Symbolic Math Toolbox* discussed in Chapter 6, *Symbolic Math Toolbox Tutorial* offers many capabilities for manipulating polynomials.
- Polynomials are represented in MATLAB by row vectors of their coefficients in descending order.
- `roots(p)` computes the roots of the polynomial `p`.
- `poly(r)` finds the polynomial associated with the roots `r`.
- `conv(a,b)` multiplies the two polynomials `a` and `b`.
- `deconv(c,b)` divides the polynomial `b` into `c`.
- MATLAB has no built-in function to add polynomials. However, it is easy to build a function M-file that does so.
- `polyder(p)` computes the derivative of the polynomial `p`.
- `polyval(p,x)` evaluates the polynomial `p` at all the values in `x`.
- `residue(n,d)` computes the partial fraction expansion of the ratio of `n` to `d` where `n` and `d` are polynomials.

- `polyder(n,d)` computes the derivative of the ratio of `n` to `d` where `n` and `d` are polynomials.

5.17 Curve Fitting and Interpolation

In numerous application areas, one is faced with the task of describing data, often measured, with an analytic function. There are two approaches to this problem. In *interpolation*, the data is assumed to be correct and what is desired is some way to describe what happens between the data points. This approach is discussed in the next subsection. In the method to be discussed here, *curve fitting* or *regression*, one seeks to find some smooth curve that "best fits" the data, but does not necessarily pass through any data points. The figure shown below illustrates these two approaches. The 'o' marks are the data points; the solid lines connecting them depict linear interpolation, and the dashed curve is a "best fit" to the data.

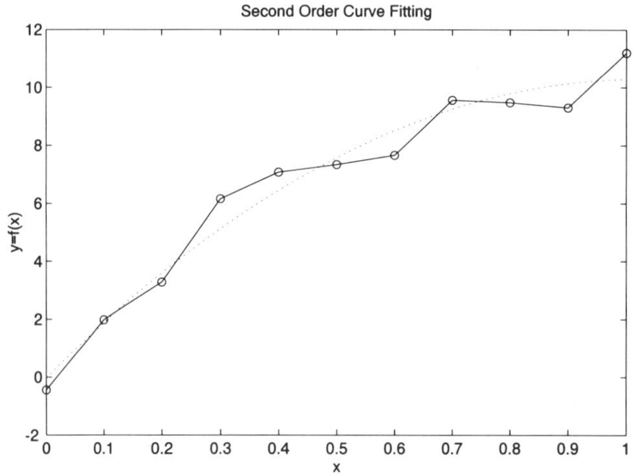

5.17.1 Curve Fitting

Curve fitting involves answering two fundamental questions: What is meant by "best fit"? And, what kind of a curve should be used? "Best fit" can be defined in many different ways and there are an infinite number of curves. So where do we go from here? As it turns out, when "best fit" is interpreted as minimizing the sum of the squared error at the data points and the curve used is restricted to polynomials, curve fitting is fairly straightforward.

Mathematically, this is called least squares curve fitting to a polynomial. If this description is confusing to you, study the above figure again. The vertical distance between the dashed curve and a marked data point is the error at that point. Squaring this distance at each data point and adding the squared distances together is the "sum of the squared error." The dashed curve is the curve that makes this sum of squared error as small as it can be, i.e., it is a "best fit." The term "least squares" is just an abbreviated way of say "minimizing the sum of the squared error."

In MATLAB, the function `polyfit` solves the least squares curve fitting problem. To illustrate the use of this function, let's start with the data in the above plot:

```
» x=[0 .1 .2 .3 .4 .5 .6 .7 .8 .9 1];

» y=[-.447 1.978 3.28 6.16 7.08 7.34 7.66 9.56 9.48 9.30 11.2];
```

To use `polyfit`, we must give it the above data and the order or degree of the polynomial we wish to best fit to the data. If we choose `n=1` as the order, the best straight line approximation will be found. This is often called linear regression. On the other hand, if we choose `n=2` as the order, a quadratic polynomial will be found. For now, let's choose a quadratic polynomial:

```
» n=2;
» p=polyfit(x,y,n)
p =
   -9.8108    20.1293    -0.0317
```

The output of `polyfit` is a row vector of the polynomial coefficients. Here the solution is $y = -9.8108x^2 + 20.1293x - 0.0317$. To compare the curve fit solution to the data points, let's plot both:

```
» xi=linspace(0,1,100);
```

creates x-axis data for plotting the polynomial.

```
» z=polyval(p,xi);
```

calls the MATLAB function `polyval` to evaluate the polynomial `p` at the data points in `xi`.

```
» plot(x,y,'o',x,y,xi,z,':')
```

plots the original data `x` and `y` marking the data points with `'o'`, plots the original data again drawing straight lines between the data points, and plots the polynomial data `xi` and `z` using a dotted line.

```
» xlabel('x'),ylabel('y=f(x)'),
» title('Second Order Curve Fitting')
```

labels the plot. The result of these steps is shown in the plot at the beginning of this section.

The choice of polynomial order is somewhat arbitrary. It takes two points to define a straight line or first order polynomial. (If this isn't clear to you, mark two points and draw a straight line between them.) It takes three points to define a quadratic or second order polynomial. Following this progression, it takes $n + 1$ data points to uniquely specify an nth order polynomial. Thus, in the above case where there are 11 data points we could choose up to a 10th order polynomial. However, given the poor numerical properties of higher order polynomials, one should not choose a polynomial order any higher than necessary. In addition, as the polynomial order increases, the approximation becomes less smooth since higher order polynomials can be differentiated more times before they become zero. For example, choosing a 10th order polynomial:

```
» pp=polyfit(x,y,10)
» format short e  % change display format
» pp.'  % display polynomial coefficients as a column
ans =
  -4.6436e+05
   2.2965e+06
  -4.8773e+06
   5.8233e+06
  -4.2948e+06
   2.0211e+06
  -6.0322e+05
   1.0896e+05
  -1.0626e+04
   4.3599e+02
  -4.4700e-01
```

Note the size of the polynomial coefficients in this case compared to those of the earlier quadratic fit. Note also the seven orders of magnitude difference between the smallest (–4.4700e–01) and largest (5.8233e+06) coefficients. How about plotting this solution and comparing it to the original data and quadratic curve fit?

```
» zz=polyval(pp,xi);  % evaluate 10th order polynomial
» plot(x,y,'o',xi,z,':',xi,zz) % plot data
» xlabel('x'),ylabel('y=f(x)'),
» title('2nd and 10th Order Curve Fitting')
```

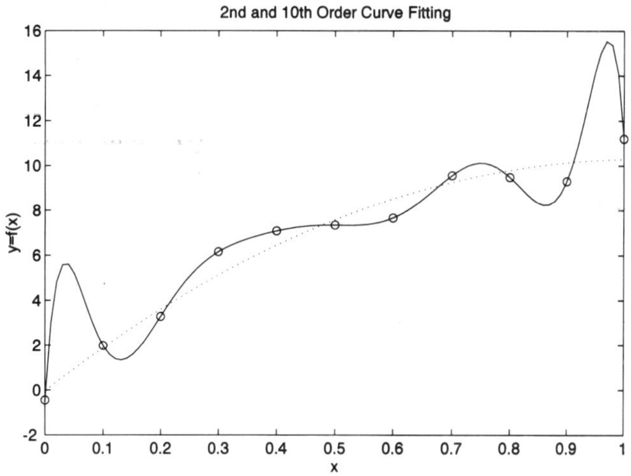

In the above plot, the original data is marked with `'o'`, the quadratic curve fit is dotted, and the 10th order fit is solid. Note the wave-like ripples that appear between the data points at the left and right extremes in the 10th order fit. Based on the above plot it is clear that the "more is better" philosophy does not necessarily apply here.

5.17.2 One-Dimensional Interpolation

As described in the above subsection on curve fitting, interpolation is defined as a way of estimating values of a function between those given by some set of data points. Interpolation is a valuable tool when one cannot quickly evaluate the function at the desired intermediate points. For example, this is true when the data points are the result of some experimental measurements or lengthy computational procedure.

Perhaps the simplest example of interpolation is MATLAB plots. By default, MATLAB draws straight lines connecting the data points used to make a plot. This *linear* interpolation guesses that intermediate values fall on a straight line between the entered points. Certainly as the number of data points increases and the distance between them decreases, linear interpolation becomes more accurate. For example:

```
» x1=linspace(0,2*pi,60);
» x2=linspace(0,2*pi,6);
» plot(x1,sin(x1),x2,sin(x2),'--')
» xlabel('x'),ylabel('sin(x)'),title('Linear Interpolation')
```

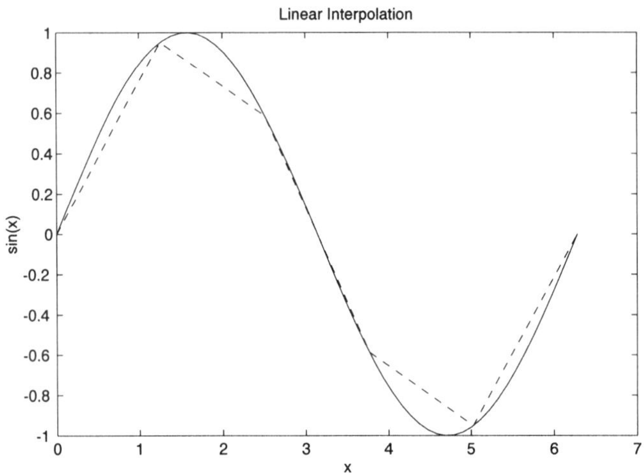

Of the two plots of the sine function, the one using 60 points is much smoother and more accurate between the data points than the one using only 6 points.

As with curve fitting, there are decisions to be made. There are multiple approaches to interpolation depending on the assumptions made. Moreover, it is possible to interpolate in more than one dimension. That is, if you have data reflecting a function of two variables, $z = f(x,y)$, you can interpolate between values of both x and y to find intermediate values of z. MATLAB provides a number of interpolation options in the one-dimensional function `interp1` and in the two-dimensional function `interp2`. Each of these functions will be illustrated in the following subsections.

To illustrate one-dimensional interpolation, consider the following scenario:

As part of a science project, Lisa records the official temperature in Springfield every hour for twelve hours so that she can use the information to report on the local climate.

Lisa analyzes her data:

```
» hours=1:12;
```

is an index for the hours when the data was taken, and

```
» temps=[5 8 9 15 25 29 31 30 22 25 27 24];
```

are the official temperature readings in Celsius.

```
» plot(hours,temps,hours,temps,'+')  % view temperatures
» title('Springfield Temperature')
» xlabel('Hour'),ylabel('Degrees Celsius')
```

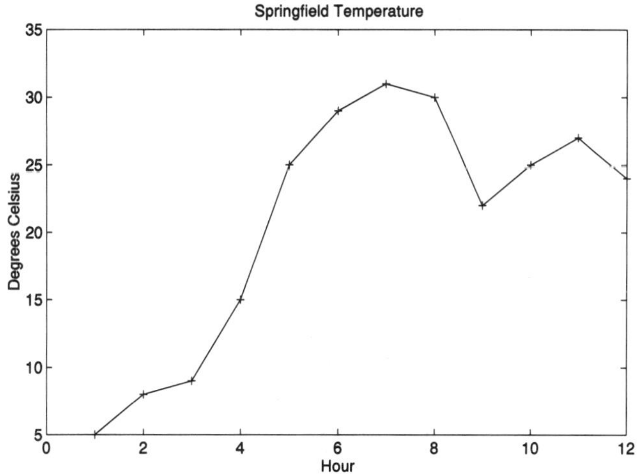

As shown in the plot, MATLAB draws lines that linearly interpolate the data points. To estimate the temperature at any given time, Lisa could try to interpret the plot visually. Alternatively, she could use the function `interp1` to do so:

```
» t=interp1(hours,temps,9.3)  % estimate at hour=9.3
t =
   22.9000

» t=interp1(hours,temps,4.7)  % estimate at hour=4.7
t =
    22

» t=interp1(hours,temps,[3.2 6.5 7.1 11.7])
t =
   10.2000
   30.0000
   30.9000
   24.9000
```

This default usage of `interp1` is described by `interp1(x,y,xo)` where `x` is the independent variable (abscissa), `y` is the dependent variable (ordinate), and `xo` is an array of the values to interpolate. In addition, this default usage assumes *linear* interpolation.

Rather than assume that a straight line connects the data points, we can assume that some smoother curve fits the data points. The most common assumption is that a third order polynomial, i.e., a cubic polynomial, is used to model each segment between consecutive data points and that the slope of each cubic polynomial matches at the data points. This type of interpolation is called *cubic splines* or just splines. Lisa finds this solution:

```
» t=interp1(hours,temps,9.3,'spline')  % estimate at hour=9.3
t =
   21.8577

» t=interp1(hours,temps,4.7,'spline')  % estimate at hour=4.7
t =
   22.3143

» t=interp1(hours,temps,[3.2 6.5 7.1 11.7],'spline')
t =
    9.6734
   30.0427
   31.1755
   25.3820
```

Note that the answers for *spline* interpolation are different from the linear interpolation results shown above. Since interpolation is a process of estimating or guessing values, it makes sense that applying different guessing rules leads to different results.

One of the most common uses of spline interpolation is to smooth data. That is, given a set of data, use spline interpolation to evaluate the data at a finer interval. For example:

```
» h=1:0.1:12;  % estimate temperature every 1/10 hour
» t=interp1(hours,temps,h,'spline');
» plot(hours,temps,'--',hours,temps,'+',h,t)
» title('Springfield Temperature')
» xlabel('Hour'),ylabel('Degrees Celsius')
```

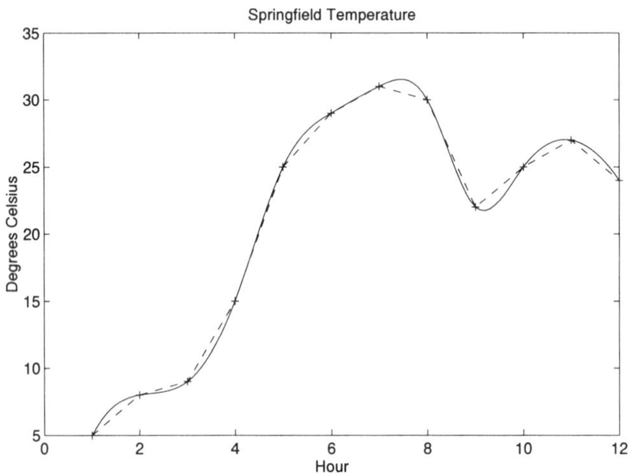

In the above plot, the dashed line is linear interpolation, the solid line is the smoothed spline interpolation, and the original data is marked with `'+'`. By asking for a finer resolution on the `hours` axis and using spline interpolation, we have a smoother, but not necessarily more accurate, estimate of the temperature. In particular, note how the slope of the spline solution does not change abruptly at the data points.

Before discussing two-dimensional interpolation, it is important to recognize the two major restrictions enforced by `interp1`. First, one cannot ask for results outside the range of the independent variable, e.g., `interp1(hours,temps,13.5)` leads to an error since `hours` varies between 1 and 12. Second, the independent variable must be *monotonic*. That is, the independent variable must always increase or must always decrease. In our example, `hours` is monotonic. However, if we had defined the independent variable to be the actual time of day:

```
» time_of_day=[7:12 1:6]  % start at 7AM, end at 6PM
time_of_day =
     7    8    9   10   11   12    1    2    3    4    5    6
```

the independent variable would not be monotonic since `time_of_day` increases until 12, then drops to 1, then increases again. If `time_of_day` were used instead of `hours` in `interp1`, an error would be returned. By the same reasoning, one cannot interpolate `temps` to find the hour when some temperature occurred because `temps` is not monotonic.

5.17.3 Two-Dimensional Interpolation

Two-dimensional interpolation is based on the same underlying ideas as one-dimensional interpolation. However, as the name implies, two-dimensional interpolation interpolates functions of two variables, $z = f(x,y)$. To illustrate this added dimension, consider the following problem:

Barney got a job in the research laboratory at Tastesewgood, Inc. There he is trying to perfect a new microwave brownie recipe. To test the uniformity of baked test recipes, he pulls them out of the microwave oven when "done" and measures the internal brownie temperature on a 3-by-5 grid around the pan as shown below. From this data Barney wishes to determine the temperature distribution throughout the pan.

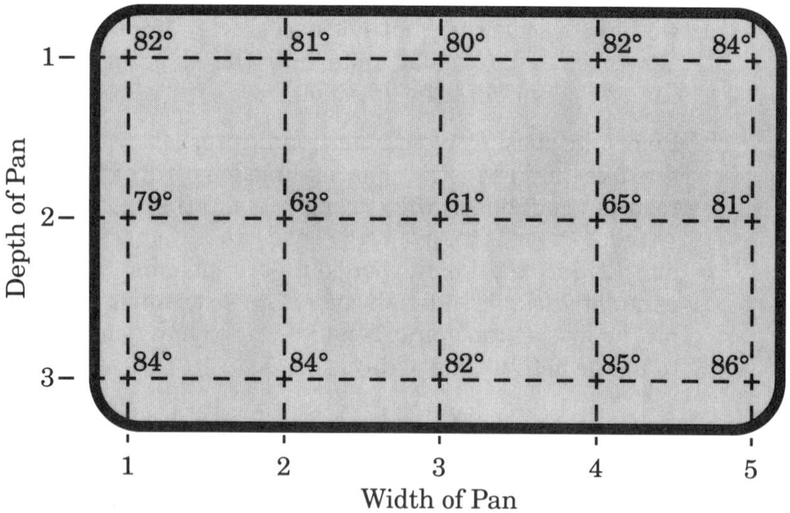

Barney uses MATLAB to solve his problem by creating the script M-files:

```
% middepth.m
% analyze test data from batch no. 51483.
% slice the pan at half depth, look across width

width=1:5;  % width of pan
depth=1:3;  % depth of pan
temps=[82 81 80 82 84;79 63 61 65 81;84 84 82 85 86];

wi=1:0.2:5;  % choose resolution for width
d=2;  % center of pan

zl=interp2(width,depth,temps,wi,d);  % linear interpolate
zc=interp2(width,depth,temps,wi,d,'cubic');  % cubic

plot(wi,zl,'--',wi,zc)  % plot linear and cubic
xlabel('Width of Pan')
ylabel('Degrees Celsius')
title('Temperature at Depth = 2')
```

which produces the plot

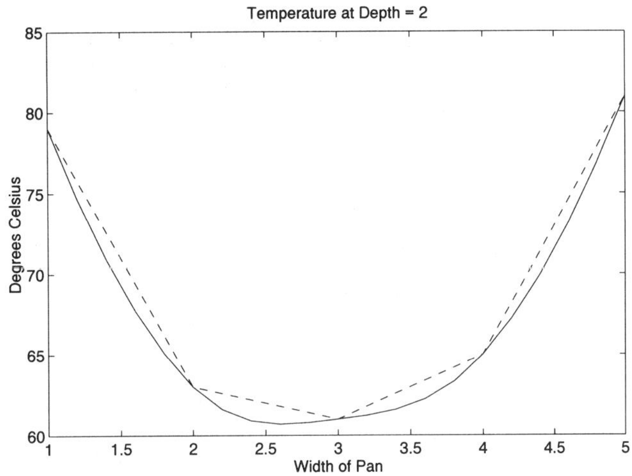


```
% midwidth.m
% analyze test data from batch no. 51483.
% slice the pan at its half width and look across depth

width=1:5;  % width of pan
depth=1:3;  % depth of pan
temps=[82 81 80 82 84;79 63 61 65 81;84 84 82 85 86];  % data

di=1:0.2:3;  % choose resolution for depth
w=3;  % center of pan

zl=interp2(width,depth,temps,w,di);  % linear interpolate
zc=interp2(width,depth,temps,w,di,'cubic');  % cubic

plot(di,zl,'--',di,zc)  % plot linear and cubic
xlabel('Depth of Pan')
ylabel('Degrees Celsius')
title('Temperature at Width = 3')
```

which produces the plot

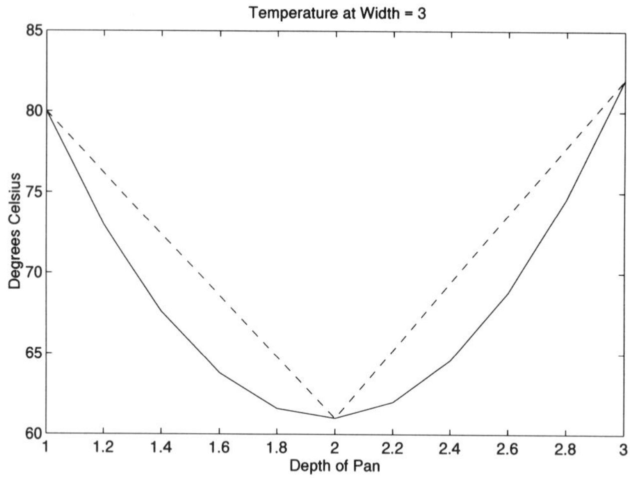


```
% overpan.m
% analyze test data from batch no. 51483.
% view temperature distribution over the pan

width=1:5;  % width of pan
depth=1:3;  % depth of pan
temps=[82 81 80 82 84;79 63 61 65 81;84 84 82 85 86];  % data

di=1:0.2:3;  % resolution for depth
wi=1:0.2:5;  % resolution for width
zc=interp2(width,depth,temps,wi,di,'cubic');  % cubic

mesh(wi,di,zc)  % create 3-D plot (See Section 5.20)
xlabel('Width of Pan')
ylabel('Depth of Pan')
zlabel('Degrees Celsius')
title('Temperature of Batch no. 51483')
axis('ij')  % (See Section 5.20)
grid
```

which produces the plot

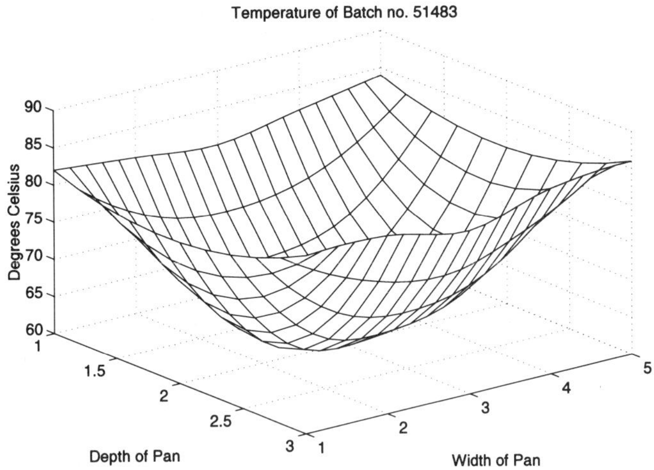

The above example clearly demonstrates that two-dimensional interpolation is more complicated simply because there's more to keep track of. The basic form of `interp2` is `interp2(x,y,z,xi,yi,method)`. Here `x` and `y` are the two independent variables and `z` is a matrix of the dependent variable having the size `length(y)` rows and `length(x)` columns. `xi` is an array of the values to interpolate along the x-axis; `yi` is an array of values to interpolate along the y-axis. The optional parameter `method` can be `'linear'`, `'cubic'`, or `'nearest'`. In this case, `cubic` does not mean cubic splines, but rather another algorithm using cubic polynomials. For more information regarding these methods ask for online help, e.g., `help interp2`.

5.17.4 Summary

- Interpolation is the process of estimating intermediate values of a function given a set of data points describing the function.
- Curve fitting or regression is the process of finding a curve that "best fits" a set of data points. The "best fit" generally does not pass through any of the data points.
- Least squares curve fitting to a polynomial finds polynomial coefficients that minimize the sum of the error squared at the data points, where error is defined as the difference between the known data points and the polynomial evaluated at the data points.

- The function `polyfit(x,y,n)` returns the polynomial coefficients of the nth order polynomial that minimizes the least square error given the data in `x` and `y` $= f($`x`$)$.
- The function `polyval(p,x)` evaluates the polynomial specified by `p` at the point given in `x`.
- At least $n+1$ points are required for an nth order polynomial curve fit.
- The order of the polynomial chosen for curve fitting should remain small.
- Interpolation is the process of estimating values of a function between those given by some set of data points.
- Interpolation is generally not useful for estimating values of a function outside those given by some set of data points.
- The independent variable(s) used for interpolation must be monotonic.
- One-dimensional interpolation estimates $y = f(x)$ for values of x given knowledge of data pairs $\{x_i, y_i\}$ describing $y = f(x)$.
- Two-dimensional interpolation estimates $z = f(x,y)$ for values of x and y given knowledge of data triplets $\{x_i,y_i,z_i\}$ describing $z = f(x,y)$.

5.18 Numerical Analysis

Whenever it is difficult to integrate, differentiate, or determine some specific value of a function analytically, a computer can be called upon to numerically approximate the desired solution. This area of computer science and mathematics is known as numerical analysis. As you could have guessed by now, MATLAB provides tools to solve these problems. In this subsection, the use of these tools will be illustrated.

5.18.1 Plotting

Up to this point, plots of a function have been generated by simply evaluating the function over some range and plotting the resulting vectors. For many cases this is sufficient. However, sometimes a function is flat and unexciting over some range and then acts wildly over another. Using the traditional plotting approach in this case can lead to a plot that misrepresents the true nature of a function. As a result MATLAB provides a smart plotting function called `fplot`. This function carefully evaluates the function to be plotted and makes sure that all its oddities are represented in the output plot. As its input, this function needs to know the name of the function as a character string and the plotting range as a two-element array. For example:

```
» fplot('humps',[0 2])
» title('FPLOT of humps')
```

evaluates the function `humps` between `0` and `2` and displays the plot:

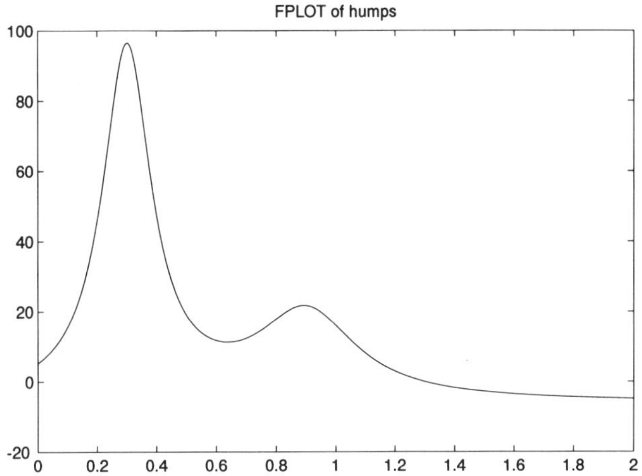

In this example, `'humps'` is the M-file function:

```
function y = humps(x)
% HUMPSA function used by QUADDEMO, ZERODEMO and FPLOTDEMO.
% HUMPS(X) is a function with strong maxima near x=.3 and x=.9.
% See QUADDEMO, ZERODEMO and FPLOTDEMO.

% Copyright (c) 1984–93 by The MathWorks, Inc.

y = 1 ./ ((x–.3).^2 + .01) + 1 ./ ((x–.9).^2 + .04) – 6;
```

`fplot` works for any function M-file with one vector input and one vector output. That is, as in `humps` above, the output variable `y` returns an array the same size as the input `x`, with `y` and `x` associated in the array-to-array sense. The most common mistake made in using `fplot` (as well as other numerical analysis functions) is forgetting to put the name of the function in quotes. That is, `fplot` needs to know the name of the function as a character string. If `fplot(humps,[0 2])` is typed, MATLAB thinks `humps` is a variable in the workspace rather than the name of a function. Note that this problem can be avoided by defining the variable humps to be the desired character string:

```
» humps='humps';
» fplot(humps,[0 2])
```

Now MATLAB evaluates the variable `humps` to find the string `'humps'`.

For simple functions that can be expressed as a single character string, such as $y = 2e^{-x}\sin(x)$, `fplot` can plot the function without creating an M-file by simply writing the function to be plotted as a complete character string using x as the independent variable:

```
» f='2*exp(-x).*sin(x)';
```

Here, the function $f(x) = 2e^{-x}\sin x$ is defined using array multiplication.

```
» fplot(f,[0 8])
» title(f),xlabel('x')
```

plots the function over the range $0 \le x \le 8$, producing the plot:

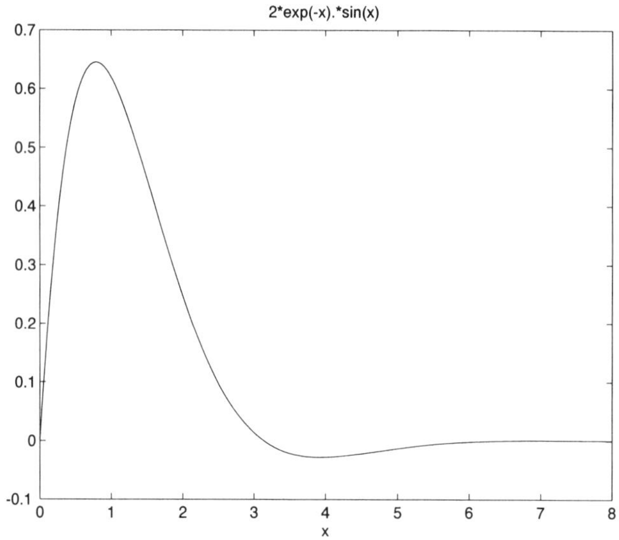

Beyond these basic features, the function `fplot` has many more powerful capabilities. For more information see Part 3, *Reference,* or online help.

5.18.2 Minimization

In addition to the visual information provided by plotting, it is often necessary to determine other more specific attributes of a function. Of particular interest in many applications are function extremes, i.e., its *maxima* (peaks) and its *minima* (valleys). Mathematically, these extremes are found analytically by determining where the derivative (slope) of a function is zero. This fact can be readily understood by inspecting the slope of the peaks plot at its peaks and valleys. Clearly, when a function is simply defined this process often works. However, even for many simple functions that can be differentiated readily, it is often not possible to find where the derivative is zero. In these cases and in cases where it is difficult or impossible to find the derivative analytically, it is necessary to search for function extremes numerically. MATLAB provides two functions that perform this task, `fmin` and `fmins`. These two functions find minima of one-dimensional and n-dimensional functions respectively. Only `fmin` will be discussed here. Further information regarding `fmins` can be found in Section 8.1, *Reference Tables*. Since a maximum of $f(x)$ is equal to a minimum of $-f(x)$, `fmin` and `fmins` can be used to find both minima and maxima. If this fact is not clear, visualize the preceding plot flipped upside down. In the upside down state, the peaks become valleys and the valleys become peaks.

To illustrate one-dimensional minimization and maximization, consider the preceding example once again. From the figure, there is a maximum near $x_{max} = 0.7$ and a minimum near $x_{min} = 4$. Analytically, these points can be shown to be $x_{max} = \pi/4 \approx .785$ and $x_{min} = 5\pi/4 \approx 3.93$. Writing a script M-file using a text editor for convenience and using `fmin` to find them numerically gives:

```
% ex_fmin.m
fn='2*exp(-x)*sin(x)';        % function for min
xmin=fmin(fn,2,5)              % search over range 2<x<5
emin=5*pi/4-xmin               % find error
x=xmin;                        % define x for eval of fn
ymin=eval(fn)                  % evaluate at xmin
fx='-2*exp(-x)*sin(x)';       % function for max: note minus sign
xmax=fmin(fx,0,3)              % search over range 0<x<3
emax=pi/4-xmax                 % find error
x=xmax;                        % define x for eval of fn
ymax=eval(fn)                  % evaluate at xmax
```

Running this M-file results in the following:

```
» ex_fmin
xmin =
    3.9270
emin =
   1.4523e-06
ymin =
   -0.0279
xmax =
    0.7854
emax =
  -1.3781e-05
ymax =
    0.6448
```

These results agree well with the preceding plot. Note that `fmin` works a lot like `fplot`. The function to be evaluated can be expressed in a function M-file or just given as a character string with x being the independent variable. The latter was done here. This example also introduces the function `eval` which takes a character string and interprets it as if the string was typed at the

MATLAB prompt. Since the function to be evaluated was given as a character string with an independent variable `x`, setting `x` equal to `xmin` and `xmax` allows `eval` to evaluate the function to find `ymin` and `ymax`.

Finally, it is important to note that numerical minimization involves searching for a minimum; `fmin` evaluates the function over and over looking for a minimum. This searching can take a significant amount of time if evaluating the function requires a lot of computations or if the function has more than one minimum within the search range. In some cases the searching process does not find a solution at all! When `fmin` cannot find a minimum, it stops and provides an explanation.

5.18.3 Zero Finding

Just as one is interested in finding function extremes, it is sometimes important to find out where a function crosses zero or some other constant value. Trying to find this point analytically is often difficult and many times impossible. In the preceding `humps` function plot, repeated below, the function crosses zero near $x = 1.2$.

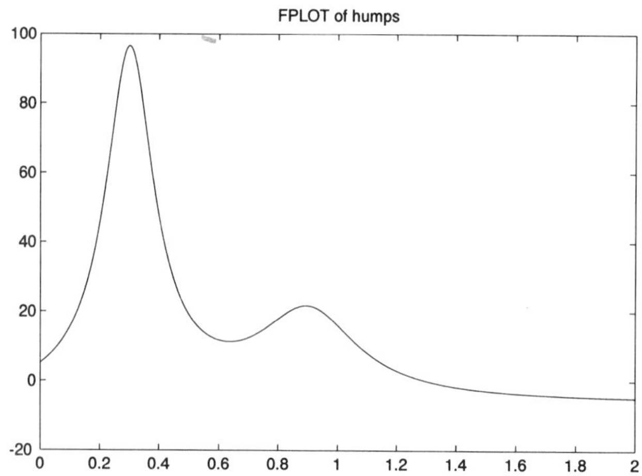

Once again MATLAB provides a numerical solution to this problem. The function fzero searches for the zero of a one-dimensional function. To illustrate the use of this function, let's use the humps example again:

```
» xzero=fzero('humps',1.2)  % look for zero near 1.2
xzero =
    1.2995

» yzero=humps(xzero)  % evaluate at xzero
yzero =
   3.5527e-15
```

So the zero actually occurs close to 1.3. As before, the zero searching process may not find a solution. If fzero does not find one, it will stop and provide an explanation.

The function fzero must be given the name of a function when it's called. For some reason it was never given the capability to accept a function described by a character string using x as the independent variable. Thus even though this feature is available in both fplot and fmin, it does not work with fzero.

While fzero finds where a function is zero, it can also be used to find where a function is equal to any constant. All that's required is a simple redefinition. For example, to find where the function $f(x)$ equals the constant c, define the function $g(x)$ as $g(x) = f(x) - c$. Then using $g(x)$ in fzero will find the value of x where $g(x)$ is zero, which occurs when $f(x) = c$!

5.18.4 Integration

The integral or the area under a function is yet another useful attribute. MATLAB provides three functions for numerically computing the area under a function over a finite range: trapz, quad, and quad8. The function trapz approximates the integral under a function by summing the area of trapezoids formed from the data points as shown below using the function `humps`.

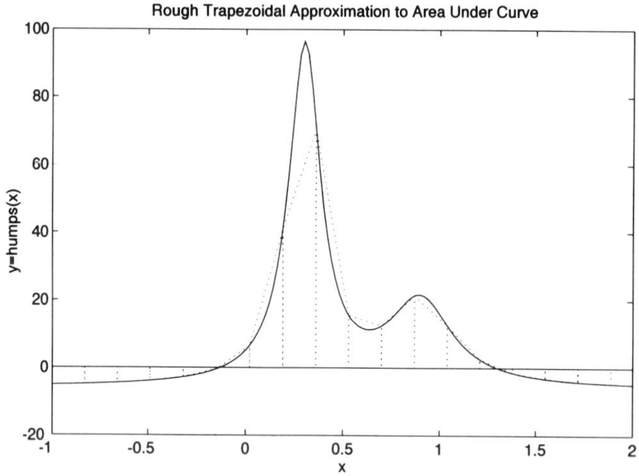

As is apparent from the figure, the area of individual trapezoids underestimates the true area in some segments and overestimates it in others. As with linear interpolation, this approximation gets better as the number of trapezoids increases. For example, if we roughly double the number of trapezoids used in the above figure we get a much better approximation as shown below.

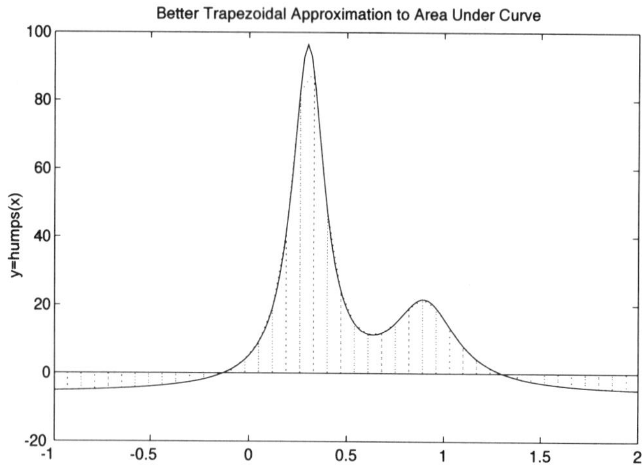

To compute the area under y = humps(x) over the range $-1 < x < 2$ using `trapz` for each of the two plots shown above:

```
» x=-1:.17:2;          % rough approximation
» y=humps(x);
» area=trapz(x,y)      % call trapz just like plot command
area =
   25.9174

» x=-1:.07:2;          % better approximation
» y=humps(x);
» area=trapz(x,y)
area =
   26.6243
```

Naturally the solutions are different. Based on inspection of the plots, the rough approximation probably underestimates the area. Nothing certain can be said about the better approximation except that it is likely to be much more accurate. Clearly, if one could somehow change individual trapezoid widths to match the characteristics of the function, i.e., make them narrower where the function changes more rapidly, much greater accuracy can be achieved.

The MATLAB functions `quad` and `quad8`, which are based on the mathematical concept of quadrature, take this approach. These integration functions both operate in the same way. Both evaluate the function to be integrated at whatever intervals are necessary to achieve accurate results. Moreover, both functions make higher order approximations than a simple trapezoid, with `quad8` being more rigorous than `quad`. These functions are called in the same way that `fzero` is:

```
» area=quad('humps',-1,2)  % area between -1 and 2
area =
   26.3450

» area=quad8('humps',-1,2)
area =
   26.3450
```

Note that both of these functions return essentially the same estimate of the area and that estimate is between the two `trapz` estimates.

For more information on MATLAB's integration functions, see Section 8.2, *Matlab Commands and Functions* and online help.

5.18.5 Differentiation

As opposed to integration, numerical differentiation is much more difficult. Integration describes an overall or macroscopic property of a function, whereas differentiation describes the slope of a function at a point, which is a microscopic property of a function. As a result, integration is not sensitive to minor changes in the shape of a function, whereas differentiation is. Any small change in a function can easily create large changes in its slope in the neighborhood of the change.

Because of this inherent difficulty with differentiation, numerical differentiation is avoided whenever possible, especially if the data to be differentiated is obtained experimentally. In this case, it is best to perform a least squares curve fit to the data, then differentiate the resulting polynomial. For example, reconsider the curve fitting example from the *Curve Fitting and Interpolation* section:

```
» x=[0 .1 .2 .3 .4 .5 .6 .7 .8 .9 1];
» y=[-.447 1.978 3.28 6.16 7.08 7.34 7.66 9.56 ...
    9.48 9.30 11.2]; % data
» n=2; % order of fit
» p=polyfit(x,y,n)  % find polynomial coefficients
p =
   -9.8108   20.1293   -0.0317

» xi=linspace(0,1,100);
» z=polyval(p,xi); % evaluate polynomial
» plot(x,y,'o',x,y,xi,z,':')
» xlabel('x'),ylabel('y=f(x)'),
» title('Second Order Curve Fitting')
```

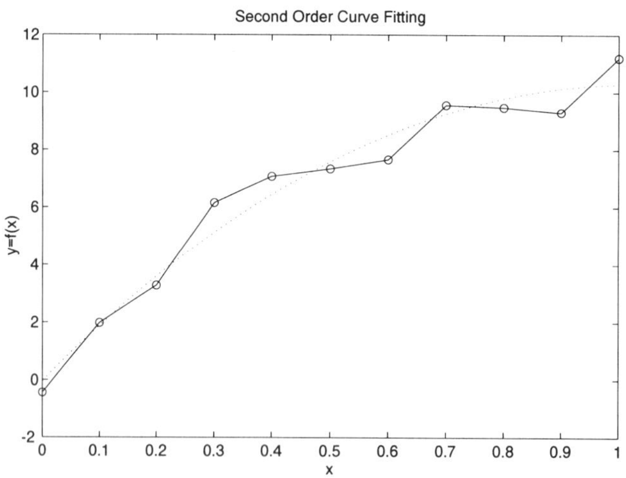

The derivative in this case is found by using the polynomial derivative function polyder:

```
» pd=polyder(p)
pd =
   -19.6217   20.1293
```

The derivative of $y = -9.8108x^2 + 20.1293x - 0.0317$ is $dy/dx = -19.6217x + 20.1293$. Since the derivative of a polynomial is yet another polynomial of the next lowest order, the derivative can also be evaluated and plotted:

```
» z=polyval(pd,xi);  % evaluate derivative
» plot(xi,z)
» xlabel('x'),ylabel('dy/dx'),
» title('Derivative of a Curve Fit Polynomial')
```

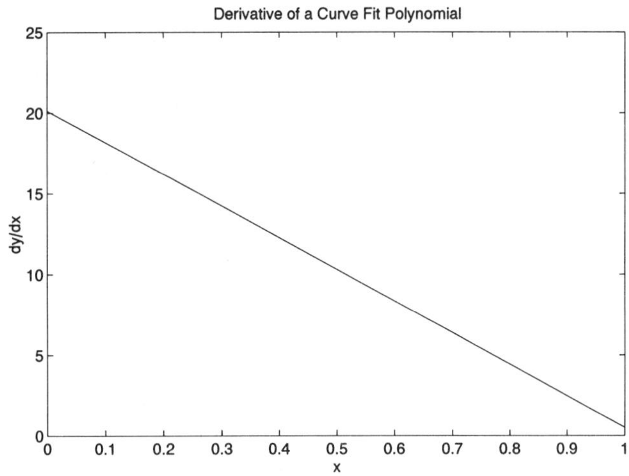

In this case, the polynomial fit was second order, making the resulting derivative first order. As a result, the derivative is a straight line, meaning that it changes linearly with x.

MATLAB provides one function for computing a very rough derivative given the data describing some function. This function, named `diff`, computes the difference between elements in an array. Since differentiation is defined as

$$\frac{dy}{dx} = \lim_{h \to 0} \frac{f(x+h) - f(x)}{(x+h) - (x)}$$

the derivative of $y = f(x)$ can be approximated by

$$\frac{\partial y}{\partial x} \approx \frac{f(x+h) - f(x)}{(x+h) - (x)} \quad \text{where } h > 0$$

which is the *finite difference of y divided by the finite difference in x*. Since `diff` computes differences between array elements, differentiation can be approximated in MATLAB. Continuing with the prior example:

```
» dy=diff(y)./diff(x);  % compute differences and divide
» xd=x(1:length(x)-1);  % new x axis for shorter dy
» plot(xd,dy)
» title('Approximate Derivative Using DIFF')
» ylabel('dy/dx'),xlabel('x')
```

Since `diff` computes the difference between elements of an array, the resulting output contains one less element than the original array. Thus, to plot the derivative, one element of the x array must be thrown out. Comparing the last two plots, it is overwhelmingly apparent that approximating the derivative by finite differences often leads to poor results.

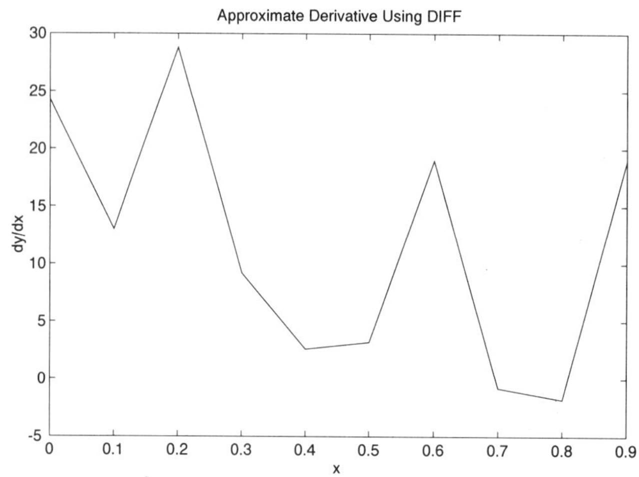

5.18.6 Differential Equations

You may already be familiar with the fact that the actions of many physical systems can be modeled by ordinary differential equations. If you are, then this section may be of interest to you.

Ordinary differential equations describe how the rate of change of variables within a system are influenced by variables within the system and by external stimuli, i.e., inputs. When ordinary differential equations can be solved analytically, features in the *Symbolic Math Toolbox* can be used to find exact solutions. This toolbox is described in Chapter 6, *Symbolic Math Toolbox Tutorial*.

In those cases where the equations cannot be readily solved analytically, it is convenient to solve them numerically. For illustration purposes, consider the classical Van der Pol differential equation which describes an oscillator:

$$\frac{d^2x}{dt^2} - \mu\left(1 - x^2\right)\frac{dx}{dt} + x = 0$$

As with all numerical approaches to solving differential equations, higher order differential equations must be rewritten in terms of an equivalent set of first order differential equations. For the above differential equation, this is accomplished by defining two new variables:

$$\text{let } y_1 = x, \text{ and } y_2 = \frac{dx}{dt}$$

$$\text{then} \quad \frac{dy_1}{dt} = y_2$$

$$\frac{dy_2}{dt} = \mu\left(1 - y_1^{\,2}\right) - y_1$$

From this set of equations, the MATLAB functions `ode23` and `ode45` can be used to find the motion of this system as time evolves. Doing so requires that we write a function M-file that returns the above derivatives given the current time and the current values of y_1 and y_2. In MATLAB the derivatives are given by a column vector called `yprime` in this case. Similarly y_1 and y_2 are written as a column vector y. The resulting function M-file is:

```
function yprime=vdpol(t,y);
%VDPOL(t,y) returns the state derivatives of the
%Van der Pol equation:
%
%  x''-mu*(1-x^2)*x'+x = 0 (' = d/dx, '' = d^2/dx^2)
%
%  let y(1) = x   and y(2) = x'
%
%  then  y(1)' = y(2)
%        y(2)' = mu*(1-y(1)^2)*y(2) -y(1)

mu=2;  % choose 0< mu < 10

yprime=[y(2)
      mu*(1-y(1)^2)*y(2)-y(1)];  % yprime is a column
```

Given this function that completely describes the differential equation, the solution is computed as:

```
» [t,y]=ode23('vdpol',0,30,[1;0]);
» y1=y(:,1);  % first column is y(1) versus time points in t
» y2=y(:,2);  % second column is y(2)
» plot(t,y1,t,y2,'--')
» xlabel('Time, seconds'), ylabel('Y(1) and Y(2)')
» title('Van der Pol Solution for mu=2')
```

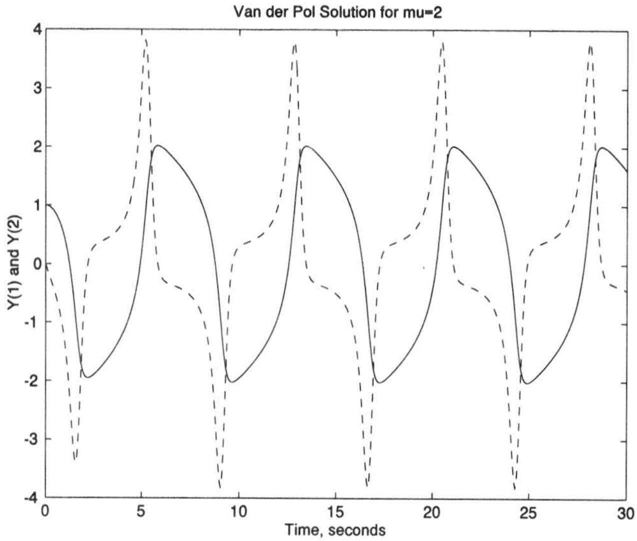

In this plot y_2 (dashed) is the derivative of y_1 (solid). The parameters passed to `ode23` are described by `ode23(function_name, to,tf,yo)` where `function_name` is the character string name of the M-file function that computes the derivatives, `to` is the initial time, `tf` is the final time, and `yo` is the vector of initial conditions. In the above example, the solution starts at time equal to zero, ends at time equal to 30 seconds, and begins with the initial condition `y=[1;0]`. The two output parameters are a column vector `t` containing the time points where the response was estimated, and the matrix `y` that has as many columns as there are differential equations (two in this case) and as many rows as there are in `t`.

The function `ode45` is used exactly the same way as `ode23`. The difference between the two functions has to do with the internal algorithm used. `ode45` uses a more complicated algorithm that does not need to compute as many points over a given time range but needs to make more computations per time point. In those cases where more output data points are desired, `ode23` is the better algorithm.

5.18.7 Summary

- `fplot(function_name,[lower upper])` plots a function between the values of `lower` and `upper`. This function is useful for plotting functions that are not easily plotted using the standard technique of evaluating over a linearly spaced range.
- `fmin(function_name,lower,upper)` finds the minimum of a one-dimensional function between the values `lower` and `upper`.
- `fzero(function_name,guess)` finds a zero of a function near the initial estimate `guess`.
- `trapz(x,y)` approximates the area under the function $y = f(x)$ given the data x and y by using the trapezoidal method.
- `quad(function_name,lower,upper)` and `quad8(function_name,lower,upper)` compute the area under a function from `lower` to `upper` using the method of quadrature.
- `diff(y)./diff(x)` approximates the derivative of $y = f(x)$ using finite differences.
- `ode23(function_name,to,tf,yo)` and `ode45(function_name,to,tf,yo)` integrate a set of differential equations described in the function `function_name` from an initial time `to`, to a final time `tf`, starting with the initial condition `yo`.
- Differentiation based on curve fitting often leads to better derivative estimates.
- All numerical analysis functions approximate the actual solution sought. The only way to get the exact solution is to solve the problem at hand analytically.
- Minimization and zero finding do not always converge to viable solutions. As a result, it is always good to check results.
- The *Symbolic Math Toolbox* discussed in Chapter 6, *Symbolic Math Toolbox Tutorial* offers many capabilities for performing zero finding, integration, differentiation, and solving ordinary differential equations. In particular, it solves these problems analytically whenever possible.

5.19 2-D Graphics

Throughout this Tutorial, a number of MATLAB's graphics features were introduced. In this and the next section, these and the many other graphics features in MATLAB will be more rigorously illustrated.

5.19.1 Using the plot Command

As you have seen in earlier examples, the most common command for plotting two-dimensional data is the `plot` command. This versatile command plots sets of data arrays on appropriate axes, and connects the points with straight lines. Here is an example you have seen before:

```
» x=linspace(0,2*pi,30);
» y=sin(x);
» plot(x,y)
```

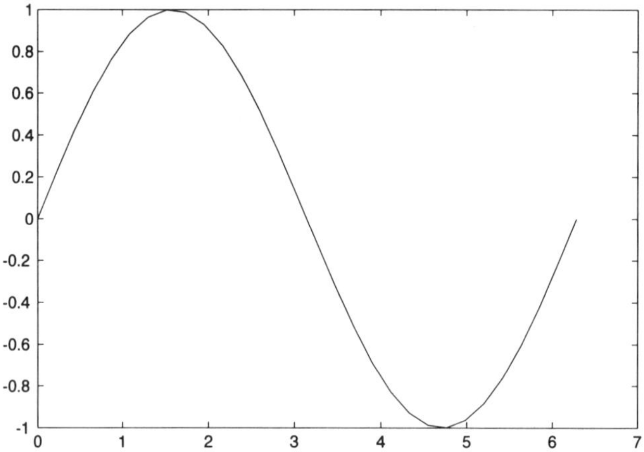

This example creates 30 data points `0 ≤ x ≤ 2π` to form the horizontal axis of the plot, and creates another vector `y` containing the sine of the data points in `x`. The `plot` command opens a graphics window, called a figure window, scales the axes to fit the data, plots the points, and then connects the points with a straight line. It also adds a numerical scale and tick marks to both axes automatically. If a figure window already exists, `plot` clears the current figure window and draws a new plot.

Let's plot a sine and cosine on the same plot:

```
» z=cos(x);
» plot(x,y,x,z)
```

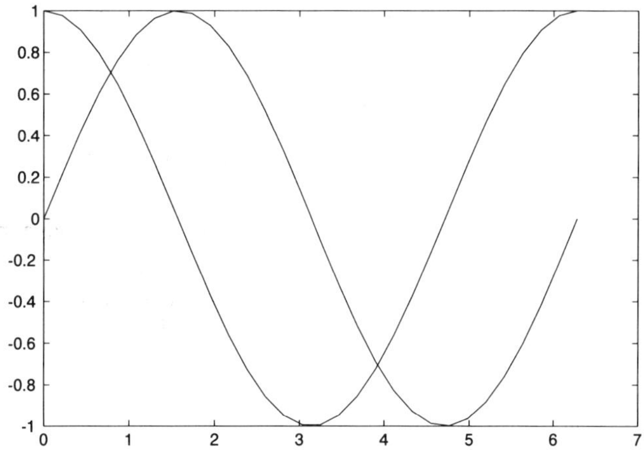

This example shows that you can plot more than one set of arrays at the same time, just by giving `plot` another pair of arguments. This time `sin(x)` vs. `x`, and `cos(x)` vs. `x` were plotted on the same plot. `plot` automatically drew the second curve in a different color on the screen. Many curves may be plotted at one time by supplying additional pairs of arguments to `plot`.

If one of the arguments is a matrix and the other a vector, the plot command plots each column of the matrix versus the vector:

```
» W=[y;z]; % create a matrix of the sin and cosine
» plot(x,W) % plot the columns of W vs. x
```

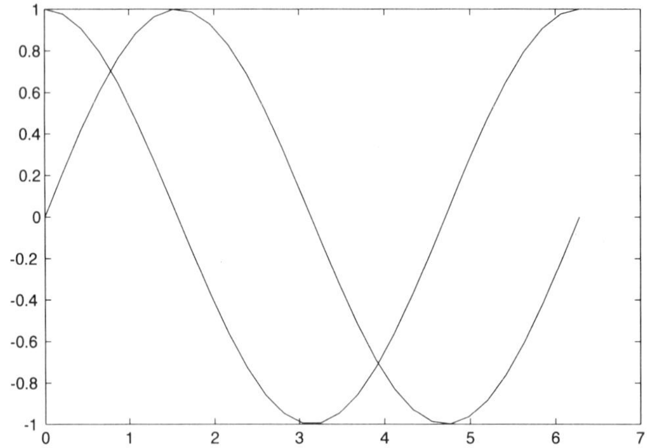

If you change the order of the arguments, the plot will rotate 90 degrees:

```
» plot(W,x) % plot x vs. the columns of W
```

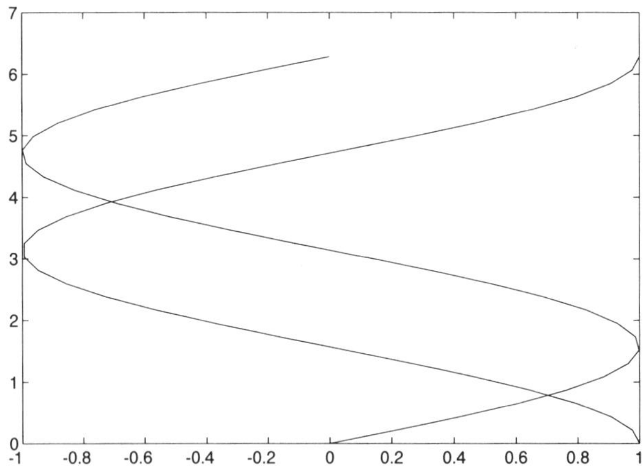

5.19.2 Linestyles, Markers, and Colors

In the previous examples, MATLAB chose the solid linestyle, and the colors yellow and magenta for the plots. You can specify the colors and linestyles you want by giving `plot` an additional argument after each pair of data arrays. The optional additional argument is a character string consisting of 1, 2, or 3 characters from the following table:

Symbol	Color	Symbol	Linestyle
y	yellow	.	point
m	magenta	o	circle
c	cyan	x	x-mark
r	red	+	plus
g	green	*	star
b	blue	-	solid line
w	white	:	dotted line
k	black	-.	dash-dot line
		--	dashed line

If you do not specify a color, MATLAB starts with yellow and cycles through the first six colors in the table for each additional line. The default linestyle is the solid line unless you specify a different linestyle. Use of the point, circle, *x*-mark, plus, and star symbols places the chosen symbol at each data point, but does not connect the data points with a straight line.

Here is an example using different linestyles, colors, and point markers:

```
» plot(x,y,'g:',x,z,'r--',x,y,'wo',x,z,'c+')
```

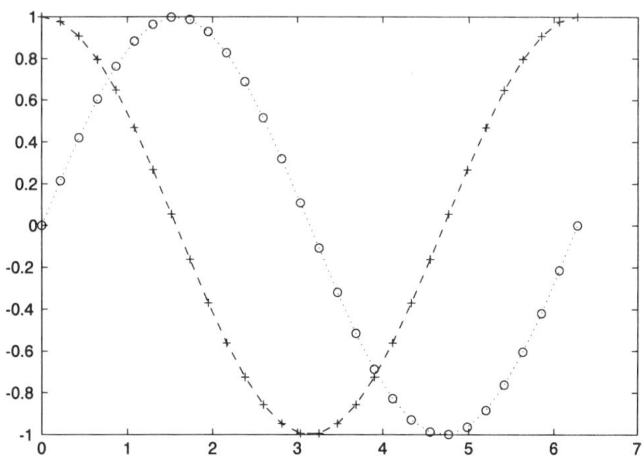

As with many of the plots in this section, your computer displays color but the figures printed here do not. Those figures that are printed in color are shown in the color plates in the center of this text.

5.19.3 Adding Grids and Labels

The `grid on` command adds grid lines to the current plot at the tick marks. The `grid off` command removes the grid. `grid` with no arguments alternately turns them on and off, i.e., toggles them. Horizontal and vertical axes can be labeled with the `xlabel` and `ylabel` commands, respectively. The `title` command adds a line of text at the top of the plot. Let's use the sine and cosine plot again as an example.

```
» x=linspace(0,2*pi,30);
» y=sin(x);
» z=cos(x);
» plot(x,y,x,z)
```

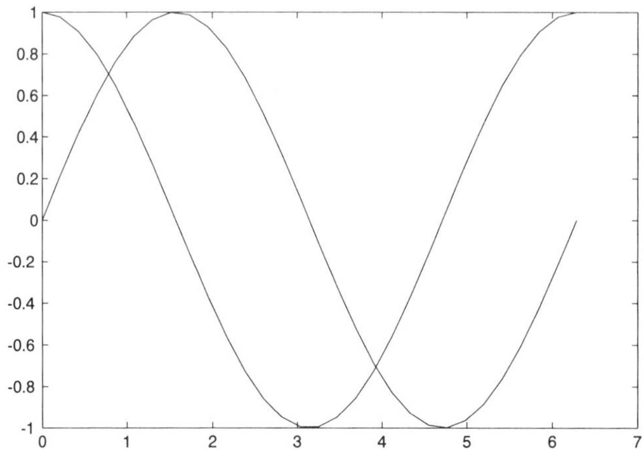

Now add a grid, title, and axis labels:

```
» grid % turn on grid lines
» xlabel('Independent Variable X')        % x-axis label
» ylabel('Dependent Variables Y and Z') % y-axis label
» title('Sine and Cosine Curves')        % title the plot
```

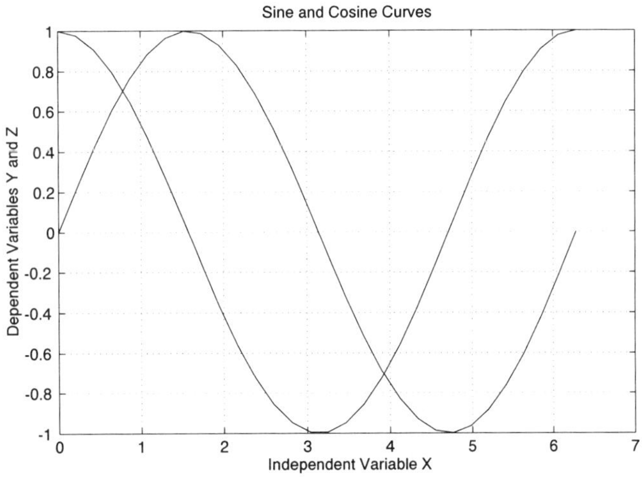

You can add a label or any other text string to any specific location on your plot with the `text` command. The format is `text(x,y,'string')`, where `(x,y)` represents the coordinates of the center left edge of the text string in units taken from the plot axes. To add a label identifying the sine curve at the location `(2.5,0.7)`:

```
» text(2.5,0.7,'sin(x)')
```

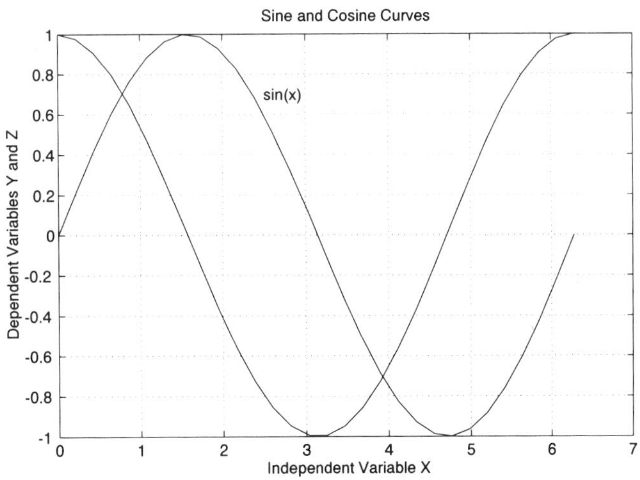

If you want to add a label, but don't want to stop to figure out the coordinates to use, you can place a text string with the mouse. The `gtext` command switches to the current figure window, puts up a cross-hair that follows the mouse, and waits for a mouse click or keypress. When either one occurs, the text is placed with the lower left corner of the first character at that location. Try labeling the second curve in the plot:

```
» gtext('cos(x)')
```

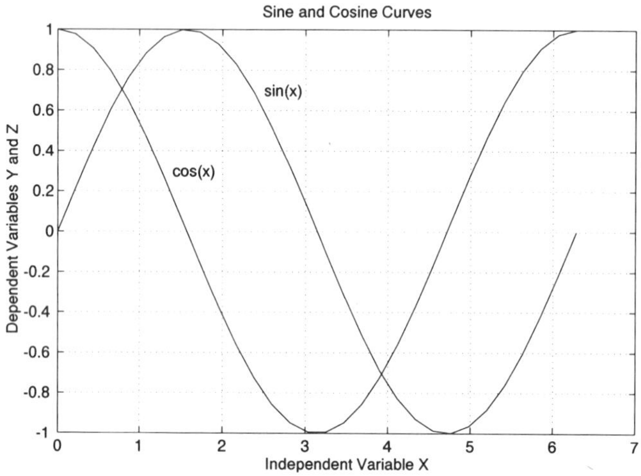

5.19.4 Customizing Axes

If you are not satisfied with the scaling and appearance of both the horizontal and vertical axes of your plot, MATLAB gives you complete control over them with the `axis` command. Because this command has so many features, only the most useful will be described here. For a more complete description see Section 8.2, *MATLAB Commands and Functions* or use online help. The primary features of the `axis` command are given in the table below.

Commands	Description
`axis([xmin xmax ymin ymax])`	Set the maximum and minimum values of the axes using values given in the row vector.
`axis auto` `axis('auto')`	Return the axis scaling to its automatic defaults: `xmin=min(x)`, `xmax=max(x)`, etc.

Commands (Continued)	Description (Continued)
`axis(axis)`	Freeze scaling at the current limits, so that if `hold` is turned on, subsequent plots use the same axis limits.
`axis xy` `axis('xy')`	Use the (default) Cartesian coordinate form, where the *system origin* (the smallest coordinate pair) is at the lower left corner. The horizontal axis increases left to right, and the vertical axis increases bottom to top.
`axis ij` `axis('ij')`	Use the *matrix* coordinate form, where the system origin is at the top left corner. The horizontal axis increases left to right, but the vertical axis increases top to bottom.
`axis square` `axis('square')`	Set the current plot to be a square rather than the default rectangle.
`axis equal` `axis('equal')`	Set the scaling factors for both axes to be equal.
`axis normal` `axis('normal')`	Turn off `axis equal` and `axis square`
`axis off` `axis('off')`	Turn off all axis labeling, grid, and tick marks. Leave the title and any labels placed by the `text` and `gtext` commands.
`axis on` `axis('on')`	Turn on axis labeling, tick marks, and grid.

Try out some of the `axis` commands on your plots. Using the example plot from above produces the following results:

```
» axis off % turn off the axes
```

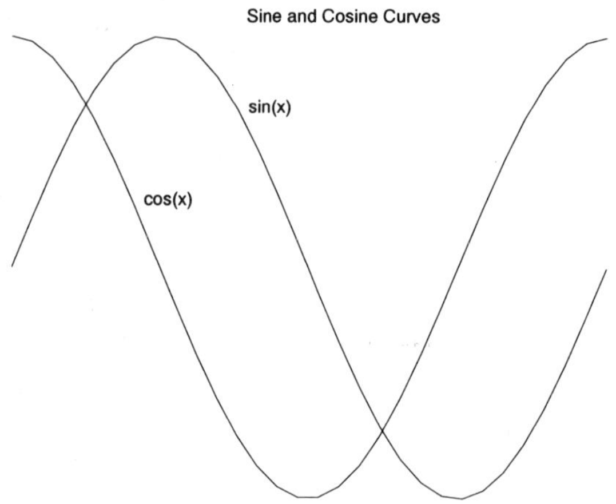


```
» axis on, grid off
```

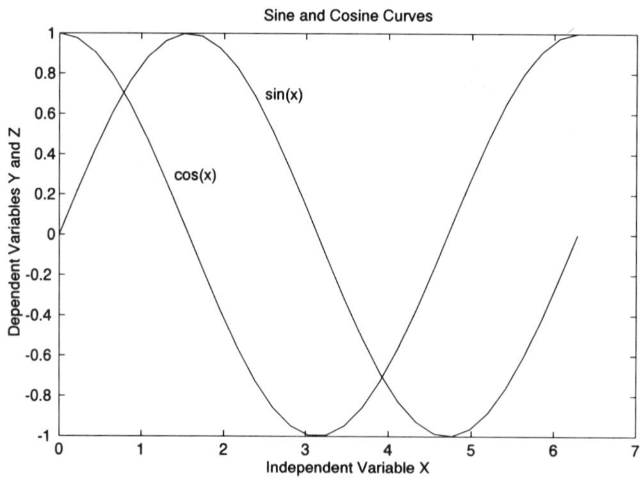

```
» axis ij % turn the plot upside down
```

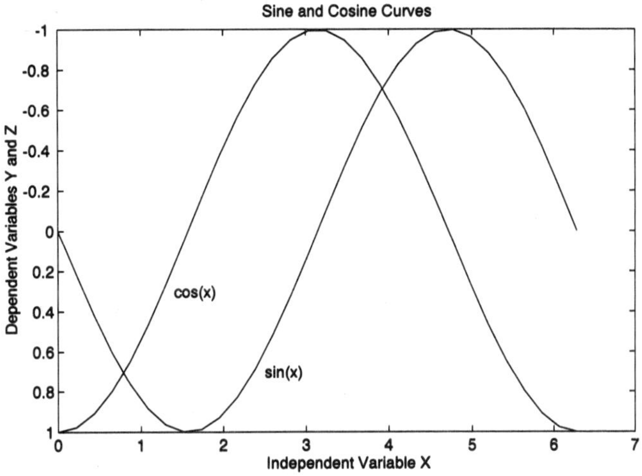


```
» axis('square','equal') % give axis two commands at once
```

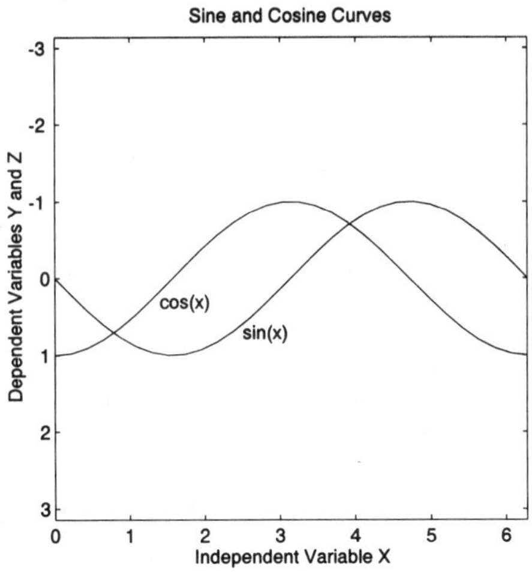

```
» axis('xy','normal') % return to the defaults
```

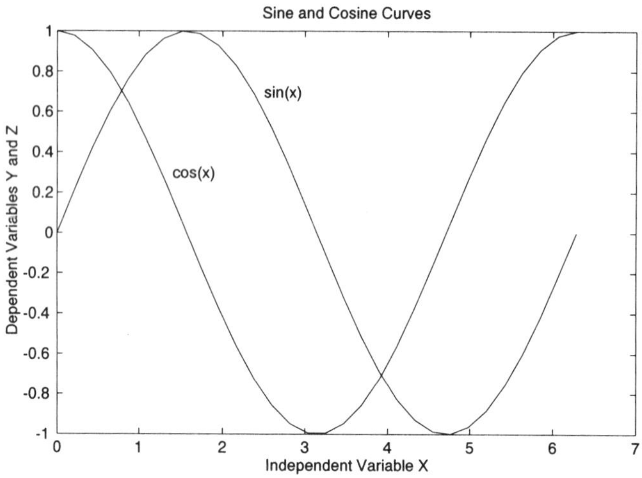

5.19.5 Printing Figures

What good is a plot if you can't print it? Printing of plots can be accomplished by using a command from the menu bar or MATLAB commands.

To print a plot using commands from the menu bar, make the figure window the active window by clicking on it with the mouse. Then use the **Print** menu item from the **File** menu (on the figure window menu bar on the MS Windows version). Using the parameters set in the **Print Setup** or **Page Setup** menu item, the current plot is sent to the printer.

MATLAB has its own printing commands that can be used from the command window. To print a figure window, click on it with the mouse or use the `figure(n)` command to make it active, and then use the `print` command.

```
» print % prints the current plot to your printer
```

The `orient` command changes the print orientation mode. The default *portrait* mode prints vertically in the middle of the page. *Landscape* mode prints horizontally and fills the page. *Tall* mode prints vertically but fills the page. The chosen printing mode remains unchanged until you change it or end your MATLAB session.

```
» orient % what is the current orientation?
ans =
      portrait

» orient landscape % print sideways on the page
» orient tall % stretch to fill the vertical page
```

5.19.6 Manipulating Plots

You can add lines to an existing plot using the `hold` command. When you set `hold on`, MATLAB does not remove the existing curves when new `plot` commands are issued. Instead, it adds new curves to the current axes. However, if the new data does not fit within the current axes limits, the axes are rescaled. Setting `hold off` releases the current figure window for new plots. The `hold` command without arguments toggles the `hold` setting.

Going back to our previous example:

```
» x=linspace(0,2*pi,30);
» y=sin(x);
» z=cos(x);
» plot(x,y)
```

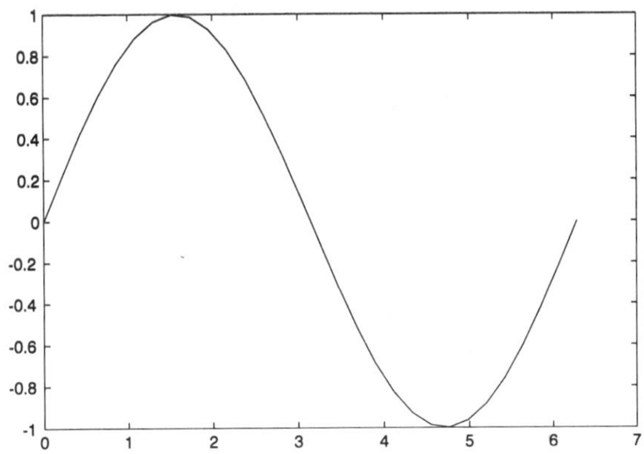

Now hold the plot and add a cosine curve.

```
» hold on
» plot(x,z,'m')
» hold off
```

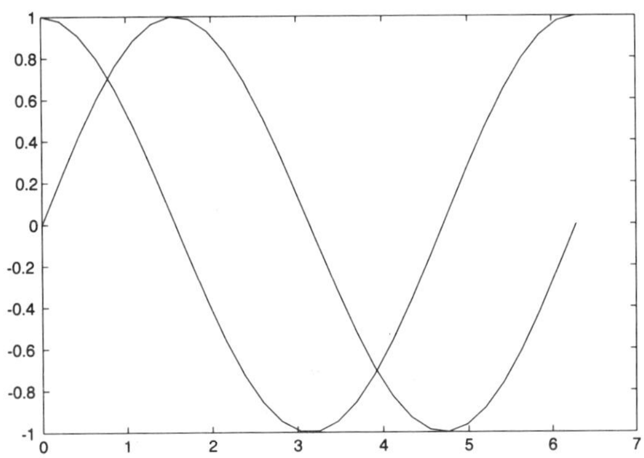

Notice that this example specified the color of the second curve. Since there is only one set of data arrays in each plot command, the line color for each plot command would otherwise default to yellow, resulting in two yellow lines on the plot.

If you want two or more plots in different figure windows, use the figure command in the command window, or the **New Figure** selection from the **File** menu in the command or figure windows. figure with no arguments creates a new figure window. You can choose a specific figure window to be the default by selecting it with the mouse, or by using figure(n), where n is the number of the window to be made active for subsequent plotting commands.

One figure window, on the other hand, can hold more than one set of axes. The subplot(m,n,p) command subdivides the current figure window into an m-by-n matrix of plotting areas, and chooses the *p*th area to be active. The subplots are numbered left to right along the top row, then the second row, etc. For example:

```
» x=linspace(0,2*pi,30);
» y=sin(x);
» z=cos(x);
» a=2*sin(x).*cos(x);
» b=sin(x)./(cos(x)+eps);
» subplot(2,2,1) % pick the upper left of 4 subplots
» plot(x,y),axis([0 2*pi -1 1]),title('sin(x)')
» subplot(2,2,2) % pick the upper right of 4 subplots
» plot(x,z),axis([0 2*pi -1 1]),title('cos(x)')
» subplot(2,2,3) % pick the lower left of 4 subplots
» plot(x,a),axis([0 2*pi -1 1]),title('2sin(x)cos(x)')
» subplot(2,2,4) % pick the lower right of 4 subplots
» plot(x,b),axis([0 2*pi -20 20]),title('sin(x)/cos(x)')
```

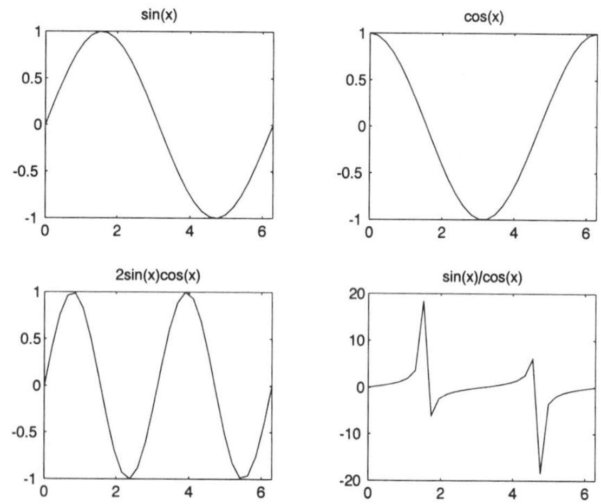


```
» subplot(1,1,1) % return to a single plot in the figure window
```

Use the command `subplot(1,1,1)` to return to the default mode and use the entire figure window for a single set of axes. If you print a figure window containing multiple plots, all of them will be printed on the same page. For example, if the current figure window contains four subplots and the orientation is landscape mode, each of the plots will use one quarter of the printed page.

MATLAB provides an interactive tool to expand sections of a 2-D plot to see more detail, or to *zoom in* on a region of interest. The command `zoom on` turns on zoom mode. Clicking the mouse button on a Macintosh or left mouse button on an MS-Windows computer within the figure window expands the plot by a factor of 2 centered around the point under the mouse pointer. Each time you click, the plot expands. Click the right mouse button on an MS-Windows computer or shift-click on the Macintosh to zoom out by a factor of 2. You can also click-and-drag to zoom into a specific area. `zoom out` returns the plot to its initial state. `zoom off` turns off zoom mode. `zoom` with no arguments toggles the zoom state of the active figure window.

Try zooming in and out on a plot created using the M-file called `peaks.m`. This is an interesting function that generates a square matrix of data. The data is based on a function of two variables, and contains data points for `x` and `y` in the range –3 to 3. The function is:

$$f(x,y) = 3(1-x)^2 e^{-(y+1)^2-x^2} - 10\left(\frac{x}{5} - x^3 - y^5\right)e^{-x^2-y^2} - \frac{1}{3}e^{-(x+1)^2-y^2}$$

You can specify the size of the square matrix `peaks` generates by passing it an argument. If you omit the argument, it defaults to 31. Try this example:

```
» M=peaks(25)        % create a 25-by-25 matrix of data
» plot(M)            % plot the columns of M
» title('Peaks Plot for ZOOM Practice')
» zoom on
```

The command `plot(M)` where `M` is a matrix, plots each column of `M` vs. its index. The example above plotted 25 lines on the plot. Zoom in and out with the mouse to experiment with zooming.

5.19.7 Other 2-D Plotting Features

- `loglog` is the same as `plot`, except that logarithmic scales are used for both axes.
- `semilogx` is the same as `plot`, except that the x-axis uses a logarithmic scale, and the y-axis uses a linear scale.
- `semilogy` is the same as `plot`, except that the y-axis uses a logarithmic scale, and the x-axis uses a linear scale.

• Plots in polar coordinates can be created using the `polar(t,r,S)` command, where `t` is the angle vector in radians, `r` is the radius vector, and `S` is an optional character string describing color, marker symbol, and/or linestyle. See `plot` for a description of appropriate string values.

```
» t=0:.01:2*pi;
» r=sin(2*t).*cos(2*t);
» polar(t,r)
» title('Polar Plot of sin(2t)cos(2t)')
```

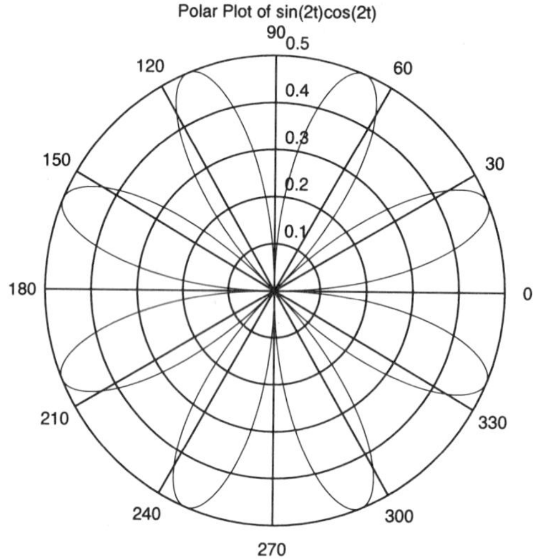

- Bar and stair plots can be generated using the `bar` and `stairs` plotting commands. Here are examples of a bell curve:

```
» x=-2.9:0,2:2.9;
» y=exp(-x.*x);
» bar(x,y)
» title('Bar Chart of a Bell Curve')
```

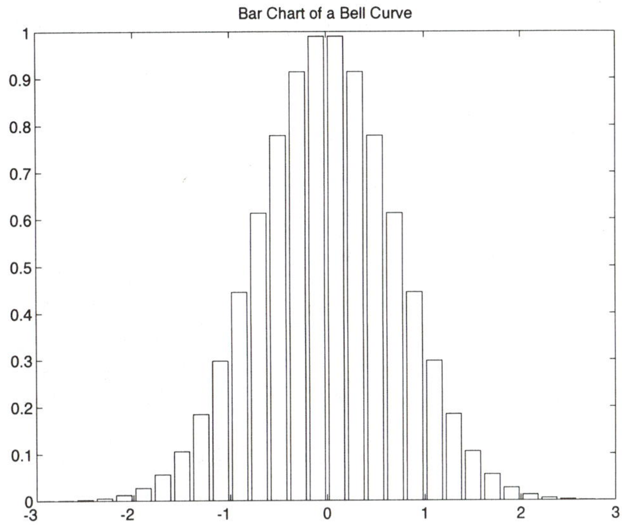


```
» stairs(x,y)
» title('Stair Chart of a Bell Curve')
```

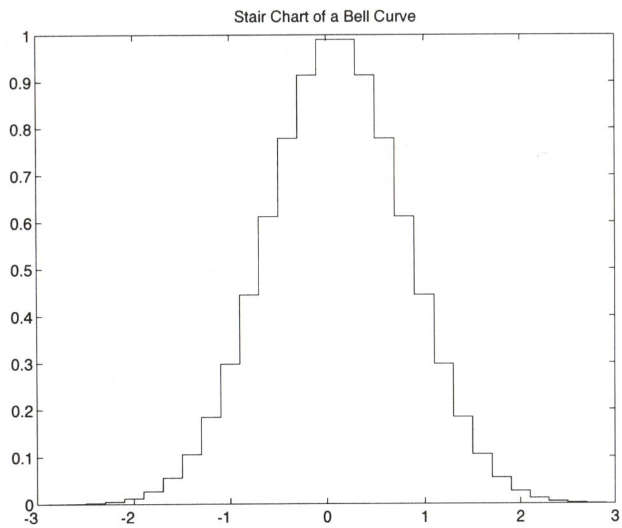

- `hist(y)` draws a 10-bin histogram for the data in vector `y`. `hist(y,n)` where `n` is a scalar draws a histogram with `n` bins. `hist(y,x)` where `x` is a vector draws a histogram using the bins specified in `x`. Here is an example of a bell-curve histogram from Gaussian data:

```
» x=-2.9:0.2:2.9; % specify the bins to use
» y=randn(5000,1); % create 5000 random points
» hist(y,x) % draw the histogram
» title('Histogram of Gaussian Data')
```

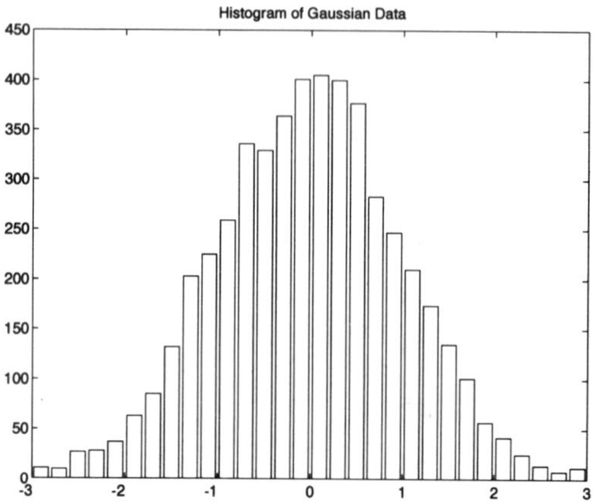

- Discrete sequence data can be plotted using the `stem` function. `stem(y)` creates a plot of the data points in vector `y` connected to the horizontal axis by a line. An optional character string argument can be used to specify linestyle. `stem(x,y)` plots the data points in `y` at the values specified in `x`.

```
» y=randn(50,1); % create some random data
» stem(y,':') % draw a stem plot with dotted line
» title('Stem Plot of Random Data')
```

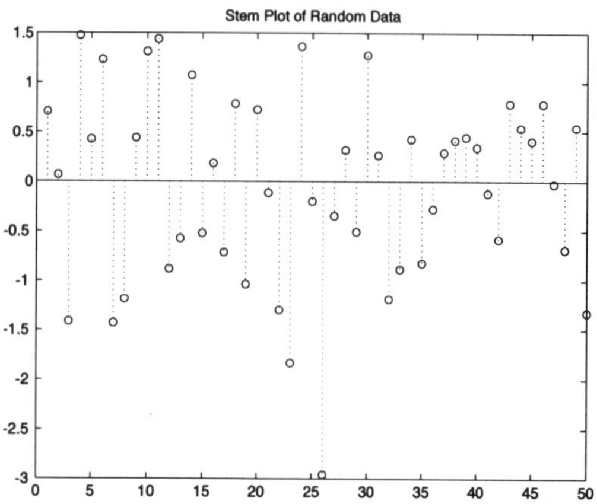

- A plot can include error bars at the data points. `errorbar(x,y,e)` plots the graph of vector `x` vs. vector `y` with error bars specified by vector `e`. All vectors must be the same length. For each data point (x_i, y_i), an error bar is drawn a distance e_i above and e_i below the data point.

```
» x=0:0.1:2; % create a vector
» y=erf(x); % y is the error function of x
» e=rand(size(x))/10; % generate random error values
» errorbar(x,y,e) % create the plot
» title('Errorbar Plot')
```

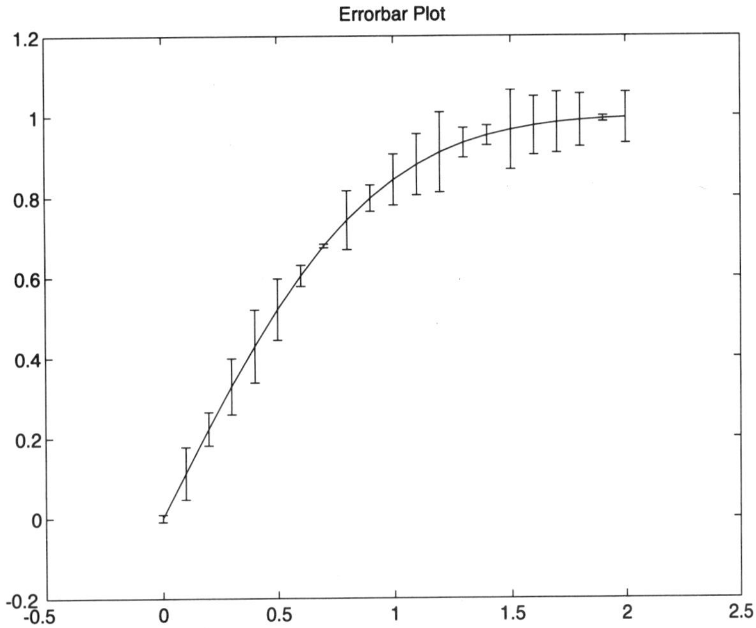

- Complex data can be plotted using `compass` and `feather`. `compass(z)` draws a plot that displays the angle and magnitude of the complex elements of `z` as arrows emanating from the origin. `feather(z)` plots the same data using arrows emanating from equally spaced points on a horizontal line. compass(x,y) and feather(x,y) are equivalent to compass(x+i*y) and feather(x+i*y).

```
» z=eig(randn(20,20));
» compass(z)
» title('Compass Plot of the Eigenvalues of a Random Matrix')
```

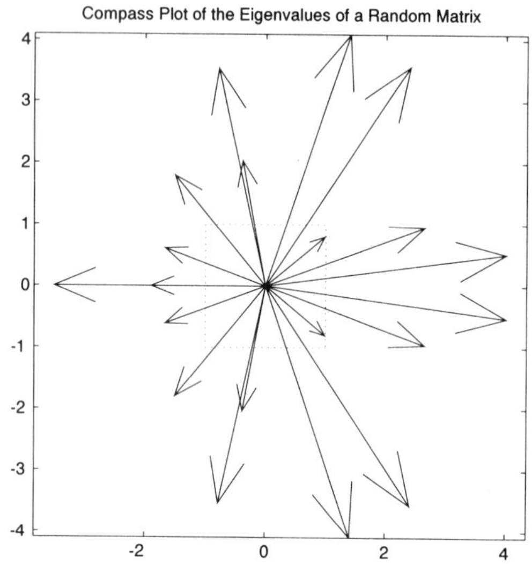


```
» feather(z)
» title('Feather Plot of Eigenvalues of a Random Matrix')
```

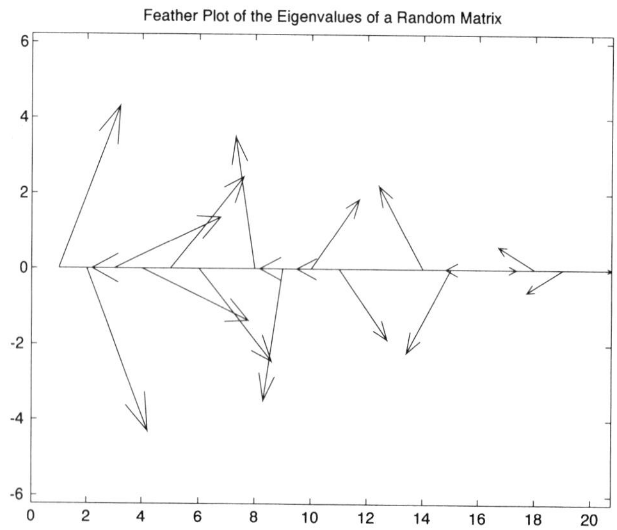

- `rose(t)` draws a 20-bin polar histogram for the angles in vector `t`. `rose(t,n)` where `n` is a scalar draws a histogram with `n` bins. `rose(t,x)` where `x` is a vector draws a histogram using the bins specified in `x`. Here is an example of an angle histogram:

```
» t=randn(1000,1)*pi;
» rose(t)
» title('Angle Histogram of Random Angles')
```

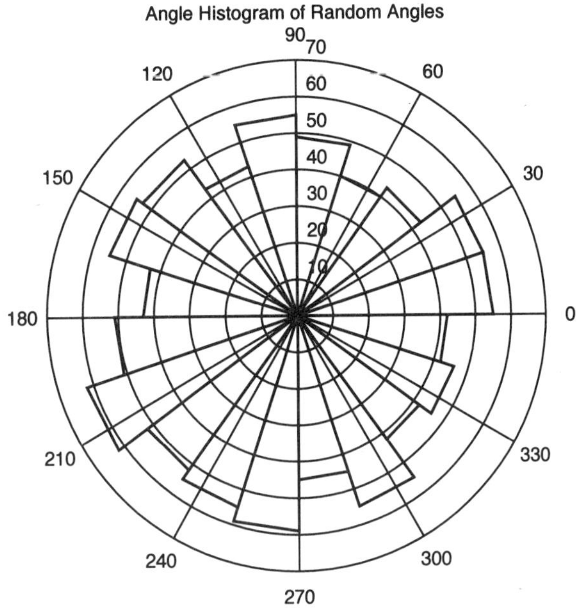

- `ginput` provides a means of selecting points from the current plot using the mouse. `[x,y]=ginput(n)` gets up to `n` points from the current axes and returns their coordinates in the column vectors `x` and `y`. If `n` is omitted, an unlimited number of points are gathered until the **Return** or **Enter** key is pressed. As an example, let's plot a function, and then plot some points with the mouse.

```
» x=linspace(-2*pi,2*pi,60);
» y=sin(x).^2./(x+eps);
» plot(x,y)
» title('Plot of sin(x)^2/x')
» [a,b]=ginput(8); % get up to 8 points
» hold on
» plot(a,b,'co') % plot the collected points
» hold off
```

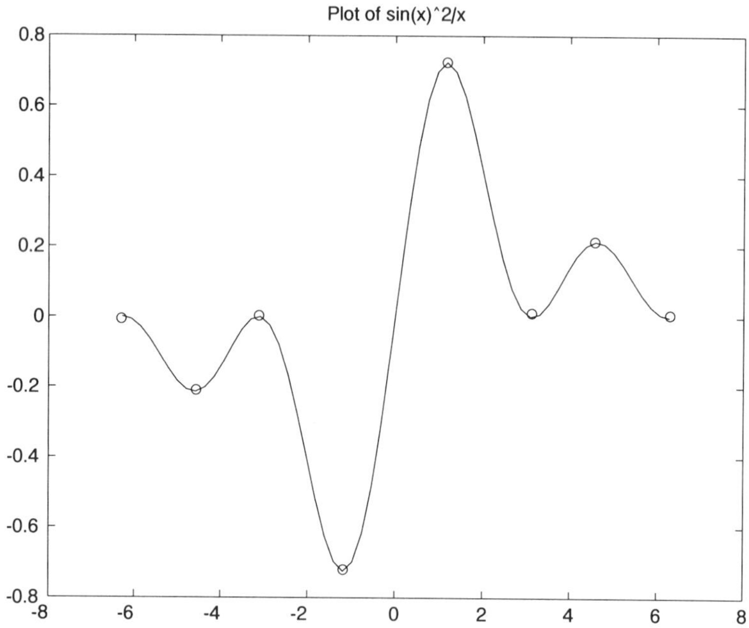

- The fplot command gives you an easy way to automatically plot a function of one variable between specified limits without creating a data set for the variable. fplot('fun',[xmin xmax]) plots the function fun over the range xmin ≤ x ≤ xmax with automatic scaling of the y-axis. fplot('fun',[xmin xmax ymin ymax]) specifies the y-axis limits as well. There are restrictions on the type of function that can be plotted, and additional arguments can be specified. See the *Numerical Analysis* section as well as Chapter 6, *Symbolic Math Toolbox Tutorial* for further information.

```
» fplot('sin(x)./x',[-20 20 -.4 1.2])
» title('Fplot of f(x)=sin(x)/x')
» xlabel('x')
» ylabel('f(x)')
```

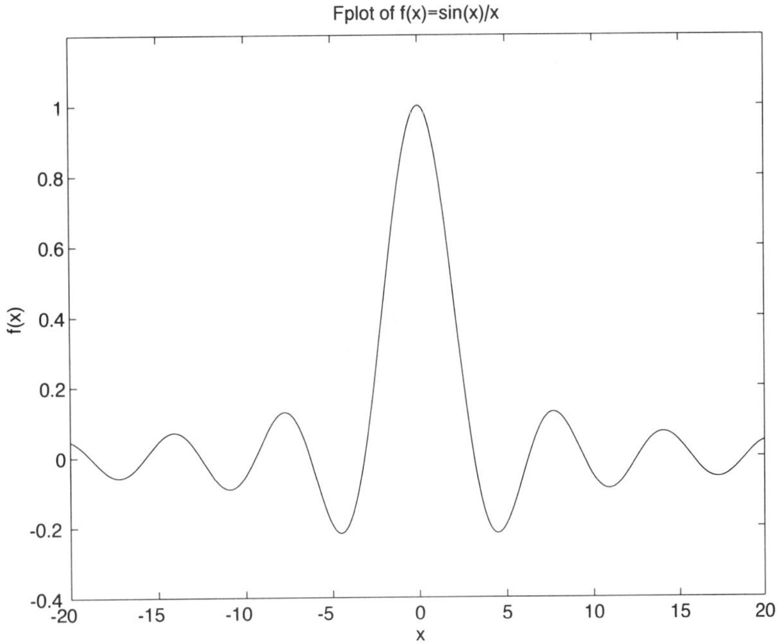

- `fill(x,y,'c')` fills the 2-D polygon defined by the column vectors `x` and `y` with the color specified by `c`. The vertices of the polygon are specified by the pairs (x_i, y_i). If necessary, the polygon is closed by connecting the last vertex to the first. See Section 8.2, *MATLAB Commands and Functions* or online help for more information. Try this example:

```
» t=(1/8:2/8:15/8)'*pi; % column vector
» x=sin(t);
» y=cos(t);
» fill(x,y,'r') % a filled red circle using only 8 points
» axis('square')
» text('-.11,0,'STOP')
» title('Red Stop Sign')
```

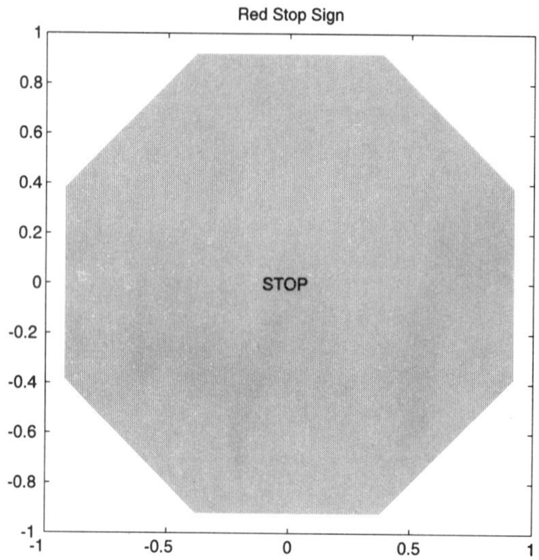

5.19.8 Summary

- The `plot` command creates a plot of vectors or columns of matrices. The form of the command is `plot(`x_1`,`y_1`,`S_1`,`x_2`,`y_2`,`S_2`,...)` where (x_n, y_n) are sets of data arrays and S_n are optional strings specifying color, marker symbols, and/or linestyles.
- `grid` turns on a grid at the tick marks of your plot.
- Titles and axis labels can be added using `title`, `xlabel`, and `ylabel`.
- The `text(x,y,S)` command adds the character string `S` to the current plot at the coordinates `(x,y)`.
- `gtext` lets you place text on your plot interactively using the mouse.
- `axis([xmin xmax ymin ymax])` scales the current plot to the values given as arguments. `axis('string')`, where `'string'` is one of a number of spe-

cific options, changes axis limits or the appearance of the plot in many ways.

- The `print` command prints the plot in the current figure window to your printer, the clipboard, or to a file.
- Orientation of printed plots can be changed with the `orient` command. The three orientations are `portrait`, `landscape`, and `tall`.
- You can add plots to your current plot by setting `hold on`. Setting `hold off` allows the next `plot` command to clear the figure window before plotting.
- Multiple figure windows can be generated by the `figure` command. `figure(n)` chooses figure window `n` to be the active figure window.
- A figure window may be subdivided and any subdivision made active with the `subplot` command.
- If you set `zoom on`, the active figure window can be expanded interactively using the mouse.

5.20 3-D Graphics

MATLAB provides a variety of functions to display 3-D data. Some functions plot lines in three dimensions, while others draw surfaces and wire frames. Pseudocolor can be used to represent a fourth dimension.

5.20.1 Line Plots

The `plot` command from the 2-D world can be extended into three dimensions with `plot3`. The format is the same as the 2-D `plot`, except the data is in triples, rather than pairs. The generalized format of `plot3` is `plot3(`$x_1, y_1, z_1, S_1, x_2, y_2, z_2, S_2, \ldots$`)`, where x_n, y_n, and z_n are vectors or matrices, and S_ns are optional character strings specifying color, marker symbol, and/or linestyle. Here is an example of a 3-D helix:

```
» t=0:pi/50:10*pi;
» plot3(sin(t),cos(t),t)
» title('Helix'),xlabel('sin(t)'),ylabel('cos(t)'),zlabel('t')
```

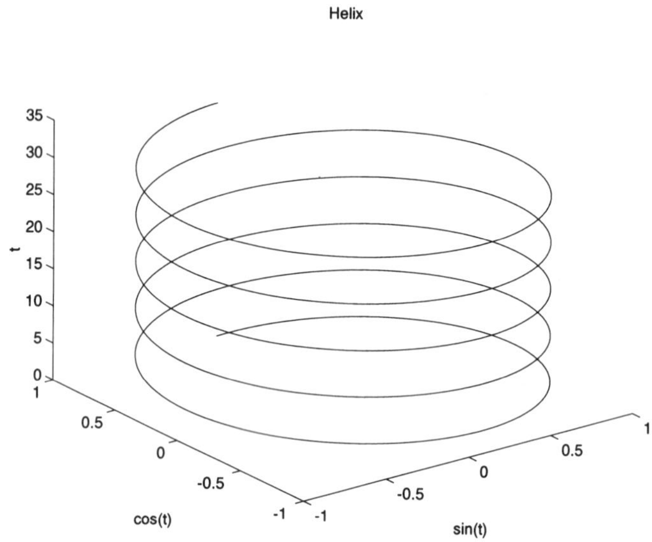

Notice that there is a zlabel function corresponding to the 2-D xlabel and ylabel functions. In the same way, the axis command has a 3-D form:

`axis([xmin xmax ymin ymax zmin zmax])` sets the limits of all three axes. Other axis commands also work in 3-D. Let's change the plot origin:

```
» axis('ij')          % make the y–axis increase back–to–front
```

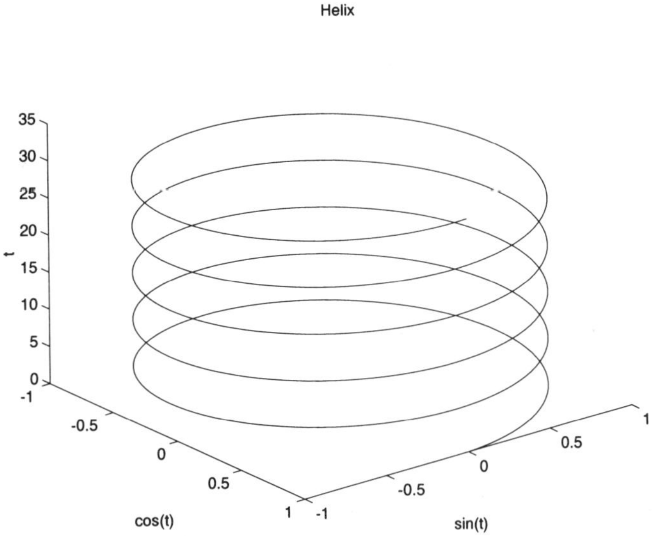

The `text` function also has a 3-D form: `text(x,y,z,'string')` will place the text `'string'` beginning at the coordinate `(x,y,z)` on the current plot.

5.20.2 Mesh and Surface Plots

MATLAB defines a mesh surface by the z-coordinates of points above a rectangular grid in the x-y plane. It forms a plot by joining adjacent points with straight lines. The result looks like a fishing net with the knots at the data points. Mesh plots are very useful for visualizing large matrices or plotting functions of two variables.

The first step in generating the mesh plot of a function of two variables, $z=f(x,y)$, is to generate `X` and `Y` matrices consisting of repeated rows and columns respectively, over some range of the variables x and y. MATLAB provides the function `meshgrid` for this purpose. `[X,Y]=meshgrid(x,y)` creates a matrix `X` whose rows are copies of the vector `x`, and a matrix `Y` whose columns are copies of the vector `y`. This pair of matrices can then be used to evaluate functions of the two variables using MATLAB's array mathematics features.

Here is an example of using meshgrid to generate evenly-spaced data points in the x-y plane between –7.5 and 7.5 in both x and y.

```
» x=-7.5:.5:7.5;
» y=x;
» [X,Y]=meshgrid(x,y);
```

X and Y are a pair of matrices representing a rectangular grid of points in the x-y plane. Any function $z = f(x,y)$ can be generated using these points.

```
» R=sqrt(X.^2+Y.^2)+eps;      % distance from the origin (0,0)
» Z=sin(R)./R;                % calculate sin(r)/r
```

The matrix R contains the radius of each point in [X,Y]. This is the distance from each point to the center of the matrix, which is the origin. Adding eps prevents dividing by zero and generating NaNs (Not-a-Numbers) in the data. The matrix Z now contains the sine of the radius divided by the radius for each point in the plane. The following command generates the mesh plot shown on **Color Plate 1** in the center of this text.

```
» mesh(X,Y,Z)
```

Notice that the line colors are related to the height of the mesh above the x-y plane. mesh will accept an optional argument to control the colors and color ranges used in the plot. This ability to change how MATLAB uses color will be discussed later in the sections on **Color Maps**.

In this example, mesh mapped the values of the matrix elements to the points (X_{ij}, Y_{ij}, Z_{ij}) in 3-D space. mesh can also take a single matrix as an argument; mesh(Z) uses the points (i,j,Z_{ij}). That is, Z is plotted versus its subscripts or indices. In this particular case, mesh(Z) just changes the scale of the x and y axes to the indices of matrix Z. Try it for yourself.

A *surface* plot of the same matrix Z looks like the mesh plot previously generated, except that the spaces between the lines (called *patches*) are filled in. Plots of this type are generated using the surf function, which has all of the same arguments as the mesh function. **Color Plate 2** in the center of this text illustrates the following example.

```
» surf(X,Y,Z)
```

To help illustrate the next several topics, let's return to the peaks function discussed earlier. A 3-D mesh plot of this function can be generated by:

```
» mesh(peaks)
» title('Mesh Plot of the Peaks Function')
```

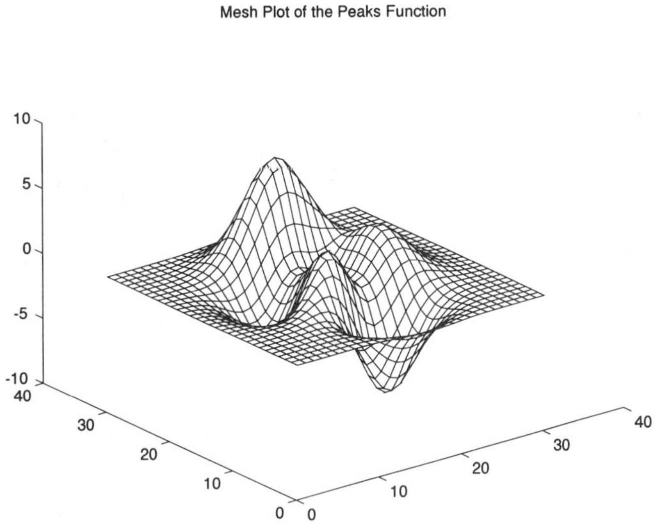

Contour plots show lines of constant elevation or height. Maybe you've seen these before. Certainly, if you've ever seen a topographical map, you know what a contour plot looks like. In MATLAB, contour plots in 2-D and 3-D are generated using the contour and contour3 functions respectively. For example, contour plots of the peaks function using the following commands are illustrated in **Color Plates 3** and **4**.

```
» [x,y,z]=peaks;
» contour(x,y,z,20)          % generate 20 2–D contours
» contour3(x,y,z,20)         % the same contour plot in 3–D
» axis([–3 3 –3 3 –6 8]      % adjust scale
```

Another interesting way to visualize contour information is by using color to represent height. The pcolor function maps height to a set of colors and presents the same information as the contour plot at the same scale. Here is the peaks function again:

```
» z=peaks;
» pcolor(z)
» title('Peaks Pseudocolor Plot')
```

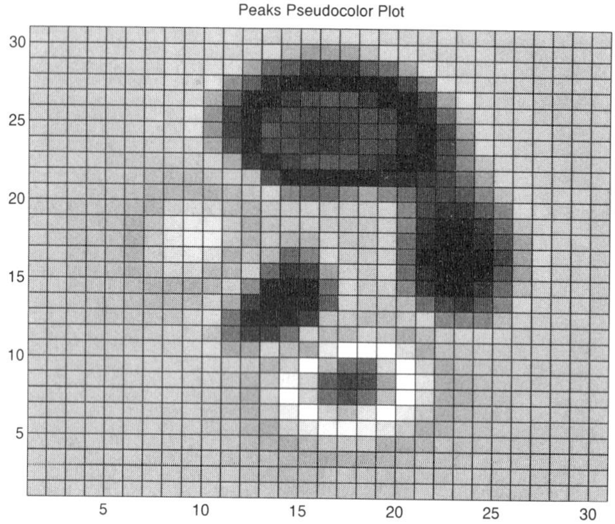

Since both pcolor and contour show the same information at the same scale, it is often useful to superimpose the two. The next example uses colormap and shading to change the appearance of the plot. Both functions will be discussed later in this tutorial. **Color Plate 5** illustrates this example.

```
» [x,y,z]=peaks;
» colormap(hot)    % select a set of colors
» pcolor(x,y,z)    % generate the psudocolor plot
» shading flat     % remove the grid lines
» hold on
» contour(x,y,z,20,'k') % plot 20 contour lines
» hold off
```

5.20.3 Manipulating Plots

MATLAB allows you to specify the angle from which to view a 3-D plot. The function view(azimuth,elevation) sets the angle of view by specifying your *azimuth* and *elevation*. Elevation describes the angle in degrees at which you observe the plot above the *x-y* plane. Azimuth describes the angle within the *x-y* plane where you stand. These concepts are illustrated below.

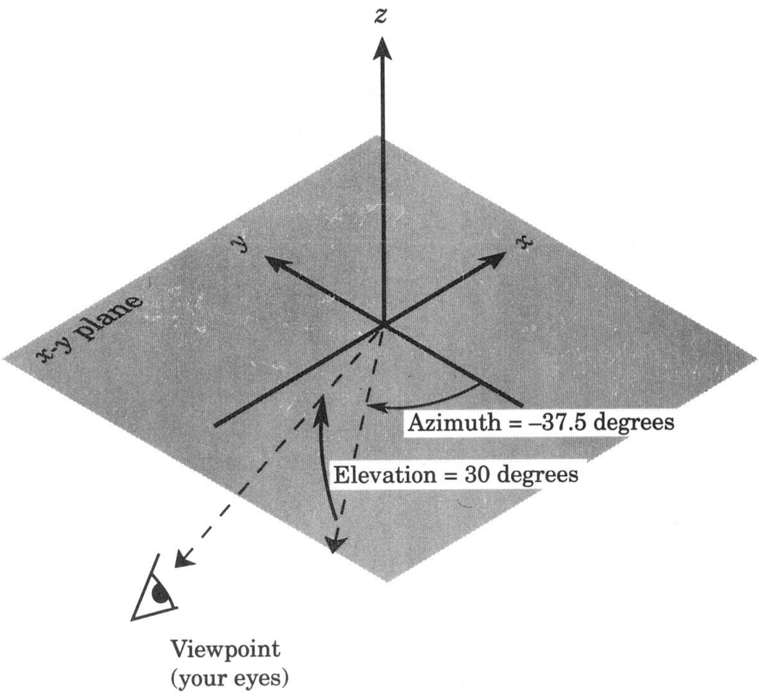

Azimuth is measured in degrees from the negative *y*-axis, with the above figure showing the default MATLAB azimuth of –37.5 degrees, i.e., the negative *y*-axis is rotated counterclockwise away from you 37.5 degrees. Elevation is the angle at which your eyes view the *x-y* plane. The above figure attempts to illustrate the default MATLAB elevation of 30 degrees, i.e., you are looking down at the *x-y* plane at an angle of 30 degrees. Using `view` to set various viewpoints allows you to inspect a figure from any direction. For example, if elevation is set negative, you view the figure from the bottom. If azimuth is set positive, the figure turns clockwise from its default view. You can even look at the figure from directly above by setting view to `view(0,90)`. In fact, this is the default 2-D viewpoint, where the *x*-axis increases from left to right and the *y*-axis increases from bottom to top.

Here are a few examples of the peaks mesh from different viewpoints:

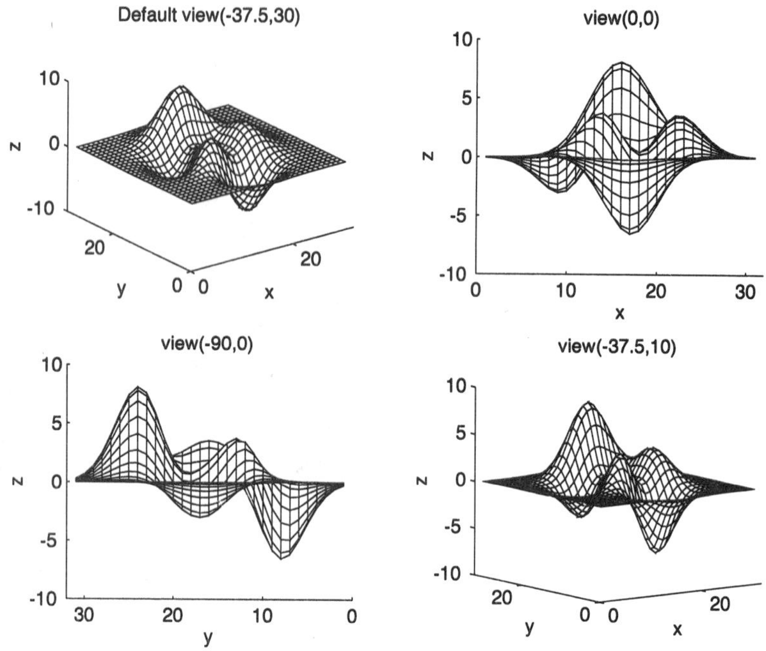

The view command has another form that may be more useful in some instances. view([x y z]) places your view on a vector containing the Cartesian coordinate (x,y,z) in 3-D space. The distance you are from the origin is not affected. For example, view([0 –10 0]), view([0 –1 0]), and view(0,0) all produce the same view. In addition, the azimuth and elevation of the current view can be obtained using [az,el]=view. For example:

```
» view([–7 –9 7]) % view through (–7,–9,7) to the origin
» [az,el]=view % find the azimuth and elevation
az =
       –37.8750
el =
        31.5475
```

The hidden command controls hidden line removal. When you plot a mesh that overlaps itself from your viewpoint (like peaks, or the sinc function), the part of the mesh that is behind another part, i.e., the hidden lines, are removed.

You only see the parts that are in your line of sight. If you turn `hidden off`, you can look right through the mesh. Here is an example:

```
» mesh(peaks(20)+7) % coarse (20) mesh, shifted up
» hold on
» pcolor(peaks(20)) % add a pseudocolor plot
» hold off
» title('Mesh With Hidden On')
```

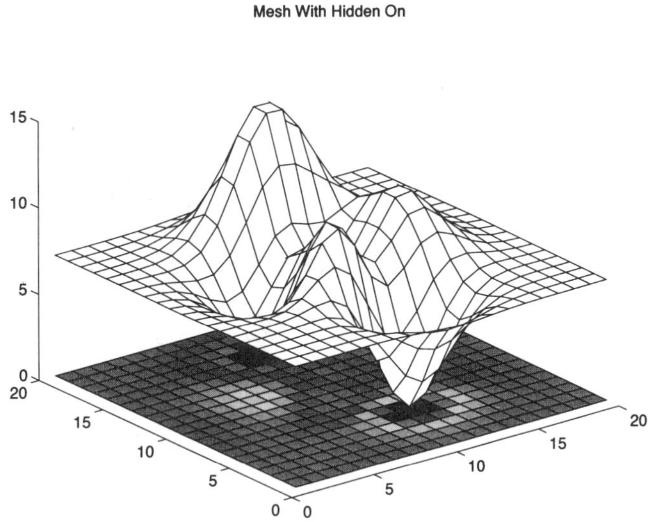

Now turn off hidden line removal and see the difference:

```
» hidden off
» title('Mesh With Hidden Off')
```

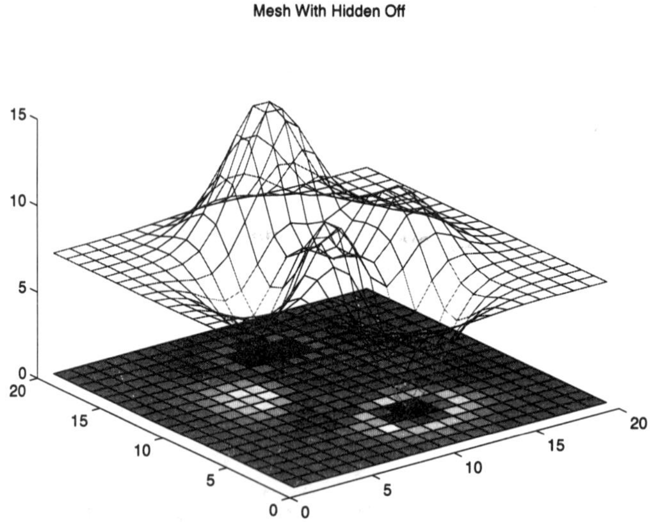

Shading was mentioned earlier. You can choose between three kinds of shading for mesh, surf, pcolor, and fill plots: flat, interpolated, and faceted shading (the default). With flat shading, each mesh line segment or surface patch has a constant color. Faceted shading is flat shading with superimposed black mesh lines. Interpolated shading varies the color of the patch or line segment linearly. Try shading flat, shading interp, and shading faceted on mesh and surf plots to see the effects.

5.20.4 Other 3-D Plotting Features

- The clabel function adds height labels to contour plots. There are three forms: clabel(cs), clabel(cs,V), and clabel(cs,'manual'). clabel(cs), where cs is the contour structure returned from a contour command, i.e., cs=contour(z), labels all known contours with their heights. The label positions are selected randomly. clabel(c,V) labels just those contour levels given in vector V. clabel(c,'manual') places contour labels at the locations clicked on with a mouse, similar to the ginput command described earlier. Pressing the **Return** key terminates labeling.
- Two alternate forms of the mesh command add to the mesh plot. meshc plots the mesh and adds a contour plot beneath it. meshz plots the mesh

and draws a *curtain plot*, or reference plane, beneath the mesh. Try `meshc(peaks)` and `meshz(peaks)` to see the result.

- There are two alternate forms of the `surf` command, as well. `surfc` draws a `surf` plot and adds a contour plot beneath it. `surfl` draws a `surf` plot, but adds surface highlights from a light source. The general form is `surfl(X,Y,Z,S,K)` where `X`, `Y`, and `Z` are the same as `surf`. `S` is an optional vector in Cartesian (`S=[Sx Sy Sz]`) or spherical (`S=[az,el]`) coordinates that specifies the direction of the light source. If not specified, `S` defaults to 45 degrees counterclockwise from the current view direction. `K` is an optional vector that specifies the contribution due to ambient light, diffuse reflection, specular reflection, and the specular spread coefficient (`K=[ka,kd,ks,spread]`).

```
» colormap(gray)
» surfl(peaks),shading interp
» title('Surfl Plot of Peaks With Default Lighting')
```

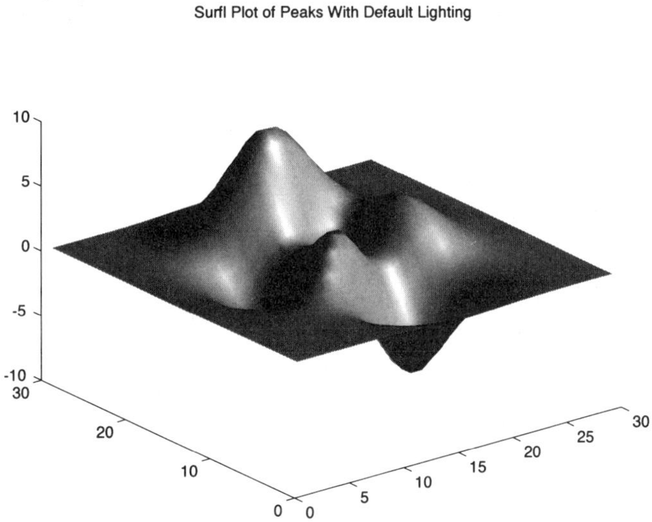

- `fill3`, the 3-D version of `fill`, draws filled 3-D polygons in 3-D space. The general form is `fill3(x,y,z,c)` where the vertices of the polygon are specified by triples of the components of `x`, `y`, and `z`. If necessary, the polygon is closed by connecting the last vertex to the first. If `c` is a character, the polygon is filled with the specified color as shown in the table for plot. `c` can also be an RGB row vector triple (`[r g b]`), where `r`, `g`, and `b` are values between 0 and 1 that represent the amount of red, green, and blue in the resulting color. If `c` is a vector or matrix, it is used as an index into a color map (discussed next). Multiple polygons may be specified by adding more arguments: `fill3(x1,y1,z1,c1,x2,y2,z2,c2,...)`. The following exercise will fill four random triangles with color:

```
» colormap(cool)
» fill3(rand(3,4),rand(3,4),rand(3,4),rand(3,4))
```

5.20.5 Understanding Color Maps

Colors and color maps were mentioned a number of times in this tutorial. This section is a brief exploration into the subject of color and color maps. MATLAB defines a color map as a three-column matrix. Each row of the matrix defines a particular color using numbers in the range 0 to 1. These numbers specify the RGB values: the intensity of the red, green, and blue components of a color. Some representative samples are given in the following table:

Red	Green	Blue	Color
0	0	0	black
1	1	1	white
1	0	0	red
0	1	0	green
0	0	1	blue
1	1	0	yellow
1	0	1	magenta
0	1	1	cyan
0.5	0.5	0.5	medium gray
0.5	0	0	dark red
1	0.62	0.4	copper
0.49	1	0.83	aquamarine

There are 10 MATLAB functions that generate predefined color maps:

Function	Color Map Description
hsv	Hue-saturation-value
hot	Black-red-yellow-white
cool	Shades of cyan and magenta
pink	Pastel shades of pink
gray	Linear gray-scale
bone	Gray-scale with a tinge of blue
jet	A variant of HSV
copper	Linear copper-tone
prism	Prism
flag	Alternating red, white, blue, and black

Each will generate a 64-by-3 matrix specifying the RGB descriptions of 64 colors. Each of these functions can be given an argument specifying the number of rows to generate. For example, `hot(m)` will generate an `m`-by-3 matrix containing the RGB values of colors which range from black, through shades of red, orange, and yellow, to white.

Most computers can handle up to 256 colors in an 8-bit color lookup table. This means that up to three or four 64-by-3 color maps can be in use at one time in different figures. If more color map entries are used, the computer may have to swap out entries in its hardware lookup table. As a result, it is usually prudent to keep the total number of different color map entries used at any one time below 256.

5.20.6 Using Color Maps

The statement `colormap(M)` installs the matrix `M` as the color map to be used by the current figure. For example, `colormap(cool)` installs a 64-entry version of the `cool` color map.

The `plot`, `plot3`, `contour`, and `contour3` functions do not use color maps; they use the colors listed in the `plot` color and linestyle table. Most other plotting functions, such as `mesh`, `surf`, `fill`, `pcolor`, and their variations, use the current color map.

Functions that take a color argument usually accept the argument in one of three forms: a character string representing one of the colors in the `plot` color and linestyle table, e.g., `'r'`, a 3-entry row vector representing a single RGB value (`[.25 .50 .75]`), or a column vector. If the color argument is a column vector or matrix, the elements are scaled and used as indices into the current color map matrix.

5.20.7 Displaying Color Maps

You can display a color map in a number of ways. One way is to view the elements in a color map matrix:

```
» hot(8)
ans =
        0.3333       0       0
        0.6667       0       0
        1.0000       0       0
        1.0000  0.3333       0
        1.0000  0.6667       0
        1.0000  1.0000       0
        1.0000  1.0000  0.5000
        1.0000  1.0000  1.0000
```

In addition, the `pcolor` function can be used to display a color map. Try this example a few times using different color map functions and varying the parameter `n`:

```
» n=8;
» colormap(jet(n))
» pcolor([1:n+1;1:n+1]')
» title('Using Pcolor to Display a Color Map')
```

Color Plate 6 shows eight color maps you can create using the standard MATLAB colormap functions.

The `colorbar` function adds a vertical or horizontal color bar (color scale) to your current figure window showing the color mappings for the current axis. `colorbar('horiz')` places a color bar horizontally beneath your current plot. `colorbar('vertical')` places a vertical color bar to the right of your plot. `colorbar` without arguments either adds a vertical color bar if no color bars exist, or updates any existing color bar. **Color Plate 7** shows the result of the following example.

```
» [x,y,z]=peaks;
» mesh(x,y,z);
» colormap(hsv)
» axis([-3 3 -3 3 -6 8])
» colorbar
```

5.20.8 Creating and Altering Color Maps

The fact that color maps are matrices means that you are able to manipulate them like other arrays. The function `brighten` takes advantage of this fact to adjust a given color map to increase or decrease the intensity of the dark colors. `brighten(n)` brightens ($0 < n \le 1$) or darkens ($-1 \le n < 0$) the current color map. `brighten(n)` followed by `brighten(-n)` restores the original color map. The command `newmap=brighten(n)` creates a brighter or darker version of the current color map without changing the current map. The command `newmap=brighten(cmap,n)` creates an adjusted version of the specified color map without affecting either the current color map or `cmap`.

You can create your own color map by generating an `m`-by-3 matrix `mymap` and installing it with `colormap(mymap)`. Each value in a color map matrix must be between 0 and 1. If you try to use a matrix with more or fewer than three columns or containing any values less than zero or greater than one, `colormap` will report an error.

You can combine color maps arithmetically, although the results are sometimes unpredictable. For example, the map called `pink` is simply:

```
» pinkmap=sqrt(2/3*gray+1/3*hot);
```

Since color maps are matrices, they can be plotted. The `rgbplot` command plots color map matrix values just the way `plot` would, but uses the colors red, green, and blue. Try `rgbplot(hot)`. This shows that the red component increases first, then the green, then the blue. `rgbplot(gray)` shows that all three columns increase linearly and equally (all three lines overlap). Try `rgbplot` with some of the other color maps, such as `jet`, `hsv`, and `prism`.

Normally, a color map is scaled to extend from the minimum to the maximum values of your data; that is, the entire color map is used to render your plot. You may occasionally wish to change the way colors are used. The `caxis` function allows you to use the entire color map for a subset of your data range, or use only a portion of the current color map for your entire data plot.

The current values of `cmin` and `cmax` are returned by `caxis` without arguments. These will normally be the minimum and maximum values of your data. `caxis([cmin cmax])` uses the entire color map for data in the range between `cmin` and `cmax`; data points greater than `cmax` will be rendered with the color associated with `cmax`, and data points smaller than `cmin` will be rendered with the color associated with `cmin`. If `cmin` is less than `min(data)` and/or `cmax` is greater than `max(data)`, the colors associated with `cmin` and/or `cmax` will never be used; only a portion of the color map will be used. `caxis('auto')` will restore the default values of `cmin` and `cmax`.

For the following examples, refer to **Color Plate 8**. The first example is illustrated in **Color Plate 8a**.

```
» pcolor([1:17;1:17]')
» title('Default Color Range')
» colormap(hsv(8))
» caxis('auto')
» colorbar
» caxis
ans =
       1    17
```

As you can see, all eight colors in the current color map are used for the entire data set, two bars for each color. If the colors are mapped to values from –3 to 23, only five colors will be used in the plot as shown in **Color Plate 8b**, which is generated by the commands:

```
» title('Extended Color Range')
» caxis([-3 23]) % extend the color range
» colorbar    % redraw the color scale
```

If the colors are mapped to values from 5 to 12, all colors are used. However, the data less than 5 or greater than 12 get the colors associated with 5 and 12, respectively. This is shown in **Color Plate 8c**, which is produced by the commands:

```
» title('Restricted Color Range')
» caxis([5 12])  % restrict the color range
» colorbar    % redraw the color scale
```

5.20.9 Summary

- `plot3` is the 3-D version of the `plot` command and is used in the same way.
- `zlabel` is used to label the z-axis on 3-D plots.
- The `axis([xmin xmax ymin ymax zmin zmax])` command sets limits for all three axes. `axis('ij')` moves the origin of a 3-D plot, and changes the y-axis to increase left-to-right in the default view.
- The 3-D form of the text function, `text(x,y,z,S)`, places the character string `S` at the location `(x,y,z)` on the current plot.
- A rectangular grid of evenly spaced points in the x-y plane can be easily generated using `meshgrid`. `[X,Y]=meshgrid(x,y)`, where `x` and `y` are vectors, generates a matrix `X`, whose rows are copies of the vector `x`, and matrix `Y`, whose columns are copies of the vector `y`. The coordinates (X_{ij}, Y_{ij}) represent a regular rectangular grid in the x-y plane.

- `mesh(Z)`, where `Z` is a matrix, plots a mesh in three dimensions with intersections at the points (i,j,Z_{ij}). `mesh(X,Y,Z)` plots a mesh with intersections at the points (X_{ij},Y_{ij},Z_{ij}). A color argument is optional.
- `surf` is the surface version of `mesh`, and uses the same arguments. `surf` draws a mesh plot and fills in the patches (holes between line segments) with appropriate colors.
- Contour plots can be generated with `contour`. This function uses the general form `contour(X,Y,Z,n,S)` where `X` and `Y` are optional vectors specifying the `x` and `y` axes, `Z` is the matrix of data points, `n` is an optional number of contour lines to draw, and `S` is a character string specifying color and/or linetype as in the `plot` command. `contour3` draws contours in 3-D space.
- Pseudocolor plots can be generated using `pcolor`. `pcolor(C)` generates a "checkerboard" plot of matrix `C`, where the values of the elements of `C` determine the color of each cell of the plot. `pcolor(X,Y,C)`, where `X` and `Y` are vectors or matrices draws a pseudocolor plot on the grid defined by `X` and `Y`. `pcolor` is really `surf` with its view set to directly above.
- The angle of view of a 3-D plot can be set with the `view` command. The usual format is `view(a,e)`, where `a` is the azimuth (angle in degrees from the y-axis), and `e` is the angle of elevation (in degrees above the x-y plane). An alternative form is `view([x y z])`, where `(x,y,z)` is the Cartesian coordinate of the observer. `view(2)` sets the default 2-D view (`view(0,90)`), and `view(3)` sets the default 3-D view (`view(-37.5,30)`).
- Hidden line removal can be controlled with `hidden off` and `hidden on`.
- Three types of shading are available using `shading flat`, `shading interp`, and `shading faceted`.
- A color map is a 3-column matrix whose rows contain the RGB values of a color in the form `[r g b]`.
- MATLAB supplies 10 functions to create different color maps.
- `colormap(M)` installs the matrix M as the current color map. `M` must have three columns and contain only values between 0 and 1.
- Functions that take a color argument usually accept either a character representing one of the colors in the plot color and linestyle table, a single RGB value in the form `[r g b]`, or a column vector or matrix whose elements are scaled and used as indices into the current color map.
- Color maps can be displayed by viewing the elements of the color map matrix, using `pcolor` to display color bars, or using the `colorbar` function to add a color scale to an existing plot.

- `brighten` adjusts the brightness of the current color map, or creates a new color map from an existing one.
- Color maps can be created by any array operation, and can be installed and used as long as they consist of three columns, and contain only values between 0 and 1.

6 Symbolic Math Toolbox Tutorial

6.1 Introduction

In the previous chapter you learned how MATLAB can be used like a very powerful, programmable, top-of-the-line, everything-including-the-kitchen-sink calculator. Even a powerful calculator, however, has its limitations. Like a calculator, *basic* MATLAB uses numbers. It takes numbers (`123/4`) or variables (`x = [1 2 3]`) and acts upon them using the commands and functions you specify (`y = cos(pi/4)` or `r = log(4/3)`) to produce numerical results. What most calculators and *basic* MATLAB lack is the ability to manipulate mathematical expressions without actually using numbers. *Basic* MATLAB must have numbers to work with. For example, asking for the sine of a variable that has not been assigned a numerical value results in:

```
» y=sin(x)  % take the sine of x
??? Undefined function or variable x.
```

The commands and functions introduced in this chapter can change all that. Now you can tell MATLAB to manipulate *expressions* that let you compute with mathematical symbols rather than numbers. This process is often called *symbolic math.* Here are some examples of symbolic expressions:

$$\cos(x^2) \qquad 3x^2 + 5x - 1 \qquad v = \frac{d}{dx}2x^2 \qquad f = \int \frac{x^2}{\sqrt{1-x}}dx$$

The *Symbolic Math Toolbox* is a collection of *tools* (functions) for MATLAB that is used for manipulating and solving symbolic expressions. There are tools to combine, simplify, differentiate, integrate, and solve algebraic and differential equations. Other tools are used in linear algebra to derive exact results for inverses, determinants, and canonical forms, and to find the eigenvalues of symbolic matrices without the error introduced by numerical computations.

Variable precision arithmetic, which calculates symbolically and returns a result to any specified accuracy, is also supported. If some of these topics are foreign to you, don't worry. As with *basic* MATLAB, you don't need to know everything to make MATLAB help you solve problems.

The tools in the *Symbolic Math Toolbox* are built upon the powerful software program called Maple®, originally developed at the University of Waterloo in Canada. When you ask MATLAB to perform some symbolic operation, it asks Maple to do it, then returns the result to the MATLAB command window. As a result, performing symbolic manipulations in MATLAB is a natural extension of the way you use MATLAB to crunch numbers.

6.2 Symbolic Expressions

Symbolic expressions are character strings, or arrays of character strings, that represent numbers, functions, operators, and variables. The variables are not required to have predefined values. *Symbolic equations* are symbolic expressions containing an equals sign. *Symbolic arithmetic* is the practice of solving these symbolic equations by applying known rules and identities to the given symbols, exactly the way you learned to solve them in algebra and calculus. *Symbolic matrices* are arrays whose elements are symbolic expressions.

6.2.1 MATLAB Representation of Symbolic Expressions

MATLAB represents symbolic expressions internally as character strings to differentiate them from numeric variables or operations; otherwise, they look almost exactly like basic MATLAB commands. Here are some examples of symbolic expressions along with their MATLAB equivalents:

Symbolic Expression	MATLAB Representation
$\frac{1}{2x^n}$	`'1/(2*x^n)'`
$y = \frac{1}{\sqrt{2x}}$	`y = '1/sqrt(2*x)'`
$\cos(x^2) - \sin(2x)$	`'cos(x^2) - sin(2*x)'`
$M = \begin{bmatrix} a & b \\ c & d \end{bmatrix}$	`M = sym('[a,b;c,d]')`
$f = \int_a^b \frac{x^3}{\sqrt{1-x}}\,dx$	`f = int('x^3/sqrt(1-x)','a','b')`

MATLAB symbolic functions let you manipulate these expressions in many ways. For example:

```
» diff('cos(x)')  % differentiate cos(x) with respect to x
ans =
     -sin(x)

» M=sym('[a,b;c,d]')  % create a symbolic matrix M
M =
     [a,b]
     [c,d]

» determ(M)  % find the determinant of the symbolic matrix M
ans =
     a*d-b*c
```

Notice that in the first example above, the symbolic expression was defined *implicitly* by using single quotes to tell MATLAB that `'cos(x)'` is a string and imply that `diff('cos(x)')` is a symbolic expression rather than a numeric expression, while in the second example the `sym` function was used to *explicitly* tell MATLAB that `M = sym('[a,b;c,d]')` is a symbolic expression. Often the explicit `sym` function is not required where MATLAB can determine the argument type on its own.

In MATLAB the form `func arg` is equivalent to `func('arg')` where `func` is a function and `arg` is a character string argument. For example, MATLAB can figure out that `diff cos(x)` and `diff('cos(x)')` both mean `diff(sym('cos(x)'))`, but the first form is certainly easier to type. However, at times the sym function *is* necessary. In the second example above:

```
» M=[a,b;c,d]  % M is a numeric matrix using values a through d
???  Undefined function or variable a.

» M='[a,b;c,d]' % M is a character string, not a symbolic matrix
M =
        [a,b;c,d]

» M=sym('[a,b;c,d]')  % M is a symbolic matrix
M =
        [a,b]
        [c,d]
```

In the example above, M was defined three ways: numerically (if a, b, c, and d had been predefined), as a character string, and as a symbolic matrix. Many symbolic functions are smart enough to convert character strings to symbolic expressions automatically. But in some cases, especially when creating a symbolic array, the sym function must be used to convert a character string specifically to a symbolic expression. The implicit form, e.g., diff cos(x), is most useful for simple tasks that do not reference earlier results. However, the simplest form (without quotes) requires an argument that is a single character string containing no embedded spaces:

```
» diff x^2+3*x+5  % the argument is equivalent to 'x^2+3*x+5'
ans =
        2*x+3

» diff x^2 + 3*x + 5  % spaces define separate strings
???  Error using ==> diff
Too many input arguments.
```

Symbolic expressions without variables are called *symbolic constants*. When symbolic constants are displayed, they are often difficult to distinguish from integers. For example:

```
» f=symop('(3*4-2)/5+1')  % reduce symbolic constant
f =
     3

» isstr(f)  % is f a string? (1=yes, 0=no)
ans =
     1
```

In this case, f represents the symbolic constant '3'; not the number 3. MATLAB stores strings as the ASCII representation of the characters. Consequently, if you perform a numeric operation on a string, it uses the ASCII value of each character in the operation. Since the number 51 is the ASCII representation of the character '3', adding 1 to f numerically will not produce the expected result:

```
» f+1
ans =
        52
```

6.2.2 Symbolic Variables

When working with symbolic expressions containing more than one variable, exactly one variable is the *independent* variable. If MATLAB is not told which variable is the independent variable, it selects one based on the following rule:

The default independent variable in a symbolic expression is the unique, lower case letter, other than *i* and *j*, that is not part of a word. If there is no such character, *x* is chosen. If the character is not unique, the one closest to *x* alphabetically is chosen. If there is a tie, the one later in the alphabet is chosen.

The default independent variable, sometimes known as the *free* variable, in the expression '1/(5+cos(x))' is 'x'; the free variable in the expression '3*y+z' is 'y'; and the free variable in the expression 'a+sin(t)' is 't'. The free symbolic variable in the expression 'sin(pi/4)–cos(3/5)' is 'x' because this expression is a symbolic constant, containing no symbolic variables. You can ask MATLAB to tell you which variable in a symbolic expression it thinks is the independent variable by using the symvar function:

```
» symvar('a*x+y')  % find the default symbolic variable
ans =
        x

» symvar('a*t+s/(u+3)')  % u is closest to 'x'
ans =
        u

» symvar('sin(omega)')  % 'omega' is not a single character
ans =
        x

» symvar('3*i+4*j')  % i and j are equal to sqrt(-1)
ans =
        x

» symvar('y+3*s','t') % find the variable closest to t
ans =
        s
```

If symvar cannot find a default symbolic variable using the rule, it will assume there is none, and return x. This will be true for expressions containing multi-character variables, such as `alpha` or `s2`, as well as symbolic constants, which contain no variables.

Most commands give you the option to specify the independent variable if desired:

```
» diff('x^n') % differentiate with respect to the default variable 'x'
ans =
        x^n*n/x

» diff('x^n','n')  % differentiate with respect to 'n'
ans =
        x^n*log(x)

» diff('sin(omega)')  % differentiate using the default variable (x)
ans =
        0

» diff('sin(omega)','omega')  % specify the independent variable
ans =
        cos(omega)
```

6.2.3 Things to Try on Your Own

- Here are some expressions to practice on. Given each symbolic expression, use MATLAB syntax to create the equivalent MATLAB symbolic expressions:

$$f = ax^2 + bx + c \qquad f' = \frac{d}{dx}\sqrt{3x^2 + 2x + 5} \qquad z = \frac{3\cos(w)}{\sin\left(\frac{2w}{w+1}\right)}$$

$$A = \begin{bmatrix} 3\sin(t) & -\cos(t^2) \\ \cos(2t) & -\sin(t) \end{bmatrix} \qquad p = \frac{3s^2 + 2s + 1}{4s - 2} \qquad r = e^{-2t}$$

Here are answers to the above expressions:

```
» f='a*x^2+b*x+c'
f =
       a*x^2+b*x+c

» f_prime=diff('sqrt(3*x^2+2*x+5)')
f_prime =
       1/2/(3*x^2+2*x+5)^(1/2)*(6*x+2)

» z='3*cos(w)/sin(2*w/(w+1))'
z =
       3*cos(w)/sin(2*w/(w+1))

» A=sym('[3*sin(t) -cos(t^2); cos(2*t) -sin(t)]')
A =
       [ 3*sin(t),-cos(t^2)]
       [ cos(2*t),  -sin(t)]

» p='(3*s^2-2*s+1)/(4*s-2)'
p =
       (3*s^2+2*s+1)/(4*s-2)

» r='exp(-2*t)'
r =
       exp(-2*t)
```

- Find the default independent variable returned by `symvar(expression)` in the following expressions:

$z = 3ac + 4b - 2$ $\quad f = s^{3nt}$ $\quad x = 4a + 3c + b^2 + 1$

$n = 3r + 2s^2 + 5$ $\quad q = r + \sqrt[3]{3k^2 + 2P}$ $\quad y = \dfrac{\tan(3z)}{\sin(2v)} - \dfrac{k}{p}$

  Answers: *c, t, c, s, r, z*.

- Find the default symbolic variable returned by `symvar(expression,'n')` in the previous expressions.

  Answers: *c, n, c, r, k, p*.

6.2.4 Summary

- Basic MATLAB commands and functions operate on numeric values and predefined variables that represent numeric values or numeric matrices.

- Symbolic commands and functions operate on symbolic expressions without requiring that the variables be predefined.
- In MATLAB, the form `func arg` is equivalent to `func('arg')` in all cases. For example, the commands `help format` and `help('format')` are equivalent.
- MATLAB represents symbolic expressions as character strings to distinguish them from numeric variables, e.g., `y=sin(x)` is a numeric expression, whereas `y='sin(x)'` is a character string that can be interpreted as a symbolic expression.
- The `sym` function is used to explicitly define a symbolic expression in cases where ambiguity can exist, e.g., `M='[a b;c d]'` is a string, whereas `M=sym('[a b;c d]')` is a symbolic expression.
- Remember, some symbolic constants may look like integers. The `isstr` function returns 1 if its argument is a symbolic expression.
- The `symvar` function can be used to find out which variable in a symbolic expression will be used by the MATLAB symbolic functions as the default independent variable.
- Most symbolic functions let you specify the independent variable in one or more of their forms.

6.3 Operations on Symbolic Expressions

Once you have created a symbolic expression, you will probably want to change it in some way. You may wish to extract part of an expression, combine two expressions, or find the numeric value of a symbolic expression. There are many symbolic tools that let you accomplish these tasks.

All symbolic functions (with a few specific exceptions discussed later) act on symbolic expressions and symbolic arrays, and return symbolic expressions or arrays. The result may sometimes look like a number, but it is a symbolic expression internally represented by a character string. As we discussed earlier, you can find out if what looks like a number is an integer or a string by using the `isstr` function from MATLAB.

6.3.1 Extracting Numerators and Denominators

If your expression is a *rational polynomial* (a ratio of two polynomials), or can be expanded to a rational polynomial (including those with a denominator of 1), you can extract the numerator and denominator using numden. For example, given the expressions:

$$m = x^2 \qquad f = \frac{ax^2}{b-x} \qquad g = \frac{3}{2}x^2 + \frac{2}{3}x - \frac{3}{5}$$

$$h = \frac{x^2+3}{2x-1} + \frac{3x}{x-1} \qquad k = \begin{bmatrix} \frac{3}{2} & \frac{2x+1}{3} \\ \frac{4}{x^2} & 3x+4 \end{bmatrix}$$

numden combines and rationalizes the expression if necessary, and returns the resulting numerator and denominator. The MATLAB statements to do this are:

```
» m='x^2'  % create a simple expression
m =
        x^2

» [n,d]=numden(m)  % get the numerator and denominator
n =
        x^2
d =
        1

» f='a*x^2/(b-x)'  % create a rational expression
f =
        a*x^2/(b-x)

» [n,d]=numden(f)  % get the numerator and denominator
n =
        a*x^2
d =
        b-x
```

The first two expressions gave the expected result.

```
» g='3/2*x^2+2/3*x-3/5'  % rationalize and extract
g =
        3/2*x^2+2/3*x-3/5

» [n,d]=numden(g)
n =
        45*x^2+20*x-18
d =
        30

» h='(x^2+3)/(2*x-1)+3*x/(x-1)' % sum of rational polynomials
h =
        (x^2+3)/(2*x-1)+3*x/(x-1)

» [n,d]=numden(h)  % rationalize and extract
n =
        x^3+5*x^2-3
d =
        (2*x-1)*(x-1)
```

These two expressions, for g and h, were *rationalized*, or turned into a single expression with a numerator and denominator, before the parts were extracted.

```
» k=sym('[3/2,(2*x+1)/3;4/x^2,3*x+4]')  % symbolic array
k =
        [  3/2,(2*x+1)/3]
        [4/x^2,    3*x+4]

» [n,d]=numden(k)
n =
        [3, 2*x+1]
        [4, 3*x+4]
d =
        [  2, 3]
        [x^2, 1]
```

This expression, k, was a symbolic array. numden returned two new arrays, n and d, where n was the array of numerators, and d was the array of denominators. If you use the form s=numden(f), numden returns only the numerator into the variable s.

6.3.2 Standard Algebraic Operations

A number of standard algebraic operations can be performed on symbolic expressions. The functions symadd, symsub, symmul, and symdiv add, subtract, multiply, and divide two expressions, and sympow raises one expression to the power of another. For example, given two functions:

$$f = 2x^2 + 3x - 5 \qquad g = x^2 - x + 7$$

```
» f='2*x^2+3*x-5'  % define the symbolic expressions
f =
      2*x^2+3*x-5

» g='x^2-x+7'
g =
      x^2-x+7

» symadd(f,g)  % find an expression for f + g
ans =
      3*x^2+2*x+2

» symsub(f,g)  % find an expression for f - g
ans =
      x^2+4*x-12

» symmul(f,g)  % find an expression for f * g
ans =
      (2*x^2+3*x-5)*(x^2-x+7)

» symdiv(f,g)  % find an expression for f / g
ans =
      (2*x^2+3*x-5)/(x^2-x+7)

» sympow(f,'3*x')  % find an expression for f^3
ans =
      (2*x^2+3*x-5)^3
```

Another general-purpose function lets you create new expressions from other symbolic variables, expressions, and operators. symop takes up to 16 comma-separated arguments, each of which can be a symbolic expression, numeric value, or operator (+, –, *, /, ^, (, or)). symop then concatenates the arguments and returns the resulting expression.

```
» f='cos(x)'  % create an expression
f =
      cos(x)

» g='sin(2*x)'  % create another expression
f =
      sin(2*x)

» symop(f,'/',g,'+',3)  % combine them
ans =
      cos(x)/sin(2*x)+3
```

All of these operations work with array arguments as well. They will be discussed again later in this chapter.

6.3.3 Advanced Operations

MATLAB has the capability to perform more advanced operations on symbolic expressions. The `compose` function combines $f(x)$ and $g(x)$ into $f(g(x))$, the `finverse` function finds the functional inverse of an expression, and the `symsum` function finds the symbolic summation of an expression.

Given the expressions:

$$f = \frac{1}{1+x^2} \qquad g = \sin(x) \qquad h = \frac{1}{1+u^2} \qquad k = \sin(v)$$

```
» f='1/(1+x^2)';  % create the four expressions
» g='sin(x)';
» h='1/(1+u^2)';
» k='sin(v)';
» compose(f,g)  % find an expression for f(g(x))
ans =
        1/(1+sin(x)^2)

» compose(g,f)  % find an expression for g(f(x))
ans =
        sin(1/(1+x^2))
```

`compose` can also be used on functions that have different independent variables:

```
» compose(h,k,'u','v')  % given h(u), k(v), find h(k(v))
ans =
        1/(1+sin(v)^2)
```

The functional inverse of an expression, say $f(x)$, is the expression $g(x)$ that satisfies the condition $g(f(x)) = x$. For example, the functional inverse of e^x is $\ln(x)$ since $\ln(e^x) = x$. The functional inverse of $\sin(x)$ is $\arcsin e(x)$, and the functional inverse of

$$\frac{1}{\tan(x)}$$

is

$$\arctan\left(\frac{1}{x}\right).$$

The `finverse` function returns the functional inverse of an expression, and warns you if the result is not unique.

```
» finverse('1/x')  % inverse of 1/x is 1/x since '1/(1/x) = x'
ans =
      1/x

» finverse('x^2')  % g(x^2)=x has more than one solution
Warning: finverse(x^2) is not unique
ans =
      x^(1/2)

» finverse('a*x+b')  % find the solution to 'g(f(x))=x'
ans =
      -(b-x)/a

» finverse('a*b+c*d-a*z','a')  % find solution to 'g(f(a))=a'
ans =
      -(c*d-a)/(b-z)
```

The `symsum` function finds the symbolic summation of an expression. There are four forms of the function: `symsum(f)` returns

$$\sum_{0}^{x-1} f(x),$$

`symsum(f,'s')` returns

$$\sum_{0}^{s-1} f(s),$$

`symsum (f,a,b)` returns

$$\sum_{a}^{b} f(x),$$

and the most general form, `symsum (f,'s',a,b)` returns

$$\sum_{a}^{b} f(s).$$

Let us try

$$\sum_{0}^{x-1} x^2.$$

which should return

$$\frac{x^3}{3}-\frac{x^2}{2}+\frac{x}{6}:$$

```
» symsum('x^2')
ans =
        1/3*x^3-1/2*x^2+1/6*x
```

How about

$$\sum_{1}^{n}(2n-1)^2$$

which should return

$$\frac{n(2n-1)(2n+1)}{3}:$$

```
» symsum('(2*n-1)^2',1,'n')
ans =
        11/3*n+8/3-4*(n+1)^2+4/3*(n+1)^3

DU» factor(ans)  % change the form
ans =
        1/3*n*(2*n-1)*(2*n+1)
```

Finally, let's try

$$\sum_{1}^{\infty}\frac{1}{(2n-1)^2}$$

which should return

$$\frac{\pi^2}{8}:$$

```
» symsum('1/(2*n-1)^2',1,inf)
ans =
        1/8*pi^2
```

6.3.4 Conversion Functions

This section presents tools to convert from symbolic expressions to numeric values, and back again. These are some of the very few symbolic functions that can return numeric values. Notice, however, that some symbolic functions automatically convert a number into its symbolic representation if it is one of a number of arguments to the function. (Refer to the symop discussion in Section 6.2 for an example.)

The sym function can take a numeric argument and convert it into a symbolic representation. The numeric function does the opposite. It converts a symbolic constant (a symbolic expression with no variables) to a numeric value.

```
» phi='(1+sqrt(5))/2'  % the 'golden' ratio
phi =
      (1+sqrt(5))/2

» numeric(phi)  % convert to a numeric value
ans =
      1.6180
```

The eval function from MATLAB passes a character string to MATLAB to evaluate. Therefore eval is another function that can be used to convert a symbolic constant into a number, or to evaluate an expression.

```
» eval(phi)  % execute the string '(1+sqrt(5))/2'
ans =
      1.6180
```

As expected, numeric and eval returned the same numeric value.

You have already worked with polynomials in MATLAB, using vectors whose elements are the coefficients of the polynomials. The symbolic function sym2poly converts a symbolic polynomial to its MATLAB equivalent coefficient vector. The function poly2sym does the reverse, and lets you specify the variable to use in the resulting expression.

```
» f='2*x^2+x^3-3*x+5'  % f is the symbolic polynomial
f =
      2*x^2+x^3-3*x+5

» n=sym2poly(f)  % extract the coefficient vector
n =
      1   2  -3   5

» poly2sym(n)  % recreate the polynomial in x
ans =
      2*x^2+x^3-3*x+5

» poly2sym(n,'s')  % recreate the polynomial in s
ans =
      s^3+2*s^2-3*s+5
```

6.3.5 Variable Substitution

Suppose you have a symbolic expression in x, and you want to change the variable to y. MATLAB gives you a tool to make substitutions in symbolic expressions, called subs. The format is subs(f,new,old), where f is a symbolic expression, and new and old are characters, character strings, or other symbolic expressions. The string new will replace each occurrence of the string old in the expression f. Here are some examples:

```
» f='a*x^2+b*x+c'  % create a function f(x)
f =
      a*x^2+b*x+c

» subs(f,'s','x')  % substitute 's' for 'x' in the expression f
ans =
      a*s^2+b*s+c

» subs(f,'alpha','a')  % substitute 'alpha' for 'a' in f
ans =
      alpha*x^2+b*x+c

» g='3*x^2+5*x–4'  % create another function
g =
      3*x^2+5*x–4

» h =subs(g,'2','x')  % substitute '2' for 'x' in g
h =
      18

» isstr(h)  % show that the result is a symbolic expression
ans =
      1
```

The last example shows how subs makes the substitution, and then tries to simplify the expression. Since the result of the substitution was a symbolic constant, MATLAB could reduce it to a single symbolic value. Notice that since subs is a symbolic function, it returns a symbolic expression, actually a symbolic constant, even though it looks like a number. To get a number, we need to use the numeric or eval function to convert the string.

```
» numeric(h)  % convert symbolic to numeric
ans =
      18

» isstr(ans)  % show that the result is a numeric value
ans =
      0
```

6.3.6 Things to Try on Your Own

- Rationalization of symbolic expressions is performed within many function M-files (like `sym.m`) by the `symrat` function. This tool examines a scalar and returns a symbolic rational approximation of the scalar that is an integer, the ratio of two integers, the ratio of two integers times π, or an integer times a power of 2. Array elements are examined individually.

```
» H=hilb(3)  % our old friend the 3x3 numeric Hilbert matrix
H =
        1.0000    0.5000    0.3333
        0.5000    0.3333    0.2500
        0.3333    0.2500    0.2000

» symrat(H)  % get the symbolic rational approximation of H
ans =
        [  1, 1/2, 1/3]
        [1/2, 1/3, 1/4]
        [1/3, 1/4, 1/5]

» symrat(eps)  % rational approximation of smallest number
ans =
        2^(-52)
```

6.3.7 Summary

- All symbolic functions return symbolic expressions, with the exception of those few functions (like `numeric`, `sym2poly`, and `eval`) that are intended to convert from symbolic notation to numeric values.
- The function `numden` extracts the numerator and denominator of a symbolic expression, and rationalizes the expression before extracting if necessary. The usual form is `[n,d]=numden(f)`. If `f` is an array, `n` and `d` will be arrays of numerators and denominators, respectively.
- The `symadd`, `symsub`, `symmul`, and `symdiv` functions are used to add, subtract, multiply, and divide two expressions. The `sympow` function will raise an expression to the power of another expression.
- It is possible to piece together a new expression out of up to 16 expressions, symbolic variables, and symbolic operators using `symop`.
- Given two symbolic expressions $f(x)$ and $g(x)$, the composite function $f(g(x))$ can be found using `compose(f,g)`. Functions of other variables use the `compose(f,g,'n')` or `compose (f,g,'u','v')` forms.

- The functional inverse of a function $f(x)$ is defined as the function $g(x)$ such that $g(f(x)) = x$. The functional inverse can be found using `finverse`. The form `finverse(f,'n')` is used to specify the independent variable.
- Symbolic summation can be performed using the `symsum` function. Given a function $f(n)$, `symsum(f)` finds an expression for

$$\sum_{0}^{n-1} f(n).$$

  The form `symsum(f,a,b)` finds an expression for

$$\sum_{a}^{b} f(n).$$

- The symbolic function `sym2poly` converts a symbolic polynomial to its MATLAB form, which is a row vector of its coefficients in descending order. `poly2sym` will do the reverse, and let you choose the independent variable in the resulting expression.
- Variable substitution in symbolic expressions can be performed with the `subs` function. This tool substitutes one string for each occurrence of a second string within an expression, using the format `subs(f,'new','old')`.

6.4 Differentiation and Integration

Differentiation and integration are central to the study and application of calculus, and are used extensively in many engineering disciplines. MATLAB symbolic tools can help solve many of these kinds of problems.

6.4.1 Differentiation

Differentiation of a symbolic expression uses the diff function in one of four forms:

```
» f='a*x^3+x^2-b*x-c'  % define a symbolic expression
f =
      a*x^3+x^2-b*x-c

» diff(f)  % differentiate with respect to default (x)
ans =
      3*a*x^2+2*x-b

» diff(f,'a')  % differentiate f with respect to a
ans =
      x^3

» diff(f,2)  % differentiate f twice with respect to x
ans =
      6*a*x+2

» diff(f,'a',2)  % differentiate f twice with respect to a
ans =
      0
```

The diff function also operates on arrays. If F is a symbolic vector or matrix, diff(F) differentiates each element in the array:

```
» F=sym('[a*x,b*x^2;c*x^3,d*s]')  % create a symbolic array
F =
      [   a*x,b*x^2]
      [ c*x^3,  d*s]

» diff(F)  % differentiate the elements with respect to x
ans =
      [       a, 2*b*x]
      [3*c*x^2,      0]
```

Note that the diff function is also used in MATLAB to compute the numerical differences of a numeric vector or matrix. For a numeric vector or matrix M, diff(M) computes the numerical differences M(2:m,:)–M(1:m–1,:) as shown:

```
» M=[(1:8).^2]  % create a vector
M =
       1     4     9    16    25    36    49    64

» diff(M)  % find the differences between elements
ans =
       3     5     7     9    11    13    15
```

If the expression or variable argument to diff is numeric, MATLAB is smart enough to compute the numerical difference; if the argument is a symbolic string or variable, MATLAB differentiates the expression.

6.4.2 Integration

The integration function `int(f)`, where `f` is a symbolic expression, attempts to find another symbolic expression `F` such that `diff(F)=f`. As you probably found from your study of calculus, integration is more complicated than differentiation. The integral or antiderivative may not exist in closed form, or it may exist but the software can't find it, or the software finds it eventually, but runs out of memory or time. When MATLAB cannot find the antiderivative, it will return the command unevaluated.

```
» int('log(x)/exp(x^2)')   % attempt to integrate
ans =
        int(log(x)/exp(x^2),x)
```

The integration function, like differentiation, has more than one form. The form `int(f)` attempts to find an antiderivative with respect to the default independent variable. The form `int(f,'s')` attempts to find an antiderivative with respect to the symbolic variable `s`. The forms `int(f,a,b)` and `int(f,'s',a,b)`, where `a` and `b` are numeric values, attempt to find symbolic expressions for the definite integral from `a` to `b`. The forms `int(f,'m','n')` and `int(f,'s','m','n')`, where `m` and `n` are symbolic variables, attempt to find symbolic expressions for the definite integral from `m` to `n`.

```
» f='sin(s+2*x)') % create a symbolic function
f =
        sin(s+2*x)

» int(f)  % integrate with respect to x
ans =
        -1/2*cos(s+2*x)

» int(f,'s')  % integrate with respect to s
ans =
        -cos(s+2*x)

» int(f,pi/2,pi)  % integrate with respect to x from pi/2 to pi
ans =
        -cos(s)

» int(f,'s',pi/2,pi)  % integrate with respect to s from pi/2
to pi
ans =
        cos(2*x)-sin(2*x)

» int(f,'m','n')  % integrate with respect to x from m to n
ans =
        -1/2*cos(s+2*n)+1/2*cos(s+2*m)
```

Like the `diff` function, the integration function `int` operates on each element of a symbolic array:

```
» f=sym('[a*x,b*x^2;c*x^3,d*s]')  % create a symbolic array
f =
      [   a*x,b*x^2]
      [c*x^3,   d*s]

» int(f)  % integrate the array elements with respect to x
ans =
      [1/2*a*x^2, 1/3*b*x^3]
      [1/4*c*x^4,     d*s*x]
```

6.4.3 Example

In this section we will use the symbolic capabilities of MATLAB to find answers to a classic calculus problem.

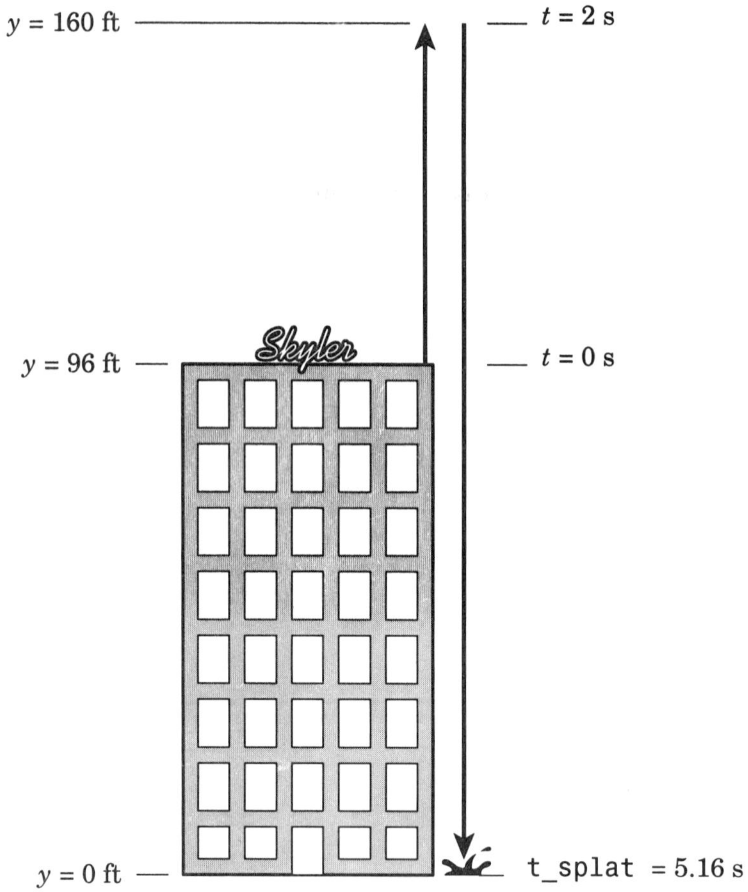

Bart is out on a field trip with his class to the top of the Skyler building in downtown Springfield. He takes a ripe tomato out of his lunch bag, leans over the edge of the roof, and hurls it up into the air. The tomato is thrown straight up, with an initial speed of $v_0 = 64$ *feet per second. The roof is* $y_0 = 96$ *feet above the ground. Where is the tomato some arbitrary* t *seconds later? When does it reach its maximum height? How high above the ground does the tomato rise? When does the tomato hit the ground? Assume there is no air resistance and that the acceleration due to gravity is a constant* a = *–32 feet per second per second in Springfield.*

Let's choose ground level to be zero height, i.e., $y = 0$ is the ground, and $y = 96$ is the top of the building. The velocity is

$$v = \frac{dy}{dt},$$

and the acceleration is

$$a = \frac{d^2y}{dt^2}.$$

Therefore, if we integrate acceleration once we get velocity and if we integrate velocity we get position or height y.

```
» a=–32  % acceleration is 32 feet/sec toward the ground
a =
       –32

» v=int(a,'t')  % find the velocity as a function of time
v =
       –32*t

» v=symadd(v,64)  % at time t=0, the velocity is 64 feet/sec.
v =
       –32*t+64

» y=int(v,'t')  % find the height y at time t by integration
y =
       –16*t^2+64*t

» y=symadd(y,96)  % the height at t=0 is 96 feet
y =
       –16*t^2+64*t+96
```

Let's check that the result is correct. Setting $t = 0$ in the above expression, we get:

```
» y0=–16*0^2+64*0+96  % the height at t=0 should be 96 feet
y0 =
       96
```

which is the correct height of the tomato before it is thrown. We could have used the `subs` function as an alternative method to check for the result at $t = 0$:

```
» y0=subs(y,0,'t')  % the height at t=0 should be 96 feet
y0 =
       96
```

We now have expressions for the velocity and position (height) as a function of time. The maximum height is reached when the tomato stops rising and starts downward. To find this point we will find the value of t when $v = 0$ by using the `solve` function. This function evaluates a symbolic string with only one variable for the value when the function is set equal to zero. In other words, solve(f) where f is a function of x solves for x when $f(x) = 0$.

```
» t_string=solve(v)  % Find the value of t when v(t)=0
t_string =
       2
```

Since solve is a symbolic function, it returns a symbolic constant (even though it looks like a number). Now we will find the maximum height which occurs at $t = 2$ seconds:

```
» ymax_string=subs(y,t_string) % substitute for default variable in y
ymax_string =
       160
```

Notice that the subs function does the same thing we did before when we checked the expression for y. subs substitutes a 2 for each t in the expression, simplifies if it can, and returns the resulting expression.

Now find the time when the tomato hits the ground

```
» t_splat=solve(y)  % the tomato hits the ground when y=0
t_splat =
       [2-10^(1/2)]
       [2+10^(1/2)]
```

Since $2-\sqrt{10}$ is negative and the tomato cannot hit the ground before it is thrown, the only meaningful solution is $2+\sqrt{10}$. Let's find the numeric value of t when the tomato hits the ground:

```
» numeric(t_splat)
ans =
       -1.1623
        5.1623
```

Thus, the height of the tomato at t seconds is given by $y = 16t^2 + 64t + 96$, the tomato rises to a maximum height of 160 feet above the ground at time $t = 2$ seconds, and hits the ground at $t = 5.1623$ seconds.

Let us introduce another factor into this problem.

Barney, on his way to the bar, is walking down the street next to the Skyler Building. If the falling tomato hits Barney, who happens to be directly under it when it arrives, and Barney is 5 feet tall, when will it hit him on the head?

For this we need to find the time t at $y = 5$. This can be done as follows:

```
» y_bonk=symadd('5=',y)  % create a symbolic equation for y=5
y_bonk =
         5 = -16*t^2+64*t+96

» t_bonk=numeric(solve(y_bonk))  % find the time t when y=5
t_bonk =
          5.1125
         -1.1125
```

Again, Barney cannot be hit before the tomato is thrown, so the only meaningful solution is $t = 5.1125$ seconds.

Notice that `solve` will solve a symbolic expression in one variable whether or not the expression is an equation containing an equals sign. If the expression contains no equals sign, `solve` sets the expression equal to zero before solving it. This implies that an expression like $3x^2 + 2x + 1 = 5x + 12$ can be solved for x:

```
» solve('(3*x^2+2*x+1)=(5*x+12)')  % solve an expression
ans =
         [1/2+1/6*141^(1/2)]
         [1/2-1/6*141^(1/2)]

» numeric(ans)  % convert to numeric values
ans =
          2.4791
         -1.4791
```

More on `solve` later in this chapter.

6.4.4 Summary

- Use the `diff` function to differentiate a symbolic expression or the elements of a symbolic array. The symbolic variable to use, and the number of differentiations to perform, can be specified.
- If the expression or variable argument to `diff` is numeric, it computes the numerical difference; if the argument is a symbolic string or variable, it differentiates the expression.
- The `int` function is used to integrate a symbolic expression or the elements of a symbolic array. Optional arguments can be used to specify the symbolic variable and the limits of a definite integral.
- Symbolic solutions of algebraic equations are generated by the `solve` function. For one equation in one unknown, `solve(f)` solves the symbolic equation f, or the equation $f = 0$, for its default variable as determined by `symvar`. Other forms of `solve` will be discussed later.

6.5 Plotting Symbolic Expressions

To get a better idea of what happened to the tomato, let's plot the result of the tomato toss:

```
» y='-16*t^2+64*t+96'   % re-create the expression for position
y =
      -16*t^2+64*t+96
» ezplot(y)  % plot the height of the tomato
```

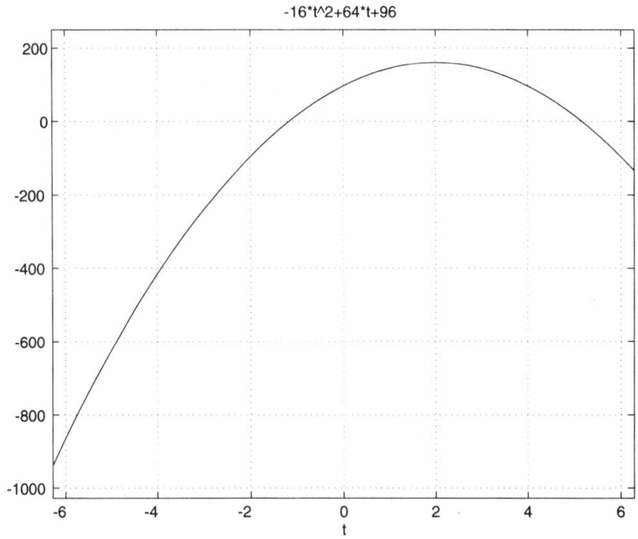

As you can see, ezplot graphs the given symbolic function over the domain $-2\pi \le t \le 2\pi$ and scales the y-axis accordingly. It also adds grids and labels. In this case, we are only interested in times between 0 and 6. Let's try again and specify the time range:

```
» ezplot(y,[0 6])  % plot y for 0<t<6
```

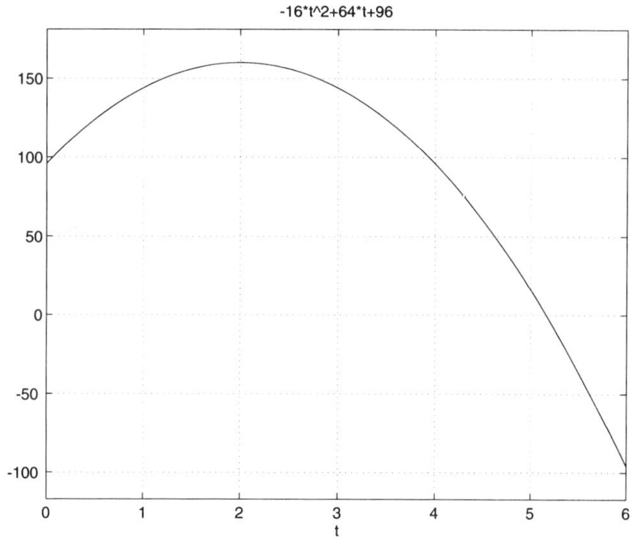

Now the region of interest shows up a little better, but there is still part of the plot below ground. The plot could be fixed by changing the time range and reissuing the `ezplot` command, e.g., `ezplot(y,[0 5.1623])`, but you have more control with some other MATLAB plotting commands. Once the plot is in the figure window, it can be modified like any other plot. Let's scale both axes of the current plot and add a title and labels:

```
» axis([0 5.5 0 165])  % scale to show the region of interest
» title('Plot of Tomato Height vs. Time')  % change the title
» ylabel('Height in feet')
» xlabel('Time in Seconds')
```

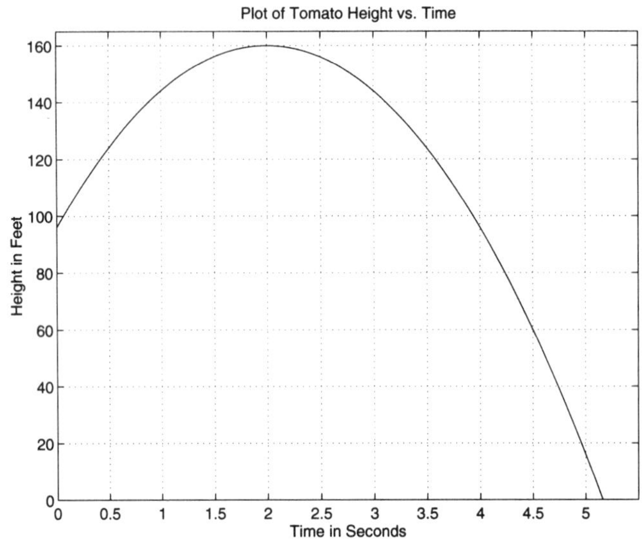

The other MATLAB plotting commands are also available to customize your plots. For example, the command `zoom on` will let you use the mouse to zoom in on any desired area of a two-dimensional plot such as the one we just generated. For more information, type `help zoom`. Other MATLAB commands to add text, identify specific points, and change scaling are also available.

6.6 Formatting and Simplifying Expressions

6.6.1 Introduction

Sometimes MATLAB will return a symbolic expression that is difficult to read. There are a number of tools available to help make the expression more readable. The first is the `pretty` function. This command attempts to display a symbolic expression in a form that resembles textbook mathematics. Let's look at a Taylor series expansion in six terms:

```
» f=taylor('log(x+1)/(x-5)')  % Six terms is the default
f =
-1/5*x+3/50*x^2-41/750*x^3+293/7500*x^4-1207/37500*x^5+0*(x^6)

» pretty(f)
                         2    41  3     293  4     1207   5        6
      - 1/5 x + 3/50 x   -  --- x  +  ---- x  -  ----- x   + 0(x  )
                             750        7500       37500
```

How about an integration:

```
» g=int('log(x)/exp(x^2)')  % a function with no antiderivative
g=
      int(log(x)/exp(x^2),x)

» pretty(g)
                                   /
                                   |   log(x)
                                   |  -------- dx
                                   |       2
                                   /  exp(x  )
```

Symbolic expressions can be presented in many equivalent forms. Some forms may be preferable to others in different situations. MATLAB uses a number of commands to simplify or change the form of symbolic expressions.

```
» f=sym('(x^2-1)*(x-2)*(x-3)')  % create a function
f =
      (x^2-1)*(x-2)*(x-3)

» collect(f)  % collect all like terms
ans =
      x^4-5*x^3+5*x^2+5*x-6

» horner(ans)  % change to Horner or nested representation
ans =
      -6+(5+(5+(-5+x)*x)*x)*x

» factor(ans)  % express as a product of polynomials
ans =
      (x-1)*(x-2)*(x-3)*(x+1)

» expand(f)  % distribute products over sums
ans =
      x^4-5*x^3+5*x^2+5*x-6
```

`simplify` is a powerful, general purpose tool that attempts to simplify an expression by the application of many different kinds of algebraic identities involving sums, integral and fractional powers, trig, exponential, and log functions, and Bessel, hypergeometric, and gamma functions. A few examples should illustrate the power of this function:

```
» simplify('log(2*x/y)')
ans =
       log(2)+log(x)-log(y)

» simplify('sin(x)^2+3*x+cos(x)^2-5')
ans =
       3*x-4

» simplify('(-a^2+1)/(1-a)')
ans =
       a+1
```

The last function to be discussed here is one of the most powerful, but least orthodox of all the simplification tools. The function `simple` tries several different simplification tools (including some within Maple itself and not described above) and then selects the form which has the fewest number of characters in the resulting expression. Let's try a cube root:

$$f = \sqrt[3]{\frac{1}{x^3} + \frac{6}{x^2} + \frac{12}{x} + 8}$$

```
» f='(1/x^3+6/x^2+12/x+8)^(1/3)'  % create the expression
f =
       1/x^3+6/x^2+12/x+8)^(1/3)

» simple(f)  % simplify it
simplify:
       (2*x+1)/x
ans =
       (2*x+1)/x

» simple(ans)  % try again-another method may help
combine(trig):
       2+1/x
ans =
       2+1/x
```

As you can see, `simple` tries a number of simplifications that may help reduce the expression, and lets you see the result of each try. Sometimes it helps to apply `simple` more than once to try a different simplification operation on the result of the first, as it did above. simple is especially useful for expressions containing trig functions. Let's try one:

$$\cos(x) + \sqrt{-\sin(x)^2}$$

```
» simple('cos(x)+sqrt(-sin(x)^2)')
simplify:
       cos(x)+(cos(x)^2-1)^(1/2)
radsimp:
       cos(x)+i*sin(x)
combine(trig):
       cos(x)+(-1/2+1/2*cos(2*x))^(1/2)
factor:
       cos(x)+(-sin(x)^2)^(1/2)
expand:
       cos(x)+(-sin(x)^2)^(1/2)
convert(exp):

1/2*exp(i*x)+1/2/exp(i*x)+1/4*4^(1/2)*exp(exp(i*x)-1/exp(i*x))
convert(tan):
(1-tan(1/2*x)^2)/(1+tan(1/2*x)^2+(-4*tan(1/2*x)^2/(1+tan(1/2*x)^2)^2)^(1/2)
ans =
       cos(x)+i*sin(x)

» simple(ans)   % one more time for luck...
convert(exp):
       exp(i*x)
convert(tan):
(1-tan(1/2*x)^2)/(1+tan(1/2*x)^2+2*i*(tan(1/2*x)/(1+tan(1/2*x)^2)
ans =
       exp(i*x)
```

6.6.2 Summary and Other Features

- A complex symbolic expression in MATLAB syntax can be presented in a form that may be easier to read using the pretty function.
- There may be many equivalent forms of a symbolic expression, some of which are more useful than others in different situations. MATLAB gives

you many tools to change the form of these expressions. Among these tools are:

Tool	Description
`collect`	Collect like terms.
`horner`	Change to Horner or nested representation.
`factor`	Attempt to factor the expression.
`expand`	Expand all terms.
`simplify`	Simplify the expression using identities.
`simple`	Attempt to find the equivalent expression yielding the shortest character string.

- It is often useful to apply `simple` more than once; the result of the first try may in itself be a candidate for further simplification.
- MATLAB symbolic functions can be used to convert a symbolic expression to its partial fraction representation. Given a rational polynomial `f`, `int(f)` will integrate the function and usually separate terms. Then `diff(ans)` will differentiate each term to produce the original expression `f` in a sum-of-terms form that is the partial fraction representation of `f`. For example:

```
» Y='(10*s^2+40*s+30)/(s^2+6*s+8)'
Y =
       (10*s^2+40*s+30)/(s^2+6*s+8)

» diff(int(Y))  % find partial fraction representation of Y
ans =
       10-15/(s+4)-5/(s+2)

» pretty(ans)
                                   15       5
                           10 - ----- - -----
                                s + 4   s + 2
```

This technique is also useful for reducing a rational polynomial where the numerator is of a higher order than the denominator:

```
» g='(x^3+5)/(x^2-1)'
g =
         (x^3+5)/(x^2-1)

» diff(int(g))  %
ans =
         x+3/(x-1)-2/(x+1)

» pretty(ans)
                                   3       2
                           x + ----- - -----
                               x - 1   x + 1
```

It is left as an exercise for the reader to create a function M-file, called `pfd.m`, that will return the partial fraction representation of an expression argument.

- Another tool that can be useful if you use the LaTeX program for word processing or publishing is `latex`. This function returns the LaTeX code required to recreate the expression you supply, and can even save the LaTeX code to a file. See Chapter 9, *Symbolic Math Toolbox Reference* or use online help for more information.

6.7 Variable Precision Arithmetic

6.7.1 Introduction

Round-off error can be introduced in any operation on numeric values, since numeric precision is limited by the number of digits preserved by each numeric operation. Repeated or multiple numeric operations can therefore accumulate error. Operations on symbolic expressions, however, are highly accurate since they do not perform numeric computations and there is no round-off error. Using `eval` or `numeric` on the result of a symbolic operation can introduce round-off error only in the converted result.

MATLAB relies exclusively on the computer's floating point arithmetic for number crunching. Although fast and easy on the computer's memory, floating point operations are limited by the number of digits supported and can introduce round-off error in each operation; they cannot produce exact results. The relative accuracy of individual arithmetic operations in MATLAB is about 16 digits. The symbolic capabilities of Maple, on the other hand, can carry out operations to any arbitrary number of digits. As the default number of digits is increased, however, additional time and computer storage are required for each computation.

Maple defaults to 16 digits of accuracy unless told differently. The function `digits` returns the current value of the global `Digits` parameter. The default number of digits of accuracy for Maple functions can be changed using `digits(n)`, where `n` is the number of digits of accuracy desired. The downside of increasing accuracy this way is that every Maple function will subsequently carry out computations to the new accuracy, increasing computation time. The display of results will not change; only the default accuracy of the underlying Maple functions will be affected.

Another function *is* available, however, that will let you perform a single computation to any arbitrary accuracy while leaving the global `Digits` parameter unchanged. The variable precision arithmetic function, or `vpa` evaluates a single symbolic expression to the default or any specified accuracy and displays a numeric result to the same accuracy:

```
» format long  % let's see all the usual digits
» pi  % how about pi to numeric accuracy
ans =
      3.14159265358979

» digits  % display the default 'Digits' value
Digits = 16

» vpa('pi')  % how about pi to 'Digits' accuracy
ans =
      3.141592653589793

» digits(18)  % change the default to 18 digits
» vpa('pi')  % evaluate pi to 'Digits' digits
ans =
      3.14159265358979324

» vpa('pi',20)  % how about pi to 20 digits
ans =
      3.1415926535897932385

» vpa('pi',50)  % how about pi to 50 digits
ans =
      3.1415926535897932384626433832795028841971693993751

» vpa('2^(1/3)',200)  % the cube root of 2 to 200 digits
ans =

1.2599210498948731647672106072782283505702514647015079800819751
121552996765139594837293965624362550941543102560356156652593990
240406137372284591103042693552469606426166250009774745265654803
068671854055
```

The vpa function applied to a symbolic matrix evaluates each element to the number of digits specified as well:

```
» A=sym('[1/4,log(sqrt(2));exp(1),3/7]')
A =
      [   1/4,log(sqrt(2))]
      [exp(1),         3/7]
» vpa(A,20)  % evaluate to 20 digits
ans =
      [.25000000000000000000, .34657359027997265471]
      [2.7182818284590452354, .42857142857142857143]
```

6.7.2 Things to Try on Your Own

- Find the value of $e^{\pi\sqrt{163}}$ to 18, 25, 30, and 40 digits. Notice that the result is close to an integer value, but is not exactly an integer.

  ```
  » vpa('exp(pi*sqrt(163))',18)
  ```

- Use the `hilb(3)` function to display the 3×3 Hilbert matrix to 20 digits of accuracy.

  ```
  » vpa(hilb(3),20)
  ```

6.7.3 Summary

- The accuracy of any numeric calculation is limited by the number of digits of precision used by your computer and software. If a calculation has intermediate steps, each step has the potential for introducing round-off error. MATLAB uses 16 digits for numeric computations.
- MATLAB symbolic operations using Maple can be carried out to any desired accuracy. The global `Digits` parameter, normally set at 16, can be changed to any value; however, increasing it will trade off time and resources for additional accuracy.
- The `vpa` function lets you evaluate a symbolic expression to any desired accuracy without affecting any other operation. The format is `vpa('expression')` which uses the `Digits` parameter, or `vpa('expression','digits')` to specify the number of digits desired.
- If you remember that `func arg` and `func('arg')` mean the same thing, and that all symbolic functions return symbolic expressions, you may not be surprised to learn that the forms `vpa pi/2 23` and `vpa('pi/2','23')` both return an expression for $\pi/2$ to 23 digits of accuracy.

- The vpa function applied to a symbolic array acts upon each element of the array.

6.8 Solving Equations

Symbolic equations can be solved using symbolic tools available in MATLAB. Some of them have been introduced earlier, and more will be examined in this section.

6.8.1 Solving a Single Algebraic Equation

We have seen earlier in the tutorial that MATLAB contains tools for solving symbolic expressions. If the expression is not an equation (it does not contain an equals sign), the `solve` function sets the symbolic expression equal to zero before solving it:

```
» solve('a*x^2+b*x+c')  % roots of the quadratic equation
ans =
        [1/2/a*(-b+(b^2-4*a*c)^(1/2))]
        [1/2/a*(-b-(b^2-4*a*c)^(1/2))]
```

The result is a symbolic vector whose elements are the two solutions. If you wish to solve for something other than the default variable `x`, `solve` will let you specify it:

```
» solve('a*x^2+b*x+c','b')  % solve for b
ans =
        -(a*x^2+c)/x
```

Symbolic equations containing equals signs can also be solved:

```
» f=solve('cos(x)=sin(x)')  % solve for x
f =
        1/4*pi

» t=solve('tan(2*x)=sin(x)')
t =
        [                       0]
        [acos(1/2+1/2*3^(1/2))]
        [acos(1/2+1/2*3^(1/2))]
```

and numeric solutions found:

```
» numeric(f)
ans =
      0.7854

» numeric(t)
ans =
             0
             0 + 0.8314i
      1.9455
```

Notice that when solving equations of periodic functions, there are an infinite number of solutions. `solve` restricts its search for solutions in these cases to a limited range near zero, and returns a non-unique subset of solutions.

If a symbolic solution cannot be found, a variable precision one will be computed:

```
» x=solve('exp(x)=tan(x)')
x =
      1.306326940423079
```

6.8.2 Several Algebraic Equations

Several algebraic equations can be solved at the same time. A statement of the form `solve(S1,S2,...,Sn)` solves `n` equations for the default variables. A statement of the form `solve (S1,S2,...,Sn,'v1,v2,...,vn')` solves `n` symbolic equations for the `n` unknowns listed as `'v1,v2,...,vn'`.

Here is a familiar example:

> *Lisa wants to go to the movies, so she dumps out her piggy bank and counts her coins. She finds that:*
>
> - *The number of dimes plus half the total number of nickels and pennies is equal to the number of quarters.*
> - *The number of pennies is 10 less than the number of nickels, dimes, and quarters.*
> - *The number of quarters and dimes is equal to the number of pennies plus ¼ of the number of nickels.*
> - *The number of quarters and pennies is one less than the number of nickels plus eight times the number of dimes.*
>
> *If the movie ticket costs $3.00, popcorn is $1.00, and a candy bar is 50 cents, does she have enough to get all three?*

First, create a set of linear equations from the information given above. Let p, n, d, and q be the number of pennies, nickels, dimes, and quarters, respectively:

$$d + \frac{n+p}{2} = q \qquad p = n + d + q - 10$$

$$q + d = p + \frac{n}{4} \qquad q + p = n + 8d - 1$$

Next, create MATLAB symbolic equations, and solve for the variables:

```
» e1='d+(n+p)/2=q';
» e2='p=n+d+q-10';
» e3='q+d=p+n/4';
» e4='q+p=n+8*d-1';
» [pennies,nickels,dimes,quarters]=solve(e1,e2,e3,e4,'p,n,d,q')
pennies =
       16
nickels =
       8
dimes =
       3
quarters =
       15
```

So Lisa has 16 pennies, 8 nickels, 3 dimes, and 15 quarters. That means she has:

```
» money=.01*16+.05*8+.10*3+.25*15
money =
4.6100
```

which is enough for a ticket, popcorn, and a candy bar with 11 cents left over.

6.8.3 Single Differential Equation

Ordinary differential equations are sometimes difficult to solve. MATLAB gives you a powerful tool to help you find solutions to differential equations.

The function `dsolve` computes symbolic solutions to ordinary differential equations. Since we are working with differential equations, we need a way to include differentials in an expression. Therefore, the syntax of `dsolve` is a little different from most other functions. The equations are specified by using the letter `D` to denote differentiation, and `D2`, `D3`, etc., to denote repeated differentiation. Any letters following `Ds` are dependent variables. The equation

$$\frac{d^2y}{dx^2} = 0$$

is represented by the symbolic expression D2y=0. The independent variable can be specified, or will default to the one chosen by the symvar rule. For example, the general solution to the first order equation

$$\frac{dy}{dx} = 1 + y^2$$

can be found by:

```
» dsolve('Dy=1+y^2')  % find the general solution
ans =
        -tan(-x+C1)
```

where C1 is a constant of integration. Solving the same equation with the initial condition $y(0) = 1$ will produce:

```
» y=dsolve('Dy=1+y^2','y(0)=1')  % add an initial condition
y =
        tan(x+1/4*pi)
```

The independent variable can be specified using this form:

```
» dsolve('Dy=1+y^2','y(0)=1','v')  % find solution to dy/dv
ans =
        tan(v+1/4*pi)
```

Let's try a second order differential equation with two initial conditions:

$$\frac{d^2y}{dx^2} = \cos(2x) - y \qquad \frac{dy}{dx}(0) = 0 \qquad y(0) = 1$$

```
» y=dsolve('D2y=cos(2*x)-y','Dy(0)=0','y(0)=1')
y =
        -2/3*cos(x)^2+1/3+4/3*cos(x)

» y=simple(y)  % y looks like it can be simplified
y =
        -1/3*cos(2*x)+4/3*cos(x)
```

Often, a differential equation to be solved contains terms of more than one order, and is presented in the following form:

$$\frac{d^2y}{dx^2} - 2\frac{dy}{dx} - 3y = 0$$

The general solution is:

```
» y=dsolve('D2y-2*Dy-3*y=0')
y =
        C1*exp(-x)+C2*exp(3*x)
```

Applying the initial conditions $y(0) = 0$ and $y(1) = 1$ gives:

```
» y=dsolve('D2y−2*Dy−3*y=0','y(0)=0,y(1)=1')
y =
      1/(exp(−1)−exp(3))*exp(−x)−1/(exp(−1)−exp(3))*exp(3*x)

» y=simple(y)  % this looks like a candidate for simplification
y =
      −(exp(−x)−exp(3*x))/(exp(3)−exp(−1))

» pretty(y)  % pretty it up

                                  exp(− x) − exp(3 x)
                                − ───────────────────
                                   exp(3) − exp(−1)
```

Now plot the result in an interesting region:

```
» ezplot(y,[−6 2])
```

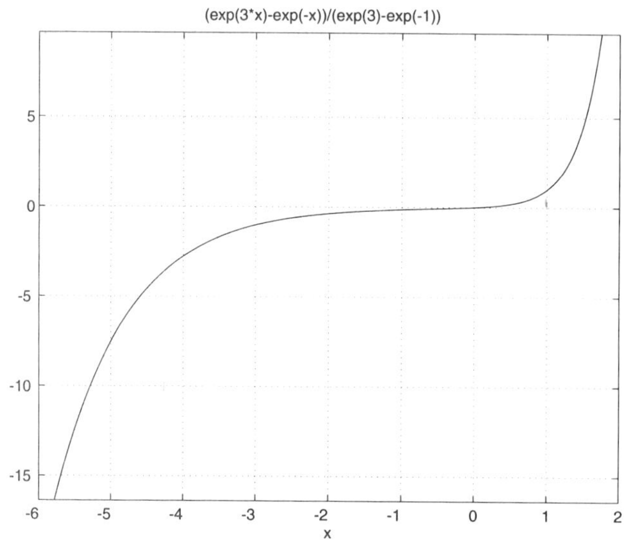

6.8.4 Several Differential Equations

The function `dsolve` can also handle several differential equations at once. Here is a pair of linear, first-order equations:

$$\frac{df}{dx} = 3f + 4g \qquad \frac{dg}{dx} = -4f + 3g$$

The general solutions are:

```
» [f,g]=dsolve('Df=3*f+4*g','Dg=-4*f+3*g')
f =
        C1*exp(3*x)*sin(4*x)+C2*exp(3*x)*cos(4*x)
g =
        -C2*exp(3*x)*sin(4*x)+C1*exp(3*x)*cos(4*x)
```

Adding initial conditions $f(0) = 0$ and $g(0) = 1$, we get:

```
» [f,g]=dsolve('Df=3*f+4*g','Dg=-4*f+3*g','f(0)=0,g(0)=1')
f =
        exp(3*x)*sin(4*x)
g =
        exp(3*x)*cos(4*x)
```

6.8.5 Examples

The implicit function $(x-3)^2 + (y-2)^2 = 5^2$ is the standard form for the equation of a circle of radius 5 with origin at (3,2). Solve the expression for y in terms of x.

```
» f='(x-3)^2+(y-2)^2=5^2'   % create the function
f =
        (x-3)^2+(y-2)^2=5^2

» y=solve(f,'y')   % solve for y
y =
        [2+(16-x^2+6*x)^(1/2)]
        [2-(16-x^2+6*x)^(1/2)]
```

There are two functions for y; the solutions are:

$$y = 2 \pm \sqrt{16 - x^2 + 6x}\ .$$

Solve the initial value problem:

$$\frac{d^2y}{dt^2} + y = 4\cos(t), \quad y\left(\frac{\pi}{2}\right) = 2, \quad \frac{dy}{dt}\left(\frac{\pi}{2}\right) = -3.$$

```
» y=dsolve('D2y+y=4*cos(t)','y(pi/2)=2*pi, Dy(pi/2)=-3')
y =
        5*cos(t)+2*sin(t)*t+pi*sin(t)
```

6.8.6 Summary and Other Features

- The `solve` function generates symbolic solutions to algebraic equations. For one equation in one unknown, `solve(f)` solves the symbolic equation f, or the equation $f=0$, for its default variable as determined by `symvar`. The form `solve(f,'v')` solves the symbolic equation for the variable `v`.

- Several symbolic equations can be solved simultaneously. The form `solve(S1,S2,...,Sn)` solves the `n` symbolic equations for the default independent variables. The form `solve (S1,S2,...,Sn,'v1,v2,...,vn')` solves the `n` symbolic equations for the variables `v1...vn`.
- The `dsolve` function computes symbolic solutions to ordinary differential equations, with or without initial conditions. MATLAB uses the convention that `D` represents

$$\frac{d}{dx},$$

`D2` represents

$$\frac{d^2}{dx^2},$$

and `D2y` represents

$$\frac{d^2y}{dx^2}.$$

The solution can contain constants of integration `C1...Cn` when initial conditions are not specified.

- Several differential equations can be solved simultaneously. The form is `dsolve(F,G)` or `dsolve(F,G,'k1,k2')` where `F` and `G` are a pair of differential equations, and `k1` and `k2` are initial conditions. As before, initial conditions are not required for general solutions containing constants of integration.
- Sometimes when MATLAB returns a solution, from `solve` for example, the solution will have the string '`RootOf(EXPR)`' in it. `RootOf` is a placeholder for all of the roots of the expression `EXPR`. This lets MATLAB return a less complicated expression than if it had included all roots. The function `allvalues(S)` will evaluate an expression `S` that contains the '`RootOf`' sub-expression, find all the roots of the `RootOf` expression, and evaluate `S`

for each of the roots. allvalues returns a vector of all possible symbolic values of the expression S.

```
» p='x^5+x^4+2'  % create an expression
p =
      x^5+x^4+2

» q=solve(p)  % solve the expression
q =
      RootOf(_Z^5+_Z^4+2)

» r=allvalues(q)  % evaluate the solution
r =
      [                          -1.451085092054719]
      [-.5515052469483773-1.018143389670579*i]
      [-.5515052469483773+1.018143389670579*i]
      [ .7770477929757369-.6512828266136642*i]
      [ .7770477929757369+.6512828266136642*i]
```

- The dsolve function uses symvar to find the default independent variable if none is specified. The exception is when x is a dependent variable. In this case t will be chosen as the independent variable. For example,

```
» dsolve('Dx=1+x^2')  % x is a dependent variable; assume Dx=dx/dt
ans =
      tan(t-C1)
```

6.9 Linear Algebra and Matrices

Linear algebra is the study and application of the properties of matrices and vectors. In this section, we present an introduction to symbolic matrices and the tools MATLAB supplies for solving problems using linear algebra.

6.9.1 Symbolic Matrices

Symbolic matrices and vectors are arrays whose elements are symbolic expressions. They can be generated with the `sym` function:

```
» A=sym('[a,b,c;b,c,a;c,a,b]')
A =
        [ a, b, c]
        [ b, c, a]
        [ c, a, b]
» G=sym('[cos(t),sin(t);-sin(t),cos(t)]')
G =
        [  cos(t), sin(t)]
        [ -sin(t), cos(t)]
```

The `sym` function can also expand a formula which specifies individual elements. *Note that in this case only, the* `i` *and* `j` *are row and column positions, respectively, and do not affect the default values of* `i` *and* `j` *(which represent* $\sqrt{-1}$*).* The following examples create 3-by-3 matrices whose elements depend on their row and column positions:

```
» S=sym(3,3,'(i+j)/(i-j+s)')  % create a matrix using a formula
S =
        [     2/s, 3/(-1+s), 4/(-2+s)]
        [3/(1+s),       4/s, 5/(-1+s)]
        [4/(2+s),  5/(1+s),       6/s]

» S=sym(3,3,'m','n','(m-n)/(m-n-t)') % use m and n in a formula
S =
        [       0, -1/(-1-t), -2/(-2-t)]
        [1/(1-t),          0, -1/(-1-t)]
        [2/(2-t),   1/(1-t),          0]
```

The `sym` function can also convert a numeric matrix to its symbolic form:

```
» M=[1.1,1.2,1.3;2.1,2.2,2.3;3.1,3.2,3.3]  % a numeric matrix
M =
        1.1000  1.2000  1.3000
        2.1000  2.2000  2.3000
        3.1000  3.2000  3.3000

» S=sym(M)  % convert to symbolic form
S =
        [11/10,  6/5, 13/10]
        [21/10, 11/5, 23/10]
        [31/10, 16/5, 33/10]
```

If the elements of the numeric matrix can be specified as the ratio of small integers, the `sym` function will use the rational (fractional) representation. If the elements are irrational, sym will represent the elements as floating point numbers in symbolic form:

```
» E=[exp(1) sqrt(2)]
E =
        2.7183    1.4142

» sym(E)
ans =
        [3060513257434036*2^(-50),  3184525836262886*2^(-51)]
```

The size (number of rows and columns) of a symbolic matrix can be found using symsize. This function returns a numeric value or vector, not a symbolic expression. The four forms of symsize are illustrated below.

```
» S=sym('[a,b,c;d,e,f]')  % create a symbolic matrix
S =
        [a,b,c]
        [d,e,f]

» d=symsize(S)  % returns the size of S in the vector d
d =
        2   3

» [m,n]=symsize(S)  % return number of rows in m, columns in n
m =
        2
n =
        3

» m=symsize(S,1)  % return the number of rows
m =
        2

» n=symsize(S,2)  % return the number of columns
n =
        3
```

Numeric arrays use the form N(m,n) to access a single element, but symbolic array elements must be referenced using symbolic functions, such as sym(S,m,n). It would be nice to be able to use the same syntax, but MATLAB representation of symbolic expressions gets in the way. A symbolic array is represented internally as an array of strings, which in turn are arrays of characters; S(m,n) returns a single character. Individual elements of symbolic arrays, therefore, must be referenced by a symbolic function such as sym, rather than directly:

```
» G=sym('[ab,cd;ef,gh]')  % a 2x2 symbolic matrix
G =
         [ab,cd]
         [ef,gh]

» G(1,2)  % the 2nd character of the 1st row of G
ans =
         a

» r=sym(G,1,2)  % the 2nd expression in the 1st row of G
r =
         cd
```

Remember that the symbolic matrix G in the example above is actually stored in the computer as a 2-by-7 array of characters. The first row is '[ab,cd]', so the second character is 'a'.

Finally, sym can be used to change an element of a symbolic array:

```
» sym(G,2,2,'pq')  % change element (2,2) in G to 'pq'
ans =
         [ab,cd]
         [ef,pq]
```

6.9.2 Algebraic Operations

A number of common algebraic operations can be performed on symbolic matrices using the symadd, symsub, symmul, and symdiv functions. Powers are computed using sympow, and the transpose of a symbolic matrix is computed using transpose:

```
» G=sym('[cos(t),sin(t);-sin(t),cos(t)]') % create matrix
G =
        [  cos(t), sin(t)]
        [ -sin(t), cos(t)]

» symadd(G,'t')  % add 't' to each element
ans =
        [  cos(t)+t, sin(t)+t]
        [ -sin(t)+t, cos(t)+t]

» symmul(G,G)  % multiply G by G; sympow(G,2) does same thing
ans =
        [cos(t)^2-sin(t)^2,    2*cos(t)*sin(t)]
        [ -2*cos(t)*sin(t), cos(t)^2-sin(t)^2]

» simple(ans)  % try to simplify
ans =
        [ cos(2*t), sin(2*t)]
        [-sin(2*t), cos(2*t)]
```

Next we'll show that G is an orthogonal matrix by showing that the transpose of G is its inverse:

```
» I=symmul(G,transpose(G))  % multiply G by its transpose
I =
        [cos(t)^2+sin(t)^2,                 0]
        [                0, cos(t)^2+sin(t)^2]

» simplify(I)  % seems to be a trig identity here
ans =
        [1, 0]
        [0, 1]
```

which is the identity matrix, as expected.

6.9.3 Linear Algebra Operations

The inverse and determinant of symbolic matrices are computed by the functions `inverse` and `determ`.

```
» H=sym(hilb(3)) % symbolic form of numeric 3x3 Hilbert matrix
H =
        [  1, 1/2, 1/3]
        [1/2, 1/3, 1/4]
        [1/3, 1/4, 1/5]

» determ(H)  % find the determinant of H
ans =
        1/2160

» J=inverse(H)  % find the inverse of H
J =
        [  9,  -36,   30]
        [-36,  192, -180]
        [ 30, -180,  180]

» determ(J)  % find the determinant of the inverse
ans =
        2160
```

The function `linsolve` is used for solving simultaneous linear equations; it is the symbolic equivalent to the MATLAB backslash operator. `linsolve(A,B)` solves the matrix equation $AX = B$ for a square matrix X. Returning to Lisa's coin problem:

> *Lisa wants to go to the movies, so she dumps out her piggy bank and counts her coins. She finds that:*
>
> - *The number of dimes plus half the total number of nickels and pennies is equal to the number of quarters.*
> - *The number of pennies is 10 less than the number of nickels, dimes, and quarters.*
> - *The number of quarters and dimes is equal to the number of pennies plus ¼ of the number of nickels.*
> - *The number of quarters and pennies is one less than the number of nickels plus eight times the number of dimes.*

Create a set of linear equations as we did last time. Let p, n, d, and q be the number of pennies, nickels, dimes, and quarters, respectively:

$$d+\frac{n+p}{2}=q \qquad p=n+d+q-10$$

$$q+d=p+\frac{n}{4} \qquad q+p=n+8d-1$$

Rearrange the expressions into the order p, n, d, q:

$$\frac{p}{2}+\frac{n}{2}+d-q=0 \qquad p-n-d-q=-10$$

$$-p-\frac{n}{4}+d+q=0 \qquad p-n-8d+q=-1$$

Next, create a symbolic array of the coefficients of the equations:

```
» A=sym('[1/2,1/2,1,-1;1,-1,-1,-1;-1,-1/4,1,1;1,-1,-8,1]')
A =
        [1/2, 1/2, 1,-1]
        [  1,  -1,-1,-1]
        [ -1,-1/4, 1, 1]
        [  1,  -1,-8, 1]

» B=sym('[0;-10;0;-1]')  % Create the symbolic column vector B
B =
        [  0]
        [-10]
        [  0]
        [ -1]

» X=linsolve(A,B)  % solve the symbolic system A*X=B for X
X =
        [16]
        [ 8]
        [ 3]
        [15]
```

The result is the same; Lisa has 16 pennies, 8 nickels, 3 dimes, and 15 quarters.

6.9.4 Other Features

- The symop function concatenates its arguments and evaluates the resulting expression:

```
» f='cos(x)'  % create an expression
f=
      cos(x)

» symop('atan(',f,'+',a,')','^2')
ans =
      atan(cos(x)+a)^2
```

Be careful if you mix arrays and scalars when using symop. For example:

```
» M = sym('[a b; c d]')
M =
      [ a,b]
      [ c,d]

» symop(M,'+','t')
ans =
      [a+t,   b]
      [  c, d+t]
```

adds t to the diagonal of M.

- The function charpoly finds the characteristic polynomial of a matrix:

```
» G=sym('[1,1/2;1/3,1/4]')  % create a symbolic matrix
G =
      [  1,1/2]
      [1/3,1/4]

» charpoly(G)  % find the characteristic polynomial of G
ans =
      x*2-5/4*x+1/12
```

- Eigenvalues and eigenvectors of symbolic matrices can be found using the eigensys function.

```
» F=sym('[1/2,1/4;1/4,1/2]')  % create a symbolic matrix
F =
        [1/2,1/4]
        [1/4,1/2]

» eigensys(F)  % find the eigenvalues of F
ans =
        [3/4]
        [1/4]

» [V,E]=eigensys(F)  % find eigenvalues E and eigenvectors V
V =
        [-1, 1]
        [ 1, 1]
E =
        [1/4]
        [3/4]
```

- The Jordan canonical form of a matrix is the diagonal matrix of eigenvalues; the columns of the transformation matrix are eigenvectors. For a given matrix A, jordan(A) attempts to find a non-singular matrix V, so that inv(V)*A*V is the Jordan canonical form. The jordan function has two forms:

```
» jordan(F)  % find the Jordan form of F, above
ans =
        [1/4,   0]
        [  0, 3/4]

» [V,J]=jordan(F)  % find the Jordan form and eigenvectors
V =
        [ 1/2, 1/2]
        [-1/2, 1/2]
J =
        [1/4,   0]
        [  0, 3/4]
```

The columns of V above are some of the possible eigenvectors of F.

- Since F, above, is nonsingular, the null space basis of F is the empty matrix, and the column space basis is the identity matrix.

```
» F=sym('[1/2,1/4;1/4,1/2]')  % recreate F
F =
       [1/2,1/4]
       [1/4,1/2]

» nullspace(F)  % nullspace of F is the empty matrix
ans =
       []

» colspace(F)  % find the column space of F
ans =
       [1, 0]
       [0, 1]
```

- Singular values of a matrix can be found using the `singvals` function. For more information see Chapter 9, *Symbolic Math Toolbox Reference* or use online help.

```
» A=sym(magic(3))  % Generate a 3x3 matrix
A =
       [8, 1, 6]
       [3, 5, 7]
       [4, 9, 2]

» singvals(A)  % find the singular value expressions
ans =
       [15.00000000000000]
       [6.928203230275511]
       [3.464101615137752]
```

- The function `jacobian(w,v)` computes the Jacobian of w with respect to v. The `(i,j)`th entry of the result is

$$\frac{df(i)}{dv(j)}.$$

Note that when f is a scalar, the Jacobian of f is the gradient of f. Use online help or see Chapter 9, *Symbolic Math Toolbox Reference* for more information.

```
» jacobian('u*exp(v)',sym('u,v'))
ans =
       [exp(v, u*exp(v)]
```

6.9.5 Summary

- Symbolic matrices and vectors are arrays whose elements are symbolic expressions. Notice that since single symbolic expressions are matrices consisting of a single element, all of the functions discussed here also apply to single expressions.
- The `sym` function can be used to create, access, or modify a symbolic matrix or expression, or to convert a numeric matrix or expression to symbolic form.
- The size of a symbolic matrix can be found using the `symsize` function, which returns numeric values.
- Other arithmetic operations can be performed on symbolic matrices and expressions using the `symadd`, `symsub`, `symmul`, `symdiv`, and `sympow` functions.
- The inverse of a symbolic matrix can be found using the `inverse` function. The symbolic determinant of a symbolic matrix uses the `determ` function.
- `linsolve(A,B)` solves the symbolic matrix equation $AX = B$ for a square symbolic matrix X.

6.10 Transforms

Transforms are used extensively in engineering to change the frame of reference between the time domain and the s-domain, the frequency domain, or the z-domain. Many techniques exist for analyzing steady-state and smoothly changing systems in the time domain, but complex systems often can be more easily analyzed in other domains.

6.10.1 Step and Impulse Functions

Engineering problems often make use of the step function $u(t)$ and impulse function $\delta(t)$ when describing systems. The step function $Ku(t\text{-}a)$, where K is a constant, is defined as $Ku(t\text{-}a) = 0$ for $t < a$, and $Ku(t\text{-}a) = K$ for $t > a$. Here is a plot of the step function $Ku(t\text{-}a)$:

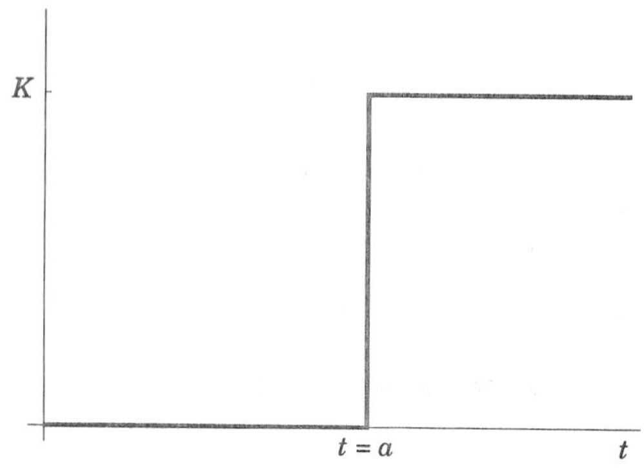

The impulse function $\delta(t)$ is the derivative of the step function $u(t)$. The impulse function $K\delta(t\text{-}a)$ is defined as $K\delta(t\text{-}a) = 0$ for $t < a$, and

$$\int_{-\infty}^{\infty} K\delta(t - a)dt = K \text{ for } t = a.$$

When graphed, it is commonly drawn as an arrow of amplitude K at $t = a$. Here is a plot of $K\delta(t\text{-}a)$:

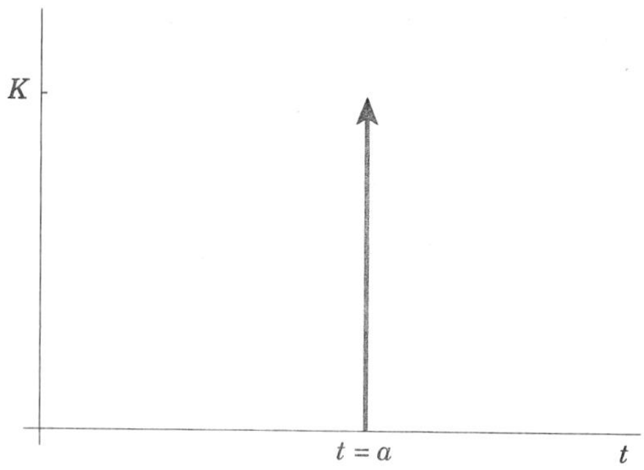

The step and impulse function notation uses the names of famous mathematicians who used these functions extensively in their work. The step function $u(t)$ is called `Heaviside(t)`, and the impulse function $\delta(t)$ is called `Dirac(t)`.

```
» u='k*Heaviside(t-a)'  % create a step function
u =
       k*Heaviside(t-a)

» d=diff(u)  % find the derivative
d =
       k*Dirac(t-a)

» int(d)  % integrate the impulse function
ans =
       k*Heaviside(t-a)
```

6.10.2 The Laplace Transform

The Laplace transform performs the operation

$$F(s) = \int_0^\infty f(t)e^{-st}dt$$

to transform $f(t)$ in the time domain to $F(s)$ in the s-domain.

The Laplace transform of the damped cosine function $e^{-at}cos(wt)$ is found using the `laplace` function:

```
» f=sym('exp(-a*t)*cos(w*t)')  % create an expression f(t)
f =
       exp(-a*x)*cos(w*x)

» F=laplace(f)  % find the Laplace transform of f
F =
       (s+a)/((s+a)^2+w^2)

» pretty(F)

                                s + a
                          ----------------
                                 2    2
                          (s + a)  + w

» laplace('Dirac(t)')  % the transform of the impulse function
ans =
       1

» laplace('Heaviside(t)')  % the transform of the step function
ans =
       1/s
```

Expressions can be transformed back to the time domain using the inverse Laplace transform, invlaplace, which performs the operation:

$$f(t) = \frac{1}{2\pi j}\int_{c+j0}^{c+j\infty} F(s)e^{st}ds$$

Using F from above, we get:

```
» invlaplace(F)  % transform F back to f
ans =
        exp(-a*t)*cos(w*t)
```

The default variable in f, above, is t, while the default variable in F, above, is s. The Laplace transform uses the form symvar(f,'t') to find the variable closest to t in the alphabet rather than using x, since f is usually a function of time. The inverse transform uses symvar(F,'s') to determine the independent variable if it is not specified.

6.10.3 The Fourier Transform

The Fourier transform and inverse Fourier transform are used extensively in circuit analysis to determine the characteristics of a system in both the time and frequency domains. MATLAB uses the fourier and invfourier functions to transform expressions between domains. The Fourier and inverse Fourier transforms are defined as:

$$F(\omega) = \int_{-\infty}^{\infty} f(t)e^{-j\omega t}dt \qquad f(t) = \frac{1}{2\pi}\int_{-\infty}^{\infty} F(\omega)e^{j\omega t}d\omega$$

MATLAB uses a w to represent ω in symbolic expressions.

```
» f='t*exp(-t^2)'  % create a function
f =
        t*exp(-t^2)/t

» F=fourier(f)  % transform using the usual w and t parameters
F =
        -1/2*i*pi^(1/2)*w*exp(-1/4*w^2)

» invfourier(F)  % find the inverse Fourier transform
ans =
        t*exp(-t^2)
```

Since both i and j represent $\sqrt{-1}$, MATLAB has to choose between them when returning an expression containing this value. By default, MATLAB will return i, as shown above.

Often when using the Fourier transform to solve engineering problems, expressions can include a step function $u(t)$ and/or an impulse function (t).

Consider the function $f(t) = -e^{-t}u(t) + 3\delta(t)$. The Fourier transform can be found by:

```
» fourier('-exp(-t)*Heaviside(t)+3*Dirac(t)')
ans =
      -1/(1+i*w)+3

» invfourier(ans)  % find the inverse
ans =
      -exp(-t)*Heaviside(t)+3*Dirac(t)
```

The default variables in Fourier transforms are `t` and `w`. The Fourier transform uses the form `symvar(f,'t')` to find the variable, while the inverse Fourier transform uses `symvar(F,'w')` to choose the independent variable if it is not specified.

6.10.4 Things to Try on Your Own

- The Fourier and inverse Fourier transforms convert between the time domain and the frequency domain as $f(t) \leftrightarrow F(\omega)$, where t represents time, and ω represents frequency in radians per second. Some engineering fields and some instructors tend to use the transform pair $g(t) \leftrightarrow G(f)$, where t represents time, and f represents frequency in cycles per second (Hz). The relationship used is $\omega = 2\pi f$ or $f = \omega / 2\pi$. It would be convenient to have functions that would do the conversion for you so that you could work with either convention. For example, to convert $g = e^{-t^2}$ to the frequency domain using ω, and then do the same using f, you could:

```
» g='exp(-t^2)'  % create the function
g =
      exp(-t^2)

» Gw=fourier(g)  % get the transform in terms of w
Gw =
      pi^(1/2)*exp(-1/4*w^2)

» invfourier(Gw)  % get the inverse to check the result
ans =
      exp(-t^2)

» Gf=ftf(g)  % get the transform in terms of f
Gf =
      pi^(1/2)*exp(-pi^2*f^2)

» invftf(Gf)  % get the inverse to check it out
ans =
      exp(-t^2)
```

This was done using a function M-file called ftf.m for the $g(t) \to G(f)$ transform, and another function M-file called invftf.m for the inverse. Here is an example of an M-file for ftf.m:

```
function G = ftf(g)
%
%  Fourier transform g(t) => G(f) where f = w/(2*pi)
%
G=fourier(g,'w','t');           % get the transform
G=subs(G,'(2*pi*f)','w');  % change the variable
```

- It is left as an exercise for the reader to write the M-file invftf.m.

6.10.5 The *z*-Transform

The Laplace and Fourier transforms are used to analyze continuous-time systems. z-transforms, on the other hand, are used to analyze discrete-time systems. The z-transform is defined as:

$$F(z) = \sum_{n=0}^{\infty} f(n) z^{-n}$$

where z is a complex number.

z-transforms and inverse z-transforms are obtained using the ztrans and invztrans functions. The format is similar to the fourier and laplace transform functions:

```
» f='2^n/7-(-5)^n/7'  % create a function
f =
      2^n/7-(-5)^n/7

» G=ztrans(f)  % transform using the usual z and n parameters
G =
      z/(z-2)/(z+5)

» pretty(G)
                              z
                       ---------------
                       (z - 2) (z + 5)

» invztrans(G)  % find the inverse z-transform
ans =
      1/7*2^n-1/7*(-5)^n
```

The default variables in z-transforms are n and z. The z-transform uses the form symvar(f,'n') to find the variable, while the inverse Fourier transform uses symvar(F,'z') to choose the independent variable if it is not specified.

Note that all of these transforms including the Laplace, Fourier, and z-transform pairs have forms that let you specify different independent variables.

6.10.6 Summary

- The step function $u(t)$ is represented in MATLAB symbolic notation by `Heaviside(t)`; the impulse function $\delta(t)$ is represented by `Dirac(t)`. These functions will often appear in transforms.
- Laplace transforms and inverse Laplace transforms `laplace` and `invlaplace`, transform $f(t) \leftrightarrow F(s)$. Use `laplace(f)` to transform $f(t) \rightarrow F(s)$, and `invlaplace(F)` to transform $F(s) \rightarrow f(t)$.
- Fourier transforms are found using `fourier` and `invfourier`. These functions are used to transform $f(t) \leftrightarrow F(\omega)$, and have the same forms as the `laplace` transform pair.
- z-transforms are found using `ztrans` and `invztrans`. These functions are used to transform $f(n) \leftrightarrow F(z)$, and have the same forms as the `laplace` and `fourier` functions.

6.11 Interactive Symbolic Tools

As you have seen as you worked your way through this Tutorial, the *Symbolic Math Toolbox* contains many powerful functions to solve both simple and complex problems using symbolic expressions. As a reward for all of your hard work, we will now present two unique "fun" tools that are included in MATLAB.

6.11.1 Riemann Sums

The integral of a function can represent the area under the curve of the function, and can be approximated over a closed interval using Riemann sums. MATLAB supplies an interactive tool that lets you look at different approximations of the integral using this technique.

The form of the function is `rsums(f)` where `f` is a symbolic function. This will bring up a graphics window where the area under the curve of the function from 0 to 1 will be approximated by 10 rectangles. The title displays the function and the total calculated area of the rectangles. Underneath the graph is a horizontal slider that lets you change the number of rectangles used to approximate the curve from as few as 2 to as many as 256 using the mouse. Here is an example:

```
» f='10*x*exp(-5*x^2)'  % create an interesting function
f =
      10*x*exp(-5*x^2)

» ezplot(f)  % view the graph from -2*pi to 2*pi
```

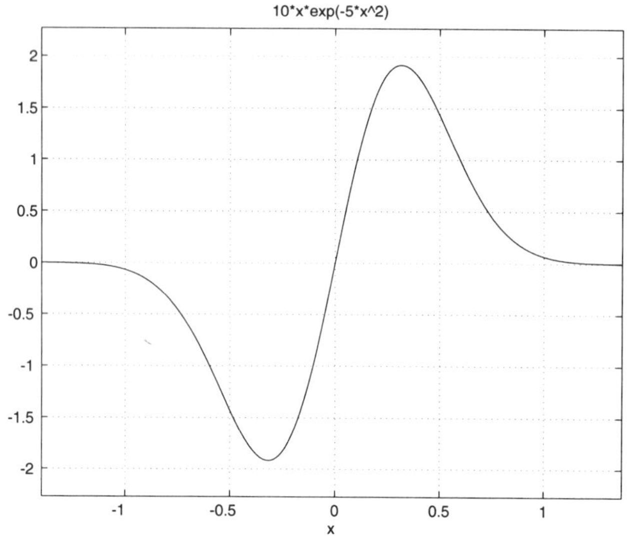

```
» vpa(int(f,0,1),6)  % the value of the integral from 0 to 1
ans =
      .993262

» rsums(f)  % the Riemann approximation from 0 to 1
```

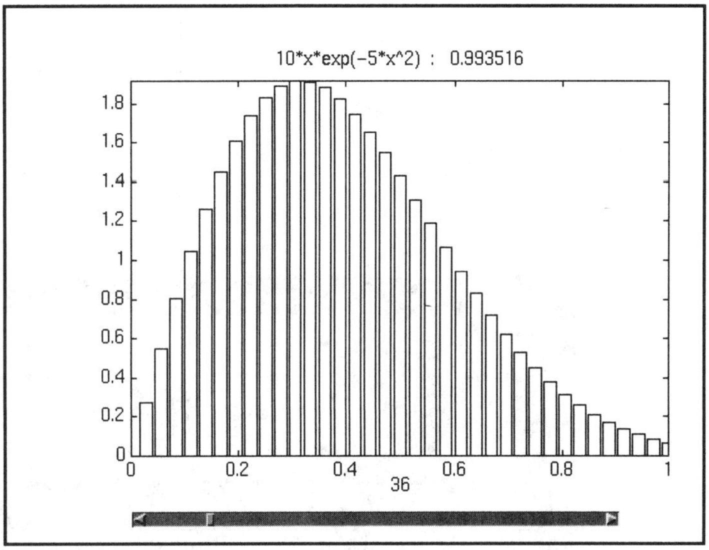

Use the mouse to change the setting of the slider and notice how the total area under the rectangles (shown to the right of the title) approaches the value that we just obtained from the symbolic integral. It will never match, because the maximum number of rectangles that can be used is 256, while the integral is calculated in the limit.

Try some of the other functions on your own.

6.11.2 Function Calculator

This tool is an interactive graphical function calculator, called `funtool`, that uses mouse clicks to perform operations on symbolic expressions. `funtool` manipulates two functions of a single variable, `f(x)` and `g(x)`, and graphs them in two windows. A third window controls the calculator, and contains the symbolic functions `f` and `g`, the domain of the graphs `x`, and a constant expression a that you can change by typing in the text boxes. There are three rows of buttons under the text boxes that invoke MATLAB symbolic functions, and a fourth row of buttons to control the calculator itself.

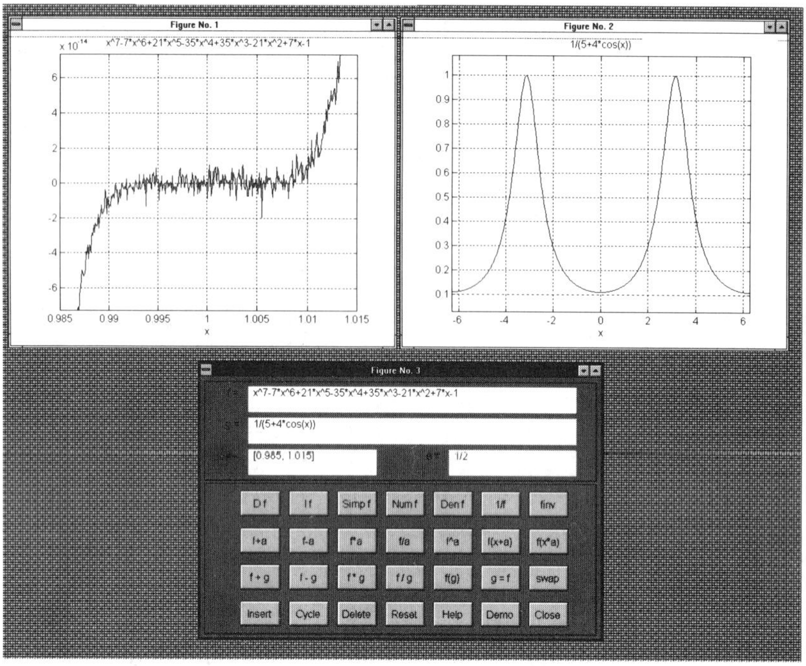

The buttons in the first row operate on `f` alone. There are buttons to differentiate, integrate, simplify, extract the numerator or denominator, calculate `1/f`, and replace `f` by its inverse function. The buttons in the second row translate and scale `f` by the parameter `a`. There are buttons to add, subtract, multiply, and divide `f` by `a`, and to raise `f^a`. The last two buttons replace `f(x)` by `f(x+a)` or `f(x*a)`. The buttons in the third row operate on both `f(x)` and `g(x)`. The first four buttons add, subtract, multiply, and divide `f(x)` by `g(x)`. The fifth button replaces `f(x)` by `f(g(x))`, and the last two buttons copy `f(x)` to `g(x)`, or swap `f(x)` and `g(x)`.

The last row of buttons controls the calculator itself. The `funtool` calculator maintains a list of functions called `fxlist`. The first three buttons add the active function to the list, cycle through the list, and delete the active function from the list. The **Reset** button sets `f`, `g`, `x`, `a`, and `flist` to their initial values. The **Help** button gives you some more detailed information about `funtool`. The **Demo** button shows how to generate `f=sin(x)` from the initial values in nine clicks, and the **Close** button closes all `funtool` windows.

Try out `funtool`. Experience the possibilities of many of MATLAB's symbolic functions.

7 Signals and Systems Toolbox Tutorial

7.1 Introduction

The preceding material clearly demonstrates the power of MATLAB to solve numerous problems. Given the ability to write function M-files, it is possible to write functions for specific academic disciplines. Because matrix analysis has widespread application in signal processing and linear systems, *The Student Edition of MATLAB* offers a number of tools tailored to these areas. These tools form the *Signals and Systems Toolbox*. If you are not involved in the study of signal processing or linear systems, the functions in this toolbox may not be of much use to you at this time. However, these tools embody important concepts that underlie applications such as speech recognition, image processing, financial analysis, and process control. If you are studying any of these areas, you will find that the power of this toolbox simplifies many of the computations required to solve common problems.

Because of the large number and broad scope of the functions in the *Signals and Systems Toolbox*, it is impractical to illustrate each function. The utility of individual toolbox functions becomes obvious when the underlying academic material is understood. As a result, we will describe how MATLAB represents systems, organizes the available functions for easy reference, and provides several examples.

7.2 System Representations in MATLAB

Continuous and discrete (or digital) systems are represented in a consistent manner in MATLAB. Systems can be written in transfer function, zero-pole-gain, or state-space form. Since transfer functions are described by ratios of polynomials, they are described in MATLAB by row vectors of their coefficients. The zero-pole-gain form is closely associated with the transfer function form. In this form, the roots of the numerator and denominator polynomials and the overall gain of the system are described with MATLAB arrays. The state-space form is perhaps the most natural form for MATLAB since the state-space representation is described by the matrices **A**, **B**, **C**, and **D**. The following tables illustrate these forms.

Transfer Function	
Continuous	$$H(s) = \frac{N(s)}{D(s)} = \frac{N_1 s^m + N_2 s^{m-1} + \ldots + N_m s + N_{m+1}}{D_1 s^n + D_2 s^{n-1} + \ldots + D_n s + D_{n+1}}$$ where $m \le n$ MATLAB: `num=[N1 N2 ... Nm+1], den=[D1 D2 ... Dn+1]`
Discrete (z^1 form)	$$H(z) = \frac{N(z)}{D(z)} = \frac{N_1 z^m + N_2 z^{m-1} + \ldots + N_m z + N_{m+1}}{D_1 z^n + D_2 z^{n-1} + \ldots + D_n z + D_{n+1}}$$ where $m \le n$ MATLAB: `num=[N1 N2 ... Nm+1], den=[D1 D2 ... Dn+1]`
(z^{-1} form)	$$H(z) = \frac{N(z)}{D(z)} = \frac{N_1 + N_2 z^{-1} + \ldots + N_n z^{-n+1} + N_{n+1} z^{-n}}{D_1 + D_2 z^{-1} + \ldots + D_n z^{-n+1} + D_{n+1} z^{-n}}$$ MATLAB: `num=[N1 N2 ... Nn+1], den=[D1 D2 ... Dn+1]`

Zero-Pole-Gain	
Continuous	$H(s) = \frac{N(s)}{D(s)} = K\frac{(s-z_1)(s-z_2)\ldots(s-z_m)}{(s-p_1)(s-p_2)\ldots(s-p_n)}$ where $m \le n$ MATLAB: `K, Z=[z1; z2; ... ;zm] P=[p1; p2; ... ;pn]`
Discrete	$H(z) = \frac{N(z)}{D(z)} = K\frac{(z-z_1)(z-z_2)\ldots(z-z_m)}{(z-p_1)(z-p_2)\ldots(z-p_n)}$ where $m \le n$ MATLAB: `K, Z=[z1; z2; ... ;zm] P=[p1; p2; ... ;pn]`

State-Space	
Continuous	$\dot{x} = Ax + Bu$ $y = Cx + Du$ MATLAB: `A, B, C, D`
Discrete	$x[n+1] = Ax[n] + Bu[n]$ $y[n] = Cx[n] + Du[n]$ MATLAB: `A, B, C, D`

Most of the tools in the *Signals and Systems Toolbox* transparently handle both the transfer function and state-space representations. In addition, the state-space representation naturally handles the multiple-input, multiple-output (MIMO) case. On the other hand, the transfer function and zero-pole-gain representations can only be extended to the single-input, multiple-output (SIMO) case. When a system has multiple outputs, the denominator of its transfer function remains the same for all outputs, but it has a different numerator associated with each output. In MATLAB the numerator is represented by a matrix with the ith row containing the numerator associated with the ith output. For example,

$$\begin{bmatrix} Y_1(s) \\ Y_2(s) \end{bmatrix} = \begin{bmatrix} \dfrac{3s+2}{3s^3+5s^2+2s+1} \\ \dfrac{8s^2+2s+2}{3s^3+5s^2+2s+1} \end{bmatrix} U(s)$$

```
num = [0 3 2
       8 2 2]

den = [3 5 2 1]
```

7.3 Signals and Systems Toolbox Organization

The *Signals and Systems Toolbox* offers a wide variety of functions. To help you find functions associated with a given problem type, they are organized into tables below. The first column in each table is the generic MATLAB function name. The second column shows typical syntax options. If you are familiar with the underlying academic material for particular functions, the notation in these syntax options may be intuitively obvious to you. In any case, more specific information about individual functions can be found in Chapter 10, *Signals and Systems Toolbox Reference*.

Frequency Response		

`bode`	`bode(num,den)`
	`bode(num,den,W)`
	`bode(A,B,C,D,iu)`
	`bode(A,B,C,D,iu,W)`
		Bode plots

`freqs`	`[H,W]=freqs(num,den)`
	`H=freqs(num,den,W)`
	`[H,W]=freqs(num,den,n)`
		continuous-time frequency response

`freqz`	`[H,W]=freqz(num,den)`
	`[H,W]=freqz(num,den,n)`
	`H=freqz(num,den,W)`
	`[H,f]=freqz(num,den,n,Fs)`
		discrete-time frequency response

`nyquist`	`nyquist(num,den)`
	`nyquist(num,den,W)`
	`nyquist(A,B,C,D,iu)`
	`nyquist(A,B,C,D,iu)`
		Nyquist plots

Frequency Domain Analysis		
abs	abs(x)	magnitude of a complex variable
angle	angle(x)	phase angle of a complex variable
grpdelay	[Gd,W]=grpdelay(num,den) [Gd,W]=grpdelay(num,den,n) Gd=freqz(num,den,W) [Gd,f]=freqz(num,den,n,Fs)	discrete-time group delay
psd	psd(x) psd(x,nfft) psd(x,nfft,Fs) psd(x,nfft,Fs,window)	power spectral density estimation
specgram	specgram(x)	spectrogram
unwrap	unwrap(x)	unwrap phase response

Prototype Filter Design		
buttap	[Z,P,K]=buttap(n)	Butterworth analog lowpass prototype filter design
cheb1ap	[Z,P,K]=cheb1ap(n,Rp)	Chebyshev I analog lowpass prototype filter design
cheb2ap	[z,p,k]=cheb2ap(n,Rs)	Chebyshev II analog lowpass prototype filter design
ellipap	[Z,P,K]=ellipap(n,Rp,Rs)	elliptic analog lowpass prototype filter design

Filter Design		

`butter`	`[num,den]=butter(n,Wn)`
	`[num,den]=butter(n,Wn,[Wn1 Wn2])`
	`[num,den]=butter(n,Wn,'high')`
	`[num,den]=butter(n,[Wn1 Wn2],'stop')`
	`[num,den]=butter(n,Wn,'s')`
	`[num,den]=butter(n,Wn,[Wn1 Wn2],'s')`
	`[num,den]=butter(n,Wn,'high','s')`
	`[num,den]=butter(n,[Wn1 Wn2],'stop','s')`
		Butterworth digital or analog filter design

`cheby1`	`[num,den]=cheby1(n,Rs,Wn)`
	`[num,den]=cheby1(n,Rs,Wn,[Wn1 Wn2])`
	`[num,den]=cheby1(n,Rs,Wn,'high')`
	`[num,den]=cheby1(n,Rs,[Wn1 Wn2],'stop')`
	`[num,den]=cheby1(n,Rs,Wn,'s')`
	`[num,den]=cheby1(n,Rs,[Wn1 Wn2],'s')`
	`[num,den]=cheby1(n,Rs,Wn,'high','s')`
	`[num,den]=cheby1(n,Rs,[Wn1 Wn2],'stop','s')`
		Chebyshev I digital or analog filter design

`cheby2`	`[num,den]=cheby2(n,Rp,Wn)`
	`[num,den]=cheby2(n,Rp,Wn,[Wn1 Wn2])`
	`[num,den]=cheby2(n,Rp,Wn,'high')`
	`[num,den]=cheby2(n,Rp,[Wn1 Wn2],'stop')`
	`[num,den]=cheby2(n,Rp,Wn,'s')`
	`[num,den]=cheby2(n,Rp,[Wn1 Wn2],'s')`
	`[num,den]=cheby2(n,Rp,Wn,'high','s')`
	`[num,den]=cheby2(n,Rp,[Wn1 Wn2],'stop','s')`
		Chebyshev II digital or analog filter design

`ellip`	`[num,den]=ellip(n,Rp,Rs,Wn)`
	`[num,den]=ellip(n,Rp,Rs,[Wn1 Wn2])`
	`[num,den]=ellip(n,Rp,Rs,Wn,'high')`
	`[num,den]=ellip(n,Rp,Rs,[Wn1 Wn2],'stop')`
	`[num,den]=ellip(n,Rp,Rs,Wn,'s')`
	`[num,den]=ellip(n,Rp,Rs,[Wn1 Wn2],'s')`
	`[num,den]=ellip(n,Rp,Rs,Wn,'high','s')`
	`[num,den]=ellip(n,Rp,Rs,[Wn1 Wn2],'stop','s')`
		elliptic digital or analog filter design

`remez`	`num=remez(n,F,M)`
	`num=remez(n,F,M,W)`
	`num=remez(n,F,M,'Hilbert')`
	`num=remez(n,F,M,'differentiator')`
		Parks-McClellan optimal digital FIR filter design

`yulewalk`	`[num,den]=yulewalk(n,F,M)`	Yule-Walker digital IIR filter design

Frequency Domain Transformations		
lp2bp	[numt,dent]=lp2bp(num,den,Wo,Bw) [At,Bt,Ct,Dt]=lp2bp(A,B,C,D,Wo,Bw)	lowpass to bandpass transformation
lp2bs	[numt,dent]=lp2bs(num,den,Wo,Bw) [At,Bt,Ct,Dt]=lp2bs(A,B,C,D,Wo,Bw)	lowpass to bandstop transformation
lp2hp	[numt,dent]=lp2hp(num,den,Wo) [At,Bt,Ct,Dt]=lp2hp(A,B,C,D,Wo)	lowpass to highpass transformation
lp2lp	[numt,dent]=lp2lp(num,den,Wo) [At,Bt,Ct,Dt]=lp2lp(A,B,C,D,Wo)	lowpass to lowpass transformation
polystab	dent=polystab(den)	discrete-time stabilization

Fast Fourier Transform		
fft	fft(x) fft(x,n)	fast Fourier transform
fft2	fft2(X) fft2(X,nr,nc)	2-D fast Fourier transform
ifft	ifft(x) ifft(x,n)	inverse fast Fourier transform
ifft2	ifft(X) ifft(X,nr,nc)	2-D inverse fast Fourier transform

Window Functions		
boxcar	boxcar(n)	boxcar or rectangular window
hamming	hamming(n)	Hamming window
hanning	hanning(n)	Hanning window
kaiser	kaiser(n,beta)	Kaiser window
triang	triang(n)	triangular window

Continuous to Discrete Conversion		
bilinear	[Zd,Pd,Kd]=bilinear(Z,P,K,Fs) [numd,dend]=bilinear(num,den,Fs) [Ad,Bd,Cd,Dd]=bilinear(A,B,C,D,Fs) [Zd,Pd,Kd]=bilinear(Z,P,K,Fs,Fp)	bilinear transformation
c2d	[Ad,Bd]=c2d(A,B,Ts)	zero-order hold continuous to discrete conversion
c2dm	[Ad,Bd,Cd,Dd]=c2dm(A,B,C,D,Ts,'zoh') [Ad,Bd,Cd,Dd]=c2dm(A,B,C,D,Ts,'foh') [Ad,Bd,Cd,Dd]=c2dm(A,B,C,D,Ts,'tustin') [Ad,Bd,Cd,Dd]=c2dm(A,B,C,D,Ts,'prewarp') [Ad,Bd,Cd,Dd]=c2dm(A,B,C,D,Ts,'matched') [numd,dend]=c2dm(num,den,Ts,'zoh') [numd,dend]=c2dm(num,den,Ts,'foh') [numd,dend]=c2dm(num,den,Ts,'tustin') [numd,dend]=c2dm(num,den,Ts,'prewarp') [numd,dend]=c2dm(num,den,Ts,'matched')	continuous to discrete conversion given a specific method
impinvar	[numd,dend]=impinvar(num,den) [numd,dend]=impinvar(num,den,Fs)	impulse invariant transformation

Linear System Representations		
residue	[R,P,k]=residue(num,den) [num,den]=residue(R,P,k)	continuous-time partial fraction expansion and inverse
residuez	[R,P,k]=residuez(num,den) [num,den]=residuez(R,P,k)	discrete-time partial fraction expansion and inverse
sos2ss	[A,B,C,D]=sos2ss(sos)	second-order section to state-space
sos2tf	[num,den]=sos2tf(sos)	second-order section to transfer function
sos2zp	[Z,P,K]=sos2zp(sos)	second-order section to zero-pole
ss2sos	sos=ss2sos(A,B,C,D) sos=ss2sos(A,B,C,D,iu)	state-space to second-order section
ss2tf	[num,den]=ss2tf(A,B,C,D,iu)	state-space to transfer function
ss2zp	[Z,P,K]=ss2zp(A,B,C,D,iu)	state-space to zero-pole
tf2ss	[A,B,C,D]=tf2ss(num,den)	transfer function to state-space
tf2zp	[Z,p,k]=tf2zp(num,den)	transfer function to zero-pole
zp2sos	sos=zp2sos(Z,P,K)	zero-pole to second-order section

Linear System Representations (Continued)		
zp2ss	[A,B,C,D]=zp2ss(Z,P,K)	zero-pole to state-space
zp2tf	[num,den]=zp2tf(Z,P,K)	zero-pole to transfer function

Linear System Manipulation		
append	[A,B,C,D]=append(A1,B1,C1,D1,A2,B2,C2,D2)	combine two state-space systems
canon	[At,Bt,Ct,Dt]=canon(A,B,C,D,'modal') [At,Bt,Ct,Dt,T]=canon(A,B,C,D,'modal') [At,Bt,Ct,Dt]=canon(A,B,C,D,'companion') [At,Bt,Ct,Dt,T]=canon(A,B,C,D,'companion')	canonical state transformations
cloop	[Ac,Bc,Cc,Dc]=cloop(A,B,C,D,sign) [numc,denc]=cloop(num,den,sign)	close loop around system
feedback	[A,B,C,D]=feedback(A1,B1,C1,D1,A2,B2,C2,D2,sign) [num,den]=feedback(num1,den1,num2,den2,sign)	form feedback connection of two systems
series	[A,B,C,D]=series(A1,B1,C1,D1,A2,B2,C2,D2) [num,den]=series(num1,den1,num2,den2)	series connection of two systems
ssselect	[Ae,Be,Ce,De]=ssselect(A,B,C,D,inputs,outputs)	select subsystem from state-space system

Linear System Analysis		
ctrb	co=ctrb(A,B)	compute controllability matrix
damp	[Wn,Z]=damp(A) [Wn,Z]=damp(num) [Wn,Z]=damp(P)	compute damping of continuous-time poles
ddamp	[mag,Wn,Z]=ddamp(A,Ts) [mag,Wn,Z]=ddamp(num,Ts) [mag,Wn,Z]=ddamp(P,Ts)	compute damping of discrete-time poles
obsv	ob=obsv(A,B)	compute observability matrix
rlocus	rlocus(num,den) rlocus(num,den,k) rlocus(A,B,C,D) rlocus(A,B,C,D,k)	Evans root locus
tzero	Z=tzero(A,B,C,D) [Z,gain]=tzero(A,B,C,D)	find transmission zeros

Linear System Design		
acker	K=acker(A,B,P)	state feedback using Ackermann's formula
lqe	[L,P,E]=lqe(A,G,C,Q,R) [L,P,E]=lqe(A,G,C,Q,R,N)	linear quadratic estimator design
lqr	[K,S,E]=lqr(A,B,Q,R) [K,S,E]=lqr(A,B,Q,R,N)	linear quadratic regulator design
place	K=place(A,B,P)	robust state feedback

Time Domain Response		
dimpulse	dimpulse(A,B,C,D,iu) dimpulse(A,B,C,D,iu,n) dimpulse(num,den) dimpulse(num,den,n)	discrete-time unit impulse response
dstep	dstep(A,B,C,D,iu) dstep(A,B,C,D,iu,n) dstep(num,den) dstep(num,den,n) dstep(A,B,C,D,iu,n)	discrete-time unit step response
filter	y=filter(num,den,u) [y,Zf]=filter(num,den,u,Zi)	discrete-time filter response
impulse	impulse(A,B,C,D,iu) impulse(A,B,C,D,iu,t) impulse(num,den) impulse(num,den,t)	continuous-time unit impulse response
lsim	lsim(A,B,C,D,U,t) lsim(A,B,C,D,U,t,Xo) lsim(num,den,u,t)	arbitrary input continuous linear system response
step	step(A,B,C,D,iu) step(A,B,C,D,iu,t) step(num,den) step(num,den,t)	continuous-time unit step response

Miscellaneous		
`chop`	`chop(X,n)` `chop(X,n,unit)`	round to n significant digits
`detrend`	`detrend(x)` `detrend(x,0)`	remove linear trend or average value from data
`filtdemo`	`filtdemo`	discrete-time filtering demo
`sinc`	`sinc(x)`	sinc function
`sosdemo`	`sosdemo`	second-order section demo
`strips`	`strips(x)`	strip chart plot
`zplane`	`zplane(z,p)` `zplane(num,den)`	discrete-time zero-pole plot

7.4 Examples

This section illustrates some of the most common functions in the *Signals and Systems Toolbox.*

7.4.1 Continuous-Time Systems

Continuous-time systems are systems whose dynamics and signals vary continuously over time. Typical examples include electrical circuits and the motion of the planets in our solar system. The numerical examples to follow will use the system described by the transfer function:

$$H(s) = \frac{100}{s^2 + 4s + 100}$$

The frequency response of this system can be plotted by using the function bode:

```
» nc=100; dc=[1 4 100];
» bode(nc,dc)
```

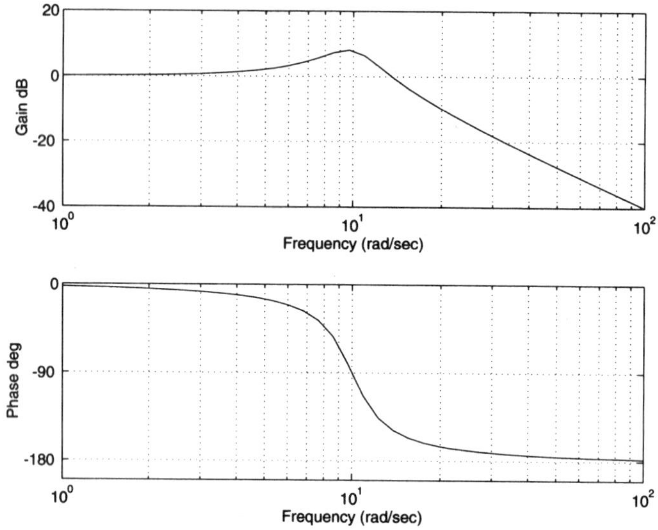

Its step response can be plotted with the function step:

```
» step(nc,dc)
```

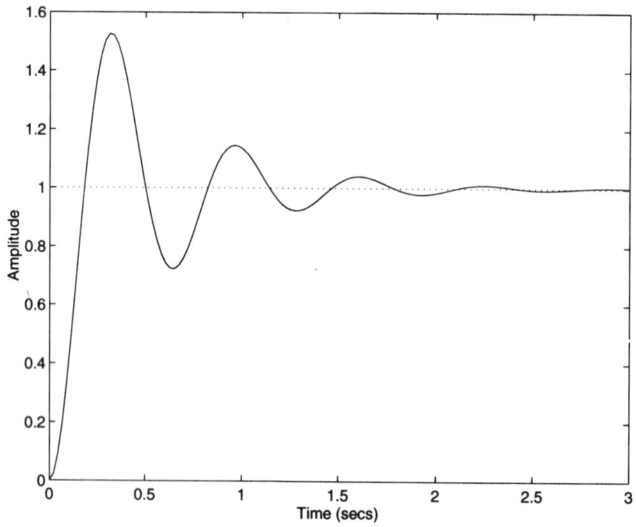

A state-space description of this system is given by:

```
» [A,B,C,D]=tf2ss(nc,dc)
A =
      -4   -100
       1      0
B =
       1
       0
C =
       0    100
D =
       0
```

Finally, converting this system to an equivalent discrete-time system using a zero-order-hold equivalence at a sampling period of 0.01 seconds gives:

```
» [n,d]=c2d(nc,dc,0.01)
n =
    2.7183
d =
    0.0172      0.0687      1.7183
```

7.4.2 Discrete-Time Systems

Discrete-time systems are systems whose dynamics and signals change only at distinct points in time. In between these points, the system is commonly undefined. The following examples will utilize the discrete-time transfer function description given below:

$$H(z) = \frac{z + 0.37}{z^2 - 0.2z + 0.37}$$

The frequency response of this system can be plotted by using the function `freqz`:

```
» nd=[1 0.37]; dd=[1 -0.2 0.37];
» freqz(nd,dd)
```

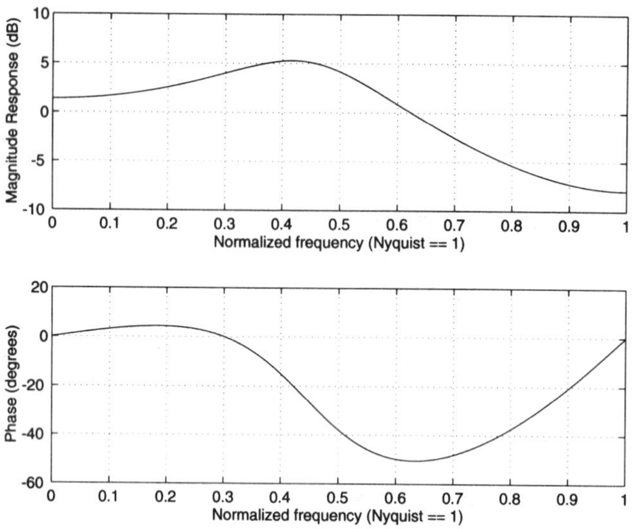

Here the frequency axis is normalized so that 1.0 is the Nyquist frequency, i.e., 1.0 equals one half the sampling frequency.

The step response can be plotted with the function dstep:

```
» dstep(nd,dd)
```

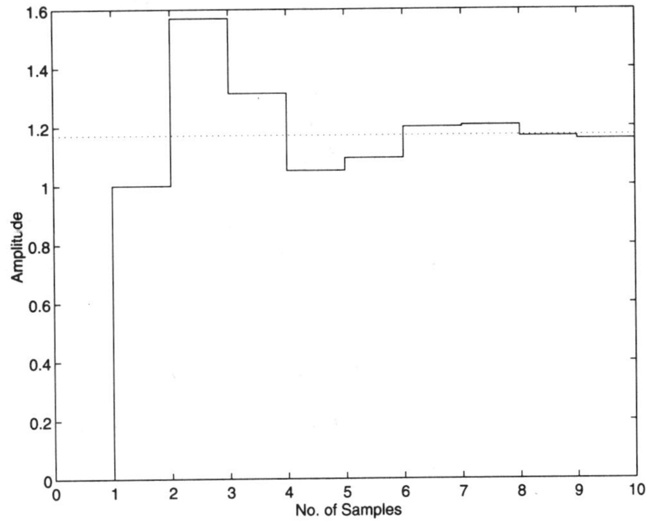

This system can also be converted to a state representation using tf2ss, provided the numerator is padded with leading zeros so that it has the same length as the denominator:

```
» ndd=[0 nd]
ndd =
           0     1.0000     0.3700

» [A,B,C,D]=tf2ss(ndd,dd)
A =
      0.2000   −0.3700
      1.0000         0
B =
       1
       0
C =
      1.0000     0.3700
D =
       0
```

7.4.3 Filtering

The *Signals and Systems Toolbox* offers a variety of functions for the design of analog and digital filters. An nth order analog lowpass Butterworth filter with unity 3dB bandwidth is found by using `buttap(n)`. For example, the third-order filter is given by:

```
» [Z,P,K]=buttap(3) % zeros, poles, gain form
Z =
     []
P =
  -0.5000 + 0.8660i
  -1.0000 + 0.0000i
  -0.5000 - 0.8660i
K =
     1

» [n,d]=zp2tf(Z,P,K) % convert to transfer function
n =
     0       0       0       1
d =
    1.0000     2.0000     2.0000     1.0000

» bode(n,d) % find and plot frequency response
```

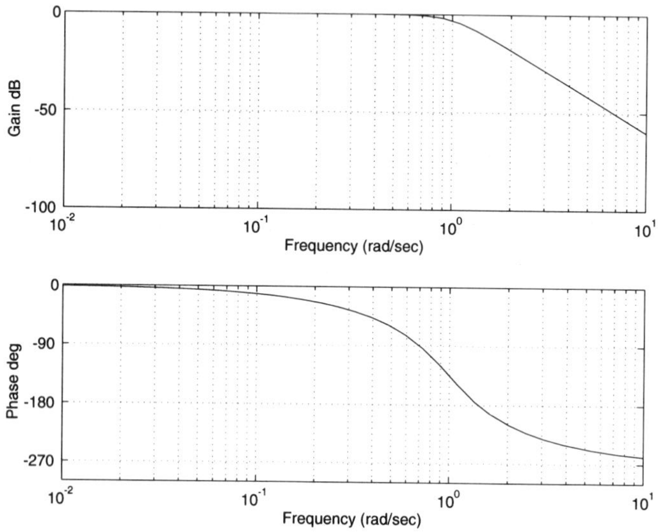

This filter can be converted to a bandpass filter using the function lp2bp:

```
» [nbp,dbp]=lp2bp(n,d,2,1); % form is: lp2bp(n,d,Wo,Bw)
» bode(nbp,dbp,logspace(-1,2));
```

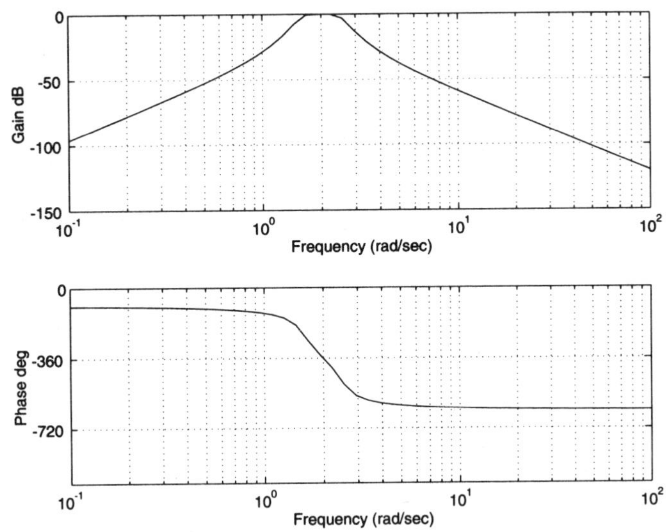

Here the requested center frequency is 2 rad/s and the bandwidth is 1 rad/s. Rather than use the default frequency axis computed by bode, the frequency axis is specified by the points in logspace(−1,2), i.e., 50 points between 10^{-1} and 10^2 rad/s.

A digital Butterworth lowpass filter can be designed from the Butterworth filter above by using the bilinear transformation:

```
» Fs=100/(2*pi)  % choose sampling frequency
Fs =
   1.5915e+01

» [nd,dd]=bilinear(n,d,Fs)  % perform bilinear transformation
nd =
   2.9118e-05   8.7355e-05   8.7355e-05   2.9118e-05
dd =
   1.0000e+00  -2.8744e+00   2.7566e+00  -8.8193e-01

» freqz(nd,dd,128,Fs)  % plot response at 128 points
```

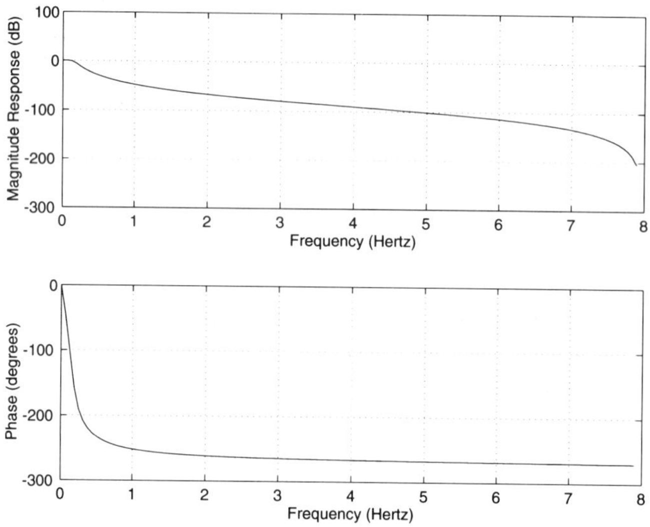

7.4.4 Spectral Analysis

The FFT algorithm finds wide application in digital signal processing. Its uses include filtering, convolution, computation of frequency response and group delay, and power spectrum estimation.

In MATLAB, the FFT algorithm is embodied in the function `fft`. One example illustrating the use of this function is given under the `fft` listing in Chapter 10, *Signals and Systems Toolbox Reference*. For further illustration, consider the problem of estimating the continuous Fourier transform of the signal

$$f(t) = \begin{cases} 12e^{-3t} & t \geq 0 \\ 0 & t < 0 \end{cases}$$

Analytically, this Fourier transform is given by

$$F(\omega) = \frac{12}{3 + j\omega}$$

Although using the FFT in this case has little real value since the analytic solution is known, this example illustrates an approach to estimating the Fourier transform of less common signals. The following MATLAB statements estimate $|F(\omega)|$ using the FFT and graphically compare it to the analytic expression above:

```
» N=128;
```

makes the number of points be a power of 2, which speeds the FFT computation,

```
» t=linspace(0,3,N);
» f=2*exp(-3*t);
```

identifies the sample points and evaluates $f(t)$ at those samples. At the final sample point the function $f(t)$ is approximately zero so aliasing is minimized.

```
» Ts=t(2)-t(1);
» Ws=2*pi/Ts;
```

identifies the sampling period and sampling frequency in rad/sec,

```
» F=fft(f);
```

finds the FFT of $f(t)$,

```
» Fp=F(1:N/2+1)*Ts;
```

extracts only the positive frequency components from `F` and multiplies them by sampling period to estimate $F(\omega)$,

```
» W=Ws*(0:N/2)/N;
```

creates the continuous frequency axis, which starts at zero and ends at the Nyquist frequency `Ws/2`,

```
» Fa=2./(3+j*W);
```

evaluates the analytical Fourier transform,

```
» plot(W,abs(Fa),W,abs(Fp),'+')
» xlabel('Frequency, Rad/s'),ylabel('|F(w)|')
```

generates the following plot where the solid line is the analytical solution and the + marks are the FFT estimates.

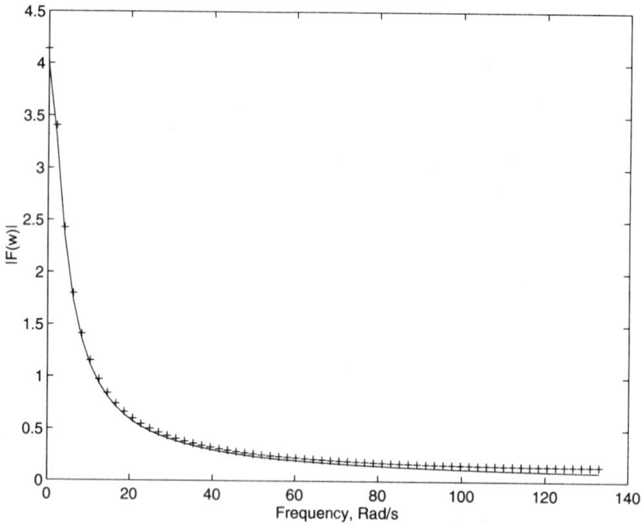

Part Three

Reference

8

MATLAB Reference

8.1 Reference Tables

MATLAB provides 20 main categories of functions. Some of MATLAB's functions are built into the interpreter, while others take the form of M-files. The M-file functions, and in the case of the built-in functions, M-files containing only help text, are organized into 20 directories, each containing the files associated with a category. The MATLAB command `help` displays an online table of these main categories.

MATLAB's Main Categories of Functions	
`color`	Color control and lighting model functions.
`datafun`	Data analysis and Fourier transform functions.
`demos`	Demonstrations and samples.
`elfun`	Elementary math functions.
`elmat`	Elementary matrices and matrix manipulation.
`funfun`	Function functions – nonlinear numerical methods.
`general`	General purpose commands.
`graphics`	General purpose graphics functions.
`iofun`	Low-level file I/O functions.
`lang`	Language constructs and debugging.
`matfun`	Matrix functions – numerical linear algebra.
`ops`	Operators and special characters.
`plotxy`	Two-dimensional graphics.
`plotxyz`	Three-dimensional graphics.
`polyfun`	Polynomial and interpolation functions.
`sparfun`	Sparse matrix functions.
`specfun`	Specialized math functions.
`specmat`	Specialized matrices.
`sounds`	Sound processing functions.
`strfun`	Character string functions.

The following pages contain tables of functions within each of these specific areas. If you execute `help` on one of the directory names listed on the left side of this table, MATLAB displays an online version of the tables within that area.

General Purpose Commands

Managing Commands and Functions	
`demo`	Run demos.
`expo`	Run MATLAB EXPO demonstration program.
`help`	Online documentation.
`info`	Information about MATLAB and The MathWorks.
`lasterr`	Last error message generated.
`lookfor`	Keyword search through the help entries.
`path`	Control MATLAB's search path.
`subscribe`	Become a subscribing MATLAB user.
`type`	List M-file.
`ver`	Current MATLAB and toolbox versions.
`version`	Current MATLAB version number.
`what`	Directory listing of M-, MAT- and MEX-files.
`whatsnew`	Display README files for MATLAB and toolboxes.
`which`	Locate functions and files.

Managing Variables and the Workspace	
`clear`	Clear variables and functions from memory.
`disp`	Display matrix or text.
`length`	Length of vector.
`load`	Retrieve variables from disk.
`pack`	Consolidate workspace memory.
`save`	Save workspace variables to disk.
`size`	Size of matrix.
`who`	List current variables.
`whos`	List current variables, long form.

Working with Files and the Operating Environment	
`cd`	Change current working directory.
`cedit`	Set command line editing parameters (UNIX only).
`delete`	Delete file.
`diary`	Save text of MATLAB session.
`dir`	Directory listing.
`getenv`	Get environment value.
`hostid`	MATLAB server host identification number.
`ls`	Directory listing.
`matlabroot`	Root directory of MATLAB installation.
`pwd`	Show current working directory.
`tempdir`	Name of system temporary directory.
`tempname`	Unique name for temporary file.
`terminal`	Set graphics terminal type.
`unix`	Execute operating system command; return result.
`!`	Execute operating system command.

Controlling the Command Window	
`clc`	Clear command window.
`echo`	Echo commands inside script files.
`format`	Set output format.
`home`	Send cursor home.
`more`	Control paged output in command window.

Starting and Quitting from MATLAB	
`matlabrc`	Master startup M-file.
`quit`	Terminate MATLAB.
`startup`	M-file executed when MATLAB is invoked.

Operators and Special Characters

<table>
<tr><th colspan="2">Operators and Special Characters</th></tr>
<tr><td>+</td><td>Plus.</td></tr>
<tr><td>–</td><td>Minus.</td></tr>
<tr><td>*</td><td>Matrix multiplication.</td></tr>
<tr><td>.*</td><td>Array multiplication.</td></tr>
<tr><td>^</td><td>Matrix power.</td></tr>
<tr><td>.^</td><td>Array power.</td></tr>
<tr><td>kron</td><td>Kronecker tensor product.</td></tr>
<tr><td>\</td><td>Backslash or left division.</td></tr>
<tr><td>/</td><td>Slash or right division.</td></tr>
<tr><td>./</td><td>Array division.</td></tr>
<tr><td>:</td><td>Colon.</td></tr>
<tr><td>( )</td><td>Parentheses.</td></tr>
<tr><td>[ ]</td><td>Brackets.</td></tr>
<tr><td>.</td><td>Decimal point.</td></tr>
<tr><td>..</td><td>Parent directory.</td></tr>
<tr><td>...</td><td>Continuation.</td></tr>
<tr><td>,</td><td>Comma.</td></tr>
<tr><td>;</td><td>Semicolon.</td></tr>
<tr><td>%</td><td>Comment.</td></tr>
<tr><td>!</td><td>Exclamation point.</td></tr>
<tr><td>'</td><td>Transpose and quote.</td></tr>
<tr><td>.'</td><td>Nonconjugated transpose.</td></tr>
<tr><td>=</td><td>Assignment.</td></tr>
<tr><td>==</td><td>Equality.</td></tr>
<tr><td><></td><td>Relational operators.</td></tr>
<tr><td>&</td><td>Logical AND.</td></tr>
<tr><td>|</td><td>Logical OR.</td></tr>
<tr><td>~</td><td>Logical NOT.</td></tr>
<tr><td>xor</td><td>Logical EXCLUSIVE OR.</td></tr>
</table>

Logical Functions	
`all`	True if all elements of vector are true.
`any`	True if any element of vector is true.
`exist`	Check if variables or functions exist.
`find`	Find indices of nonzero elements.
`finite`	True for finite elements.
`isempty`	True for empty matrix.
`ishold`	True if `hold` is on.
`isieee`	True for IEEE floating-point arithmetic.
`isinf`	True for infinite elements.
`isletter`	True for alphabetic character.
`isnan`	True for Not-A-Number.
`isreal`	True if all matrix elements are real.
`issparse`	True for sparse matrix.
`isstr`	True for text string.

Language Constructs and Debugging

MATLAB as a Programming Language	
`eval`	Execute string with MATLAB expression.
`feval`	Execute function specified by string.
`function`	Add new function.
`global`	Define global variable.
`nargchk`	Validate number of input arguments.

Control Flow	
`break`	Terminate execution of loop.
`else`	Used with `if`.
`elseif`	Used with `if`.
`end`	Terminate the scope of `for`, `while` and `if` statements.
`error`	Display message and abort function.
`for`	Repeat statements a specific number of times.
`if`	Conditionally execute statements.
`return`	Return to invoking function.
`while`	Repeat statements an indefinite number of times.

Interactive Input	
`input`	Prompt for user input.
`keyboard`	Invoke keyboard as if it were a script file.
`menu`	Generate menu of choices for user input.
`pause`	Wait for user response.

Debugging	
`dbclear`	Remove breakpoint.
`dbcont`	Resume execution.
`dbdown`	Change local workspace context.
`dbquit`	Quit debug mode.
`dbstack`	List who called whom.
`dbstatus`	List all breakpoints.
`dbstep`	Execute one or more lines.
`dbstop`	Set breakpoint.
`dbtype`	List M-file with line numbers.
`dbup`	Change local workspace context.

Matrices and Matrix Manipulation

Elementary Matrices

Elementary Matrices	
`eye`	Identity matrix.
`gallery`	Test matrices - matrix condition and eigenvalues.
`linspace`	Linearly spaced vector.
`logspace`	Logarithmically spaced vector.
`meshgrid`	X and Y arrays for 3-D plots.
`ones`	Ones matrix.
`rand`	Uniformly distributed random numbers.
`randn`	Normally distributed random numbers.
`zeros`	Zeros matrix.
`:`	Regularly spaced vector.

Special Variables and Constants	
`ans`	Most recent answer.
`computer`	Computer type.
`eps`	Floating-point relative accuracy.
`flops`	Count of floating-point operations.
`i, j`	Imaginary unit.
`inf`	Infinity.
`NaN`	Not-a-Number.
`nargin`	Number of function input arguments.
`nargout`	Number of function output arguments.
`pi`	3.1415926535897....
`realmax`	Largest floating-point number.
`realmin`	Smallest floating-point number

Time and Dates	
`clock`	Wall clock.
`cputime`	Elapsed CPU time.
`date`	Calendar.
`etime`	Elapsed time function.
`tic, toc`	Stopwatch timer functions.

Matrix Manipulation	
`diag`	Create or extract diagonals.
`fliplr`	Flip matrix in the left/right direction.
`flipud`	Flip matrix in the up/down direction.
`isreal`	True for matrix containing real elements only.
`reshape`	Change size.
`rot90`	Rotate matrix 90 degrees.
`tril`	Extract lower triangular part.
`triu`	Extract upper triangular part.
`:`	Index into matrix, rearrange matrix.

Specialized Matrices

Specialized Matrices	
compan	Companion matrix.
hadamard	Hadamard matrix.
hankel	Hankel matrix.
hilb	Hilbert matrix.
invhilb	Inverse Hilbert matrix.
magic	Magic square.
pascal	Pascal matrix.
rosser	Classic symmetric eigenvalue test problem.
toeplitz	Toeplitz matrix.
vander	Vandermonde matrix.
wilkinson	Wilkinson's eigenvalue test matrix.

Math Functions

Elementary Math Functions

Elementary Math Functions	
abs	Absolute value.
acos	Inverse cosine.
acosh	Inverse hyperbolic cosine.
acot	Inverse cotangent.
acoth	Inverse hyperbolic cotangent.
acsc	Inverse cosecant.
acsch	Inverse hyperbolic cosecant.
angle	Phase angle.
asec	Inverse secant.
asech	Inverse hyperbolic secant.
asin	Inverse sine.
asinh	Inverse hyperbolic sine.
atan	Inverse tangent.
atan2	Four quadrant inverse tangent.
atanh	Inverse hyperbolic tangent.
ceil	Round towards plus infinity.
conj	Complex conjugate.
cos	Cosine.
cosh	Hyperbolic cosine.
cot	Cotangent.
coth	Hyperbolic cotangent.
csc	Cosecant.

Elementary Math Functions (Continued)	
`csch`	Hyperbolic cosecant.
`exp`	Exponential.
`fix`	Round towards zero.
`floor`	Round towards minus infinity.
`gcd`	Greatest common divisor.
`imag`	Complex imaginary part.
`lcm`	Least common multiple.
`log`	Natural logarithm.
`log10`	Common logarithm.
`real`	Complex real part.
`rem`	Remainder after division.
`round`	Round towards nearest integer.
`sec`	Secant.
`sech`	Hyperbolic secant.
`sign`	Signum function.
`sin`	Sine.
`sinh`	Hyperbolic sine.
`sqrt`	Square root.
`tan`	Tangent.
`tanh`	Hyperbolic tangent.

Specialized Math Functions

Specialized Math Functions	
bessel	Bessel functions.
besseli	Modified Bessel functions of the first kind.
besselj	Bessel functions of the first kind.
besselk	Modified Bessel functions of the second kind.
bessely	Bessel functions of the second kind.
beta	Beta function.
betainc	Incomplete beta function.
betaln	Logarithm of beta function.
ellipj	Jacobian elliptic functions.
ellipke	Complete elliptic integral.
erf	Error function.
erfc	Complementary error function.
erfcx	Scaled complementary error function.
erfinv	Inverse error function.
expint	Exponential integral.
gamma	Gamma function.
gammainc	Incomplete gamma function.
gammaln	Logarithm of gamma function.
legendre	Associated Legendre functions.
log2	Dissect floating point numbers.
pow2	Scale floating point numbers.
rat	Rational approximation.
rats	Rational output.

Matrix Functions—Numerical Linear Algebra

Matrix Analysis	
cond	Matrix condition number.
det	Determinant.
etree	Elimination tree of a matrix.
norm	Matrix or vector norm.
null	Null space.
orth	Orthogonalization.
rcond	LINPACK reciprocal condition estimator.
rank	Number of linearly independent rows or columns.
rref	Reduced row echelon form.
subspace	Angle between two subspaces.
trace	Sum of diagonal elements.

Linear Equations	
chol	Cholesky factorization.
inv	Matrix inverse.
lscov	Least squares in the presence of known covariance.
lu	Factors from Gaussian elimination.
nnls	Non-negative least-squares.
pinv	Pseudoinverse.
qr	Orthogonal-triangular decomposition.
\ and /	Linear equation solution.

Eigenvalues and Singular Values	
balance	Diagonal scaling to improve eigenvalue accuracy.
cdf2rdf	Complex diagonal form to real block diagonal form.
eig	Eigenvalues and eigenvectors.
hess	Hessenberg form.
poly	Characteristic polynomial.
qz	Generalized eigenvalues.
rsf2csf	Real block diagonal form to complex diagonal form.
schur	Schur decomposition.
svd	Singular value decomposition.

Matrix Functions	
expm	Matrix exponential.
funm	Evaluate general matrix function.
logm	Matrix logarithm.
sqrtm	Matrix square root.

Low Level Functions	
qrdelete	Delete columns from QR factorization.
qrinsert	Insert columns into QR factorization.

Data Analysis and Fourier Transform Functions

Basic Operations	
`cumprod`	Cumulative product of elements.
`cumsum`	Cumulative sum of elements.
`max`	Largest component.
`mean`	Average or mean value.
`median`	Median value.
`min`	Smallest component.
`prod`	Product of elements.
`sort`	Sort in ascending order.
`std`	Standard deviation.
`sum`	Sum of elements.
`trapz`	Numerical integration using trapezoidal method.

Finite Differences	
`del2`	Five-point discrete Laplacian.
`diff`	Difference function and approximate derivative.
`gradient`	Approximate gradient.

Correlation	
`corrcoef`	Correlation coefficients.
`cov`	Covariance matrix.

Filtering and Convolution	
`conv`	Convolution and polynomial multiplication.
`conv2`	Two-dimensional convolution.
`deconv`	Deconvolution and polynomial division.
`filter`	One-dimensional digital filter.
`filter2`	Two-dimensional digital filter.

Fourier Transforms	
abs	Magnitude.
angle	Phase angle.
cplxpair	Sort numbers into complex conjugate pairs.
fft	Discrete Fourier transform.
fft2	Two-dimensional discrete Fourier transform.
fftshift	Move zeroth lag to center of spectrum.
ifft	Inverse discrete Fourier transform.
ifft2	Two-dimensional inverse discrete Fourier transform.
nextpow2	Next higher power of 2.
unwrap	Remove phase angle jumps across 360° boundaries.

Vector Functions	
cross	Vector cross product.
dot	Vector dot product.

Polynomial and Interpolation Functions

Polynomials	
conv	Multiply polynomials.
deconv	Divide polynomials.
poly	Construct polynomial with specified roots.
polyder	Differentiate polynomial.
polyeig	Solve polynomial eigenvalue problem.
polyfit	Fit polynomial to data.
polyval	Evaluate polynomial.
polyvalm	Evaluate polynomial with matrix argument.
residue	Partial-fraction expansion (residues).
roots	Find polynomial roots.

Data Interpolation	
griddata	Data gridding.
interp1	One-dimensional interpolation (1-D table lookup).
interp2	Two-dimensional interpolation (2-D table lookup).
interpft	One-dimensional interpolation using FFT method.

Function Functions

Function Functions - Nonlinear Numerical Methods	
fmin	Minimize function of one variable.
fmins	Minimize function of several variables.
fplot	Plot function.
fzero	Find zero of function of one variable.
ode23	Solve differential equations, low order method.
ode45	Solve differential equations, high order method.
quad	Numerically evaluate integral, low order method.
quad8	Numerically evaluate integral, high order method.

Sparse Matrix Functions

Elementary Sparse Matrices	
spdiags	Sparse matrix formed from diagonals.
speye	Sparse identity matrix.
sprandn	Sparse random matrix.
sprandsym	Sparse symmetric random matrix.

Full to Sparse Conversion	
find	Find indices of nonzero entries.
full	Convert sparse matrix to full matrix.
sparse	Create sparse matrix from nonzeros and indices.
spconvert	Convert from sparse matrix external format.

Working with Nonzero Entries of Sparse Matrices	
issparse	True if matrix is sparse.
nnz	Number of nonzero entries.
nonzeros	Nonzero entries.
nzmax	Amount of storage allocated for nonzero entries.
spalloc	Allocate memory for nonzero entries.
spfun	Apply function to nonzero entries.
spones	Replace nonzero entries with ones.

Visualizing Sparse Matrices	
`gplot`	Plot graph, as in "graph theory."
`spy`	Visualize sparsity structure.

Reordering Algorithms	
`colmmd`	Column minimum degree.
`colperm`	Order columns based on nonzero count.
`dmperm`	Dulmage-Mendelsohn decomposition.
`randperm`	Random permutation vector.
`symmmd`	Symmetric minimum degree.
`symrcm`	Reverse Cuthill-McKee ordering.

Norm, Condition Number, and Rank	
`condest`	Estimate 1-norm condition.
`normest`	Estimate 2-norm.
`sprank`	Structural rank.

Miscellaneous	
`spaugment`	Form least squares augmented system.
`spparms`	Set parameters for sparse matrix routines.
`symbfact`	Symbolic factorization analysis.

Two-Dimensional Graphics

Elementary X-Y Graphs	
`fill`	Draw filled 2-D polygons.
`loglog`	Log-log scale plot.
`plot`	Linear plot.
`semilogx`	Semi-log scale plot, x-axis logarithmic.
`semilogy`	Semi-log scale plot, y-axis logarithmic.

Specialized X-Y Graphs	
bar	Bar graph.
comet	Animated comet plot.
compass	Compass plot.
errorbar	Error bar plot.
feather	Feather plot.
fplot	Plot function.
hist	Histogram plot.
polar	Polar coordinate plot.
rose	Angle histogram plot.
stairs	Stairstep plot.
stem	Stem plot for discrete sequence data.

Graph Annotation	
grid	Grid lines.
gtext	Mouse placement of text.
legend	Add legend to plot.
text	Text annotation.
title	Graph title.
xlabel	*X*-axis label.
ylabel	*Y*-axis label.

Coordinate System Conversion	
cart2pol	Cartesian to polar coordinates.
pol2cart	Polar to Cartesian coordinates.

Miscellaneous	
zoom	Zoom in and out of a two-dimensional plot.

Three-Dimensional Graphics

Line and Area Fill Commands	
fill3	Draw filled three-dimensional polygons in 3-D space.
plot3	Plot lines and points in 3-D space.

Contour and Other 2-D Plots of 3-D Data	
`clabel`	Contour plot elevation labels.
`comet3`	Three-dimensional animated comet plot.
`contour`	Contour plot.
`contour3`	Three-dimensional contour plot.
`contourc`	Contour plot computation (used by `contour`).
`image`	Display image.
`imagesc`	Scale data and display as image.
`pcolor`	Pseudocolor (checkerboard) plot.
`quiver`	Quiver plot.
`slice`	Volumetric slice plot.

Surface and Mesh Plots	
`mesh`	Three-dimensional mesh surface.
`meshc`	Combination `mesh`/`contour` plot.
`meshz`	Three-dimensional mesh with zero plane.
`slice`	Volumetric visualization plot.
`surf`	Three-dimensional shaded surface.
`surfc`	Combination `surf`/`contour` plot.
`surfl`	Three-dimensional shaded surface with lighting.
`waterfall`	Waterfall plot.

Graph Appearance	
`axis`	Axis scaling and appearance.
`caxis`	Pseudocolor axis scaling.
`colormap`	Color lookup table.
`hidden`	Mesh hidden line removal mode.
`shading`	Color shading mode.
`view`	Three-dimensional graph viewpoint specification.
`viewmtx`	View transformation matrices.

Graph Annotation	
grid	Grid lines.
legend	Add legend to plot.
text	Text annotation.
title	Graph title.
xlabel	X-axis label.
ylabel	Y-axis label.
zlabel	Z-axis label for three-dimensional plots.

3-D Objects	
cylinder	Generate cylinder.
sphere	Generate sphere.

Coordinate System Conversion	
cart2sph	Cartesian to polar coordinates.
sph2cart	Polar to Cartesian coordinates.

Graphics Functions

Figure Window Creation and Control	
capture	Screen capture of current figure (UNIX only).
clf	Clear current figure.
close	Close figure.
figure	Create figure (graph window).
gcf	Get handle to current figure.
graymon	Set default figure properties for grayscale monitors.
newplot	Determine correct axes and figure for new graph.
refresh	Redraw current figure window.
whitebg	Toggle figure background color.

Axis Creation and Control	
`axes`	Create axes in arbitrary positions.
`axis`	Control axis scaling and appearance.
`caxis`	Control pseudocolor axis scaling.
`cla`	Clear current axes.
`gca`	Get handle to current axes.
`hold`	Hold current graph.
`ishold`	True if `hold` is on.
`subplot`	Create axes in tiled positions.

Handle Graphics Objects	
`axes`	Create axes.
`figure`	Create figure window.
`image`	Create image.
`line`	Create line.
`patch`	Create patch.
`surface`	Create surface.
`text`	Create text.
`uicontrol`	Create user interface control.
`uimenu`	Create user interface menu.

Handle Graphics Operations	
`delete`	Delete object.
`drawnow`	Flush pending graphics events.
`findobj`	Find object with specified properties.
`gco`	Get handle of current object.
`get`	Get object properties.
`reset`	Reset object properties.
`rotate`	Rotate an object.
`set`	Set object properties.

Dialog Boxes	
`uigetfile`	Retrieve name of file to open through dialog box.
`uiputfile`	Retrieve name of file to write through dialog box.

Hardcopy and Storage	
`orient`	Set paper orientation.
`print`	Print graph or save graph to file.
`printopt`	Configure local printer defaults.

Movies and Animation	
`getframe`	Get movie frame.
`movie`	Play recorded movie frames.
`moviein`	Initialize movie frame memory.

Miscellaneous	
`ginput`	Graphical input from mouse.
`ishold`	Return hold state.
`rbbox`	Rubberband box for region selection.
`waitforbuttonpress`	Wait for key/button press over figure.

Color Control and Lighting Model Functions

Color Controls	
`caxis`	Pseudocolor axis scaling.
`colormap`	Color lookup table.
`shading`	Color shading mode.

Colormaps	
bone	Gray-scale with a tinge of blue colormap.
contrast	Contrast-enhancing grayscale colormap.
cool	Shades of cyan and magenta colormap.
copper	Linear copper-tone colormap.
flag	Alternating red, white, blue, and black colormap.
gray	Linear grayscale colormap.
hsv	Hue-saturation-value colormap.
hot	Black-red-yellow-white colormap.
jet	Variation of HSV colormap (no wrap).
pink	Pastel shades of pink colormap.
prism	Prism colors colormap.
white	All white monochrome colormap.

Colormap Related Functions	
brighten	Brighten or darken colormap.
colorbar	Display colormap as color scale.
hsv2rgb	Hue-saturation-value to red-green-blue conversion.
rgb2hsv	Red-green-blue to hue-saturation-value conversion.
rgbplot	Plot colormap.
spinmap	Spin colormap.

Lighting Models	
diffuse	Diffuse reflectance.
specular	Specular reflectance.
surfl	Three-dimensional shaded surface with lighting.
surfnorm	Surface normals.

Sound Processing Functions

General Sound Functions	
saxis	Sound axis scaling.
sound	Convert vector into sound.

SPARCstation-specific Sound Functions	
auread	Read Sun audio file.
auwrite	Write Sun audio file.
lin2mu	Linear to mu-law conversion.
mu2lin	Mu-law to linear conversion.

.WAV Sound Functions	
wavread	Load MS-Windows 3.1 .WAV format sound file.
wavwrite	Save MS-Windows 3.1 .WAV format sound file.

Character String Functions

General	
abs	Convert string to numeric values.
blanks	Create string of blanks.
deblank	Remove trailing blanks and null spaces from string.
eval	Execute string with MATLAB expression.
findstr	Find one string within another.
isstr	True for string.
setstr	Convert numeric values to string.
str2mat	Form text matrix from individual strings.
string	About character strings in MATLAB.
strrep	String search and replace.
strtok	First token in string.

String Comparison	
isletter	True for alphabetic character.
lower	Convert string to lower case.
strcmp	Compare strings.
upper	Convert string to upper case.

String to Number Conversion	
`int2str`	Convert integer to string.
`num2str`	Convert number to string.
`sprintf`	Convert number to string under format control.
`sscanf`	Convert string to number under format control.
`str2num`	Convert string to number.

Hexadecimal to Number Conversion	
`dec2hex`	Convert decimal integer to hex string.
`hex2dec`	Convert hex string to decimal integer.
`hex2num`	Convert hex string to IEEE floating-point number.

Low-level File I/O Functions

File Opening and Closing	
`fclose`	Close file.
`fopen`	Open file.

Unformatted I/O	
`fread`	Read binary data from file.
`fwrite`	Write binary data to file.

Formatted I/O	
`fgetl`	Read line from file, discard newline character.
`fgets`	Read line from file, keep newline character.
`fprintf`	Write formatted data to file.
`fscanf`	Read formatted data from file.

File Positioning	
`feof`	Test for end-of-file.
`ferror`	Inquire file I/O error status.
`frewind`	Rewind file.
`fseek`	Set file position indicator.
`ftell`	Get file position indicator.

String Conversion	
sprintf	Write formatted data to string.
sscanf	Read string under format control.

Specialized File I/O	
csvread	Read a file of comma-separated values.
csvwrite	Write a file of comma-separated values.
uigetfile	Retrieve name of file to open through dialog box.
uiputfile	Retrieve name of file to write through dialog box.
wk1read	Read a Lotus 1-2-3 WK1 spreadsheet file.
wk1write	Write a Lotus 1-2-3 WK1 spreadsheet file.

8.2 MATLAB Commands and Functions

This section of the guide contains a description of the operators in MATLAB, followed by detailed descriptions of all MATLAB commands and functions, in alphabetical order. Online help is also available – if you execute `help` on one of the function names, MATLAB displays an abbreviated version of the reference entry.

Each reference entry contains a purpose, synopsis, description, examples, algorithm, references, and other entries to see.

Arithmetic Operators + - * / \ ^ '

Purpose Matrix and array arithmetic.

Synopsis

```
A+B
A–B
A*B    A.*B
A/B    A./B
A\B    A.\B
A^B    A.^B
A'     A.'
```

Description MATLAB has two different types of arithmetic operations. Matrix arithmetic operations are defined by the rules of linear algebra. Array arithmetic operations are carried out element-by-element. The period or decimal point character (.) distinguishes the array operations from the matrix operations. However, since the matrix and array operations are the same for addition and subtraction, the character pairs `.+` and `.–` are not used.

`+` Addition. `A+B` adds `A` and `B`. `A` and `B` must have the same dimension, unless one is a scalar. A scalar can be added to a matrix of any dimension.

`–` Subtraction. `A–B` subtracts `B` from `A`. `A` and `B` must have the same dimensions, unless one is a scalar. A scalar can be subtracted from a matrix of any dimension.

`*` Matrix multiplication. `A*B` is the linear algebraic product of the matrices `A` and `B`. The number of columns of `A` must equal the number of rows of `B`, unless one of them is a scalar. A scalar can multiply a matrix of any dimension.

`.*` Array multiplication. `A.*B` is the element-by-element product of the arrays `A` and `B`. `A` and `B` must have the same dimension, unless one of them is a scalar.

`\` Backslash or matrix left division. If `A` is a square matrix, `A\B` is roughly the same as `inv(A)*B`, except it is computed in a different way. If `A` is an `n`-by-`n` matrix and `B` is a column vector with `n` components, or a matrix with several such columns, then `X = A\B` is the solution to the equation $AX = B$ computed by Gaussian elimination (see "Algorithm" for details). A warning message prints if `A` is badly scaled or nearly singular.

If `A` is an `m`-by-`n` matrix with `m ~= n` and `B` is a column vector with `m` components, or a matrix with several such columns, then `X = A\B` is the solution in the least squares sense to the under- or overdetermined system of equations $AX = B$. The effective rank, `k`, of `A`, is determined from the QR decomposition with pivoting (see "Algorithm" for details). A solution `X` is computed which has at most `k` nonzero components per column. If `k < n`, this is usually not be the same solution as `pinv(A)*B`, which is the least squares solution with the smallest `residual norm`, $\|AX-B\|$.

Array left division. `A.\B` is the matrix with elements `B(i,j)/A(i,j)`. `A` and `B` must have the same dimensions, unless one of them is a scalar.

Slash or matrix right division. `B/A` is roughly the same as `B*inv(A)`. More precisely, `B/A = (A'\B')'`. See `\`.

Array right division. `A./B` is the matrix with elements `A(i,j)/B(i,j)`. `A` and `B` must have the same dimensions, unless one of them is a scalar.

Matrix power. `X^p` is `X` to the power `p`, if `p` is a scalar. If `p` is an integer, the power is computed by repeated multiplication. If the integer is negative, `X` is inverted first. For other values of `p`, the calculation involves eigenvalues and eigenvectors, such that if `[V,D] = eig(X)`, then `X^p = V*D.^p/V`.

If `x` is a scalar and `P` is a matrix, `x^P` is `x` raised to the matrix power `P` using eigenvalues and eigenvectors. `X^P`, where `X` and `P` are both matrices, is an error.

Array power. `A.^B` is the matrix with elements `A(i,j)` to the `B(i,j)` power. `A` and `B` must have the same dimensions, unless one of them is a scalar.

Matrix transpose. `A'` is the linear algebraic transpose of `A`. For complex matrices, this involves the complex conjugate transpose.

Array transpose. `A.'` is the array transpose of `A`. For complex matrices, this does not involve conjugation.

Example Here are two vectors, and the results of various matrix and array operations on them, printed with `format rat`.

Matrix Operations		Array Operations	
x	1 2 3	y	4 5 6
x'	1 2 3	y'	4 5 6
x+y	5 7 9	x–y	–3 –3 –3
x + 2	3 4 5	x–2	–1 0 1
x * y	Error	x.*y	4 10 18
x'*y	32	x'.*y	Error
x*y'	4 5 6 8 10 12 12 15 18	x.*y'	Error
x*2	2 4 6	x.*2	2 4 6
x\y	16/7	x.\y	4 5/2 2
2\x	1/2 1 3/2	2./x	2 1 2/3
x/y	0 0 1/6 0 0 1/3 0 0 1/2	x./y	1/4 2/5 1/2
x/2	1/2 1 3/2	x./2	1/2 1 3/2
x^y	Error	x.^y	1 32 729
x^2	Error	x.^2	1 4 9
2^x	Error	2.^x	2 4 8

Matrix Operations		Array Operations
(x+i*y)'	1 − 4i 2 − 5i	3 − 6i
(x+i*y).'	1 + 4i 2 + 5i	3 + 6i

Algorithm The specific algorithm used for solving the simultaneous linear equations denoted by `X = A\B` and `X = B/A` depends upon the structure of the coefficient matrix `A`.

- If `A` is a triangular matrix, or a permutation of a triangular matrix, then `X` can be computed quickly by a permuted backsubstitution algorithm. The check for triangularity is done for full matrices by testing for zero elements and for sparse matrices by accessing the sparse data structure. Most nontriangular matrices are detected almost immediately, so this check requires a negligible amount of time.
- If `A` is symmetric, or Hermitian, and has positive diagonal elements, then a Cholesky factorization is attempted (see `chol`). If `A` is sparse, a symmetric minimum degree preordering is applied (see `symmmd` and `spparms`). If `A` is found to be positive definite, the Cholesky factorization attempt is successful and requires less than half the time of a general factorization. Nonpositive definite matrices are usually detected almost immediately, so this check also requires little time. If successful, the Cholesky factorization is

  ```
  A = R'*R
  ```

  where `R` is upper triangular. The solution `X` is computed by solving two triangular systems,

  ```
  X = R\(R'\B)
  ```

- If `A` is square, but not a permutation of a triangular matrix, or is not Hermitian with positive elements, or the Cholesky factorization fails, then a general triangular factorization is computed by Gaussian elimination with partial pivoting (see `lu`). If `A` is sparse, a nonsymmetric minimum degree preordering is applied (see `colmmd` and `spparms`). This results in

  ```
  A = L*U
  ```

  where `L` is a permutation of a lower triangular matrix and `U` is an upper triangular matrix. Then `X` is computed by solving two permuted triangular systems.

  ```
  X = U\(L\B)
  ```

- If `A` is not square and is full, then Householder reflections are used to compute an orthogonal-triangular factorization.

  ```
  A*P = Q*R
  ```

  where `P` is a permutation, `Q` is orthogonal and `R` is upper triangular (see `qr`). The least squares solution `X` is computed with

  ```
  X = P*(R\(Q'*B)
  ```

- If `A` is not square and is sparse, then the augmented matrix

  ```
  S = [c*I A; A' 0]
  ```

is formed (see `spaugment`). The default value of the residual scaling factor is `c = max(max(abs(A)))/1000` (see `spparms`). The least squares solution `X` and the residual

```
R = B–A*X
```

are computed by solving

```
S * [R/c; X] = [B; 0]
```

with minimum degree preordering and sparse Gaussian elimination with numerical pivoting.

The various matrix factorizations are computed by MATLAB implementations of the algorithms employed by LINPACK routines `ZGECO`, `ZGEFA` and `ZGESL` for square matrices and `ZQRDC` and `ZQRSL` for rectangular matrices. See the *LINPACK User's Guide* for details.

Diagnostics From matrix division, if a square `A` is singular:

```
Matrix is singular to working precision.
```

From element-wise division, if the divisor has zero elements:

```
Divide by zero.
```

On machines without IEEE arithmetic, like the VAX, the above two operations generate the error messages shown. On machines with IEEE arithmetic, only warning messages are generated. The matrix division returns a matrix with each element set to `Inf`; the element-wise division produces `NaN`s or `Inf`s where appropriate.

If the inverse was found, but is not reliable:

```
Warning: Matrix is close to singular or badly scaled.
    Results may be inaccurate.  RCOND = xxx
```

From matrix division, if a nonsquare `A` is rank deficient:

```
Warning: Rank deficient, rank = xxx tol = xxx
```

See Also `det`, `inv`, `lu`, `orth`, `qr`, `rcond`, `rref`

References

[1] Dongarra, J.J., J.R. Bunch, C.B. Moler, and G.W. Stewart, *LINPACK User's Guide*, SIAM, Philadelphia, 1979.

Relational Operators < ≤ > ≥ == ~=

Purpose Relational operations.

Synopsis

```
A < B
A > B
A <= B
A >= B
A == B
A ~= B
```

Description The relational operators are <, ≤, >, ≥, ==, and ~=. Relational operators perform element-by-element comparisons between two matrices. They return a matrix of the same size, with elements set to 1 where the relation is true, and elements set to 0 where it is not.

The operators <, ≤, >, and ≥ use only the real part of their operands for the comparison. The operators == and ~= test real and imaginary parts.

The relational operators have precedence midway between the logical operators (except ~) and the arithmetic operators.

To test if two strings are equivalent, use `strcmp`, which allows vectors of dissimilar length to be compared.

Examples If one of the operands is a scalar and the other a matrix, the scalar expands to the size of the matrix. For example, the two pairs of statements:

```
X = 5; X >= [1 2 3; 4 5 6; 7 8 10]
X = 5*ones(3,3); X >= [1 2 3; 4 5 6; 7 8 10]
```

produce the same result:

```
X =
     1     1     1
     1     1     0
     0     0     0
```

See Also `&`, `|`, `~`, `all`, `any`, `find`, `strcmp`

Logical Operators & | ~

Purpose Logical operations.

Synopsis

```
A & B
A | B
~A
```

Description The symbols `&`, `|`, and `~` are the logical operators AND, OR, and NOT. They work element-wise on matrices, with 0 representing FALSE and anything nonzero representing TRUE. `A & B` does a logical AND, `A | B` does a logical OR, and `~A` complements the elements of `A`. The function `xor(A,B)` implements the exclusive OR operation.

Inputs		AND	OR	XOR
0	0	0	0	0
0	1	0	1	1
1	0	0	1	1
1	1	1	1	0

The logical operators `&` and `|` have the lowest precedence, with arithmetic operators and relational operators being higher. The logical operator `~` has the same precedence as the arithmetic operators.

The precedence for the logical operators with respect to each other is:

1. NOT has the highest precedence.

2. AND and OR have equal precedence, and are evaluated from left to right.

Examples Here are two scalar expressions that illustrate precedence relationships for arithmetic, relational, and logical operators:

```
1 & 0 + 3
3 > 4 & 1
```

They evaluate to 1 and 0 respectively, and are equivalent to:

```
1 & (0 + 3)
(3 > 4) & 1
```

Here are two examples that illustrate the precedence of the logical operators to each other:

```
1 | 0 & 0 = 0
0 & 0 | 1 = 1
```

See Also <, <=, >, >=, ==, ~=, `all`, `any`, `find`, `xor`

Special Characters [ ] ( ) = ' . , ; % !

Purpose Special characters.

Synopsis `[ ] ( ) = ' . , ; % !`

Description

`[ ]` Brackets are used to form vectors and matrices. `[6.9 9.64 sqrt(-1)]` is a vector with three elements separated by blanks. `[6.9, 9.64, i]` is the same thing. `[1+j 2-j 3]` and `[1 +j 2 -j 3]` are not the same. The first has three elements, the second has five.

`[11 12 13; 21 22 23]` is a 2-by-3 matrix. The semicolon ends the first row.

Vectors and matrices can be used inside `[ ]` brackets. `[A B;C]` is allowed if the number of rows of `A` equals the number of rows of `B` and the number of columns of `A` plus the number of columns of `B` equals the number of columns of `C`. This rule generalizes in a hopefully obvious way to allow fairly complicated constructions.

`A = [ ]` stores an empty matrix in `A`.

For the use of `[` and `]` on the left of an "=" in multiple assignment statements, see `lu`, `eig`, `svd`, and so on.

`( )` Parentheses are used to indicate precedence in arithmetic expressions in the usual way. They are used to enclose arguments of functions in the usual way. They are also used to enclose subscripts of vectors and matrices in a manner somewhat more general than usual. If `X` and `V` are vectors, then `X(V)` is `[X(V(1)), X(V(2)), ..., X(V(n))]`. The components of `V` are rounded to nearest integers and used as subscripts. An error occurs if any such subscript is less than 1 or greater than the dimension of X. Some examples are

- X(3) is the third element of X.
- X([1 2 3]) is the first three elements of X.
- X([sqrt(2) sqrt(3) 4*atan(1)]) is also the first three elements of X.

If X has n components, X(n:-1:1) reverses them. The same indirect subscripting works in matrices. If V has m components and W has n components, then A(V,W) is the m-by-n matrix formed from the elements of A whose subscripts are the elements of V and W. For example, A([1,5],:) = A([5,1],:) interchanges rows 1 and 5 of A.

= Used in assignment statements. == is the relational EQUALS operator. See the Relational Operators.

' Matrix transpose. X' is the complex conjugate transpose of X. X.' is the nonconjugate transpose.

Quote. 'any text' is a vector whose components are the ASCII codes for the characters. A quote within the text is indicated by two quotes.

. Decimal point. 314/100, 3.14 and .314e1 are all the same.

Element-by-element operations are obtained using .* , .^ , ./, or .\. See the Arithmetic Operators.

Three or more points at the end of a line indicate continuation.

, Comma. Used to separate matrix subscripts and function arguments. Used to separate statements in multistatement lines. For multi-statement lines, the comma can be replaced by a semicolon to suppress printing.

; Semicolon. Used inside brackets to end rows. Used after an expression or statement to suppress printing or separate statements.

% Percent. The percent symbol denotes a comment; it indicates a logical end of line. Any following text is ignored.

! Exclamation point. Indicates that the rest of the input line is issued as a command to the operating system.

See Also Arithmetic, relational, and logical operators.

Colon :

Purpose Create vectors, matrix subscripting, and `for` iterations.

Description The colon is one of the most useful operators in MATLAB. It can create vectors, subscript matrices, and specify `for` iterations.

The colon operator uses the following rules to create regularly spaced vectors:

`j:k`	is the same as `[j,j+1,...,k]`
`j:k`	is empty if `j > k`
`j:i:k`	is the same as `[j,j+i,j+2i, ...,k]`
`j:i:k`	is empty if `i > 0` and `j > k` or if `i < 0` and `j < k`

Below are the definitions that govern the use of the colon to pick out selected rows, columns, and elements of vectors and matrices:

`A(:,j)`	is the `j`-th column of `A`
`A(i,:)`	is the `i`-th row of `A`
`A(:,:)`	is the same as `A`
`A(j:k)`	is `A(j), A(j+1),...,A(k)`
`A(:,j:k)`	is `A(:,j), A(:,j+1),...,A(:,k)`
`A(:)`	is all the elements of `A`, regarded as a single column. On the left side of an assignment statement, `A(:)` fills `A`, preserving its shape from before.

Examples Using the colon with integers,

```
D = 1:4
```

results in

```
D =
     1     2     3     4
```

Using two colons to create a vector with arbitrary real increments between the elements,

```
E = 0:.1:.5
```

results in

```
E =
     0    0.1000    0.2000    0.3000    0.4000    0.5000
```

See Also `for`, `linspace`, `logspace`, `reshape`

abs

Purpose Absolute value and string to numeric conversion.

Synopsis `Y = abs(X)`

Description `abs(X)` returns the absolute value for each element of `X`.

If `X` is complex, `abs(X)` returns the complex modulus (magnitude):

```
abs(X) = sqrt(real(X).^2+imag(X).^2)
```

If `X` is a MATLAB string, `abs(X)` returns the numeric values of the ASCII characters in the string. The way the string prints changes; the internal representation does not.

Examples

```
abs(-5) = 5
abs(3+4i) = 5
abs('3+4i') = [51  32  43  32  52  105]
```

See Also `angle`, `setstr`, `sign`, `strings`, `unwrap`

acos, acosh

Purpose Inverse cosine and inverse hyperbolic cosine.

Synopsis

```
Y = acos(X)
Y = acosh(X)
```

Description `acos` and `acosh` operate element-wise on matrices. Their domains and ranges include complex values. All angles are in radians.

`acos(X)` returns the inverse cosine for each element of `X`. For real elements of `X` in the range [-1,1], `acos(X)` is real and in the range [0,`pi`]. For real elements of `X` outside the range [-1,1], `acos(X)` has nonzero imaginary parts.

`acosh(X)` returns the inverse hyperbolic cosine for each element of `X`.

Examples Graph the inverse cosine function for the range $-1 \le x \le 1$.

```
x = -1:.05:1;
plot(x, acos(x))
```

Algorithm

$$\cos^{-1}(z) = -i \log\left[z + i\left(1 - z^2\right)^{1/2}\right]$$

$$\cosh^{-1}(z) = \log\left[z + \left(z^2 - 1\right)^{1/2}\right]$$

See Also `atan2`, `cos`, `exp`, `funm`, `log`, `sin`, `sqrt`, `tan`

acot, acoth

Purpose Inverse cotangent and inverse hyperbolic cotangent.

Synopsis
```
Y = acot(X)
Y = acoth(X)
```

Description `acot` and `acoth` operate element-wise on matrices. Their domains and ranges include complex values. All angles are in radians.

`acot(X)` returns the inverse cotangent for each element of `X`.

`acoth(X)` returns the inverse hyperbolic cotangent for each element of `X`.

Algorithm

$$\cot^{-1}(z) = \tan^{-1}\left(\frac{1}{z}\right)$$

$$\coth^{-1}(z) = \tanh^{-1}\left(\frac{1}{z}\right)$$

See Also `cot`, `coth`

---

acsc, acsch

Purpose Inverse cosecant and inverse hyperbolic cosecant.

Synopsis
```
Y = acsc(X)
Y = acsch(X)
```

Description `acsc` and `acsch` operate element-wise on matrices. Their domains and ranges include complex values. All angles are in radians.

`acsc(X)` returns the inverse cosecant for each element of `X`.

`acsch(X)` returns the inverse hyperbolic cosecant for each element of `X`.

Algorithm

$$\csc^{-1}(z) = \sin^{-1}\left(\frac{1}{z}\right)$$

$$\operatorname{csch}^{-1}(z) = \sinh^{-1}\left(\frac{1}{z}\right)$$

See Also `csc`, `csch`

---

all

Purpose Test arrays for logical conditions.

Synopsis
```
Y = all(X)
```

Description `all(X)`, where `X` is a vector, returns 1 if *all* of the elements of `X` are nonzero. It returns 0 if any single element is zero. For matrices, `all(X)` operates on the columns of `X`, returning a row vector of 1s and 0s.

Examples This function is particularly useful in `if` statements

```
if all(A < 0.5)
    do something
end
```

because an `if` wants to respond to a single condition, not a vector of possibly conflicting suggestions. Applying the function twice, as in `all(all(A))`, always reduces the matrix to a scalar condition.

See Also `&`, `|`, `~`, `any`

---

angle

Purpose Phase angle.

Synopsis `P = angle(Z)`

Description `angle(Z)` returns the phase angles, in radians, of the elements of complex matrix `Z`. The angles lie between `–pi` and `pi`.

For complex `Z`, the magnitude and phase are given by

```
r = abs(Z)
theta = angle(Z)
```

and the statement

```
Z = r.*exp(i*theta)
```

converts back to the original complex `Z`.

Algorithm `angle` can be expressed as:

```
angle(Z) = imag(log(Z)) = atan2(imag(Z),real(Z))
```

See Also `abs`, `unwrap`

---

ans

Purpose The most recent answer.

Synopsis `ans`

Description `ans` is the variable created automatically when no output argument is specified.

Examples The statement

```
2+2
```

is the same as

```
ans = 2+2
```

---

any

Purpose Test arrays for logical conditions.

Synopsis `Y = any(X)`

Description `any(X)`, where `X` is a vector, returns 1 if *any* of the elements of `X` are nonzero. It returns 0 if all the elements are zero. For matrices, `any(X)` operates on the columns of `X`, returning a row vector of 1s and 0s.

Examples This function is particularly useful in `if` statements

```
if any(A < 0.5)
    do something
end
```

because an `if` wants to respond to a single condition, not a vector of possibly conflicting suggestions. Applying the function twice, as in `any(any(A))`, always reduces the matrix to a scalar condition.

See Also `&`, `|`, `~`, `all`

---

asec, asech

Purpose Inverse secant and inverse hyperbolic secant.

Synopsis
```
Y = asec(X)
Y = asech(X)
```

Description `asec` and `asech` operate element-wise on matrices. Their domains and ranges include complex values. All angles are in radians.

`asec(X)` returns the inverse secant for each element of `X`.

`asech(X)` returns the inverse hyperbolic secant for each element of `X`.

Algorithm

$$\sec^{-1}(z) = \cos^{-1}\left(\frac{1}{z}\right)$$

$$\operatorname{sech}^{-1}(z) = \cosh^{-1}\left(\frac{1}{z}\right)$$

See Also `sec`, `sech`

---

asin, asinh

Purpose Inverse sine and inverse hyperbolic sine.

Synopsis
```
Y = asin(X)
Y = asinh(X)
```

Description `asin` and `asinh` operate element-wise on matrices. Their domains and ranges include complex values. All angles are measured in radians.

`asin(X)` returns the inverse sine for each element of `X`. For real elements of `X` in the range [-1,1], `asin(X)` is in the range [-pi/2,pi/2]. For real elements of `X` outside the range [-1,1], `asin(X)` has nonzero imaginary parts.

asinh(X) returns the inverse hyperbolic sine for each element of X.

Examples Graph the inverse sine function for the range $-1 \le x \le 1$.

```
x = -1:.05:1;
plot(x, asin(x))
```

Algorithm

$$\sin^{-1}(z) = -i \ \log\left[iz + \left(1 - z^2\right)^{½}\right]$$

$$\sinh^{-1}(z) = \log\left[z + \left(z^2 + 1\right)^{½}\right]$$

See Also atan2, cos, exp, funm, log, sin, sqrt, tan

---

atan, atanh

Purpose Inverse tangent and inverse hyperbolic tangent.

Synopsis

```
Y = atan(X)
Y = atanh(X)
```

Description atan and atanh operate element-wise on matrices. Their domains and ranges include complex values. All angles are in radians.

atan(X) returns the inverse tangent for each element of X. For real elements of X in the range [-1,1], atan(x) is in the range [–pi/4,pi/4]. atan(X) is always real.

atanh(X) returns the inverse hyperbolic tangent for each element of X.

Examples Graph the inverse tangent function for the range $-2 \le x \le 2$.

```
x = -2:.05:2;
plot(x, atan(x))
```

Algorithm

$$\tan^{-1}(z) = \log\left(\frac{i+z}{i-z}\right)$$

$$\tanh^{-1}(z) = \frac{1}{2}\log\left(\frac{i+z}{i-z}\right)$$

See Also atan2, cos, exp, funm, log, sin, sqrt, tan

---

atan2

Purpose Four-quadrant inverse tangent.

Synopsis

```
P = atan2(Y,X)
```

Description `atan2(Y,X)` returns a matrix `P` the same size as `X` and `Y` containing the element-by-element, four-quadrant arctangent of the real parts of `Y` and `X`. Any imaginary parts are ignored.

Elements of `P` are in the interval `[-pi, pi]`. The specific quartile is determined by `sign(Y)` and `sign(X)`:

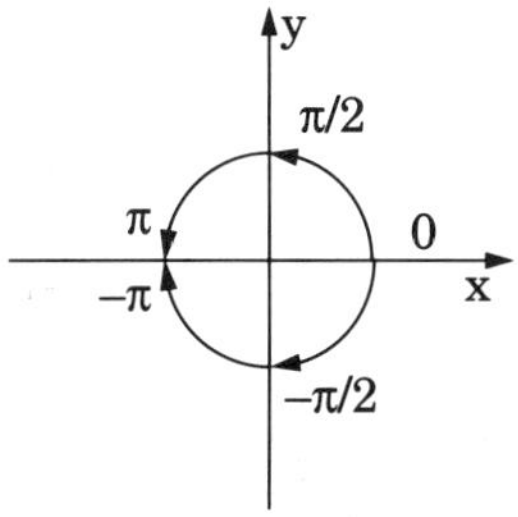

This contrasts with the result of `atan(Y/X)`, which is limited to the interval `[-pi/2, pi/2]`, or the right side of this diagram.

Examples Any complex number $z = x + iy$ can be written in polar coordinates with

```
r = abs(z)
theta = atan2(imag(z),real(z))
```

Then `z` is equal to

```
r *exp(i *theta)
```

This is a common operation, so MATLAB provides a function, `angle(z)`, that simply computes `atan2(imag(z),real(z))`.

See Also `abs`, `angle`, `cos`, `funm`, `sin`, `tan`

---

auread

Purpose Read μ-law encoded audio file.

Synopsis `Y = auread('`*`filename`*`')`

Description `Y = auread('`*`filename`*`')` reads the audio file in *`filename`* and converts the data from μ-law encoded bytes to a signal in the range $-1 \le Y \le 1$.

`auread` reads the data from the specified file as 8-bit, unsigned character data. The high-order bit is a sign bit, where 1 represents positive data and 0 represents negative data. The remaining seven bits represent signal magnitude, and are inverted (1's complement).

See Also `auwrite`

---

auwrite

Purpose Write μ-law encoded audio file.

Synopsis `auwrite(Y,'`*`filename`*`')`

Description auwrite(Y,'*filename*') converts the file Y to μ-law encoded bytes and writes it to the audio file *filename*. On UNIX machines, if *filename* is not specified, auwrite uses /dev/audio on either the local machine, or on the remote machine specified by the environment variable DISPLAY.

See Also auread

---

axes

Purpose Place axes at specified position by creating axes graphics object.

Synopsis

```
h = axes
axes(h)
h = axes('PropertyName',PropertyValue,...)
```

Description axes is a low-level function for creating axes objects. *Axes objects* are children of figure objects and parents of image, line, patch, surface, and text objects. Axes objects define a frame of reference that orients their children objects within the figure.

h = axes creates the default full-window axes and returns its handle.

axes(h) makes the axes with handle h the current axes.

h = axes('*PropertyName*',*PropertyValue*,...) is an object creation function that accepts property name/property value pairs as input arguments. These properties, which control various aspects of axes objects, are described under "Object Properties." You can also set and query property values after creation using the set and get functions.

Specify default axes properties at the axes' parent level, that is, at the figure object. To do so, call the set function, supplying as arguments the handle of the parent figure, a default name string, and the desired default value. Construct the default name string by prepending the string 'DefaultAxes' to the desired axes property name. For example

```
set(fig_handle,'DefaultAxesColor','red')
```

sets the default axes rectangle color to red for all axes created in the figure with handle fig_handle.

The axis (not axes) function provides simplified access to commonly used properties that control the scaling and appearance of axes.

Use gca (get current axes) to obtain the handle of the current axes.

Object Properties This section lists property names along with the type of values each accepts.

AspectRatio — Aspect ratio for two-dimensional axes.

A two-element vector, [axis_ratio data_ratio], where

- axis_ratio is the ratio of the length of the horizontal axis to the length of the vertical axis (width divided by height). MATLAB creates the largest axes with this ratio that fits into the rectangle defined by Position.

- `data_ratio` is the ratio of the length of a data unit along the horizontal axis to the length of a data unit along the vertical axis. To create axes with this `data_ratio`, MATLAB changes the limits of one axis while maintaining the ratio specified in `axis_ratio`. This change does not affect the corresponding limits property (`XLim` or `YLim`).

Specify the ratios as any number in the range `[0,Inf]`. The default for both is `NaN`, which specifies no ratio. By default, MATLAB changes either ratio in order create an axes that best fills the figure window.

`Box`

Axes box mode.

`on`	Enclose graphics area in box (2-D) or cube (3-D).
`off`	(Default.) Do not display enclosing box.

`ButtonDownFcn`

Callback string, object selection.

Any legal MATLAB expression, including the name of an M-file or function. When you select the object, the string is passed to the `eval` function to execute the specified function. Initially the empty matrix.

`Children`

Children of axes.

A read-only vector containing the handles of all objects displayed within the axes. The children objects of axes can be images, lines, patches, surfaces, and text.

`CLim`

Color limits.

A two-element vector `[cmin cmax]` that determines how MATLAB maps data values to the colormap. `cmin` is the data value to map to the first colormap entry, and `cmax` is the data value to map to the last colormap entry. By default `Clim` is [0 1], mapping the full range of the colormap to the full range of the data.

Data values less than `cmin` are mapped to `cmin` and data values greater than `cmax` are mapped to `cmax`. `NaN`s are clipped by making them transparent. `CLim` values outside the range of the data cause MATLAB to use a limited portion of the colormap.

`CLim` affects the rendering of surface and patch objects, but not images, lines, or text.

`CLimMode`

Color limits mode.

`auto`	(Default.) MATLAB calculates color limits that span the full range of the data of the axes' children.

	`manual`	Color limits do not automatically change (see the `CLim` property). Setting values for `CLim` sets this property to `manual`.
`Clipping`	Data clipping.	
	`on`	(Default.) No effect for axes objects.
	`off`	No effect for axes objects.
`Color`	Color of the axes rectangle.	
	`ColorSpec`	A three-element RGB vector or one of MATLAB's predefined names. See the `ColorSpec` reference page for more information on specifying color.
	`none`	(Default.) The same as the figure background color.

`ColorOrder`

Axes color order.

An `m`-by-3 matrix of RGB values. If you do not specify a line color with `plot` and `plot3`, these functions cycle through the `ColorOrder` to obtain the color for each line plotted. By default, `ColorOrder` contains the first six colors in MATLAB's predefined color palette:

1. yellow
2. magenta
3. cyan
4. red
5. green
6. blue

`CurrentPoint`

3-D coordinates for a pair of points in the axes' data space.

A 2-by-3 matrix that specifies the end points of a line, in 3-D coordinates, that extends from the front to the back of the axes volume. The format of the matrix is

```
[xback  yback  zback;  xfront  yfront  zfront]
```

These coordinates are always relative to the axes volume, even if the pointer is outside the axes. All measurements are in units specified by the `Units` property.

`DrawMode`	Draw mode.	
	`normal`	(Default.) Draw objects from back to front based on the current view.
	`fast`	Draws objects in the order in which you originally specified them. This disables the three-dimensional sorting usually per-

	formed by MATLAB, resulting in faster rendering.
`FontAngle`	Italics for axes text. `normal` (Default.) Regular font angle. `italic` Italics. `oblique` Italics on some systems.
`FontName`	Font family. A string specifying the name of the font to use for axes tick labels. To display and print properly, this must be a font that your system supports. Axes labels do not display in a new font until you manually redisplay them (by setting the `XLabel`, `YLabel`, and `ZLabel` properties). The default font is Helvetica.
`FontSize`	Font point size. An integer specifying the font size, in points, to use for axes labels and titles. The default point size is 12.
`FontWeight`	Bolding for axes text. `normal` (Default.) Regular font weight. `bold` Bold weight.
`GridLineStyle`	Grid line style. A string specifying the line type for grid lines. The string consists of a single character, in quotes, specifying solid lines (–), dashed lines (– –), dotted lines(:), or dash-dot lines (–.). The default grid line style is dotted.
`Interruptible`	Callback interruptibility. `yes` The callback specified by `ButtonDownFcn` is interruptible by other callbacks. `no` (Default.) The `ButtonDownFcn` callback is not interruptible.
`LineStyleOrder`	Axes line style order. A string specifying line styles in the order used to plot multiple lines in the axes. Specify up to four line styles in the format `'linestyle1|linestyle2...'` where each *`linestyle`* is a character string specifying solid lines (–), dashed lines (– –), dotted lines(:), or dash-dot lines(–.). The default line style order is solid lines only. Some plotting functions, such as `plot`, overrule this property with their own default line style order.

`LineWidth`	Width of *x*-, *y*-, and *z*-axis lines. The width, in points, of the lines that represent each axis. The default line width is 0.5 points.
`NextPlot`	Axes handling for subsequent plots. `new` Create a new axes before drawing. `add` Add new objects to the current axes. Setting `hold` to `on` sets `NextPlot` to `add`. `replace` (Default.) Destroy the current axes (and its contents) and create a new axes at the same position before drawing. Setting `hold` to `off` sets `NextPlot` to `replace`. This property is used by the built-in high-level graphing functions `plot`, `plot3`, `fill`, `fill3`, and the M-file graphing functions `mesh`, `surf`, `bar`, and so on. The M-file `newplot` is a preamble for handling the `NextPlot` property. M-file graphing functions like `mesh`, `surf`, and `bar` call `newplot` to control this property before drawing their respective graphs. When creating M-files that implement graphing commands, use `newplot` at the beginning of the file. See the M-file `pcolor` for an example. Also see the `NextPlot` property of figure objects.
`Parent`	Handle of the axes' parent object. A read-only property that contains the handle of the axes' parent object. The parent of an axes object is the figure in which it is displayed. The utility function `gcf` also returns the handle of the current axes' parent figure.
`Position`	Size and position of axes. A four-element vector, `[left bottom width height]`, where `left` and `bottom` are the distance from the lower-left corner of the figure window to the lower-left corner of the axes. `width` and `height` are the dimensions of the axes rectangle. All measurements are in units specified by the `Units` property.
`TickDir`	Tick mark direction `in` Tick marks are directed inward from the axis lines (default for 2-D plotting). `out` Tick marks are directed outward from the axis lines (default for 3-D plotting functions unless `hold` is `on`).
`TickLength`	Tick mark length. A two-element vector `[2Dlength 3Dlength]` that specifies the length of the axes' tick marks, where

- `2Dlength` is the length of tick marks used for two-dimensional views.
- `3Dlength` is the length of tick marks used for three-dimensional views.

Each element is in units normalized relative to the dimensions of the rectangle defined by the axes. Specifying a tick length of 0.1 draws a tick mark that is 1/10 of the width or height of the rectangle. The default vector is [0.01 0.025].

`Title` Axes title.

The handle of the text object that defines the axes title. First create the text object in order to obtain its handle. The following statement performs both steps:

```
set(gca,'title',text(0,0,'axes title'))
```

While the `text` function requires location data, it is not used to place the text. Instead, MATLAB centers the string `'axes title'` above the axes.

You can also define a text object at an arbitrary location and pass its handle directly. In this case MATLAB moves the text string to the correct location for an axes title. The `title` function provides a simpler means to create a title by encapsulating this procedure.

`Type` Type of graphics object.

A read-only string; always `'axes'` for an axes object.

`Units` Unit of measurement.

`pixels`	Screen pixels.
`normalized`	(Default.) Normalized coordinates. The lower-left corner of the figure window maps to (0,0) and the upper-right corner to (1.0,1.0).
`inches`	Inches.
`cent`	Centimeters.
`points`	Points. Each point is equivalent to 1/72 of an inch.

If you change the value of `Units`, return it to its default value after completing your computation so as not to affect other functions that assume `Units` is set to the default value.

`UserData` User-specified data.

Any matrix you want to associate with the axes object. The object does not use this data, but you can retrieve it using the `get` command.

`View` Viewpoint of axes.

A two-element vector, `[az,el]`, that establishes a viewpoint used to transform three-dimensional plots into the two-dimensional space of the screen. The *viewpoint* is the location of the eye of an observer looking at the three-dimensional plot, where

- `az` is the azimuth. The *azimuth* revolves around the z-axis, with positive values causing the viewpoint to rotate in a counter-clockwise direction.
- `el` is the elevation. The *elevation* specifies an angle above or below the object. Positive values of elevation cause the viewpoint to move above the object (that is, to look down on it); negative values cause the viewpoint to move below the object.

Specify each element in degrees. Changing the `View` property alters the `XForm` property's transformation matrix. The default `View` vector is [0 90].

`Visible` Axes visibility.

`on`	(Default.) Axes lines, tick marks, and labels are visible on the screen.
`off`	Axes lines, tick marks, and labels are not drawn.

The children of the axes are not affected by the axes' `Visible` property.

`XForm` View transformation matrix.

A 4-by-4 matrix that defines the transformation from three-dimensional plots to the two-dimensional screen. Specify the transformation matrix directly with this property, or indirectly by specifying the azimuth and elevation of the viewpoint using the `View` property.

Changing the `View` property changes `XForm` as well. `XForm`, however, can define views that you cannot define using only azimuth and elevation (such as perspective views). It is also possible to specify scaling and translations in the same matrix. Such transforms are discussed in most books on computer graphics. The default `XForm` matrix is the 4-by-4 identity matrix.

Properties That Control the x-Axis

`XColor`

Color of *x*-axis.

A three-element vector specifying an RGB triple, or a predefined MATLAB color string. This property determines the display color for the *x*-axis, tick marks, tick mark labels, and the *x*-axis grid lines.The default axis color is white. See the *ColorSpec* reference page for details on specifying colors.

`XDir`

Direction of increasing *x* values.

`normal`	(Default.) *X* values increase from left to right (right-handed coordinate system).
`reverse`	*X* values increase from right to left.

`XGrid`

x-axis gridline mode.

`on`	MATLAB draws grid lines perpendicular to the *x*-axis at each tick mark (i.e., along lines of constant *x* values).
`off`	(Default.) Grid lines are not drawn.

`XLabel`

x-axis label.

The handle of the text object used to label the *x*-axis. First create the text object in order to obtain its handle. The following statement performs both steps:

```
set(gca,'xlabel',text(0,0,'axis label'))
```

While the `text` function requires location data, it is not used to place the text. Instead, MATLAB places the string `'axis label'` appropriately for an *x*-axis label.

You can also define a text object at an arbitrary location and pass its handle directly. In this case MATLAB moves the text string to the correct location for an axis label. The `xlabel` function provides a simpler means to label the *x*-axis by encapsulating this procedure.

`XLim`

x-axis limits.

A two-element vector `[xmin xmax]` that specifies the minimum and maximum *x*-axis values, where

- `xmin` is the minimum *x*-axis value.
- `xmax` is the maximum *x*-axis value.

Changing `XLim` affects the scale of the *x*-dimension as well as the placement of labels and tick marks on the *x*-axis. The default for `XLim` is [0 1].

`XLimMode` — x-axis limits mode.

`auto`	(Default.) MATLAB calculates x-axis limits (`XLim`) that span the `XData` of the axes' children and produce round numbers for the x-axis limits.
`manual`	MATLAB takes x-axis limits from `XLim`; the limits do not depend on the `XData` in the children objects. Setting values for `XLim` sets this property to `manual`.

`XScale` — x-axis scaling.

`linear`	(Default.) Linear scaling for the x-axis.
`log`	Logarithmic scaling for the x-axis.

`XTick` — x-axis tick mark spacing.

A vector of values that correspond to the x-data values at which you want to place tick marks. If you do not want tick marks displayed, set `XTick` to the empty vector, `[]`.

`XTickLabelMode` — x-axis tick mark labeling mode.

`auto`	(Default.) MATLAB calculates x-axis tick labels (`XTickLabels`) that span the `XData` of the axes' children.
`manual`	MATLAB takes x-axis tick labels from `XTickLabels`; it does not depend on the `XData` in the children objects. Setting values for `XTickLabels` sets this property to `manual`.

`XTickLabels` — x-axis tick mark labels.

A matrix of strings to use as labels for tick marks along the x-axis. These labels replace the numeric labels generated by MATLAB. If you do not specify enough text labels for all the tick marks, MATLAB uses all of the labels specified, and then labels the remaining tick marks by reusing the specified labels. The command

```
set(gca,'XTickLabels',['Old Data';'New Data'])
```

labels the first two tick marks on the x-axis `'Old Data'` and `'New Data'` respectively. Each character string must have an equal number of characters because of the way MATLAB stores strings.

This property does not control the number of tick marks or their locations.

`XTickMode` — *x*-axis tick mark mode.

- `auto` — (Default.) MATLAB calculates *x*-axis tick mark spacing (`XTick`) that spans the `XData` of the axes' children.
- `manual` — MATLAB takes *x*-axis tick spacing from `XTick`; it does not depend on the `XData` in the children objects. Setting values for `XTick` sets this property to `manual`.

Properties That Control the y-Axis

`YColor` — Color of *y*-axis.

A three-element vector specifying an RGB triple, or a predefined MATLAB color string. This property determines the display color for the *y*-axis, tick marks, tick mark labels, and the *y*-axis grid lines. The default axis color is white. See the *`ColorSpec`* reference page for details on specifying colors.

`YDir` — Direction of increasing *y* values.

- `normal` — (Default.) *y* values increase from bottom to top (right-handed coordinate system).
- `reverse` — *y* values increase from top to bottom.

`YGrid` — *y*-axis gridline mode.

- `on` — MATLAB draws grid lines perpendicular to the *y*-axis at each tick mark (i.e., along lines of constant *y* values).
- `off` — (Default.) Grid lines are not drawn.

`YLabel` — *y*-axis label.

The handle of the text object used to label the *y*-axis. First create the text object in order to obtain its handle. The following statement performs both steps:

```
set(gca,'ylabel',text(0,0,'axis label'))
```

While the `text` function requires location data, it is not used to place the text. Instead, MATLAB places the string `'axis label'` appropriately for a *y*-axis label.

You can also define a text object at an arbitrary location and pass its handle directly. In this case MATLAB moves the text string to the correct location for an axis label. The `ylabel` function provides a simpler means to label the *y*-axis by encapsulating this procedure.

`YLim`	y-axis limits. A two-element vector `[ymin ymax]` that specifies the minimum and maximum y-axis values, where • `ymin` is the minimum y-axis value. • `ymax` is the maximum y-axis value. Changing `YLim` affects the scale of the y-dimension as well as the placement of labels and tick marks on the y-axis. The default for `YLim` is [0 1].	
`YLimMode`	y-axis limits mode.	
	`auto`	(Default.) MATLAB calculates y-axis limits (`YLim`) that span the `YData` of the axes' children and produce round numbers for the y-axis limits.
	`manual`	MATLAB takes y-axis limits from `YLim`; the limits do not depend on the `YData` in the children objects. Setting values for `YLim` sets this property to `manual`.
`YScale`	y-axis scaling.	
	`linear`	(Default.) Linear scaling for the y-axis.
	`log`	Logarithmic scaling for the y-axis.
`YTick`	y-axis tick mark spacing. A vector of values that correspond to the y-data values at which you want to place tick marks. If you do not want tick marks displayed, set `YTick` to the empty vector, `[]`.	
`YTickLabelMode`	y-axis tick mark labeling mode.	
	`auto`	(Default.) MATLAB calculates y-axis tick labels (`YTickLabels`) that span the `YData` of the axes' children.
	`manual`	MATLAB takes y-axis tick labels from `YTickLabels`; it does not depend on the `YData` in the children objects. Setting values for `YTickLabels` sets this property to `manual`.
`YTickLabels`	y-axis tick mark labels. A matrix of strings to use as labels for tick marks along the y-axis. These labels replace the numeric labels generated by MATLAB. If you do not specify enough text labels for all the tick marks, MATLAB uses all of the labels specified, and	

then labels the remaining tick marks by reusing the specified labels. The command

```
set(gca,'YTickLabels',['Old Data';'New Data'])
```

labels the first two tick marks on the y-axis `'Old Data'` and `'New Data'` respectively. Each character string must have an equal number of characters because of the way MATLAB stores strings.

This property does not control the number of tick marks or their locations.

`YTickMode` — y-axis tick mark mode.

`auto`	(Default.) MATLAB calculates y-axis tick mark spacing (`YTick`) that spans the `YData` of the axes' children.
`manual`	MATLAB takes y-axis tick spacing from `YTick`; it does not depend on the `YData` in the children objects. Setting values for `YTick` sets this property to `manual`.

Properties That Control the z-Axis

`ZColor` — Color of z-axis.

A three-element vector specifying an RGB triple, or a predefined MATLAB color string. This property determines the display color for the z-axis, tick marks, tick mark labels, and the z-axis grid lines. The default axis color is white. See the *`ColorSpec`* reference page for details on specifying colors.

`ZDir` — Direction of increasing z values.

`normal`	(Default.) z values increase pointing out of the screen (two-dimensional) or from bottom to top (three-dimensional), corresponding to a right-handed coordinate system.
`reverse`	z values increase pointing into the screen (two-dimensional) or from top to bottom (three-dimensional).

`ZGrid` — z-axis gridline mode.

`on`	MATLAB draws grid lines perpendicular to the z-axis at each tick mark (i.e., along lines of constant z values).
`off`	(Default.) Grid lines are not drawn.

`ZLabel`	z-axis label.

The handle of the text object used to label the z-axis. First create the text object in order to obtain its handle. The following statement performs both steps:

```
set(gca,'zlabel',text(0,0,'axis label'))
```

While the `text` function requires location data, it is not used to place the text. Instead, MATLAB places the string `'axis label'` appropriately for a z-axis label.

You can also define a text object at an arbitrary location and pass its handle directly. In this case MATLAB moves the text string to the correct location for an axis label. The `zlabel` function provides a simpler means to label the z-axis by encapsulating this procedure.

`ZLim` — z-axis limits.

A two-element vector `[zmin zmax]` that specifies the minimum and maximum z-axis values, where

- `zmin` is the minimum z-axis value.
- `zmax` is the maximum z-axis value.

Changing `ZLim` affects the scale of the z-dimension as well as the placement of labels and tick marks on the z-axis. The default for `ZLim` is [0 1].

`ZLimMode` — z-axis limits mode.

`auto`	(Default.) MATLAB calculates z-axis limits (`ZLim`) that span the `ZData` of the axes' children and produce round numbers for the z-axis limits.
`manual`	MATLAB takes z-axis limits from `ZLim`; the limits do not depend on the `ZData` in the children objects. Setting values for `ZLim` sets this property to `manual`.

`ZScale` — z-axis scaling.

`linear`	(Default.) Linear scaling for the z-axis.
`log`	Logarithmic scaling for the z-axis.

`ZTick` — z-axis tick mark spacing.

A vector of values that correspond to the z-data values at which you want to place tick marks. If you do not want tick marks displayed, set `ZTick` to the empty vector, `[]`.

`ZTickLabelMode`

z-axis tick mark labeling mode.

`auto`	(Default.) MATLAB calculates *z*-axis tick labels (`ZTickLabels`) that span the `ZData` of the axes' children.
`manual`	MATLAB takes *z*-axis tick labels from `ZTickLabels`; it does not depend on the `ZData` in the children objects. Setting values for `ZTickLabels` sets this property to `manual`.

`ZTickLabels`

z-axis tick mark labels.

A matrix of strings to use as labels for tick marks along the *z*-axis. These labels replace the numeric labels generated by MATLAB. If you do not specify enough text labels for all the tick marks, MATLAB uses all of the labels specified, and then labels the remaining tick marks by reusing the specified labels. The command

```
set(gca,'ZTickLabels',['Old Data';'New Data'])
```

labels the first two tick marks on the *z*-axis `'Old Data'` and `'New Data'` respectively. Each character string must have an equal number of characters because of the way MATLAB stores strings.

This property does not control the number of tick marks or their locations.

`ZTickMode`

z-axis tick mark mode.

`auto`	(Default.) MATLAB calculates *z*-axis tick mark spacing (`ZTick`) that spans the `ZData` of the axes' children.
`manual`	MATLAB takes *z*-axis tick spacing from `ZTick`; it does not depend on the `ZData` in the children objects. Setting values for `ZTick` sets this property to `manual`.

Examples

An important axes property is `Position`. It allows you to define the location of the axes within the figure window. For example,

```
h = axes('Position',rect)
```

creates an axes object at the specified position within the current figure window and returns a handle to it. You specify the location and size of the axes with a rectangle defined by a four element vector,

```
rect = [left, bottom, width, height]
```

The `left` and `bottom` elements of this vector define the distance from the lower-left corner of the figure to the lower-left corner of the rectangle. The `width` and `height` elements define the dimensions of the rectangle. You specify these values in units determined by the `Units` property. By default, MATLAB uses normalized units where

(0,0) is the lower-left corner and (1.0,1.0) is the upper-right corner of the figure window.

You can define multiple axes in a single figure window:

```
clf
axes('position',[.1  .1  .8  .6])
mesh(peaks(20));
axes('position',[.1  .7  .8  .2])
pcolor([1:10;1:10]);
```

In this example, the first plot occupies the bottom two-thirds of the figure, and the second occupies the top third.

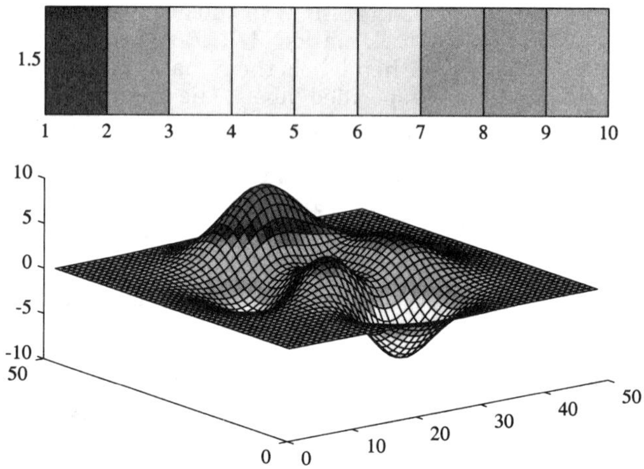

See Also axis, cla, clf, figure, gca, subplot

---

axis

Purpose Axis scaling and appearance.

Synopsis

```
axis([xmin xmax ymin ymax])
axis([xmin xmax ymin ymax zmin zmax])
axis auto
axis(axis)
v = axis;

axis ij
axis xy

axis square
axis equal
```

```
axis off
axis on
[s1,s2,s3] = axis('state')
```

Description `axis` provides an easy way to manipulate the most important properties of axes.

`axis([xmin xmax ymin ymax])` sets the scaling for the *x*- and *y*-axes on the current plot.

`axis([xmin xmax ymin ymax zmin zmax])` sets the scaling for the *x*-, *y*- and *z*-axes on the current plot.

`axis auto` returns the axis scaling to its default autoscaling mode where the best axis limits are computed automatically.

`axis(axis)` freezes the scaling at the current limits, so that if `hold` is turned on, subsequent plots use the same limits.

`v = axis` returns a row vector containing the scaling for the current plot. If the current plot is two-dimensional, `v` has four components; if it is three-dimensional, `v` has six components.

`axis ij` redraws the graph in *matrix coordinates*. The coordinate system origin is at the upper-left corner. The `i`-axis is vertical and is numbered from top to bottom. The `j`-axis is horizontal and is numbered from left to right.

`axis xy` causes the graph to return to the default Cartesian axes form. The coordinate system origin is at the lower-left corner. The *x*-axis is horizontal and is numbered from left to right. The *y*-axis is vertical and is numbered from bottom to top.

`axis square` sets the current axes region to be square.

`axis equal` indicates that the scaling factors and tic mark increments for the *x*- and *y*-axis are equal.

`axis off` turns off all axis labeling and tic marks.

`axis on` turns on axis labeling and tic marks.

`[s1,s2,s3] = axis('state')` returns three strings indicating the current setting of three axis labeling properties:

```
s1 = 'auto' or 'manual'
s2 = 'on' or 'off'
s3 = 'xy' or 'ij'
```

`axis(s1,s2,s3)` restores the axes labeling properties to the values indicated by the three strings. The default is

```
axis auto,on,xy
```

Examples The two statements

```
x = 0:.01:pi/2;
plot(x,tan(x))
```

do not produce a very satisfactory result because the automatic scaling of the *y*-axis is based on `ymax = tan(1.57)`, which is well over 1000.

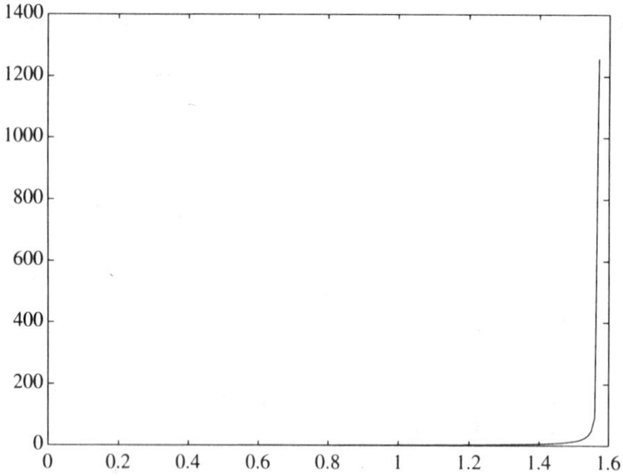

If the statement is followed by

```
axis([0  pi/2  0  10])
```

a much more satisfactory plot results.

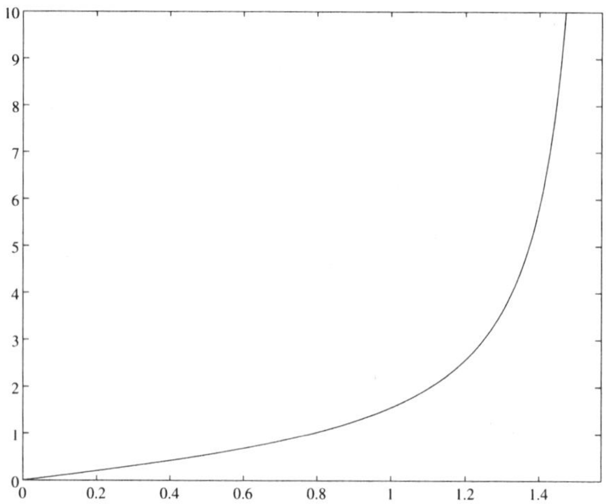

Consider the 10-by-5 matrix `Z` with elements `Z(i,j) = j/i`. The simplest way to generate this matrix is with doubly nested `for` loops:

```
m = 10;
n = 5;
for i = 1:m
    for j = 1:n
        Z(i,j) = j/i;
    end
end
```

The matrix is

```
Z =
    1.00  2.00  3.00  4.00  5.00
    0.50  1.00  1.50  2.00  2.50
    0.33  0.67  1.00  1.33  1.67
    ...
    0.11  0.22  0.33  0.44  0.56
    0.10  0.20  0.30  0.40  0.50
```

To display this in matrix coordinates, enter

```
mesh(Z), axis ij
```

which produces

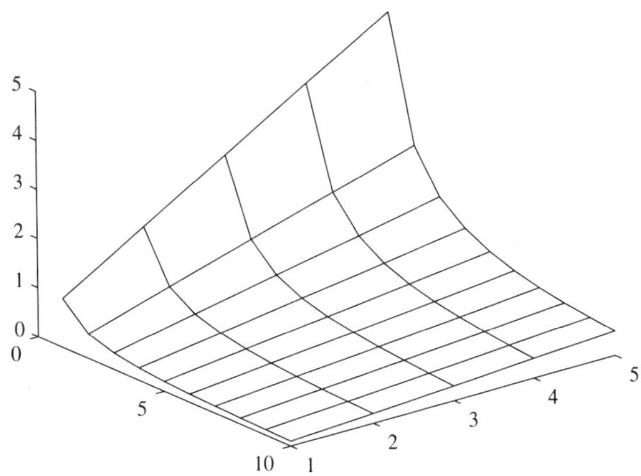

The plot shows ten rows and five columns. The first element

```
Z(1,1) = 1.0
```

is at the upper left; the largest element

```
Z(1,5) = 5.0
```

is at the upper right; and the values increase linearly with the column index, `j`.

On the other hand, consider the function $z = f(x, y) = x/y$ in Cartesian coordinates. This can be evaluated on a grid covering the rectangular region $0 \le x \le 1, 0 \le y \le 2$ with

```
m = 10;
n = 5;
x = (1:n)/n;
y = 2*(1:m)'/m;
[X,Y] = meshgrid(x,y);
Z = X ./ Y;
```

The vectors `x` and `y` are

```
x = [0.20  0.40  0.60  0.80  1.00]
```

and

```
y = [0.20  0.40  0.60 ... 1.80  2.00]'
```

Moreover, the example is arranged so that the 10-by-5 array `Z` is the same as the one given above.

Since this is an example in Cartesian coordinates, enter

```
mesh(x,y,Z)
```

which produces

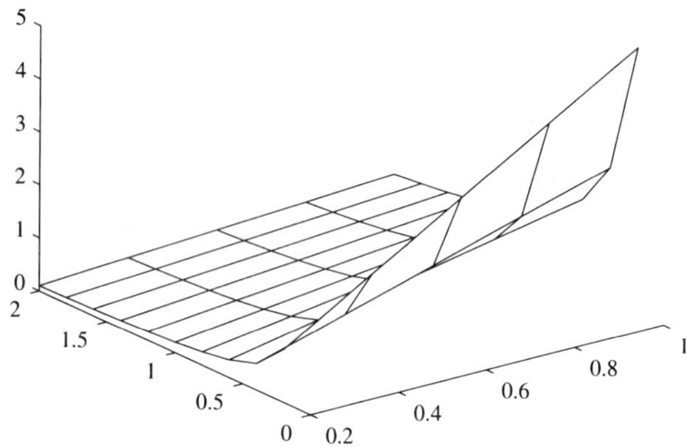

Now, even though the matrix is the same, the default Cartesian coordinate system puts the origin at the lower left and the maximum value at the lower right. Moreover, by supplying values for the *x*- and *y*- coordinates, different labeling on the axes is obtained.

Algorithm `axis` sets the axes object properties `XLim`, `YLim`, `ZLim`, `XLimMode`, `YLimMode`, `ZLimMode`, `YDir`, and `Position`.

See Also `axes`, `get`, `set`, `subplot`
All the properties of `axes` objects.

---

balance

Purpose Improve accuracy of computed eigenvalues.

Synopsis

```
[D,B] = balance(A)
B = balance(A)
```

Description Nonsymmetric matrices can have poorly conditioned eigenvalues. Small perturbations in the matrix, such as roundoff errors, can lead to large perturbations in the eigenvalues. The quantity which relates the size of the matrix perturbation to the size of the eigenvalue perturbation is the condition number of the eigenvector matrix,

```
cond(V) = norm(V)*norm(inv(V))
```

where

```
[V,D] = eig(A)
```

(The condition number of `A` itself is irrelevant to the eigenvalue problem.)

Balancing is an attempt to concentrate any ill conditioning of the eigenvector matrix into a diagonal scaling. Balancing usually cannot turn a nonsymmetric matrix into a symmetric matrix; it only attempts to make the norm of each row equal to the norm of the corresponding column. Furthermore, the diagonal scale factors are limited to powers of two so they do not introduce any roundoff error.

`[D,B] = balance(A)` returns a diagonal matrix `D` whose elements are integer powers of two, and a balanced matrix `B` so that

```
B = D\A*D
```

If `A` is symmetric, then `B == A` and `D` is the identity.

`B = balance(A)` returns just the balanced matrix `B`.

MATLAB's eigenvalue function, `eig(A)`, automatically balances `A` before computing its eigenvalues. Turn off the balancing with `eig(A,'nobalance')`.

Examples This example shows the basic idea. The matrix A has large elements in the upper right and small elements in the lower left. It is far from being symmetric.

```
A = [1   100   10000;  .01   1   100;  .0001   .01   1]
A =

   1.0e+04 *

    0.0001    0.0100    1.0000
    0.0000    0.0001    0.0100
    0.0000    0.0000    0.0001
```

Balancing produces a diagonal D matrix with elements which are powers of two and a balanced matrix B which is closer to symmetric than A.

```
[D,B] = balance(A)
D =

   1.0e+03 *

    2.0480         0         0
         0    0.0320         0
         0         0    0.0003

B =

    1.0000    1.5625    1.2207
    0.6400    1.0000    0.7812
    0.8192    1.2800    1.0000
```

To see the effect on eigenvectors, first compute the eigenvectors of A.

```
[V,E] = eig(A); V
V =

   -1.0000    0.9999   -1.0000
    0.0050    0.0100    0.0034
    0.0000    0.0001    0.0001
```

Note that all three vectors have the first component the largest. This indicates V is badly conditioned; in fact cond(V) is 1.5325e+04. Next, look at the eigenvectors of B.

```
[V,E] = eig(B); V
V =

   -0.8873    0.6933   -0.8642
    0.2839    0.4437    0.1887
    0.3634    0.5679    0.4664
```

Now the eigenvectors are well behaved and cond(V) is 14.903. The ill conditioning is concentrated in the scaling matrix; cond(D) is 8192.

This example is small and not really badly scaled, so the computed eigenvalues of A and B agree within roundoff error; balancing has little effect on the computed results.

Algorithm balance is built into the MATLAB interpreter. It uses the algorithm in [1] originally published in Algol, but popularized by the FORTRAN routines BALANC and BALBAK from EISPACK.

Successive similarity transformations via diagonal matrices are applied to `A` to produce `B`. The transformations are accumulated in the transformation matrix `D`.

The `eig` function automatically uses balancing to prepare its input matrix.

Limitations Balancing can destroy the properties of certain matrices; use it with some care. If a matrix contains small elements that are due to roundoff error, balancing may scale them up to make them as significant as the other elements of the original matrix.

Diagnostics If `A` is not a square matrix:

```
Matrix must be square.
```

See Also `eig`, `hess`, `schur`

References

[1] B.N. Parlett and C. Reinsch, "Balancing a Matrix for Calculation of Eigenvalues and Eigenvectors," *Handbook for Auto. Comp.*, Vol. II, Linear Algebra, pp. 315-326, 1971.

---

bar

Purpose Bar graph.

Synopsis

```
bar(y);
bar(x,y);
[xb,yb] = bar(...)
```

Description `bar(y)` draws a bar graph of the elements of vector `y`.

`bar(x,y)` draws a bar graph of the elements in vector `y` at the locations specified in vector `x`. The values in `x` must be evenly spaced and ascending.

If `x` and `y` are matrices of the same size, one bar graph per column is drawn.

`[xb,yb] = bar(y)` and `[xb,yb] = bar(x,y)` do not draw graphs, but return vectors `xb` and `yb` such that `plot(xb,yb)` plots the bar chart. This is useful in situations where more control is needed over the appearance of a graph; for example, to incorporate a bar chart into a more elaborate `plot` statement.

Examples Plot a bell shaped curve:

```
x = -2.9:0.2:2.9;
bar(x,exp(-x.*x))
```

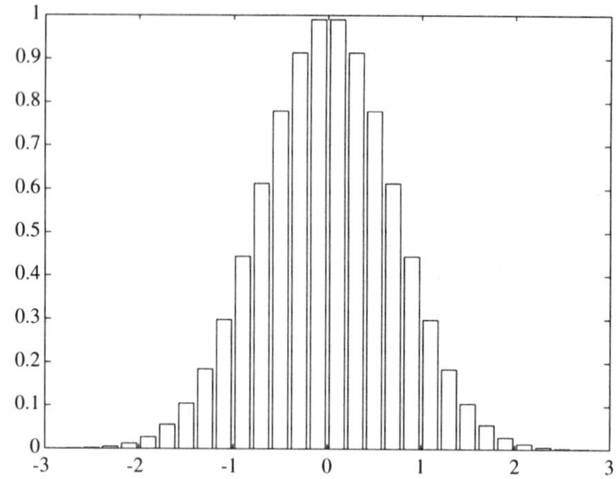

See Also `hist`, `stairs`

---

besseli

Purpose Modified Bessel functions of the first kind.

Synopsis

```
I = besseli(alpha,x)
E = besseli(alpha,x,1)
```

Description `I = besseli(alpha,x)` computes modified Bessel functions of the first kind for real, nonnegative order `alpha` and argument `x`. If `alpha` is a scalar and `x` is a vector, `I` is a vector the same length as `x`. If `x` is a vector of length `m` and `alpha` is a vector of length `n`, then `I` is an `m`-by-`n` matrix and `I(i,k)` is `besseli(alpha(k), x(i))`. The elements of `x` can be any nonnegative real values, in any order. For `alpha`, the increment between elements must be 1, and all elements must be between 0 and 1000, inclusive.

`E = besseli(alpha,x,1)` computes `besseli(alpha,x).*exp(-x)`.

The relationship between the modified Bessel function of the first kind `I` and the Bessel function of the first kind `J` is

$$I_\alpha(x) = i^{-\alpha} J_\alpha(ix)$$

Algorithm `besseli` uses three-term backward recurrence for most `x`, and an asymptotic expansion for large `x`.

See Also `bessel`, `besselj`, `besselk`, `bessely`

References [1] Abramowitz, M. and I.A. Stegun, *Handbook of Mathematical Functions*, National Bureau of Standards, Applied Math. Series #55, Dover Publications, 1965, sections 9.1.1, 9.1.89 and 9.12, formulas 9.1.10 and 9.2.5.

---

besselj

Purpose Bessel functions of the first kind.

Synopsis `J = besselj(alpha,x)`

Description `J = besselj(alpha,x)` computes Bessel functions of the first kind for real, nonnegative order `alpha` and argument `x`. If `alpha` is a scalar and `x` is a vector, `J` is a vector the same length as `x`. If `x` is a vector of length `m` and `alpha` is a vector of length `n`, then `J` is an `m`-by-`n` matrix and `J(i,k)` is `besselj(alpha(k), x(i))`. The elements of `x` can be any nonnegative real values, in any order. For `alpha`, the increment between elements must be 1, and all elements must be between 0 and 1000, inclusive.

Algorithm `besselj` uses three-term recurrence for most `x`, and an asymptotic series for large `x`.

See Also `bessel`, `besseli`, `besselk`, `bessely`

References [1] Abramowitz, M. and I.A. Stegun, *Handbook of Mathematical Functions*, National Bureau of Standards, Applied Math. Series #55, Dover Publications, 1965, sections 9.1.1, 9.1.89 and 9.12, formulas 9.1.10 and 9.2.5.

---

besselk

Purpose Modified Bessel functions of the second kind.

Synopsis

```
K = besselk(alpha,x)
K = besselk(alpha,x,1)
```

Description `K = besselk(alpha,x)` computes modified Bessel functions of the second kind for real, nonnegative order `alpha` and argument `x`. If `alpha` is a scalar and `x` is a vector, `K` is a vector the same length as `x`. If `x` is a vector of length `m` and `alpha` is a vector of length `n`, then `K` is an `m`-by-`n` matrix and `K(i,k)` is `besselk(alpha(k), x(i))`. The elements of `x` can be any nonnegative real values, in any order. For `alpha`, the increment between elements must be 1, and all elements must be between 0 and 1000, inclusive.

The relationship between `K` and the ordinary Bessel functions `J` and `Y` is

$$K_\alpha(x) = \frac{\pi}{2} i^{-\alpha} \left(J_\alpha(ix) + iY_\alpha(ix) \right)$$

`K = besselk(alpha,x,1)` computes `besselk(alpha,x).*exp(-x)`.

Algorithm The `besselk` algorithm is based on a FORTRAN program by W.J. Cody and L. Stoltz, Applied Mathematics Division, Argonnne National Laboratory, dated May 30, 1989.

See Also `bessel`, `besseli`, `besselj`, `bessely`

bessely

Purpose Bessel functions of the second kind.

Synopsis `Y = bessely(alpha,x)`

Description `Y = bessely(alpha,x)` computes Bessel functions of the second kind for real, nonnegative order `alpha` and argument `x`. If `alpha` is a scalar and `x` is a vector, `Y` is a vector the same length as `x`. If `x` is a vector of length `m` and `alpha` is a vector of length `n`, then `Y` is an `m`-by-`n` matrix and `Y(i,k)` is `bessely(alpha(k), x(i))`. The elements of `x` can be any nonnegative real values, in any order. For `alpha`, the increment between elements must be 1, and all elements must be between 0 and 1000, inclusive.

Algorithm The `bessely` algorithm is based on a FORTRAN program by W.J. Cody and L. Stoltz, Applied Mathematics Division, Argonnne National Laboratory, dated March 19, 1990. This program is available on NETLIB.

See Also `bessel`, `besseli`, `besselj`, `besselk`

beta, betainc, betaln

Purpose Beta function.

Synopsis

```
y = beta(z,w)
y = betainc(x,a,b)
y = betaln(z,w)
```

Description `y = beta(z,w)` is the beta function

$$B(z,w) = \int_0^1 t^{z-1}(1-t)^{w-1}\,dt$$

or equivalently `gamma(z)*gamma(w)./gamma(z+w)`. If both `z` and `w` are vectors or matrices, they must be the same size.

`y = betainc(x,z,w)` is the incomplete beta function

$$I_x(z,w) = \frac{1}{B(z,w)}\int_0^x t^{z-1}(1-t)^{w-1}\,dt$$

`y = betaln(z,w)` is the natural logarithm of the beta function, `log(beta(z,w))`, computed without computing `beta(z,w)`. Since the `beta` function can range over very large or very small values, its logarithm is sometimes more useful. If both `z` and `w` are vectors or matrices, they must be the same size.

Examples

```
format rat
beta((0:10)',3)
ans =
      1/0
      1/3
      1/12
      1/30
      1/60
      1/105
      1/168
      1/252
      1/360
      1/495
      1/660
```

In this case, with integer arguments,

```
beta(n,3)
= (n-1)!*2!/(n+2)!
= 2/(n*(n+1)*(n+2))
```

is the ratio of fairly small integers and the rational format is able to recover the exact result.

For `x = 510`, `betaln(x,x) = -708.8616`, which, on a computer with IEEE arithmetic, is slightly less than `log(realmin)`. Here `beta(x,x)` would underflow (or be denormal).

Algorithm

```
beta(z,w) = exp(gammaln(z)+gammaln(w)-gammaln(z+w))
betaln(z,w) = gammaln(z)+gammaln(w)-gammaln(z+w)
```

---

blanks

Purpose String of blanks.

Synopsis `blanks(n)`

Description `blanks(n)` creates a string of `n` blanks. Use `blanks` to format data and labels for use with the `disp` command.

`disp(blanks(n)')` moves the cursor down `n` lines.

Examples Use blanks and disp to display a matrix with column labels:

```
disp([blanks(5) 'Corn' blanks(6) 'Oats' blanks(6) 'Hay'])
disp(randn(5,3))

      Corn        Oats         Hay
     1.1650      1.6961     -1.4462
     0.6268      0.0591     -0.7012
     0.0751      1.7971      1.2460
     0.3516      0.2641     -0.6390
    -0.6965      0.8717      0.5774
```

See Also clc, disp, format (compact option), home

---

bone

Purpose Grayscale colormap with a tinge of blue.

Synopsis `cmap = bone(m)`

Description bone(m) returns an m-by-3 matrix containing the bone colormap. The default value for m is the length of the current colormap. The bone colormap is useful for adding an "electronic" look to gray scale images.

See Also colormap, cool, copper, flag, gray, hot, hsv, jet, pink, prism, white

---

break

Purpose Break out of flow control structures.

Synopsis `break`

Description break terminates the execution of for and while loops. In nested loops, break exits from the innermost loop only.

Examples The indented statements are repeatedly executed until nonpositive n is entered.

```
while 1
    n = input('Enter n. n <= 0 quits. n = ')
    if n <= 0,break,end
    r = rank(magic(n))
end
disp('That''s all.')
```

See Also end, error, for, if, return, while

---

brighten

Purpose Brighten or darken colormap.

Synopsis

```
brighten(beta)
map = brighten(beta)
```

`newmap = brighten(cmap,beta)`

Description `brighten(beta)` replaces the current colormap with a brighter or darker map of essentially the same colors. The modified colormap is brighter if `0 < beta < 1` and darker if `-1 < beta < 0`.

`brighten(beta)`, followed by `brighten(-beta)` restores the original map.

`map = brighten(beta)` returns a brighter or darker version of the current colormap without changing the display.

`newmap = brighten(cmap,beta)` returns a brighter or darker version of the colormap `cmap` without changing the display.

See Also `colormap`, `rgbplot`

---

capture

Purpose Screen capture of current figure (UNIX only).

Synopsis

```
capture(fig)
capture
[X,map] = capture(fig)
```

Description `capture(fig)` makes a new figure that contains a copy of the figure with handle `fig`. This screen capture contains any uimenus or uicontrols present in `fig`.

`capture`, with no input arguments, captures the current figure.

`[X,map] = capture(fig)` returns an image matrix and a colormap representing the figure `fig`. To display this information, use

```
colormap(map)
image(X)
```

`capture` produces a bitmap copy of the contents of the figure window and displays it as a single image object in a new figure window. The resolution of this copy is not as good as that obtained with the `print` command. Unlike `print`, however, `capture` includes uimenus and uicontrols. Use `capture` followed by `print` to obtain hardcopy that includes ui objects as well as an image or graphics.

See Also `image`, `print`

---

cart2pol

Purpose Transform Cartesian coordinates to polar.

Synopsis

```
[theta,rho] = cart2pol(x,y)
[theta,rho,z] = cart2pol(x,y,z)
```

Description `[theta,rho] = cart2pol(x,y)` transforms data stored in Cartesian, or *xy*, coordinates to polar coordinates. `x` and `y` must be the same size. `theta` is in radians.

`[theta,rho,z] = cart2pol(x,y,z)` transforms data stored in Cartesian coordinates to cylindrical coordinates. `x`, `y`, and `z` must be the same size.

Examples

```
t = 0:.01:2*pi;
x = cos(t).*cos(2*t).*sin(2*t);
y = sin(t).*cos(2*t).*sin(2*t);
plot(x,y), title('Cartesian')
```

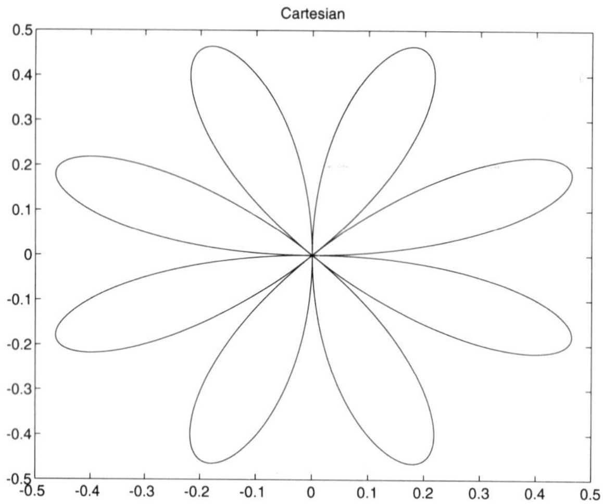


```
[theta,rho] = cart2pol(x,y);
polar(theta,rho), title('Polar')
```

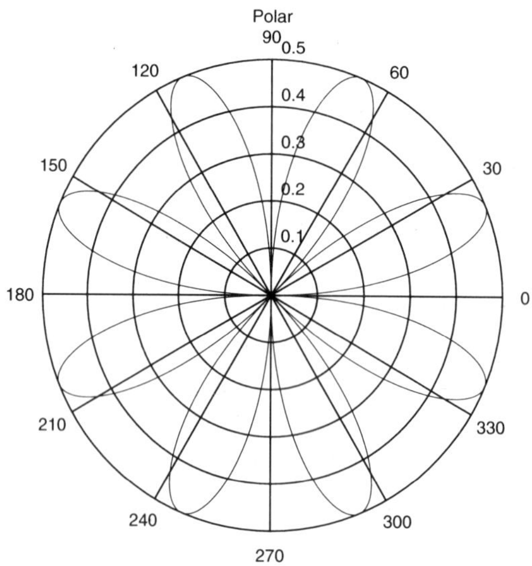

See Also cart2sph, pol2cart, sph2cart

cart2sph

Purpose Transform Cartesian coordinates to spherical.

Synopsis `[az,el,r] = cart2sph(x,y,z)`

Description `[az,el,r] = cart2sph(x,y,z)` transforms data stored in Cartesian, or *xy*, coordinates to spherical coordinates. `x`, `y`, and `z` must be the same size. `az` and `el` are in radians.

See Also `cart2pol`, `pol2cart`, `sph2cart`

caxis

Purpose Pseudocolor axis scaling.

Synopsis

```
caxis([cmin cmax])
caxis auto

v = caxis;
caxis(caxis)
```

Description `caxis` provides an easy way to manipulate the `CLim` and `CLimMode` properties of axes objects. Pseudocolor plotting functions, such as `mesh`, `surf`, and `pcolor`, create surface and patch objects which map specified color arrays into colormaps to produce color values. This mapping is controlled by the `ColorMap`, `CLim`, and `CLimMode` properties.

`caxis([cmin cmax])` sets the `Clim` color axis scaling parameters to the specified low and high values. Any values in a color array outside the interval `cmin ≤ c ≤ cmax` are clipped; that is, the corresponding patches or pixels are not plotted.

`caxis auto`, the default, sets `CLimMode` to indicate that the color axis scaling parameters are computed automatically using the minimum and maximum of the color array specified in the particular pseudocolor plotting function. In this case, only color values set to `Inf` or `NaN` are clipped.

`v = caxis` returns a two-element row vector `v` containing the `[cmin cmax]` currently in use.

`caxis(caxis)` freezes the color axis scaling at the current limits, so that if `hold` is turned on, subsequent plots use the same limits.

Algorithm Let `c` be a color array which is the last argument to the `mesh`, `surf`, or `pcolor` function. If `caxis auto` is in effect, let `cmin = min(c)` and `cmax = max(c)`. Let `m = length(map)`.

The following linear transformation maps `c` onto `k`, an array of indices in the range `1 ≤ k ≤ m`, with `c = cmin` mapping to `k = 1` and `c = cmax` mapping to `k = m`.

If `cmin ≤ c < cmax`, then

```
k = fix((c-cmin)/(cmax-cmin)*m)+1
```

If `c == cmax`, then

```
k = m
```

If `c < cmin`, `c > cmax`, or `c == NaN`, then

```
k = 'invisible'
```

Examples To obtain `(X,Y,Z)` data for a sphere of radius 1, enter

```
[X,Y,Z] = sphere(32);
C = Z;
```

In this case, values of `C` have the range [-1 1]. The statement

```
surf(X,Y,Z,C)
```

allows you to see all of the data. Values of `C` near -1 are assigned the lowest values in the color table; values of `C` near +1 are assigned the highest values in the color table.

To map the top half of the sphere to the highest value in the color table, enter:

```
caxis([-1 0])
```

To use only the bottom half of the color table, enter

```
caxis([-1 3])
```

Since the data in `C` only occupy values in the range [-1 1], the colors assigned to `C` come from the lower part of the color table.

The command

```
caxis auto
```

resets axis scaling back to autoranging and you once again see all the colors in the mesh. In this case, entering

```
z = caxis
```

makes

```
z = [-1 1]
```

See Also `axes`, `axis`, `colormap`, `get`, `set`

The `CLim` and `CLimMode` properties of `axes` objects.

The `ColorMap` property of `figure` objects.

---

cd

Purpose Change working directory.

Synopsis

```
cd
cd directory
cd ..
```

Description `cd`, by itself, prints out the current directory.

`cd` *`directory`* sets the current directory to the one specified.

`cd ..` moves to the directory above the current one.

Examples UNIX: `cd /usr/local/matlab/toolbox/demos`

DOS: `cd C:MATLAB\DEMOS`

VMS: `cd DISK1:[MATLAB.DEMOS]`

Macintosh: `cd Toolbox:Demos`

To specify a Macintosh directory name that includes spaces, enclose the name in single quotes, as in `'Toolbox:New M-Files'`.

See Also `dir`, `lookfor`, `path`, `what`

cdf2rdf

Purpose Convert complex diagonal matrix to real block diagonal form.

Synopsis `[V,D] = cdf2rdf(V,D)`

Description If the eigensystem `[V,D] = eig(X)` has complex eigenvalues appearing in complex-conjugate pairs, `cdf2rdf` transforms the system so `D` is in real diagonal form, with 2-by-2 real blocks along the diagonal replacing the complex pairs originally there. The eigenvectors are transformed so that

```
X = V*D/V
```

continues to hold. The individual columns of `V` are no longer eigenvectors, but each pair of vectors associated with a 2-by-2 block in `D` span the corresponding invariant vectors.

Examples The matrix

```
X =

     1     2     3
     0     4     5
     0    -5     4
```

has a pair of complex eigenvalues.

```
[V,D] = eig(X)

V =
     1.0000     0.4002 - 0.0191i      0.4002 + 0.0191i
          0     0.6479                0.6479
          0          0 + 0.6479i           0 - 0.6479i

D =
     1.0000          0                0
          0     4.0000 + 5.0000i      0
          0          0           4.0000 - 5.0000i
```

Converting this to real block diagonal form produces

```
[V,D] = cdf2rdf(V,D)

V =
     1.0000      0.4002      -0.0191
          0      0.6479            0
          0           0       0.6479
D =
     1     0     0
     0     4     5
     0    -5     4
```

Algorithm The real diagonal form for the eigenvalues is obtained from the complex form using a specially constructed similarity transformation.

See Also `eig`, `rsf2csf`

---

cedit

Purpose Set command line editing and recall parameters (UNIX only).

Synopsis

```
cedit('on')
cedit('off')
cedit('vms')
cedit('emacs')
```

Description `cedit('on')` enables command line edit and recall mode. In this mode, valid keystrokes are

Keystroke	Description
Ctrl-p	Previous line
Ctrl-n	Next line
Ctrl-b	One character left
Ctrl-f	One character right
Esc-b, Ctrl-l	Cursor word left (not supported on Macintosh)
Esc-f, Ctrl-r	Cursor word right (not supported on Macintosh)
Ctrl-a	Beginning of line
Ctrl-e	End of line
Ctrl-u	Cancel line
Ctrl-d	In-place delete
Ctrl-t	Insert toggle (not supported on Macintosh)
Ctrl-k	Delete to end of line

`cedit('off')` turns off the command line edit and recall facility.

`cedit('vms')` enables VMS key definitions:

Keystroke	Description
Ctrl-l	Cursor word left
Ctrl-r	Cursor word right
Ctrl-b	Beginning of line
Ctrl-e	End of line
Ctrl-u	Cancel line
Ctrl-a	Insert toggle
Ctrl-d	EOF

`cedit('emacs')` returns to the default Emacs key definitions. This option is always in effect for MS-Windows and Macintosh systems, in addition to the standard fixed key bindings.

ceil

Purpose Round towards infinity.

Synopsis `Y = ceil(X)`

Description `Y = ceil(X)` rounds the elements of `X` to the nearest integers ≥ `X`.

Examples

```
X = [-1.9  -0.2  3.4  5.6  7.0]
ceil(X) = [-1  0  4  6  7]
```

See Also `fix`, `floor`, `round`

chol

Purpose Cholesky factorization.

Synopsis

```
R = chol(X)
[R,p] = chol(X)
```

Description `R = chol(X)`, where `X` is positive definite, produces an upper triangular `R` so that `R'*R = X`. If `X` is not positive definite, an error message is printed. `chol` uses only the diagonal and upper triangle of `X`. The lower triangular is assumed to be the (complex conjugate) transpose of the upper, that is, `X` is Hermitian.

`[R,p] = chol(X)` never produces an error message. If `X` is positive definite, then `p` is 0 and `R` is the same as above. If `X` is not positive definite, then `p` is a positive integer and `R` is an upper triangular matrix of order `q = p-1` so that `R'*R = X(1:q,1:q)`.

Examples The binomial coefficients arranged in a symmetric array create an interesting positive definite matrix.

```
n = 5;
X = pascal(n)
X =
     1     1     1     1     1
     1     2     3     4     5
     1     3     6    10    15
     1     4    10    20    35
     1     5    15    35    70
```

It is interesting because its Cholesky factor consists of the same coefficients, arranged in an upper triangular matrix.

```
R = chol(X)
R =
     1     1     1     1     1
     0     1     2     3     4
     0     0     1     3     6
     0     0     0     1     4
     0     0     0     0     1
```

You can destroy the positive definiteness (and actually make the matrix singular) by subtracting 1 from the last element.

```
X(n,n) = X(n,n)-1
X =
     1     1     1     1     1
     1     2     3     4     5
     1     3     6    10    15
     1     4    10    20    35
     1     5    15    35    69
```

Now an attempt to find the Cholesky factorization fails.

Algorithm `chol` uses the algorithm from the LINPACK subroutine `ZPOFA`. For a detailed description of the use of the Cholesky decomposition, see Chapter 8 of the *LINPACK User's Guide*.

References

[1] Dongarra, J.J., J.R. Bunch, C.B. Moler, and G.W. Stewart, *LINPACK User's Guide*, SIAM, Philadelphia, 1979.

---

cla

Purpose Clear current axis.

Synopsis

```
cla
cla reset
```

Description `cla` deletes all objects (lines, text, patches, surfaces, and images) from the current axes.

`cla reset` deletes all objects within the current axes and also resets all axes properties, except `Position`, to their default values.

See Also `clf`, `hold`, `reset`

clabel

Purpose Contour plot elevation labels.

Synopsis

```
clabel(C)
clabel(C,V)
clabel(C,'manual')
```

Description `clabel(C)` adds height labels to the current contour plot using the contour structure `C` output from `contour`. The label positions are selected randomly.

`clabel(C,V)` labels only those contour levels given in vector `V`. The default labels all known contours.

`clabel(C,'manual')` places contour labels at the locations selected with a mouse. Press the **Return** key to terminate labeling. Use the space bar to enter contours.

Examples Generate, draw, and label a simple contour plot:

```
[x,y] = meshgrid(-2:.2:2);
z = x.^exp(-x.^2-y.^2);
c = contour(x,y,z);
clabel(c);
```

See Also `contour`, `contourc`

clc

Purpose Clear the command window.

Synopsis `clc`

Description `clc` clears the command window.

Examples Display a sequence of random matrices, on top of each other:

```
clc, for i = 1:25, home, A = rand(5), end
```

See Also `home`, `clf`

clear

Purpose Remove items from memory.

Synopsis

```
clear
clear name
clear name1 name2 name3...
```

```
clear functions
clear variables
clear mex
clear global
clear all
```

Description Executing `clear`, by itself, clears all variables from the workspace.

`clear` *`name`* removes just the variable or function *`name`* from the workspace. Both M-file functions and MEX-file functions can be cleared this way. *`name`* can be either a variable name or a function name.

`clear` *`name1 name2 name3`* removes *`name1`*, *`name2`*, and *`name3`* from the workspace.

`clear functions` clears all the currently compiled M-functions from memory.

`clear variables` clears all variables from the workspace.

`clear mex` clears all MEX-files from memory.

`clear global` clears all global variables.

`clear all` removes all variables, functions, and MEX-files from memory. It leaves MATLAB's workspace empty, as if MATLAB were just invoked.

If *`name`* is global, `clear` *`name`* removes *`name`* from the current workspace, but leaves it accessible to any functions declaring it global. `clear global` *`name`* completely removes the global variable *`name`*. `clear global` removes all global variables.

See Also `pack`

---

clf

Purpose Clear current figure (graph window).

Synopsis

```
clf
clf reset
```

Description `clf` deletes all objects from the current figure.

`clf reset` deletes all the objects within the current figure and also resets all figure properties, except `Position`, to their default values.

See Also `cla`, `clc`, `hold`, `reset`

---

clock

Purpose Get the current date and time.

Synopsis `c = clock`

Description `clock` returns a six-element row vector `c` containing the current time and date in decimal form:

```
[year month day hour minute seconds]
```

The first five elements are integers. The `seconds` element is accurate to several digits beyond the decimal point.

Examples Rounding to the nearest second results in an integer display:

```
fix(clock)
ans =
     1994     8     4    10    20    32
```

This represents August 4, 1994, 10:20:32.

See Also `date`, `etime`, `tic`, `toc`

---

close

Purpose Close window.

Synopsis

```
close(h)
close
```

Description The `close` command closes the window unconditionally and without confirmation.

`close(h)` closes the window with handle `h`.

`close`, by itself, closes the current figure window.

Examples Close the current window by specifying its handle:

```
close(gcf)
```

See Also `delete`, `figure`, `gcf`

---

colmmd

Purpose Minimum degree ordering for elimination sparsity.

Synopsis

```
p = colmmd(S)
```

Description `p = colmmd(S)` returns the column minimum degree ordering of `S`. For an asymmetric matrix `S`, this is a column permutation `p` such that `S(:,p)` tends to have sparser LU factors than `S`.

The `colmmd` ordering is automatically used by `\` and `/` for the solution of nonsymmetric and symmetric indefinite sparse linear systems.

Some options and parameters associated with heuristics in the algorithm can be changed with `spparms`.

Algorithm The minimum degree algorithm for symmetric matrices is described in the review paper by George and Liu [1]. For nonsymmetric matrices, MATLAB's minimum degree algorithm is new and is described in the paper by Gilbert, Moler, and Schreiber [2]. It is roughly like symmetric minimum degree for `A'*A`, but does not actually form `A'*A`.

Each stage of the algorithm chooses a vertex in the graph of `A'*A` of lowest degree (that is, a column of `A` having nonzero elements in common with the fewest other columns), eliminates that vertex, and updates the remainder of the graph by adding fill (that is, merging rows). If the input matrix `S` is of size `m`-by-`n`, the columns are all eliminated and the permutation is complete after `n` stages. To speed up the process, several heuristics are used to carry out multiple stages simultaneously.

Examples The Harwell-Boeing collection of sparse matrices includes a test matrix ABB313. It is a rectangular matrix, of order 313-by-176, associated with least squares adjustments of geodesic data in the Sudan. Since this is a least squares problem, form the augmented matrix (see `spaugment`), which is square and of order 489. The spy plot shows that the nonzeros in the original matrix are concentrated in two stripes, which are reflected and supplemented with a scaled identity in the augmented matrix. The `colmmd` ordering scrambles this structure. (Note that the following example requires professional MATLAB and the Harwell-Boeing collection of software.)

```
load('abb313.mat')
S = spaugment(A);
p = colmmd(S);
spy(S)
spy(S(:,p))
```

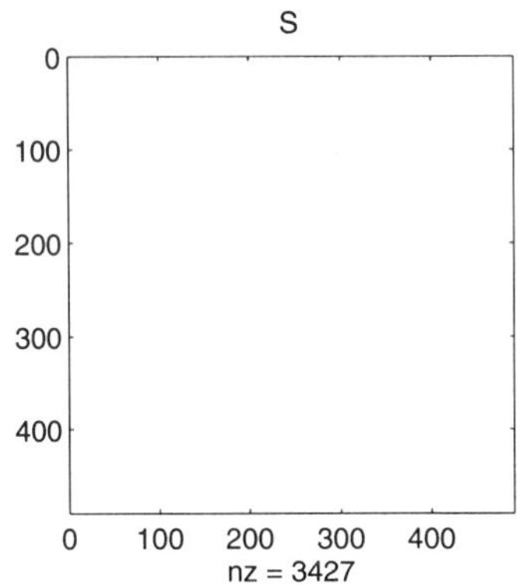

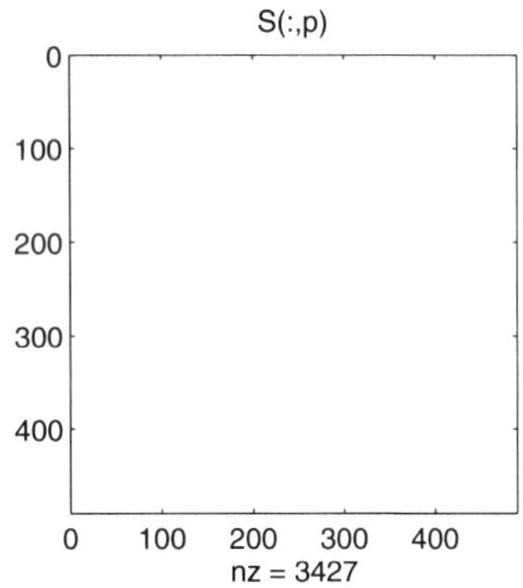

Comparing the spy plot of the LU factorization of the original matrix with that of the reordered matrix shows that minimum degree reduces the time and storage requirements by better than a factor of 2.6. The nonzero counts are 18813 and 7223, respectively.

```
spy(lu(S))
spy(lu(S(:,p)))
```

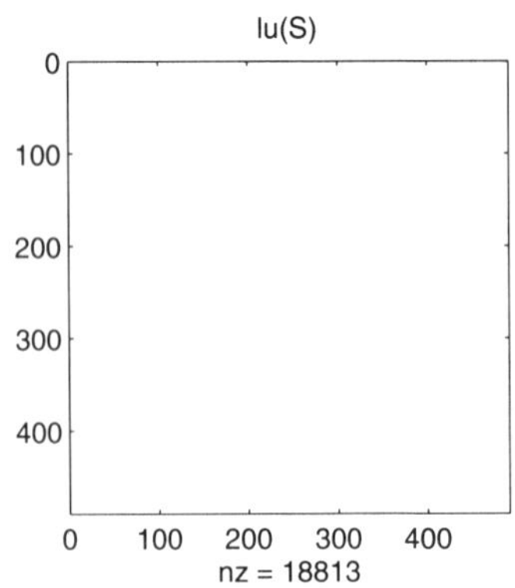

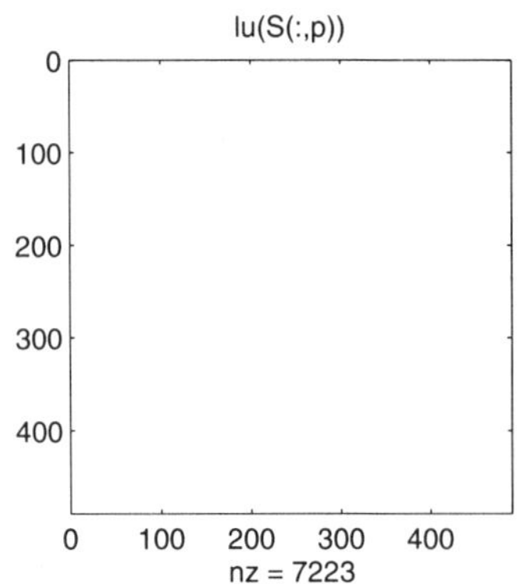

See Also `\`, `colperm`, `lu`, `spparms`, `symmmd`, `symrcm`

References
[1] George, Alan and Liu, Joseph,*"The evolution of the Minimum Degree Ordering Algorithm," SIAM Review*, 31:1-19, 1989.

[2] Gilbert, John R., Cleve Moler, and Robert Schreiber, "Sparse Matrices in MATLAB: Design and Implementation," *SIAM Journal on Matrix Analysis and Applications 13*, pp. 333-356, 1992.

---

colorbar

Purpose Display color bar (color scale).

Synopsis

```
colorbar('vert')
colorbar('horiz')
colorbar(h)
colorbar
h = colorbar(...)
```

Description `colorbar('vert')` appends a vertical color bar to the current axes, resizing the axes to make room for the color bar. `colorbar` works with both two-dimensional and three-dimensional plots.

`colorbar('horiz')` appends a horizontal color bar to the current axes.

`colorbar(h)` places the color bar in the axes `h`. The color bar is horizontal if the width of the axes is greater than its height.

`colorbar` with no arguments adds a new vertical color bar or updates an existing one.

`h = colorbar(...)` returns a handle to the color bar axes.

See Also `colormap`

---

colormap

Purpose Color lookup map.

Synopsis

```
colormap(map)
colormap('default')
colormap(hsv)
map = colormap
```

Description A colormap is an `m`-by-3 matrix of real numbers between 0.0 and 1.0. The `k`-th row of the colormap defines the `k`-th color by specifying the intensity of red, green and blue, i.e., `map(k,:) = [r(k) g(k) b(k)]`. For example, [0 0 0] is black, [1 1 1] is white, [1 0 0] is pure red, and [.5 1 .83] is aquamarine.

`colormap(map)` sets the colormap to the matrix `map`. If any values are outside the interval [0 1], an error message results.

`colormap('default')` and `colormap(hsv)` both set the current colormap to the default colormap, which varies the hue component in the hue-saturation-value color model.

`map = colormap` retrieves the current colormap. The values returned are in the interval [0 1].

A number of other colormaps, including `gray`, `hot`, `cool`, `copper`, and `pink`, are generated by M-files in the `color` directory.

Algorithm Graphics functions that use pseudocolor – `mesh`, `surf`, `pcolor`, and others – map a color matrix, `c`, whose values are in the range `[cmin, cmax]`, to an array of indices, `k`, in the range [1, `m`]. The values of `cmin` and `cmax` are either `min(min(c))` and `max(max(c))`, or are specified by `caxis`. The mapping is linear, with `cmin` mapping to index 1 and `cmax` mapping to index `m`. The indices are then used with the colormap to determine the color associated with each matrix element. See `caxis` for details.

Each figure window has its own `ColorMap` property. `colormap` is a simple M-file that sets and gets this property.

Examples An introduction to colormaps is provided by the *`Images`* and *`colormaps`* demo, `imagedemo`. Select **Color Spiral** from the menu (starts automatically on the Macintosh). This uses the `pcolor` function to display a 16-by-16 matrix whose elements vary from 0 to 255 in a rectilinear spiral. The default colormap, `hsv`, starts with red in the center, then passes through yellow, green, cyan, blue and magenta before returning to red at the outside end of the spiral. Selecting **Colormap Menu** gives access to a number of other colormaps (this is not true for the Macintosh).

The `rgbplot` function provides further information about colormaps. Try `rgbplot(hsv)`, `rgbplot(gray)`, `rgbplot(hot)`, etc.

See Also `bone`, `cool`, `copper`, `flag`, `gray`, `hot`, `hsv`, `jet`, `pink`, `prism`, `white`

`brighten`, `caxis`, *`ColorSpec`*, `image`, `mesh`, `pcolor`, `rgbplot`, `surf`

The `ColorMap` property of figure objects.

ColorSpec

Purpose Color specification.

Description `ColorSpec` is not a command, rather it refers to the three ways in which you specify color in MATLAB:

- Short name
- Long name
- RGB triple

The short names and long names are MATLAB strings that specify one of eight predefined colors. The following table lists the predefined colors and their RGB equivalents.

RGB Value	Short Name	Long Name
[1 1 0]	y	yellow
[1 0 1]	m	magenta
[0 1 1]	c	cyan
[1 0 0]	r	red
[0 1 0]	g	green
[0 0 1]	b	blue
[1 1 1]	w	white
[0 0 0]	k	black

The RGB triple is a three-element row vector whose elements specify the intensities of the red, green, and blue components of the color. The intensities must be in the range [0 1].

When you specify an RGB triple that is not one of the eight listed above, MATLAB allocates a slot in the system color table for the new color. Colors defined in this way are true colors as opposed to colors mapped to the closest entry that already exists in the color table.

The eight predefined colors and any that you specify as RGB values are not part of the figure colormap, nor are they affected by changes to the figure colormap. They are referred to a *fixed* colors, as opposed to *colormap* colors.

Examples To create a plot that uses a green line color, specify the color with either a short name, a long name, or an RGB triple. These statements are equivalent:

```
plot(xdata,ydata,'g')
plot(xdata,ydata,'green')
h = plot(xdata,ydata);
set(h,'Color',[0 1 0])
```

You can use `ColorSpec` anywhere you need to define a color. For example, this statement changes the figure background color to pink:

```
set(gcf,'Color',[1  .4  .6])
```

See Also `colormap`

colperm

Purpose Reorder columns using sparsity count.

Synopsis `j = colperm(S)`

Description `j = colperm(S)` generates a permutation `j` such that the columns of `S(:,j)` are ordered according to increasing count of nonzero entries. This is sometimes useful as a preordering for LU factorization; use `lu(S(:,j))`.

If `S` is symmetric, then `j = colperm(S)` generates a permutation `j` so that both the rows and columns of `S(j,j)` are ordered according to increasing count of nonzero entries. If S is positive definite, this is sometimes useful as a preordering for Cholesky factorization; use chol(S(j,j)).

Algorithm The algorithm involves a sort on the counts of nonzeros in each column.

Examples The n-by-n *arrowhead* matrix

```
A = [ones(1,n); ones(n-1,1) speye(n-1,n-1)]
```

has a full first row and column. Its LU factorization, lu(A), is almost completely full.

```
j = colperm(A)
```

returns j = [2:n 1]. So A(j,j) sends the full row and column to the bottom and the rear, and lu(A(j,j)) has the same nonzero structure as A itself.

On the other hand, the Bucky ball example,

```
B = bucky
```

has exactly three nonzero elements in each row and column, so

```
j = colperm(B)
```

is the identity permutation and is no help at all for reducing fill-in with subsequent factorizations.

See Also chol, colmmd, lu, symrcm

---

comet

Purpose Animated comet plot.

Synopsis

```
comet(y)
comet(x,y)
comet(x,y,p)
comet
```

Description A comet plot is an animated graph in which a circle (the comet *head*) traces the data points on the screen. A trailing segment, the comet *body*, follows the head. The *tail* traces the entire function using a solid line.

comet(y) displays an animated comet plot of the vector y.

comet(x,y) displays an animated comet plot of vector y versus vector x.

comet(x,y,p) uses a comet body of length p*length(y). The default value for p is 0.1.

comet, with no arguments, provides a demonstration of the comet plot.

Examples

```
t = 0:.01:2*pi;
x = cos(2*t).*(cos(t).^2);
y = sin(2*t).*(sin(t).^2);
comet(x,y)
```

See Also comet3

comet3

Purpose 3-D comet plot.

Synopsis

```
comet3(z)
comet3(x,y,z)
comet3(x,y,z,p)
comet3
```

Description A comet plot is an animated graph in which a circle (the comet *head*) traces the data points on the screen. A trailing segment, the comet *body*, follows the head. The *tail* traces the entire function using a solid line.

`comet3(z)` displays an animated, three-dimensional comet plot of the vector `z`.

`comet3(x,y,z)` displays an animated comet plot of the curve through the points `[x(i),y(i),z(i)]`.

`comet3(x,y,z,p)` uses a comet body of length `p*length(y)`. The default value for `p` is 0.1.

`comet3`, with no arguments, provides a demonstration of the three-dimensional comet plot.

Examples

```
t = -10*pi:pi/250:10*pi;
comet3((cos(2*t).^2).*sin(t),(sin(2*t).^2).*cos(t),t)
```

See Also `comet`

compan

Purpose Companion matrix.

Synopsis `A = compan(p)`

Description `A = compan(p)` where `p` is a vector of polynomial coefficients, returns the corresponding companion matrix whose first row is `-p(2:n)/p(1)`.

The eigenvalues of `compan(p)` are the roots of the polynomial.

Examples The polynomial

$$(x-1)(x-2)(x+3) = x^3 - 7x + 6$$

has a companion matrix given by

```
p = [1  0  -7  6]
A = compan(p)
A =
     0    7   -6
     1    0    0
     0    1    0
```

The eigenvalues are the polynomial roots:

```
eig(compan(p)) =
    -3.0000
     2.0000
     1.0000
```

This is also `roots(p)`.

See Also `eig`, `poly`, `polyval`, `roots`

---

compass

Purpose Compass plot.

Synopsis

```
compass(z)
compass(x,y)
```

Description `compass(z)` draws a graph that displays the angle and magnitude of the complex elements of `z` as arrows emanating from the origin.

`compass(x,y)` is equivalent to `compass(x+i*y)`. It displays the compass plot for the angles and magnitudes of the elements of matrices `x` and `y`.

Examples Draw a simple compass plot of the eigenvalues of a matrix:

```
z = eig(randn(20,20))
compass(z)
```

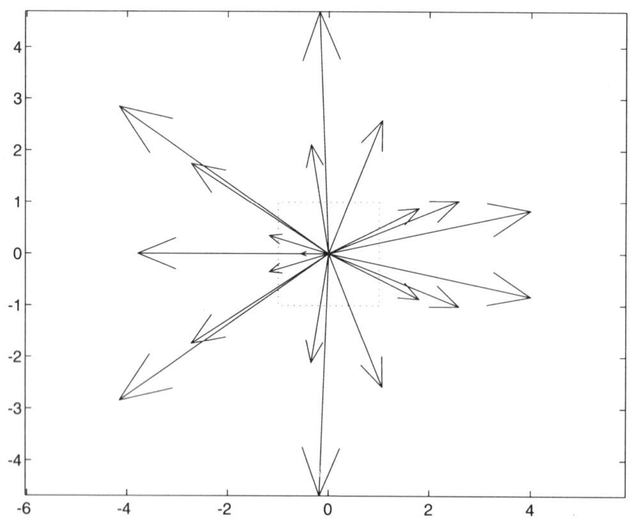

See Also `feather`, `rose`

computer

Purpose Identify the computer on which MATLAB is running.

Synopsis

```
s = computer
[s,maxsize] = computer
```

Description `s = computer` returns a string containing the type of computer on which MATLAB is running. Some computers include

String	Computer
SUN4	Sun SPARC workstation
APOLLO	Apollo workstation
PC	Personal Computer
PCWIN	PC running MS-Windows
PCAT	AT
PC386	386
MAC	Macintosh Plus or SE
MAC2	Macintosh II
HP700	HP 9000/Series 700
DECRISC	DECStation
SGI	Silicon Graphics
HP300	HP 9000/Series 300
VAXVMSD	VAX/VMS R-float
VAXVMSG	VAX/VMS G-float
CRAY	Cray X-MP or Y-MP

This list changes as new computers are added and others become obsolete.

`[s,maxsize] = computer` returns the integer `maxsize`, which contains the maximum number of elements allowed in a matrix with this version of MATLAB.

cond

Purpose Matrix condition number.

Synopsis

```
c = cond(X)
```

Description The *condition number* of a matrix measures the sensitivity of the solution of a system of linear equations to errors in the data. It gives an indication of the accuracy of the results from matrix inversion and linear equation solution.

`cond(X)` returns the 2-norm condition number, the ratio of the largest singular value of `X` to the smallest.

Algorithm `cond` uses the singular value decomposition, `svd`.

See Also `condest`, `norm`, `rank`, `rcond`, `svd`

References
[1] Dongarra, J.J., J.R. Bunch, C.B. Moler, and G.W. Stewart, *LINPACK User's Guide*, SIAM, Philadelphia, 1979.

condest

Purpose 1-norm matrix condition number estimate.

Synopsis

```
c = condest(A)
[c,v] = condest(A)
[c,v] = condest(A,trace)
```

Description `c = condest(A)` uses Higham's modification of Hager's method to estimate the condition number of a matrix. The computed `c` is a lower bound for the condition of `A` in the 1-norm.

`[c,v] = condest(A)` also computes a vector `v` such that

$$\|Av\| = \|A\|\|v\|/c$$

Thus, `v` is an approximate null vector of `A` if `c` is large.

`[c,v] = condest(A,trace)` provides further information about the estimate, where trace can be:

- 1 to print information about each step of the iteration.
- -1 to print the ratios of the estimate to the actual (computed) condition number and to `rcond`.

This function handles both real and complex matrices. It is particularly useful (without `trace = -1`) for sparse matrices.

See Also `cond`, `normest`, `rcond`

Reference

[1] Higham, N.J., FORTRAN codes for estimating the one-norm of a real or complex matrix, with applications to condition estimation, *ACM Trans. Math. Soft.*, 14, pp. 381-396 (1988).

conj

Purpose Complex conjugate.

Synopsis `Y = conj(X)`

Description `conj(X)` returns the complex conjugate of the elements of `X`.

Algorithm If `X` is a complex vector or matrix:

```
conj(X) = real(X)–i*imag(X)
```

See Also `imag`, `real`

contour

Purpose Contour plots.

Synopsis

```
contour(Z)
contour(Z,n)
```

```
contour(Z,v)
contour(x,y,Z)
contour(x,y,Z,n)
contour(x,y,Z,v)
C = contour(...)
```

Description `contour(Z)` draws a contour plot of matrix `Z`. The contours are level lines in the units of array `Z`. The lower-left corner of the plot corresponds to `Z(1,1)`. The number of contour lines and their values are chosen automatically by `contour`.

`contour(Z,n)` produces a contour plot of matrix `Z` with `n` contour levels.

`contour(Z,v)` draws a contour plot of `Z` with contour levels at the values specified in vector `v`.

`contour(x,y,Z)`, `contour(x,y,Z,n)`, and `contour(x,y,Z,v)` produce contour plots of `Z` and use the data in vectors `x` and `y` to control the axis scaling on the x- and y-axes. The elements of `x` and `y` are equally spaced.

`C = contour(...)` returns the contour matrix `C` as described in the function `contourc` and used by `clabel`.

To label the levels of a contour plot, see `clabel`.

Examples To view a contour plot of the function $z = xe^{-x^2-y^2}$

over the range $-2 \le x \le 2$, $-2 \le y \le 3$, use the statements:

```
x = -2:.2:2;
y = -2:.2:3;
[X,Y] = meshgrid(x,y);
Z = X.*exp(-X.^2-Y.^2);
contour(x,y,Z)
```

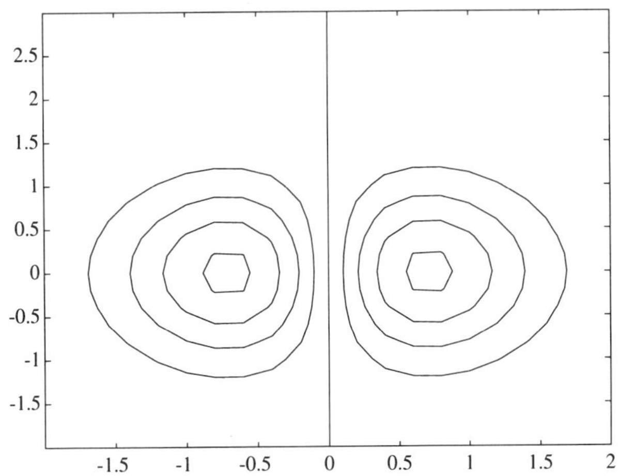

Diagnostics If the smallest dimension of `Z` is less than 2:

```
Matrix must be 2-by-2 or larger.
```

Limitations `contour` assumes that vectors `x` and `y` are monotonically increasing. If `x` and `y` are irregularly spaced, the contours are simply stretched apart, which is not the same as contouring the irregularly spaced data.

See Also `clabel`, `contour3`, `contourc`, `quiver`

---

contour3

Purpose 3-D contour plots.

Synopsis

```
contour3(Z)
contour3(Z,n)
contour3(X,Y,Z)
contour3(X,Y,Z,n)
C = contour3(...)
[C,h] = contour3(...)
```

Description `contour3` produces a three-dimensional contour plot of a surface defined on a rectangular grid.

`contour3(Z)` plots the contour lines of `Z` in three-dimensional format.

`contour3(Z,n)` plots `n` contour lines in three-dimensional format.

`contour3(X,Y,Z)` and `contour3(X,Y,Z,n)` use matrices `X` and `Y` to define the axes limits. `X` and `Y` can be vectors too, in which case they are expanded to matrices by replicating rows and columns, respectively.

`C = contour3(...)` returns the contour matrix `C` as described in the function `contourc` and used by `clabel`.

`[C,h] = contour3(...)` returns a column vector `h` of handles to the contour line objects.

Examples Plot the three-dimensional contour of the peaks function.

```
x = -3:.125:3;
y = x;
[X,Y] = meshgrid(x,y);
Z = peaks(X,Y);
contour3(X,Y,Z,20)
```

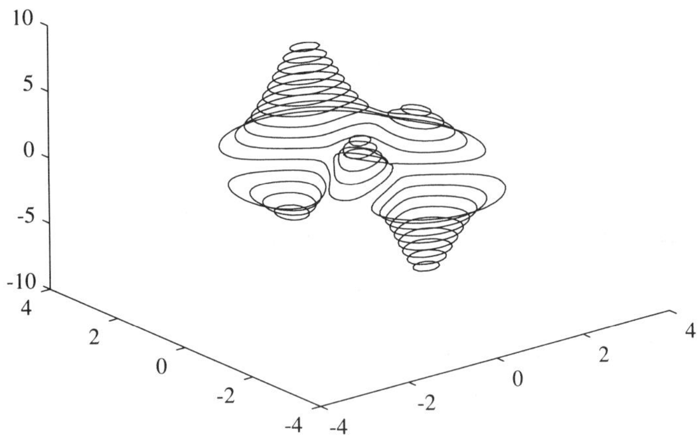

Limitations contour3 assumes that vectors x and y are montonically increasing. If x and y are irregularly spaced, the contours are simply stretched apart, which is not the same as contouring the irregularly spaced data.

See Also contour, meshc, meshgrid, surfc

---

contourc

Purpose Contour plot computation.

Synopsis

```
C = contourc(Z)
C = contourc(Z,n)
C = contourc(Z,v)
C = contourc(x,y,Z)
C = contourc(x,y,Z,n)
C = contourc(x,y,Z,v)
```

Description contourc calculates the contour matrix C used by the M-file contour to draw actual contour plots.

C = contourc(Z) computes the contour matrix for a contour plot of matrix Z. The contours are level lines in the units of Z. The number of contour lines and their corresponding values are chosen automatically by contourc.

C = contourc(Z,n) computes contours of matrix Z with n contour levels.

C = contourc(Z,v) computes contours of Z with contour levels at the values specified in vector v.

C = contourc(x,y,Z), C = contourc(x,y,Z,n), and C = contourc(x,y,Z,v) compute contours of Z and use the data in vectors x and y to control the axis scaling on the *x*- and *y*-axes. The elements of x and y are equally spaced.

The contour matrix C is a two-row matrix of contour lines. Each contiguous drawing segment contains the value of the contour, the number of (x,y) drawing pairs, and the pairs themselves. The segments are appended end-to-end,

```
C = [level1  x1  x2  x3 ... level2  x2  x2  x3 ...;
     pairs1  y1  y2  y3 ... pairs2  y2  y2  y3 ...]
```

Limitations contourc assumes that vectors x and y are monotonically increasing. If x and y are irregularly spaced, the contours are simply stretched apart, which is not the same as contouring the irregularly spaced data.

See Also contour, contour3

---

contrast

Purpose Grayscale colormap for contrast enhancement.

Synopsis

```
cmap = contrast(X,m)
cmap = contrast(X)
```

Description cmap = contrast(X,m), where X is a matrix of indices into the current colormap, returns the grayscale colormap cmap. cmap is a matrix with m rows and three identical columns such that the bins of a histogram for

```
image(X)
colormap(cmap)
```

have an approximately equal intensity distribution. The default value for m is 64.

See Also brighten, gray

---

conv

Purpose Convolution and polynomial multiplication.

Synopsis c = conv(a,b)

Description c = conv(a,b) convolves vectors a and b. Convolution can be thought of as multiplying the polynomials whose coefficients are the elements of a and b since, algebraically, this is the same operation.

Let m = length(a) and n = length(b). Then c is the vector of length m+n-1 whose k-th element is

$$c(k) = \sum_{j} a(j)b(k+1-j)$$

The sum is over all the values of j which lead to legal subscripts for a(j) and b(k+1–j), specifically j = max(1,k+1-n):min(k,m). When m = n, this gives

```
c(1) = a(1)*b(1)
c(2) = a(1)*b(2)+a(2)*b(1)
c(3) = a(1)*b(3)+a(2)*b(2)+a(3)*b(1)
...
c(n) = a(1)*b(n)+a(2)*b(n-1)+ ... +a(n)*b(1)
...
c(2*n-1) = a(n)*b(n)
```

Algorithm The convolution theorem says, roughly, that convolving two sequences is the same as multiplying their Fourier transforms. In order to make this precise, it is necessary to pad the two vectors with zeros and ignore roundoff error. Thus, if

```
X = fft([x zeros(1,length(y)-1)])
```

and

```
Y = fft([y zeros(1,length(x)-1)])
```

then

```
conv(x,y) = ifft(X.*Y)
```

Examples Let a = [1 2 3 4] and b = [10 20 30], then c = conv(a,b) is

```
c =
     10     40     100     160     170     120
```

The vectors a and b represent the polynomials

$$x^3 + 2x^2 + 3x + 4$$

$$10x^2 + 20x + 30$$

Then the product c = conv(a,b) is the polynomial

$$10x^5 + 40x^4 + 100x^3 + 160x^2 + 170x + 120$$

See Also deconv, residue, filter

convmtx, xconv2 in the *Signal Processing Toolbox*

---

conv2

Purpose 2-D convolution.

Synopsis
```
C = conv2(X,Y)
C = conv2(X,Y,'shape')
```

Description `C = conv2(X,Y)` computes the two-dimensional convolution of matrices `X` and `Y`. If one of these matrices describes a two-dimensional FIR filter, the other matrix is filtered in two dimensions.

The size in each dimension of the output matrix `C` is equal to the sum of the corresponding dimensions of the input matrices minus one. That is, if the size of `X` is`[mx, nx]` and the size of `Y` is `[my, ny]`, then the size of `C` is `[mx+my−1,nx+ny−1]`.

`C = conv2(X,Y,'`*`shape`*`')` returns a subsection of the two-dimensional convolution, as specified by the *`shape`* parameter:

- `full` returns the full two-dimensional convolution (default).
- `same` returns the central part of the convolution of the same size as `X`.
- `valid` returns only those parts of the convolution that are computed without the zero-padded edges. Using this option, `C` has size `[mx−my+1,nx−ny+1]` when `size(X) > size(Y)`.

`conv2` executes most quickly when `X` is larger than `Y`.

Examples In image processing, the Sobel edge finding operation is a two-dimensional convolution of an input array with the special matrix

```
s = [1 2 1; 0 0 0; −1 −2 −1];
```

These commands extract the horizontal edges from a raised pedestal:

```
A = zeros(10);
A(3:7,3:7) = ones(5);
H = conv2(A,s);
mesh(H)
```

These commands display first the vertical edges of `A`, then both horizontal and vertical edges.

```
V = conv2(A,s');
mesh(V)
mesh(sqrt(H.^2+V.^2))
```

See Also `conv`, `deconv`, `filter2`

---

cool

Purpose Cyan and magenta shades colormap.

Synopsis `cmap = cool(m)`

Description `cool(m)` returns an `m`-by-3 matrix containing the `cool` colormap. The default value for `m` is the length of the current colormap. The `cool` colormap varies smoothly from cyan to magenta.

Examples To color the peaks function with the cool colormap:

```
peaks
colormap(cool)
```

See Also bone, colormap, copper, flag, gray, hot, hsv, jet, pink, prism, white

copper

Purpose Linear copper tones colormap.

Synopsis `cmap = copper(m)`

Description copper(m) returns an m-by-3 matrix containing the copper colormap. The default value for m is the length of the current colormap. The copper colormap varies smoothly from black to bright copper.

Examples To view data measuring the depth of a mold for a one cent coin:

```
load penny
pcolor(-del2(flipud(P)))
colormap(copper)
shading flat
```

See Also bone, colormap, cool, flag, gray, hot, hsv, jet, pink, prism, white

corrcoef

Purpose Correlation coefficients.

Synopsis `S = corrcoef(X)`

Description S = corrcoef(X) returns a matrix of correlation coefficients calculated from an input matrix whose rows are observations and whose columns are variables. The matrix S = corrcoef(x) is related to the covariance matrix C = cov(x) by

$$S(i,j) = \frac{C(i,j)}{\sqrt{C(i,i)C(j,j)}}$$

corrcoef(X) is the 0-th lag of the covariance function, that is, the 0-th lag of xcov(x,'coeff') packed into a square array.

corrcoef removes the mean from each column before calculating the results.

See Also cov, mean, std

xcorr, xcov in the Signal Processing Toolbox

cos, cosh

Purpose Cosine and hyperbolic cosine.

Synopsis
```
Y = cos(X)
Y = cosh(X)
```

Description The trigonometric functions operate element-wise on matrices. Their domains and ranges include complex values. All angles are measured in radians.

`cos(X)` is the circular cosine for each element of `X`.

`cosh(X)` is the hyperbolic cosine for each element of `X`.

Examples
```
M = magic(3)
M =
     8     1     6
     3     5     7
     4     9     2
cos(M*pi/4) =
     1.0000      0.7071     -0.0000
    -0.7071     -0.7071      0.7071
    -1.0000      0.7071      0.0000
```

`cos(pi/2)` is not exactly zero, but rather a value the size of the floating-point accuracy, `eps`, because `pi` is only a floating-point approximation to the exact value.

Algorithm

$$\cos(x+iy) = \cos(x)\cosh(y) - i\sin(x)\sin(y)$$

$$\cos(z) = \frac{e^{iz} + e^{-iz}}{2}$$

$$\cosh(z) = \frac{e^{z} + e^{-z}}{2}$$

See Also `acos`, `asin`, `atan`, `exp`, `expm`, `funm`, `sin`, `tan`

---

cot, coth

Purpose Cotangent and hyperbolic cotangent.

Synopsis
```
Y = cot(X)
Y = coth(X)
```

Description `cot` and `coth` operate element-wise on matrices. Their domains and ranges include complex values. All angles are in radians.

`cot(X)` returns the cotangent for each element of `X`.

`coth(X)` returns the hyperbolic cotangent for each element of `X`.

Algorithm

$$\cot(z) = \frac{1}{\tan(z)}$$

$$\coth(z) = \frac{1}{\tanh(z)}$$

See Also acot, acoth

cov

Purpose Covariance matrix.

Synopsis

```
C = cov(X)
C = cov(x,y)
```

Description `C = cov(X)`, where `X` is a vector, returns a scalar containing the variance. `C = cov(X)`, where `X` is a matrix with each row an observation and each column a variable, returns the covariance matrix. `diag(cov(X))` is a vector of variances for each column, and `sqrt(diag(cov(X)))` is a vector of standard deviations.

`cov(X)` is the 0-th lag of the covariance function, that is, the 0-th lag of `xcov(X)/(n–1)` packed into a square array.

`cov(x,y)`, where `x` and `y` are column vectors of equal length, is equivalent to `cov([x y])`.

`cov` removes the mean from each column before calculating the results.

Algorithm The basic algorithm for `cov` is

```
[n,p] = size(X);
X = X-ones(n,1)*mean(X);
Y = X'*X/(n-1);
```

See Also corrcoef, mean, std

cplxpair

Purpose Sort complex numbers into complex conjugate pairs.

Synopsis

```
x = cplxpair(x)
x = cplxpair(x,tol)
```

Description `x = cplxpair(x)` returns `x` with complex conjugate pairs grouped together. The conjugate pairs are ordered by increasing real part. Within a pair, the element with negative imaginary part comes first. The purely real values are returned following all the complex pairs.

The complex conjugate pairs are forced to be exact complex conjugates. A default tolerance of `100*eps` relative to `abs(x(i))` determines which numbers are real and which elements are paired complex conjugates.

`x = cplxpair(x,tol)` overrides the default tolerance.

Examples Order into complex pairs five poles evenly spaced around the unit circle:

```
cplxpair(exp(2*pi*i*(0:4)/5)')
ans =
    -0.8090 - 0.5878i
    -0.8090 + 0.5878i
     0.3090 - 0.9511i
     0.3090 + 0.9511i
     1.0000
```

Diagnostics If there are an odd number of complex numbers, or if the complex numbers cannot be grouped into complex conjugate pairs within the tolerance, `cplxpair` generates the error message:

```
Complex numbers can't be paired.
```

---

cputime

Purpose Elapsed CPU time.

Synopsis `cputime`

Description `cputime` returns the CPU time in seconds that has been used by the MATLAB process since MATLAB started.

The return value can overflow the internal representation and wrap around.

Examples For example

```
t = cputime; surf(peaks(40)); e = cputime-t
e =
    0.4667
```

returns the CPU time used to run `surf(peaks(40))`.

See Also `clock`, `etime`, `tic`, `toc`

---

cross

Purpose Vector cross product.

Synopsis `C = cross(A,B)`

Description `C = cross(A,B)`, where `A` and `B` are three-element vectors, returns the vector cross product `C = AxB`. If `A` and `B` are 3-by-`n` matrices, `C` contains the cross products of the corresponding columns.

See Also `dot`

---

csc, csch

Purpose Cosecant and hyperbolic cosecant.

Synopsis `Y = csc(X)`

```
Y = csch(X)
```

Description `csc` and `csch` operate element-wise on matrices. Their domains and ranges include complex values. All angles are in radians.

`csc(X)` returns the cosecant for each element of `X`.

`csch(X)` returns the hyperbolic cosecant for each element of `X`.

Algorithm

$$\csc(z) = \frac{1}{\sin(z)}$$

$$\text{csch}(z) = \frac{1}{\sinh(z)}$$

See Also `acsc`, `acsch`

cumprod

Purpose Cumulative product.

Synopsis `Y = cumprod(X)`

Description `Y = cumprod(X)`, where `X` is a vector, returns a vector `Y` containing the cumulative product of the elements of `X`. If `X` is a matrix, `Y` is a matrix containing the cumulative products over each column.

Examples

```
cumprod(1:5)  =  [1  2  6  24  120]
```

See Also `prod`

cumsum

Purpose Cumulative sum.

Synopsis `Y = cumsum(X)`

Description `Y = cumsum(X)`, where `X` is a vector, returns a matrix containing the cumulative sum of the elements of `X`. If `X` is a matrix, `Y` is a matrix containing the cumulative sums over each column.

Examples

```
cumsum(1:5)  =  [1  3  6  10  15]
```

See Also `sum`

cylinder

Purpose Generate cylinder.

Synopsis

```
[X,Y,Z] = cylinder(r,n)
[X,Y,Z] = cylinder(r)
[X,Y,Z] = cylinder
cylinder = (...)
```

Description `cylinder` generates the *(x,y,z)* coordinates of a unit cylinder for use with `surf` or `mesh`.

`[X,Y,Z] = cylinder(r,n)` returns the *(x,y,z)* coordinates of a cylinder based on the generator curve in the vector `r`. `r` contains the radius at equally spaced points along the unit height of the cylinder. The cylinder generated is aligned with the *z*-axis and can be plotted using either `surf` or `mesh` (see example below). The cylinder has `n` points around the circumference.

`[X,Y,Z] = cylinder(r)`, with `n` omitted, uses `n = 20`.

`[X,Y,Z] = cylinder`, with no input arguments, uses `n = 20` and `r = [1 1]`.

`cylinder = (...)`, with no output arguments, draws a `surf` plot of the cylinder on the screen.

Examples Generate a cylinder defined by the generator function `2+cos(t)`:

```
t = 0:pi/10:2*pi;
[X,Y,Z] = cylinder(2+cos(t));
surf(X,Y,Z)
```

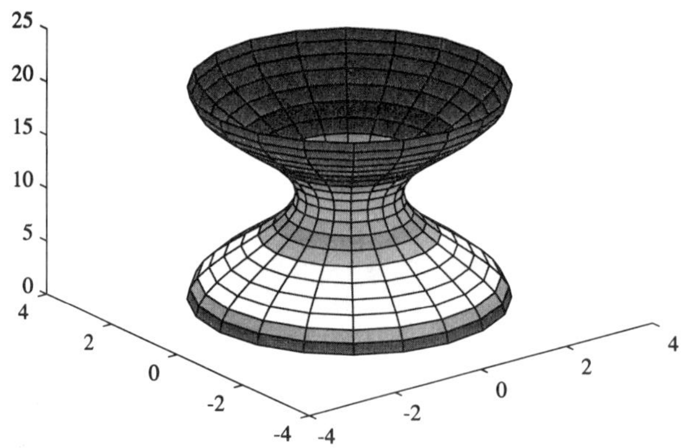

See Also `sphere`

date

Purpose Get the date.

Synopsis `s = date`

Description `s = date` returns a string containing the date in `dd-mmm-yy` format.

Examples Display the date as a string:

```
date*
ans =
       4-Mar-92
```

See Also `clock`, `etime`, `cputime`

dbclear

Purpose Clear breakpoints.

Synopsis

```
dbclear at lineno in mfilename
dbclear all in mfilename
dbclear all
dbclear in mfilename
dbclear if error
dbclear if naninf
dbclear if infnan
```

Description The `dbclear` command removes the breakpoint(s) set by a corresponding `dbstop` command.

`dbclear at` *`lineno`* `in` *`mfilename`* removes the breakpoint set at the specified line in the specified file.

`dbclear all in` *`mfilename`* removes all breakpoints in the specified M-file.

`dbclear all` removes all breakpoints in all M-file functions.

`dbclear in` *`mfilename`* removes the breakpoint set at the first executable line in the specified file.

`dbclear if error` clears the `dbstop error` statement, if set. If a runtime error occurs after this command, MATLAB terminates the current operation and returns to the base workspace.

`dbclear if naninf` and `dbclear if infnan` clear the `dbstop naninf` or `dbstop infnan` statements respectively, if set.

The `at`, `in`, and `if` keywords, familiar to users of the UNIX debugger dbx, are optional.

See Also `dbcont`, `dbdown`, `dbstack`, `dbstatus`, `dbstep`, `dbstop`, `dbtype`, `dbup`

dbcont

Purpose Resume execution.

Synopsis `dbcont`

Description The `dbcont` command resumes execution of an M-file function from a breakpoint. Execution continues until another breakpoint is reached, an error is encountered, or the entire M-file function and any calling M-files are executed.

See Also `dbclear`, `dbdown`, `dbstack`, `dbstatus`, `dbstep`, `dbstop`, `dbtype`, `dbup`

dbdown

Purpose Change local workspace context.

Synopsis `dbdown`

Description `dbdown` changes the current workspace context to the workspace of the called M-file function when a breakpoint is encountered. You must have issued the `dbup` command at least once before you can issue this command.

Multiple `dbdown` commands change the workspace context to each successively executed M-file function on the stack until the current workspace context is the current breakpoint. It is not necessary, however, to move back to the current breakpoint to continue execution or to step to the next line.

See Also `dbclear`, `dbcont`, `dbstack`, `dbstatus`, `dbstep`, `dbstop`, `dbtype`, `dbup`

dbquit

Purpose Quit debug mode.

Synopsis `dbquit`

Description The `dbquit` command immediately terminates the debugger and returns control to the base workspace prompt. The M–file being processed is *not* completed and no results are returned.

All breakpoints remain in effect.

See Also `dbclear`, `dbcont`, `dbdown`, `dbstack`, `dbstatus`, `dbstep`, `dbstop`, `dbtype`, `dbup`

dbstack

Purpose Display function call stack.

Synopsis `dbstack`

Description `dbstack` displays the line numbers and M-file names of the function calls that led to the current breakpoint, listed in the order in which they were executed. In other words, the line number of the most recently executed function call (at which the current breakpoint occurred) is listed first, followed by its calling function, which is followed by its calling function, and so on, until the base workspace is reached.

Examples

```
dbstack
In /usr/local/matlab/toolbox/matlab/cond.m at line 13
In test1.m at line 2
In test.m at line 3
```

See Also dbclear, dbcont, dbdown, dbstatus, dbstep, dbstop, dbtype, dbup

dbstatus

Purpose List all breakpoints.

Synopsis `dbstatus` *`mfilename`*

Description `dbstatus` *`mfilename`* displays a list of the line numbers for which breakpoints are set in the specified M-file function.

See Also dbclear, dbcont, dbdown, dbstack, dbstep, dbstop, dbtype, dbup

dbstep

Purpose Execute one or more lines from a breakpoint.

Synopsis

```
dbstep
dbstep nlines
dbstep in
```

Description The `dbstep` command allows you to debug an M-file function by following its execution from the current breakpoint. At a breakpoint, the `dbstep` command steps through execution of the current M-file function one line at a time or at the rate specified by `nlines`.

`dbstep` executes the next executable line of the current M-file is executed.

`dbstep nlines` executes the specified number of executable lines.

Use `dbstep in` when the next executable line is a call to another M-file function, and you want to step to the first executable line of the called function.

See Also dbclear, dbcont, dbdown, dbstack, dbstatus, dbstop, dbtype, dbup

dbstop

Purpose Set breakpoints in an M-file function.

Synopsis

```
dbstop at lineno in mfilename
dbstop in mfilename
dbstop if error

dbstop if naninf
dbstop if infnan
```

Description The `dbstop` command sets up MATLAB's debugging mode. `dbstop` sets a breakpoint at a specified location in an M-file function or causes a break in case an error occurs during execution. When the specified `dbstop` condition is met, the MATLAB prompt is displayed and you can issue any valid MATLAB command.

`dbstop at` *`lineno`* `in` *`mfilename`* stops execution just prior to execution of that line of the specified M-file function. `dbstop in` *`mfilename`* stops execution before the first executable line in the M-file function when it is called.
`dbstop if error` stops execution if a runtime error occurs in any M-file function. (The debugger does not handle compile-time syntax errors.) You can examine the local workspace and sequence of function calls leading to the error, but you cannot resume M-file execution after a runtime error. `dbstop if naninf` or `dbstop if infnan` stops execution when it detects Not-a-Number (`NaN`) or Infinity (`INF`).

Regardless of the form of the `dbstop` command, when a stop occurs, the line or error condition that caused the stop is displayed. To resume M-file function execution, you can issue a `dbcont` command or step to another line in the file with the `dbstep` command.

Any breakpoints set by the first two forms of the `dbstop` command are cleared if the M-file function is edited or cleared.

The `at`, `in`, and `if` keywords, familiar to users of the UNIX debugger dbx, are optional.

Examples Here is a short example, printed with the `dbtype` command to produce line numbers.

```
dbtype buggy
1    function z = buggy(x)
2    n = length(x);
3    z = (1:n)./x;
```

The statement

```
dbstop in buggy
```

causes execution to stop at line 2, the first executable line. The command

```
dbstep
```

then advances to line 3 and allows the value of `n` to be examined.

The example function only works on vectors; it produces an error if the input `x` is a full matrix. So the statements

```
dbstop if error
buggy(magic(3))
```

produce

```
Error using ==>./
Matrix dimensions must agree.
Error in ==> buggy.m
On line 3 ==> z = (1:n)./x;
```

Finally, if any of the elements of the input `x` are zero, a division by zero occurs. This can be illustrated with

```
dbstop if naninf
buggy(0:2)
```

which produces

```
Warning: Divide by zero
NaN/Inf debugging breakpoint hit on line 2. Stopping at next line.
2    n = length(x);
3    z = (1:n)./x;
```

See Also `dbcont`, `dbdown`, `dbstack`, `dbstatus`, `dbstep`, `dbtype`, `dbup`

dbtype

Purpose List M-file with line numbers.

Synopsis

```
dbtype mfilename
dbtype mfilename start:end
```

Description `dbtype` *`mfilename`* displays the contents of the specified M-file function with line numbers preceding each line.

`dbtype` *`mfilename start:end`* displays the portion of the file specified by a range of line numbers.

See Also `dbclear`, `dbcont`, `dbdown`, `dbstack`, `dbstatus`, `dbstep`, `dbstop`, `dbup`

dbup

Purpose Change local workspace context.

Synopsis

```
dbup
```

Description At a breakpoint, the `dbup` command changes the current workspace context to the workspace of the calling M-file function. This command allows you to examine the calling M-file function by using any other MATLAB command. In this way, you can determine what led to the arguments being passed to the called function.

Multiple `dbup` commands change the workspace context to each previous calling M-file function on the stack until the base workspace context is reached. (It is not necessary, however, to move back to the current breakpoint to continue execution or to step to the next line.)

See Also `dbclear`, `dbcont`, `dbdown`, `dbstack`, `dbstatus`, `dbstep`, `dbstop`, `dbtype`

deblank

Purpose Strip trailing blanks from the end of a string.

Synopsis

```
deblank(s)
```

Description `deblank(s)` removes the trailing blanks from the end of a character string `s`. It also removes any *null characters*, characters with an absolute value of zero, from within the string.

See Also `blanks`

dec2hex

Purpose Decimal to hexadecimal number conversion.

Synopsis `s = dec2hex(n)`

Description `dec2hex(n)` converts the decimal integer `n` to its hexadecimal representation stored in a MATLAB string.

Examples `dec2hex(1023)` is the string `'3ff'`.

See Also `format`, `hex2dec`, `hex2num`

deconv

Purpose Deconvolution and polynomial division.

Synopsis `[q,r] = deconv(b,a)`

Description `[q,r] = deconv(b,a)` deconvolves vector `a` out of vector `b`, using long division. The quotient is returned in vector `q` and the remainder in vector `r` such that `b = conv(q,a)+r`.

If `a` and `b` are vectors of polynomial coefficients, convolving them is equivalent to multiplying the two polynomials, and deconvolution is polynomial division. The result of dividing `b` by `a` is quotient `q` and remainder `r`.

Examples If

```
a = [1   2   3   4]
b = [10   20   30]
```

the convolution is

```
c = conv(a,b)
c =
    10     40     100     160     170     120
```

Use deconvolution to divide `a` back out:

```
[q,r] = deconv(c,a)
q =
    10     20     30
r =
    0     0     0     0     0     0
```

gives a quotient equal to `b` and a zero remainder.

Algorithm `deconv` uses the `filter` primitive.

See Also `conv`, `residue`

`convmtx`, `conv2`, and `filter` in the Signal Processing Toolbox

del2

Purpose Five-point discrete Laplacian.

Synopsis `V = del2(U)`

Description `V = del2(U)` returns a matrix `V` the same size as `U` with each element equal to the difference between an element of `U` and the average of its four neighbors. Elements in the interior of the matrix have four neighbors, while elements on the edges and corners have three or two neighbors.

If the matrix `U` is regarded as a function $u(x,y)$ evaluated at the point on a square grid, then `4*del2(U)` is a finite difference approximation of Laplace's differential operator applied to u, that is

$$\nabla^2 u = \frac{\partial^2 u}{\partial x^2} + \frac{\partial^2 u}{\partial y^2}$$

Algorithm where

$$v_{ij} = \left(u_{i+1,j} + u_{i-1,j} + u_{i,j+1} + u_{i,j-1}\right)/4 - u_{i,j}$$

$$u_{0,j} = u_{1,j}, \quad u_{i,n+1} = u_{i,n}, \quad etc.$$

Examples The function

$$u(x,y) = x^2 + y^2$$

has

$$\nabla^2 u = 4$$

For this function, `4*del2(U)` is also 4.

```
[x,y] = meshgrid(-4:4,-3:3);
U = x.*x+y.*y
U =
    25    18    13    10     9    10    13    18    25
    20    13     8     5     4     5     8    13    20
    17    10     5     2     1     2     5    10    17
    16     9     4     1     0     1     4     9    16
    17    10     5     2     1     2     5    10    17
    20    13     8     5     4     5     8    13    20
    25    18    13    10     9    10    13    18    25
V = 4*del2(U)
V =
     4     4     4     4     4     4     4     4     4
     4     4     4     4     4     4     4     4     4
     4     4     4     4     4     4     4     4     4
     4     4     4     4     4     4     4     4     4
     4     4     4     4     4     4     4     4     4
     4     4     4     4     4     4     4     4     4
     4     4     4     4     4     4     4     4     4
```

See Also `diff`

---

delete

Purpose Delete files and graphics objects.

Synopsis

```
delete filename
delete(h)
```

Description `delete` *`filename`* deletes the specified file.

`delete(h)` deletes the graphics object with handle `h`. The function deletes the object without requesting verification even if the object is a window.

See Also `!`, `cd`, `close`, `dir`, `type`, `who`, `what`

---

demo

Purpose Start the MATLAB Expo demo facility.

Synopsis `demo`

Description `demo` runs the MATLAB Expo.

---

det

Purpose Matrix determinant.

Synopsis `d = det(X)`

Description `d = det(X)` is the determinant of the square matrix `X`. If `X` contains only integer entries, the result `d` is also an integer.

Using `det(X) == 0` as a test for matrix singularity is appropriate only for matrices of modest order with small integer entries. Testing singularity using `abs(det(X)) <= tolerance` is rarely a good idea because it is difficult to choose the correct tolerance. The function `rcond(X)` is intended to check for singular and nearly singular matrices. See `rcond` for details.

Algorithm The determinant is computed from the triangular factors obtained by Gaussian elimination

```
[L,U] = lu(A)
s = +1 or -1 = det(L)
det(A) = s*prod(diag(U))
```

Examples The statement

```
A = [1  2  3;  4  5  6;  7  8  9]
```

produces

```
A =
     1     2     3
     4     5     6
     7     8     9
```

This happens to be a singular matrix, so

```
d = det(A)
```

produces

```
d =
    0
```

Changing `A(3,3)` with

```
A(3,3) = 0;
```

turns `A` into a nonsingular matrix, so now

```
d = det(A)
```

produces

```
d =
    27
```

See Also `\`, `/`, `inv`, `lu`, `rcond`, `rref`

---

diag

Purpose Diagonal matrices and diagonals of a matrix.

Synopsis `X = diag(v,k)`

```
X = diag(v)
v = diag(X,k)
v = diag(X)
```

Description `X = diag(v,k)`, where `v` is a vector with `n` components, returns s a square matrix `X` of order `n+abs(k)` with the elements of `v` on the `k`-th diagonal. `k` = 0 is the main diagonal, `k` > 0 is above the main diagonal, and `k` < 0 is below the main diagonal.

`diag(v)` simply puts `v` on the main diagonal.

`v = diag(X,k)`, where `X` is a matrix, returns a column vector `v` formed from the elements of the `k`-th diagonal of `X`.

`diag(X)` is the main diagonal of `X`.

Examples `diag(diag(X))` is a diagonal matrix.

`sum(diag(X))` is the trace of `X`.

The statement

```
diag(-m:m)+diag(ones(2*m,1),1)+diag(ones(2*m,1),-1)
```

produces a tridiagonal matrix of order `2*m+1`.

See Also `tril`, `triu`

---

diary

Purpose Save session in a disk file.

Synopsis

```
diary
diary filename
diary off
diary on
```

Description `diary`, by itself, toggles `diary` on and off.

`diary` *`filename`* writes a copy of all subsequent keyboard input and most of the resulting output (but not graphs) to the named file. If the file already exists, output is appended to the end of the file.

`diary off` suspends the diary.

`diary on` resumes diary mode, using the current filename, or the default filename `diary` if none has yet been specified.

The output of `diary` is an ASCII file, suitable for printing or for inclusion in reports and other documents.

Limitations You cannot put a diary into the files named `off` and `on`.

See Also `disp`, `fprintf`, `save`

---

diff

Purpose Differences and approximate derivatives.

Synopsis
```
y = diff(x)
y = diff(x,n)
```

Description `diff` calculates differences. If `x` is a row or column vector

```
x = [x(1) x(2) ... x(n)]
```

then `y = diff(x)` returns a vector of differences between adjacent elements

```
[x(2)-x(1) x(3)-x(2) ... x(n)-x(n-1)]
```

The output vector is one element shorter than the input vector. If `x` is a matrix, the differences are calculated down each column:

```
diff(X) = X(2:m,:)-X(1:m-1,:)
```

`y = diff(x,n)` is the `n`-th difference function:

```
diff(x,n) = diff(diff(x,n-1))
```

Examples `diff(y)./diff(x)` is an approximate derivative.

See Also `del2`, `prod`, `sum`

---

diffuse

Purpose Reflectance of a diffuse surface (Lambert's Law).

Synopsis `r = diffuse(Nx,Ny,Nz,S)`

Description `diffuse` returns the reflectance of a diffuse (Lambertian) surface from the normal vector components. The reflectance is the fraction of light that reflects from the surface toward the viewer. The reflectance varies from 0.0 (no light reflected) to 1.0 (all the light reflected).

`r = diffuse(Nx,Ny,Nz,S)` returns the reflectance of the surface with normal vector components `[Nx,Ny,Nz]`. The normal vector components can be matrices so that for the normal `n`,

```
n(i,j) = [Nx(i,j), Ny(i,j), Nz(i,j)]
```

These normal components can be calculated using `surfnorm`. The light source vector `S` is a three-element vector, `[Sx,Sy,Sz]`, that specifies the direction from the surface to the light source. `S` can be a two-element vector with azimuth and elevation.

`diffuse` is used as part of the function `surfl` to produce shaded surfaces.

Algorithm `diffuse` implements Lambert's law for diffuse surfaces, $r = \cos(t)$, where t is the angle between the surface normal and the light source direction.

See Also `specular`, `surfl`, `surfnorm`

---

dir

Purpose Directory listing.

Synopsis
```
dir
dir dirname
```

Description `dir` lists the files in the current directory.

`dir` *`dirname`* lists the files in the specified directory. You can use pathnames, wildcards, and any options available in your operating system.

See Also `!`, `cd`, `delete`, `type`, `what`, `who`

---

disp

Purpose Display text or matrix.

Synopsis `disp(X)`

Description `disp(X)` displays a matrix, without printing the matrix name. If `X` contains a text string, the string is displayed.

Another way to display a matrix on the screen is to type its name, but this prints a leading "`X =`," which is not always desirable.

Examples One use of `disp` in an M-file is to display a matrix with column labels:

```
disp('         Corn          Oats          Hay')
disp(rand(5,3))
```

which results in

```
         Corn          Oats          Hay
        0.2113        0.8474        0.2749
        0.0820        0.4524        0.8807
        0.7599        0.8075        0.6538
        0.0087        0.4832        0.4899
        0.8096        0.6135        0.7741
```

This shows the use of `disp` on both data and text strings.

See Also `num2str`, `setstr`, `sprintf`

---

dmperm

Purpose Dulmage-Mendelsohn decomposition of a matrix.

Synopsis

```
p = dmperm(A)
[p,q,r] = dmperm(A)
[p,q,r,s] = dmperm(A)
```

Description `p = dmperm(A)` returns a maximum matching; if `A` has full column rank then `A(p,:)` is square with nonzero diagonal.

If `A` is a reducible matrix, the linear system $Ax = b$ can be solved by permuting `A` to a block upper triangular form, with irreducible diagonal blocks, and then performing block backsubstitution. Only the diagonal blocks of the permuted matrix need to be factored, saving fill and arithmetic in the above diagonal blocks.

`[p,q,r] = dmperm(A)`, where `A` is a square matrix, finds a row permutation `p` and a column permutation `q` so that `A(p,q)` is in block upper triangular form. The third output argument `r` is an integer vector describing the boundaries of the blocks: the `k`-th block of `A(p,q)` has indices `r(k):r(k+1)-1`.

`[p,q,r,s] = dmperm(A)`, where `A` is not square, finds permutations `p` and `q` and index vectors `r` and `s` so that `A(p,q)` is block upper triangular. The blocks have indices.

```
(r(i):r(i+1)-1, s(i):s(i+1)-1).
```

In graph theoretic terms, the diagonal blocks correspond to strong Hall components of the adjacency graph of `A`.

See Also `sprank`

dot

Purpose Vector dot product.

Synopsis `C = dot(A,B)`

Description `C = dot(A,B)`, where `A` and `B` are vectors, returns the dot product of `A` and `B`. `A` and `B` must be the same size. If `A` and `B` are matrices, `C` is a row vector containing the dot product for the corresponding columns of `A` and `B`.

See Also `cross`

drawnow

Purpose Complete any pending drawing.

Synopsis `drawnow`

Description `drawnow` flushes the event queue and forces MATLAB to update the screen.

While running an M-file, MATLAB does *not* render the screen with each graphics statement. For example, executing the M-file

```
plot(x,y)
axis([0   10   0   10])
title('A short title')
grid
```

does not draw the screen until after the final statement when MATLAB returns to its prompt. This allows for efficient execution of sequences of graphics statements. In the example above, the axis is drawn once; if typed at the command line, it would be drawn twice.

Four events cause MATLAB to flush the event queue and draw the screen:

- a return to the MATLAB prompt
- a `pause` statement
- execution of a `getframe` command
- execution of a `drawnow` command

echo

Purpose Echo M-files during execution.

Synopsis

```
echo on
echo off
echo
echo fcnname on
echo fcnname off
echo fcnname
echo on all
echo off all
```

Description `echo` controls the echoing of M-files during execution. Normally, the commands in M-files do not display on the screen during execution. Command echoing can be enabled for debugging, or for demonstrations, allowing the commands to be viewed as they execute.

`echo` behaves slightly differently, depending on whether a *script file* or a *function file* is being considered. For script files, the use of `echo` is simple; echoing can be either `on` or `off`, in which case any script used is affected:

- `echo on` turns on the echoing of commands in all script files.
- `echo off` turns off the echoing of commands in all script files.
- `echo`, by itself, toggles the echo state.

With function files, the use of `echo` is more complicated. If `echo` is enabled on a function file, the file is interpreted, rather than compiled, so that each input line can be viewed as it is executed. Since this results in inefficient execution, use `echo` only for debugging.

- `echo` *`fcnname`* `on` turns on echoing of the named function file.
- `echo` *`fcnname`* `off` turns off echoing of the named function file.
- `echo` *`fcnname`* toggles the echo state of the named function file.
- `echo on all` and `echo off all` set echoing for all function files.

eig

Purpose Eigenvalues and eigenvectors.

Synopsis

```
d = eig(A)
[V,D] = eig(A)
[V,D] = eig(A,'nobalance')
d = eig(A,B)
[V,D] = eig(A,B)
```

Description The eigenvalue problem is to determine the nontrivial solutions of the equation:

$$Ax = sx$$

where A is an n-by-n matrix, x is a length n column vector, and s is a scalar. The n values of s that satisfy the equation are the *eigenvalues* and the corresponding values of x are the *right eigenvectors*. In MATLAB, the function `eig` solves for the eigenvalues s and optionally the eigenvectors x.

`d = eig(A)` returns a vector d containing the eigenvalues of matrix A.

`[V,D] = eig(A)` produces a diagonal matrix D of eigenvalues and a full matrix V whose columns are the corresponding eigenvectors so that `A*V = V*D`.

The eigenvectors are scaled so that the norm of each is 1.0.

The *left eigenvectors* can be computed with the two statements:

```
[W,D] = eig(A')
W = W'
```

The eigenvectors in D computed from A and A' are the same, although they can occur in different orders. The left eigenvectors satisfy `W*A = D*W`.

`[V,D] = eig(A,'nobalance')` finds eigenvalues and eigenvectors without a preliminary balancing step. Ordinarily, balancing improves the conditioning of the input matrix, enabling more accurate computation of the eigenvectors and eigenvalues. However, if a matrix contains small elements that are really due to roundoff error, balancing may scale them up to make them as significant as the other elements of the original matrix, leading to incorrect eigenvectors. Use the `nobalance` option in this event. See `balance` for more details.

When a matrix has no repeated eigenvalues, the eigenvectors are always independent and the eigenvector matrix V *diagonalizes* the original matrix A if applied as a similarity transformation. However, if a matrix has repeated eigenvalues, it is not similar to a diagonal matrix unless it has a full (independent) set of eigenvectors. If the eigenvectors are not independent then the original matrix is said to be *defective*. Even if a matrix is defective, the solution from `eig` satisfies `A*X = X*D`.

The *generalized* eigenvalue problem is to determine the nontrivial solutions of the equation

$$Ax = sBx$$

where both A and B are n-by-n matrices and s is a scalar. The values of s that satisfy the equation are the *generalized eigenvalues* and the corresponding values of x are the *generalized right eigenvectors*.

If B is nonsingular, the problem could be solved by reducing it to a standard eigenvalue problem

$$B^{-1}Ax = sx$$

Because B can be singular, an alternative algorithm, called the QZ method, is necessary.

In MATLAB, the function `eig` solves for the generalized eigenvalues and eigenvectors when used with two input arguments:

`d = eig(A,B)`, if A and B are square matrices, returns a vector containing the generalized eigenvalues.

`[V,D] = eig(A,B)` produces a diagonal matrix `D` of generalized eigenvalues and a full matrix `V` whose columns are the corresponding eigenvectors so that `A*V = B*V*D`.

The eigenvectors are scaled so that the norm of each is 1.0.

Examples The matrix

```
B = [3 -2 -.9 2*eps;-2 4 -1 -eps;-eps/4 eps/2 -1 0;-.5  -.5 .1 1];
```

has elements on the order of roundoff error. It is an example for which the `nobalance` option is necessary to compute the eigenvectors correctly. Try the statements

```
[VB,DB] = eig(B)
B*VB - VB*DB
[VN,DN] = eig(B, 'nobalance')
B*VN - VN*DN
```

Algorithm For real matrices, `eig(X)` uses the EISPACK routines `BALANC`, `BALBAK`, `ORTHES`, `ORTRAN`, and `HQR2`. `BALANC` and `BALBAK` balance the input matrix. `ORTHES` converts a real general matrix to Hessenberg form using orthogonal similarity transformations. `ORTRAN` accumulates the transformations used by `ORTHES`. `HQR2` finds the eigenvalues and eigenvectors of a real upper Hessenberg matrix by the QR method. The EISPACK subroutine `HQR2` is modified to make computation of eigenvectors optional.

When `eig` is used with two arguments, the EISPACK routines `QZHES`, `QZIT`, `QZVAL`, and `QZVEC` solve for the generalized eigenvalues via the QZ algorithm. Modifications handle the complex case.

When `eig` is used with one complex argument, the solution is computed using the QZ algorithm as `eig(X,eye(X))`. Modifications to the QZ routines handle the special case $B = I$.

For detailed descriptions of these algorithms, see the *EISPACK Guide*.

Diagnostics If the limit of 30n iterations is exhausted while seeking an eigenvalue:

```
Solution will not converge.
```

See Also `balance`, `hess`, `qz`, `schur`

References

[1] Smith, B. T., J. M. Boyle, J. J. Dongarra, B. S. Garbow, Y. Ikebe, V. C. Klema, and C. B. Moler, *Matrix Eigensystem Routines – EISPACK Guide*, Lecture Notes in Computer Science, volume 6, second edition, Springer-Verlag, 1976.

[2] Garbow, B. S., J. M. Boyle, J. J. Dongarra, and C. B. Moler, *Matrix Eigensystem Routines – EISPACK Guide Extension, Lecture Notes in Computer Science*, volume 51, Springer-Verlag, 1977.

[3] Moler, C. B. and G.W. Stewart, *An Algorithm for Generalized Matrix Eigenvalue Problems*, SIAM J. Numer. Anal., Vol. 10, No. 2, April 1973.

---

ellipj

Purpose Jacobian elliptic functions.

Synopsis `[sn,cn,dn] = ellipj(u,m)`

Description `[sn,cn,dn] = ellipj(u,m)` returns the Jacobian elliptic functions `sn`, `cn`, and `dn` for the values in `u` evaluated at the corresponding parameters in `m`. Inputs `u` and `m` can contain scalars or matrices, but if they are both matrices, they must be the same size.

The Jacobian elliptic functions are defined in terms of the integral:

$$u = \int_0^{\phi} \frac{d\theta}{\left(1 - m\sin^2\theta\right)^{\frac{1}{2}}}$$

Then

$$sn(u) = \sin\phi,\ cn(u) = \cos\phi,\ dn(u) = \left(1 - m\ \sin^2\phi\right)^{\frac{1}{2}},\ am(u) = \phi$$

Sometimes the elliptic functions are defined differently, which can lead to some confusion. The alternate definitions are usually in terms of the modulus `k`, which is related to `m` by:

$$k^2 = m = \sin^2(\alpha)$$

The Jacobian elliptic functions obey many mathematical identities; for a good sample, see [1].

The accuracy of the result is `eps`; the value of `eps` can be changed for a less accurate, but more quickly computed answer.

Algorithm Compute the Jacobian elliptic functions using the method of the arithmetic-geometric mean [1]. Start with the triplet of numbers:

$$a_0 = 1,\ b_0 = (1 - m)^{\frac{1}{2}},\ c_0 = (m)^{\frac{1}{2}}$$

Compute successive iterates with

$$a_i = \frac{1}{2}(a_{i-1} + b_{i-1})$$

$$b_i = (a_{i-1}b_{i-1})^{\frac{1}{2}}$$

$$c_i = \frac{1}{2}(a_{i-1} - b_{i-1})$$

Next, calculate the amplitudes in radians using

$$\sin(2\phi_{n-1} - \phi_n) = \frac{c_n}{a_n}\sin(\phi_n)$$

being careful to unwrap the phases correctly. The Jacobian elliptic functions are then simply

$$sn(u) = \sin\phi_0$$
$$cn(u) = \cos\phi_0$$
$$dn(u) = \left(1 - m\ sn(u)^2\right)^{\frac{1}{2}}$$

Limitations `ellipj` is limited to the input domain $0 \le$ `m` ≤ 1. Other values of `m` can be mapped into this range using the transformations described in [1], equations 16.10 and 16.11. `u` is limited to real values.

See Also `ellipke`

References [1] Abramowitz, M. and I.A. Stegun, *Handbook of Mathematical Functions*, Dover Publications, 1965, 17.6.

---

ellipke

Purpose Complete elliptic integrals of the first and second kind.

Synopsis

```
K = ellipke(m)
[K,E] = ellipke(m)
```

Description `[K,E] = ellipke(m)` returns the complete elliptic integral of the first and second kinds for the elements of m. The complete elliptic integral of the first kind, `K(m) = F(pi/2|m)`, is defined by

$$K(m) = \int_0^1 \left[\left(1 - t^2\right)\left(1 - mt^2\right)\right]^{-\frac{1}{2}} dt = \int_0^{\frac{\pi}{2}} \left(1 - m\ \sin^2\theta\right)^{-\frac{1}{2}} d\theta$$

The complete elliptic integral of the second kind,

`E(m) = E(K(m)) = E(pi/2|m)`

is defined by

$$E(m) = \int_0^1 \left(1 - t^2\right)^{-\frac{1}{2}} \left(1 - mt^2\right)^{\frac{1}{2}} dt = \int_0^{\frac{\pi}{2}} \left(1 - m\ \sin^2\theta\right)^{\frac{1}{2}} d\theta$$

Some definitions of `K` and `E` use the modulus k instead of the parameter m. They are related by

$$k^2 = m = \sin^2\alpha$$

The accuracy of the result is `eps`; the value of `eps` can be changed for a less accurate, but more quickly computed answer.

Algorithm The complete elliptic integral is computed using the method of the arithmetic-geometric mean described in [1], section 17.6. Start with the triplet of numbers:

$$a_0 = 1,\ b_0 = (1-m)^{\frac{1}{2}},\ c_0 = (m)^{\frac{1}{2}}$$

Compute successive iterations of *ai*, *bi*, and *ci* with

$$a_i = \frac{1}{2}(a_{i-1} + b_{i-1})$$
$$b_i = (a_{i-1} b_{i-1})^{\frac{1}{2}}$$
$$c_i = \frac{1}{2}(a_{i-1} - b_{i-1})$$

stopping at iteration n when $cn \approx 0$, within the tolerance specified by `eps`. The complete elliptic integral of the first kind is then

$$K(m) = \frac{\pi}{2a_n}$$

Limitations `ellipke` is limited to the input domain $0 \le m \le 1$.

See Also `ellipj`

References [1] Abramowitz, M. and I.A. Stegun, *Handbook of Mathematical Functions*, Dover Publications, 1965, 17.6.

---

else

Purpose Conditionally execute statements.

Synopsis

```
if expression
     statements
else
     statements
end
```

Description `else` is used to delineate an alternate block of statements.

```
if expression
     statements
else
     statements
end
```

The second set of *`statements`* is executed if the *`expression`* has any zero elements. The expression is usually the result of

```
expression rop expression
```

where *`rop`* is `==`, `<`, `>`, `<=`, `>=`, or `~=`.

See Also `elseif`, `if`

elseif

Purpose Conditionally execute statements.

Synopsis

```
if expression
      statements
elseif expression
      statements
end
```

Description `elseif` conditionally executes statements.

```
if expression
      statements
elseif expression
      statements
end
```

The second block of *statements* executes if the first `expression` has any zero elements and the second *expression* has all nonzero elements. The expression is usually the result of

```
expression rop expression
```

where *rop* is ==, <, >, <=, >=, or ~=.

`else if`, with a space between the `else` and the `if`, differs from `elseif`, with no space. The former introduces a new, nested, `if`, which must have a matching end. The latter is used in a linear sequence of conditional statements with only one terminating `end`.

The following two segments produce identical results. Exactly one of the four assignments to `x` is executed, depending upon the values of the three logical expressions, `A`, `B`, and `C`.

```
if A                    if A
   x = a                   x = a
else                    elseif B
   if B                    x = b
      x = b             elseif C
   else                    x = c
       if C             else
          x = c            x = d
       else             end
          x = d
       end
   end
end
```

See Also `break`, `else`, `end`, `for`, `if`, `return`, `while`

end

Purpose Terminate `for`, `while`, and `if` statements.

Synopsis

```
while expression
      statements
end

if expression
      statements
end

for expression
      statements
end
```

Description `end` terminates the scope of `for`, `while`, and `if` statements. Without an `end` statement, `for`, `while`, and `if` wait for further input. Each `end` is paired with the closest previous unpaired `for`, `while`, or `if` and serves to delimit its scope.

Examples This example shows `end` used with `for` and `if`. Indentation provides easier readability.

```
for i = 1:n
   for j = 1:n
      if i == j
         a(i,j) = 2;
      elseif  abs(i;j) == 1
         a(i,j) = 1;
      else
         a(i,j) = 0;
      end
   end
end
```

See Also `break`, `for`, `if`, `return`, `while`

eps

Purpose Floating-point relative accuracy.

Synopsis `eps`

Description `eps` is a permanent variable whose value is initially the distance from 1.0 to the next largest floating-point number. It can be assigned any value, including 0. `eps` is a default tolerance for `pinv` and `rank`, as well as several other MATLAB functions, so changing the value of `eps` changes the way some functions operate. On machines with IEEE floating-point arithmetic, `eps = 2^(-52)`, which is roughly 2.22e–16.

See Also `isieee`, `realmax`, `realmin`

erf, erfc, erfcx

Purpose Error functions.

Synopsis

```
y = erf(x)
y = erfc(x)
y = erfcx(x)
```

Description `y = erf(x)`, the error function, is the integral of the Gaussian distribution function from `0` to `x`:

$$erf(x) = \frac{2}{\sqrt{\pi}} \int_0^x e^{-t^2} dt$$

`y = erfc(x)` returns the value of the complementary error function:

$$erfc(x) = \frac{2}{\sqrt{\pi}} \int_x^\infty e^{-t^2} dt = 1 - erf(x)$$

`y = erfcx(x)` returns the value of the scaled complementary error function:

$$erfcx(x) = e^{x^2} erfc(x)$$

For large `x`, `erfcx(x)` is approximately

$$\left(\frac{1}{\sqrt{\pi}}\right)\frac{1}{x}$$

Algorithm The MATLAB code is a translation of a FORTRAN program by W. J. Cody, Argonne National Laboratory, NETLIB/SPECFUN, March 19, 1990. The main computation evaluates near-minimax rational approximations from "Rational Chebyshev Approximations for the Error Function," by W. J. Cody, *Math. Comp.,* 1969, pgs. 631-638.

See Also `erfinv`

erfinv

Purpose Inverse of the error function.

Synopsis

```
x = erfinv(y)
```

Description `x = erfinv(y)`, where `−1 < y < 1` returns a value in the range `−Inf < x < Inf` and satisfies `y = erf(x)` to within roundoff error. `erfinv(1)` is `Inf`, `erfinv(-1)` is `-Inf`. For `abs(y) > 1`, `erfinv(y)` is `NaN`.

Algorithm Rational approximations accurate to approximately six significant digits are used to generate an initial approximation, which is then improved to full accuracy by two steps of Newton's method. The M-file can easily be modified to eliminate the Newton improvement. The resulting code is about three times faster in execution, but is considerably less accurate.

See Also `erf`

---

error

Purpose Display error messages.

Synopsis `error('text')`

Description `error('text')` displays the text in the quoted string and causes an error message to return to the keyboard.

Examples `error` provides an error return from M-files.

```
function foo(x,y)
if nargin ~= 2
     error('Wrong number of input arguments')
end
```

See Also `break`, `disp`, `return`

---

errorbar

Purpose Errorbar plot.

Synopsis `errorbar(x,y,e)`

Description `errorbar(x,y,e)` plots the graph of vector `x` versus vector `y` with errorbars specified by vector `e`. Vectors `x`, `y`, and `e` must be the same length. The error bars are each drawn a distance of `e(i)` above and below the points in `(x,y)` so that each bar is `2*e(i)` long.

If `x`, `y`, and `e` are matrices of the same size, one errorbar graph per column is drawn.

Examples This example shows how error bars can be added to an existing plot.

```
x = 0: 0.1: 2;
y = erf(x);
e = rand(size(x))/10;
errorbar(x,y,e)
```

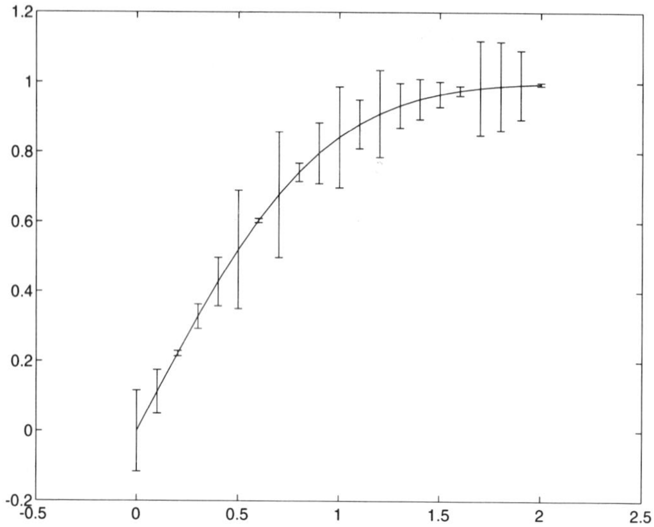

See Also bar, plot, std

etime

Purpose Elapsed time.

Synopsis `e = etime(t2,t1)`

Description `etime(t2,t1)` returns the time in seconds between `clock` output vectors `t1` and `t2`.

Examples Calculate how long a 2048-point real FFT takes.

```
x = rand(2048,1);
t = clock; fft(x); etime(clock,t)
ans =
        0.4167
```

Limitations `etime` fails across month and year boundaries. It can be fixed with some effort; see the `etime` M-file.

See Also clock, cputime, date, tic, toc

etree

Purpose Elimination tree of a matrix.

Synopsis

```
p = etree(A)
p = etree(A,'col')
[p,q] = etree(...)
```

Description `p = etree(A)` returns an elimination tree for the square symmetric matrix whose upper triangle is that of `A`. `p(j)` is the parent of column `j` in the tree, or 0 if `j` is a root.

`p = etree(A,'col')` returns the elimination tree of `A'*A`.

`[p,q] = etree(...)` returns a postorder permutation `q` on the tree.

Examples

```
etreeplot(sprandsym(50, 0.1))
```

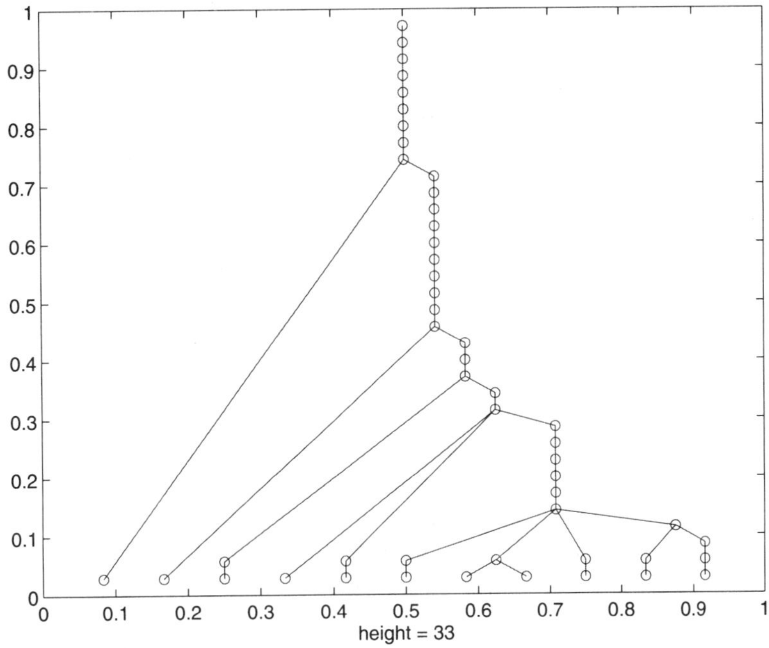

References

[1] Gilbert, John R., Cleve Moler and Robert Schreiber, "Sparse Matrices in MATLAB: Design and Implementation," *SIAM Journal on Matrix Analysis and Applications 13*, pp. 333-356, Jan. 1992.

eval

Purpose Interpret strings containing MATLAB expressions.

Synopsis

```
x = eval('string')
eval('string')
```

Description `eval('string')`, where *string* is a MATLAB text string, causes MATLAB to interpret the string as an expression or statement.

`x = eval('string')`, where *string* is an expression, returns the value of that expression. If *string* represents a statement, `eval('string')` executes that statement. *string* is often created by concatenating substrings and variables inside square brackets.

Examples The statements

```
s = '4*atan(1)';
pi = eval(s)
```

simply reset the value of `pi`.

The following loop generates a sequence of 12 matrices named `M1` through `M12`:

```
for n = 1:12
        eval(['M',int2str(n),' = magic(n)'])
end
```

The next example runs a selected M-file script. Note that the strings making up the rows of matrix `D` must all have the same length.

```
D = ['odedemo '
     'quaddemo'
     'zerodemo'
     'fitdemo '];
n = input('Select a demo number: ');
eval(D(n,:))
```

The final example reads and processes the data in several files with the names `data1.dat`, `data2.dat`, and so on.

```
k = 0;
while 1
   k = k+1;
   datak = ['data' int2str(k)];
   filename = [datak '.dat'];
   if ~exist(filename), break, end
   eval(['load ' filename]);
   X = eval(datak);
   % Process data in matrix X.
end
```

See Also `feval`

---

exist

Purpose Check if a variable or file exists.

Synopsis `e = exist('item')`

Description `exist('item')`, where *item* is either a variable or a file, returns

0 if A is none of the following

1 if the variable A exists in the workspace

2 if either A or A.m is the name of a file on disk

3 if A is a MEX-file

4 if A is a compiled SIMULINK function

5 if A is a built-in MATLAB function

See Also `dir`, `help`, `lookfor`, `what`, `which`, `who`

---

exp

Purpose Exponential.

Synopsis `Y = exp(X)`

Description `exp` is an elementary function that operates element-wise on matrices. Its domain includes complex numbers, which can lead to unexpected results if used unintentionally.

`Y = exp(X)` returns the exponential for each of the elements of X. For complex $z = x + i*y$, it returns the complex exponential,

$$e^z = e^x(\cos(y) + i\sin(y))$$

Use `expm` for the matrix exponential.

See Also `expm`, `log`, `log10`

---

expint

Purpose Exponential integral.

Synopsis `Y = expint(X)`

Description `Y = expint(X)` evaluates the exponential integral

$$\int_x^\infty \frac{e^{-t}}{t}\,dt$$

for each element of X.

Algorithm For elements of X in the range [–38,2], `expint` uses a series expansion representation (equation 5.1.11 in [1]):

$$E_i(x) = -\gamma - \ln x - \sum_{n=1}^{\infty} \frac{(-1)^n x^n}{nn!}$$

For all other elements of `X`, `expint` uses a continued fraction representation (equation 5.1.22 in [1]):

$$E_n(x) = e^{-x}\left(\frac{1}{x+}\frac{n}{1+}\frac{1}{x+}\frac{n+1}{1+}\frac{2}{x+}\cdots\right), \quad n = 1$$

References [1] Abramowitz, M. and I. A. Stegun. *Handbook of Mathematical Functions.* Chapter 5, New York: Dover Publications, 1965.

---

expm

Purpose Matrix exponential.

Synopsis `Y = expm(X)`

Description `expm(X)` is the matrix exponential of `X`. Complex results are produced if `X` has nonpositive eigenvalues. Use `exp` for the element-by-element exponential.

Algorithm `expm` is a built-in function, but it uses the Padé approximation with scaling and squaring algorithm expressed in the file `expm1.m`.

A second method of calculating the matrix exponential is via a Taylor series approximation. This method can be found in the file `expm2.m`. This method is not recommended as a general purpose method. It is often slow and inaccurate.

A third way of calculating the matrix exponential, found in the file `expm3.m`, is to diagonalize the matrix, apply the function to the individual eigenvalues, and then transform back. This method fails if the input matrix does not have a full set of linearly independent eigenvalues.

References [1] and [2] describe and compare many algorithms for computing `expm(X)`. The built-in method, `expm1`, is essentially method 3 of [2].

Examples Suppose `A` is the 3-by-3 matrix

```
1     1     0
0     0     2
0     0    –1
```

then `X = expm(A)` is

```
2.7183    1.7183    1.0862
     0    1.0000    1.2642
     0         0    0.3679
```

while `exp(A)` is

```
2.7183    2.7183    1.0000
1.0000    1.0000    7.3891
1.0000    1.0000    0.3679
```

Notice that the diagonal elements of the two results are equal; this would be true for any triangular matrix. But the off-diagonal elements, including those below the diagonal, are different.

See Also `exp`, `funm`, `logm`, `sqrtm`

References [1] G. H. Golub and C. F. Van Loan, *Matrix Computation*, p. 384, Johns Hopkins University Press, 1983.

[2] C. B. Moler and C. F. Van Loan, "Nineteen Dubious Ways to Compute the Exponential of a Matrix," *SIAM Review 20*, pp. 801-836, 1979.

---

expo

Purpose Begin MATLAB Expo.

Synopsis

```
expo
```

Description `expo` displays the initial screen for the MATLAB Expo software. MATLAB Expo is a collection of demos presented in a simple, user-driven format. Use `expo` to find out more about the capabilities of MATLAB, SIMULINK, and the applications toolboxes.

---

eye

Purpose Identity matrix.

Synopsis

```
Y = eye(n)
Y = eye(m,n)
Y = eye(size(A))
```

Description `Y = eye(n)` is the `n`-by-`n` identity matrix.

`Y = eye(m,n)` or `eye([m,n])` is an `m`-by-`n` matrix with 1s on the diagonal and 0s elsewhere.

`Y = eye(size(A))` is the same size as `A`.

See Also `ones`, `rand`, `zeros`

---

fclose

Purpose Close one or more open files.

Synopsis

```
status = fclose(fid)
status = fclose('all')
```

Description `fclose(fid)` closes the specified file, if it is open, and returns 0 if successful and -1 if unsuccessful. `fid` is a file identifier associated with an open file. (See `fopen` for a complete description of `fid`.)

`fclose('all')` closes all open files, (except standard input, output, and error) and returns 0 if successful and -1 if unsuccessful.

See Also `ferror`, `fopen`, `fprintf`, `fread`, `fscanf`, `fseek`, `ftell`, `fwrite`, `sprintf`, `sscan`

feather

Purpose Feather plot.

Synopsis

```
feather(z)
feather(x,y)
```

Description `feather(z)` draws a graph that displays the angle and magnitude of the complex elements of z as arrows emanating from equally spaced points along a horizontal axis.

`feather(x,y)` is equivalent to `feather(x+yi)`. It displays the feather plot for the angles and magnitudes of the elements of matrices `x` and `y`.

Examples Draw a simple feather plot with some random data:

```
z = randn(3,3)+randn(3,3)*j;
feather(z)
```

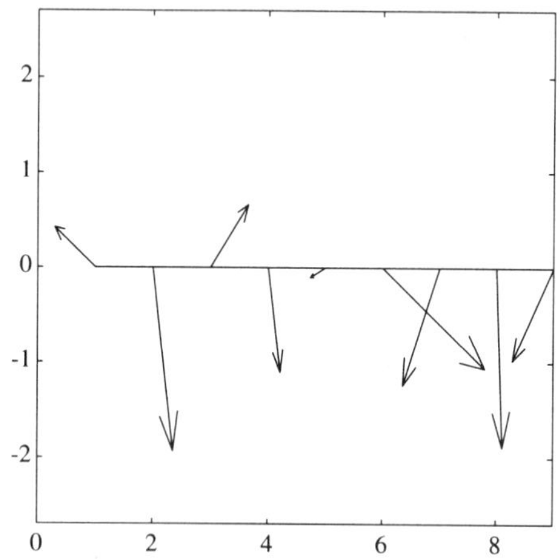

See Also `compass`, `rose`

feof

Purpose Test for end-of-file.

Synopsis `eofstat = feof(fid)`

Description `feof(fid)` tests whether the end-of-file indicator is set for the file with identifier `fid`. It returns 1 if the end-of-file indicator is set, or 0 if it is not.

See Also `fopen`

ferror

Purpose Query MATLAB about errors in file input or output.

Synopsis

```
message = ferror(fid)
message = ferror(fid,'clear')
[message, errnum] = ferror(...)
```

Description `message = ferror(fid)` returns the error message `message`. `fid` is a file identifier associated with an open file. (See `fopen` for a complete description of `fid`.)

`message = ferror(fid,'clear')` clears the error indicator for the specified file.

`[message, errnum] = ferror(...)` returns the error status number `errnum` of the most recent file I/O operation associated with the specified file.

If the most recent I/O operation performed on the specified file was successful, `ferror` returns an `errnum` value of 0 and the value of `message` is empty.

A nonzero `errnum` indicates an error occurred in the most recent file I/O operation. Two ways that you can learn more about the nature of the error are

- The value of `message` is a string that may contain information about the nature of the error.
- If you have access to a C language reference manual, you can look at the error codes generated by file I/O operations provided by the C language library `stdio`.

See Also `fclose`, `fopen`, `fprintf`, `fread`, `fscanf`, `fseek`, `ftell`, `fwrite`, `sprintf`, `sscanf`

feval

Purpose Function evaluation.

Synopsis

```
feval('function',x1,...,xn)
[y1,y2, ...] = feval('function',x1,...,xn)
```

Description `feval('`*`function`*`',x1, ...,xn)` passes the arguments `(x1, ...,xn)` to the named function.

`[y1,y2...] = feval('`*`function`*`',x1,...,xn)` returns multiple output arguments.

Examples The statements

```
[V,D] = feval('eig',A)
[V,D] = eig(A)
```

are equivalent. `feval` is useful in functions that accept string arguments specifying function names. For example, the function

```
function plotf(fun,x)
y = feval(fun,x);
plot(x,y)
```

can be used to graph other functions.

See Also `eval`, `fmin`, `funm`, `ode23`

---

fft

Purpose 1-D fast Fourier transform.

Synopsis

```
y = fft(X)
y = fft(X,n)
```

Description `y = fft(X)` is the discrete Fourier transform of vector `X`, computed with a fast Fourier transform (FFT) algorithm. If `X` is a matrix, `fft(X)` is the FFT of each column of the matrix.

`y = fft(X,n)` is the `n`-point FFT. If the length of `X` is less than `n`, `X` is padded with trailing zeros to length `n`. If the length of `X` is greater than `n`, the sequence `X` is truncated. When `X` is a matrix, the length of the columns are adjusted in the same manner.

The `fft` function employs a radix-2 fast Fourier transform algorithm if the length of the sequence is a power of two, and a slower mixed-radix algorithm if it is not.

`X = fft(x)` and `x = ifft(X)` implement the transform and inverse transform pair given for vectors of length `N` by

$$X(k) = \sum_{j=1}^{N} x(j)\omega_N{}^{(j-1)(k-1)}$$

$$x(j) = (1/N)\sum_{k=1}^{N} X(k)\omega_N{}^{-(j-1)(k-1)}$$

where

$$\omega_N = e^{-2\pi i/N}$$

is an `n`-th root of unity.

Examples A common use of Fourier transforms is to find the frequency components of a signal buried in a noisy time domain signal. Consider data sampled at 1000 Hz. Form a signal containing 50 Hz and 120 Hz and corrupt it with some zero-mean random noise:

```
t = 0:0.001:0.6;
x = sin(2*pi*50*t)+sin(2*pi*120*t);
y = x + 2*randn(size(t));
plot(y(1:50))
```

It is difficult to identify the frequency components from looking at the original signal. Converting to the frequency domain, the discrete Fourier transform of the noisy signal `y` is found by taking the 512-point fast Fourier transform (FFT):

```
Y = fft(y,512);
```

The power spectral density, a measurement of the energy at various frequencies, is

```
Pyy = Y.* conj(Y) / 512;
```

The first 256 points (the other 256 points are symmetric) can be graphed on a meaningful frequency axis with

```
f = 1000*(0:255)/512;
plot(f,Pyy(1:256))
```

Algorithm When the sequence length is a power of two, a high-speed radix-2 fast Fourier transform algorithm is employed. The radix-2 FFT routine is optimized to perform a real FFT if the input sequence is purely real, otherwise it computes the complex FFT. This causes a real power-of-two FFT to be about 40% faster than a complex FFT of the same length.

When the sequence length is not an exact power of two, an alternate algorithm finds the prime factors of the sequence length and computes the mixed-radix discrete Fourier transforms of the shorter sequences.

The time it takes to compute an FFT varies greatly depending upon the sequence length. The FFT of sequences whose lengths have many prime factors is computed quickly; the FFT of those that have few is not. Sequences whose lengths are prime numbers are reduced to the raw (and slow) discrete Fourier transform (DFT) algorithm. For this reason it is generally better to stay with power-of-two FFTs unless other circumstances dictate that this cannot be done. For example, on one machine a 4096-point real FFT takes 2.1 seconds and a complex FFT of the same length takes 3.7 seconds. The FFTs of neighboring sequences of length 4095 and 4097, however, take 7 seconds and 58 seconds, respectively.

See Also `fft2`, `fftshift`, `ifft`

`dftmtx`, `filter`, `freqz`, `specplot`, and `spectrum` in the Signal Processing Toolbox

---

fft2

Purpose 2-D fast Fourier transform.

Synopsis

```
Y = fft2(X)
Y = fft2(X,m,n)
```

Description `Y = fft2(X)` performs the two-dimensional FFT. The result `Y` is the same size as `X`.

`Y = fft2(X,m,n)` either truncates `X` or pads `X` with 0s to create an `m`-by-`n` array before doing the transform. The result is `m`-by-`n`.

Algorithm `fft2(X)` is simply

```
fft(fft(X).').'
```

This computes the one-dimensional FFT of each column `X`, then of each row of the result. The time required to compute `fft2(X)` depends strongly on the number of prime factors in `[m,n] = size(X)`. It is fastest when `m` and `n` are powers of 2.

See Also `fft`, `fftshift`, `ifft2`

---

fftshift

Purpose Rearrange the outputs of `fft` and `fft2`.

Synopsis `Y = fftshift(X)`

Description `Y = fftshift(X)` rearranges the outputs of `fft` and `fft2` by moving the zero frequency component to the center of the spectrum, which is sometimes a more convenient form.

If `X` is a vector, `Y` is a vector with the left and right halves swapped.

If `X` is a matrix, `Y` is a matrix with quadrants one and three swapped with quadrants two and four.

Examples For any matrix `X`

```
Y = fft2(X)
```

has `Y(1,1) = sum(sum(X))`; the DC component of the signal is in the upper-left corner of the two-dimensional FFT. For

```
Z = fftshift(Y)
```

this DC component is near the center of the matrix.

See Also `fft`, `fft2`

---

fgetl

Purpose Return the next line of a file as a string without newlines.

Synopsis `line = fgetl(fid)`

Description `line = fgetl(fid)` returns the next line of the file with identifier `fid`. `line` is a MATLAB string that does not include the newline and carriage return associated with the text line (to obtain those characters, use `fgets`). If `fgetl` encounters an end-of-file indicator, it returns -1.

`fgetl` is intended for use with text files only. Given a binary file with no newline characters, `fgetl` may require long execution time.

See Also `fgets`

---

fgets

Purpose Return the next line of a file as a string with newlines.

Synopsis `line = fgets(fid)`

Description `line = fgets(fid)` returns the next line for the file with identifier `fid`. `line` is a MATLAB string that includes the newline character associated with the text line (to obtain the string without the newline, use `fgetl`). If `fgets` encounters an end-of-file indicator, it returns -1.

`fgets` is intended for use with text files only. Given a binary file with no newline characters, `fgets` may require long execution time.

See Also `fgetl`

figure

Purpose Open new graphics window by creating figure object.

Synopsis

```
figure
h = figure
figure(h)
h = figure('PropertyName',PropertyValue,...)
```

Description Figure objects are children of the root object and parents of axes, uimenus, and uicontrols. They are the individual windows on the screen in which MATLAB displays graphical output. When you create a figure, you create a new window whose characteristics are controlled by a number of factors including the windowing system and the figure properties.

`figure`, by itself, opens a new figure (graph window).

`h = figure` opens a new figure and returns the next available figure number, also known as the *figure handle*. Figure handles are integers numbered sequentially starting at 1 and are displayed in the border at the top of the window.

`figure(h)` makes the figure with handle `h` the current figure for subsequent plotting commands.

`h = figure('PropertyName',PropertyValue,...)` accepts property name/property value pairs as input arguments. These properties are described under "Object Properties." You can also set and query property values after creating the object using the `set` and `get` functions.

Use `gcf` (**g**et **c**urrent **f**igure) to obtain the handle of the current figure. The current figure is the window in which any drawing command is directed.

Specify default figure properties at the figure's parent level, that is, at the root object. To do so, call the `set` function, supplying as arguments the handle 0 (for the root), a default name string, and the desired default value. Construct the default name string by prepending the string `'DefaultFigure'` to the desired figure property name. For example

```
set(0,'DefaultFigureColor','blue')
```

sets the default figure background color to blue.

Object Properties

This section lists property names along with the type of values each accepts.

`BackingStore` Store copy of figure window for fast refresh.

`on` (Default.) Store copy of figure. When obscured parts of the figure are uncovered, the system restores them without rerendering. This causes faster screen redraws, but uses more memory.

	`off`	Redraw previously obscured portions of the figure.

`ButtonDownFcn` — Callback string, object selection.

Any legal MATLAB expression, including the name of an M-file or function. When you click the mouse button with the cursor over the figure, the string is passed to the `eval` function to execute the specified function. Initially the empty matrix.

`Children` — Children of figure.

A read-only vector containing the handles of all objects displayed within the figure. The children objects of figures can be axes, uicontrols, and uimenus.

`Clipping` — Data clipping.

`on`	(Default.) No effect for figure objects.
`off`	No effect for figure objects.

`Color` — Figure background color.

A three-element RGB vector or one of MATLAB's predefined names, specifying the background color for the figure. The default `Color` is black. See the `ColorSpec` reference page for more information on specifying color.

`ColorMap` — Figure colormap.

An `m`-by-3 matrix of RGB values. Graphics commands index the colors in the colormap by their row number. For example, an index of 1 specifies the first RGB triplet, an index of 2 specifies the second RGB triplet, and so on. The maximum colormap length is determined by your system (usually slightly less than 256 colors), but the colormap must be three columns wide. The default figure colormap contains 64 predefined colors.

You can replace the default colormap, `hsv`, with your own or with one of MATLAB's predefined colormaps. To define a colormap, specify the intensity of the red component in the first column, the green component in the second column and the blue component in the third column. Color intensities are floating-point numbers in the range [0 1]. A value of 0 indicates no intensity and a value of 1 indicates full intensity. An RGB triplet of [0 0 0] specifies black and [1 1 1] specifies white.

Colormaps affect the rendering of surface, image, and patch objects, but generally do not affect other objects. See the `colormap` and `ColorSpec` reference pages for more information.

`CurrentAxes`	Figure's current axes. The handle of the figure's current axes. This is the handle returned by `gca` when this figure is the current figure. If axes children exist, there is always a current axes for the figure. If there are no axes in a figure, `get(gcf,'CurrentAxes')` creates one and returns its handle.
`CurrentCharacter`	Last key pressed. The last key pressed in the current figure window.
`CurrentMenu`	Figure's current menu. The handle of the figure's current (most recently selected) menu. If you use a single menu handler for multiple menu items, query this property to determine which menu called the handler.
`CurrentObject`	Figure's current object. The handle of the figure's current object, the object located at the `CurrentPoint` (see below). This object is the front-most object in the stacking order. Use this property to determine which object you have selected.
`CurrentPoint`	Location of last mouse press or release in the figure. A two-element vector `[x y]` containing the *x*- and *y*-coordinates of the location in the figure window where you last pressed or released the mouse button. This is measured from the lower-left corner of the figure window in units specified by the `Units` property. Initially [0 0].
`FixedColors`	Fixed colors allotted for figure. An `n`-by-3 matrix of RGB values that define the fixed colors for the figure. *Fixed colors* take up slots in the system color table, reducing the number of slots available for colormaps. Initial fixed colors are white and black.
`Interruptible`	Callback interruptibility. `yes` — Any of the figure's callbacks are interruptible by other callbacks. `no` — (Default.) The callbacks are not interruptible.
`InvertHardCopy`	Change hardcopy from white on black to black on white. `on` — (Default.) For printed output, change black figure background to white, and lines, text, and so on to black. Surface and patch colors are not affected. If the output device does

		not support color, lines and text in colors other than black or white are changed to either black or white (whichever provides the most contrast with the paper background).
	`off`	Color of printed output exactly matches screen display.

`KeyPressFcn`

Callback string, key press.

Any legal MATLAB expression, including the name of an M-file or function. When you press a keyboard key while the pointer is within the figure, the string is passed to the `eval` function to execute the specified function. Use the `CurrentCharacter` property to determine which key was pressed. Use the `PointerWindow` property of the root object to determine in which figure the key was pressed. (Note that pressing a keyboard key over a figure does not make that figure active.) Initially the empty matrix.

`MenuBar`

Display menu bar.

`figure`	(Default.) Display default MATLAB menus for figure. On some systems, these menus appear at the top of the figure window; on others, they appear at the top of the screen.
`none`	Do not display default MATLAB menus. Some systems still display system-specific menus.

`MinColorMap`

Minimum number of color table entries to use.

The minimum number of color table slots that MATLAB will make available at any time.

MATLAB keeps track of the indices in use by other applications, and avoids using them. With no color-intensive applications running, more than 200 slots can be available. This allows several 64-color MATLAB colormaps (the default size) and many other fixed colors to coexist with the rest of the applications on the screen. Figure windows display correctly even when not active.

If there are not enough free slots to install a desired figure colormap without causing false colors in other applications, the figure changes as many slots as needed to reach `MinColorMap` number of slots. If the resulting number of slots is smaller than the length of the colormap, individual colormap entries are grouped together piecewise with their neighbors to create as many groups as there are available slots in the color table. In this case, all members of a given group display in the color of the group's center member.

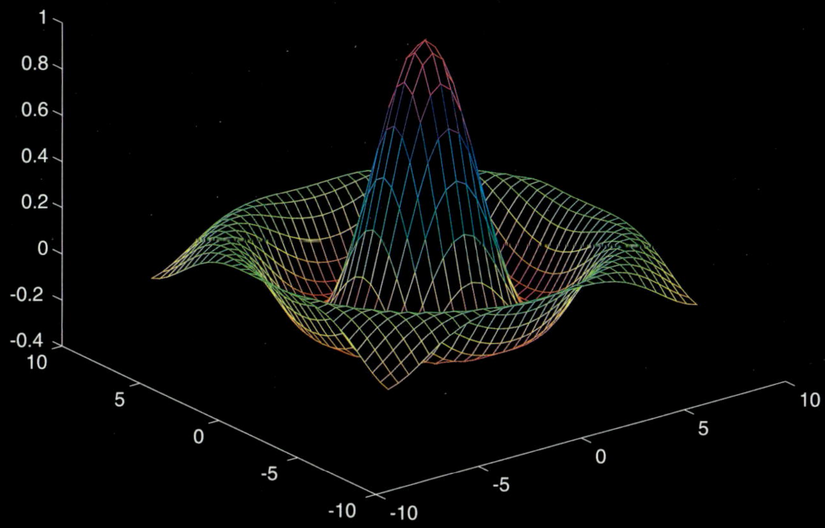

Plate 1. Mesh plot of sin(R)/(R)

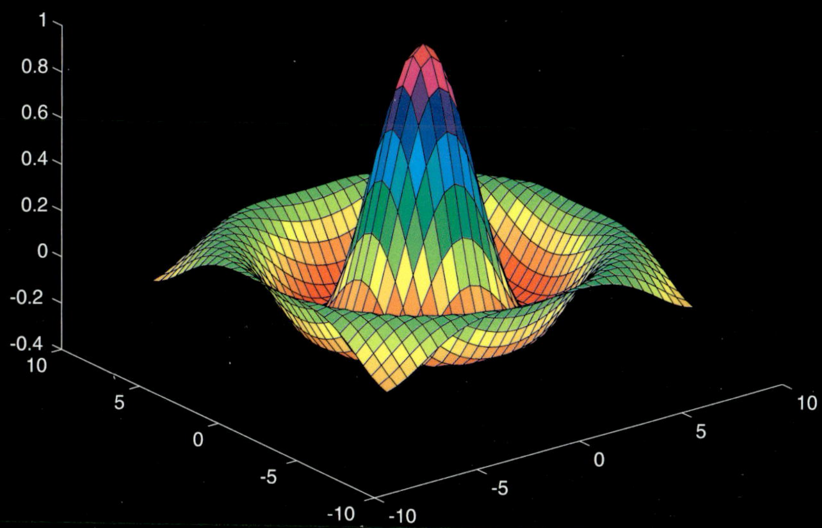

Plate 2. Surface plot of sin(R)/R

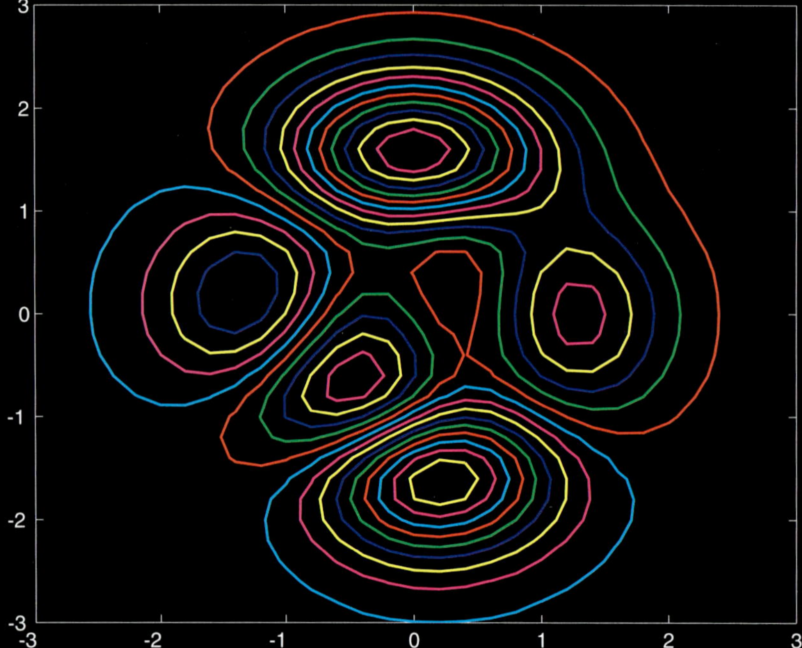

Plate 3. Two-dimensional contour plot of the peaks function

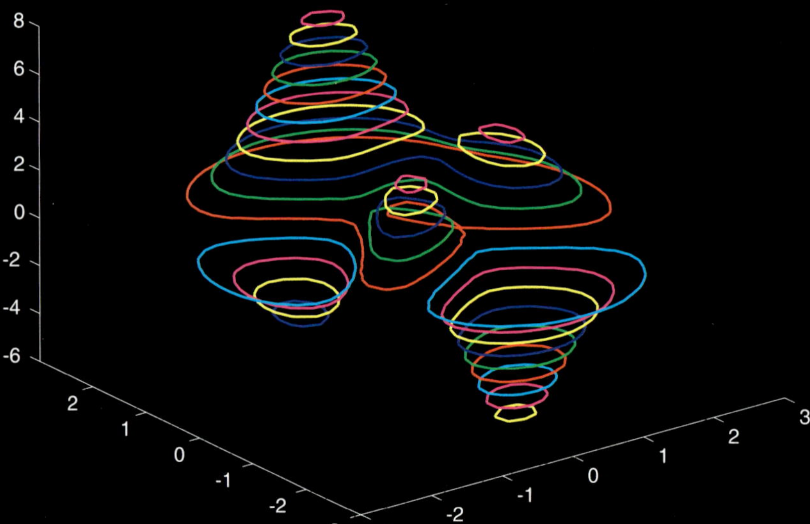

Plate 4. Three-dimensional contour plot of the peaks function

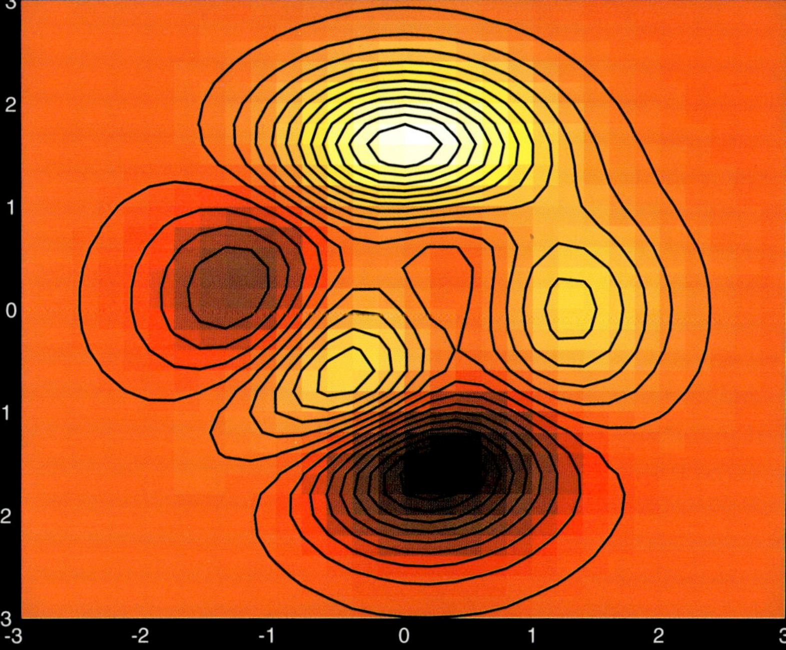

Plate 5. Combined pseudocolor and contour plots of the peaks function

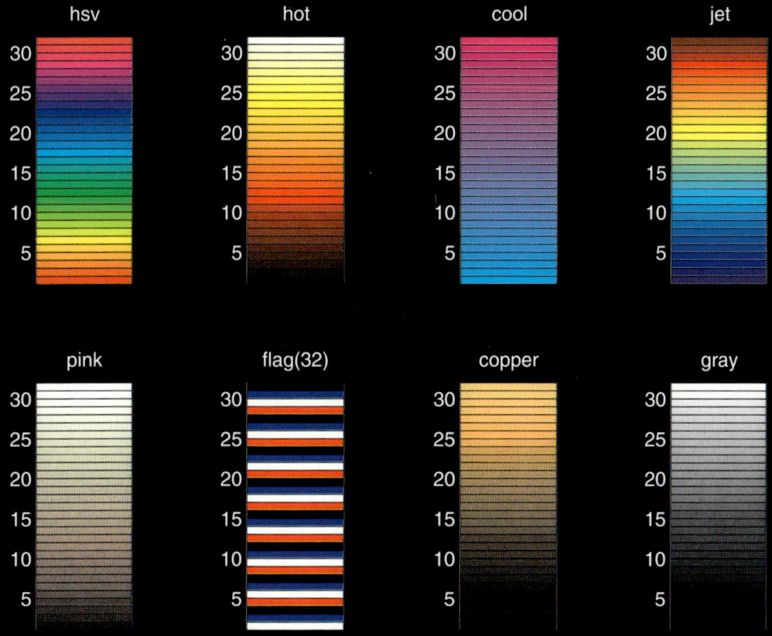

Plate 6. Eight color maps created by the indicated MATLAB functions, displayed as pseudocolor plots

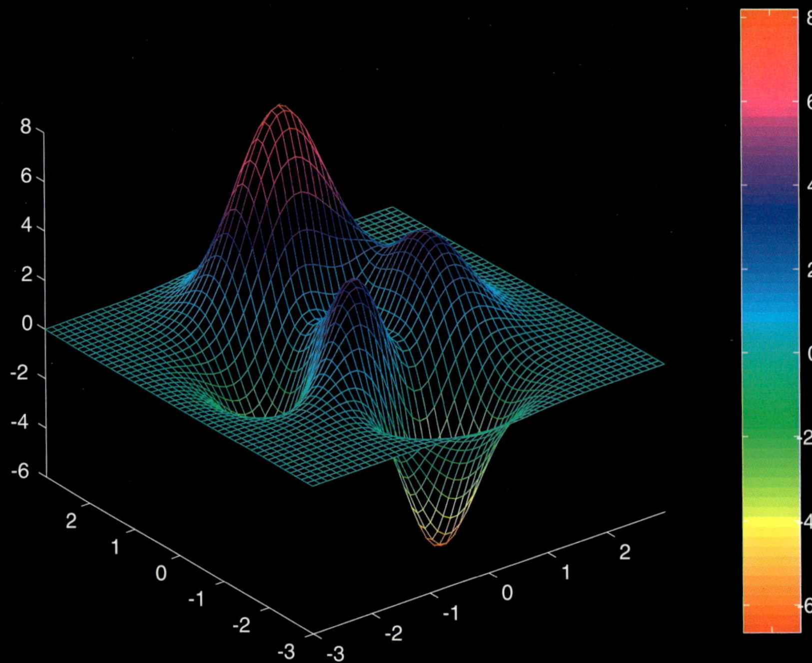

Plate 7. Mesh plot of the peaks function with a color scale

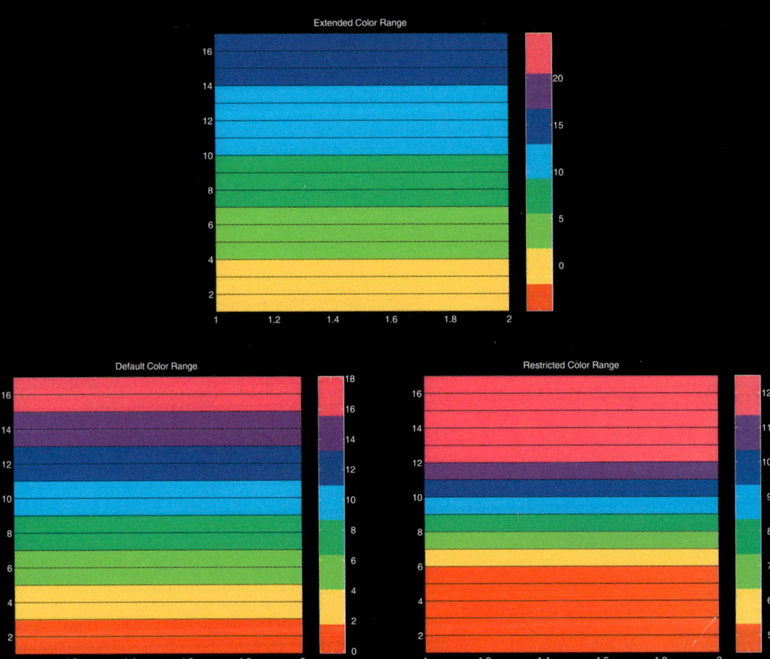

Plate 8. Pseudocolor plots illustrating the effects of extending and restricting the color range

This scheme does not interpolate lost colors, but merely subsamples the desired colormap to fit in the available number of color table slots. This works well for smoothly varying colormaps and poorly for colormaps with irregularly located colors. For images, whose colormaps often consist of irregularly located colors, it is best to set `MinColormap` to the number of distinct colors in the image to avoid displaying it in false colors, even when the figure is active.

If there are several figure windows open with similar or with small colormaps, they display simultaneously in their correct colors. If the colormaps are very large, and/or very different, it is likely that the inactive windows display in false color. Clicking on an inactive window to make it active restores the correct colors.

`Name`

Figure window title.

A string specifying a name for the window that contains the figure. By default, this property is the empty string and figure titles are `Figure No. 1`, `Figure No. 2`, and so on. If you set `Name` to *`string`*, the title becomes `Figure No. n:` *`string`*.

`NextPlot`

Figure handling for subsequent plots.

`new`	Create a new figure before drawing.
`add`	(Default.) Add new objects to the current figure.
`replace`	Reset all figure properties except `Position` and delete all figure children before drawing.

This property is used by the built-in high-level graphing functions `plot`, `plot3`, `fill`, `fill3`, and the M-file graphing functions `mesh`, `surf`, `bar`, and so on. The M-file `newplot` is a preamble for handling the `NextPlot` property. M-file graphing functions like `mesh`, `surf`, and `bar` call `newplot` to control this property before drawing their respective graphs. When creating M-files that implement graphing commands, use `newplot` at the beginning of the file. See the M-file `pcolor` for an example. Also see the `NextPlot` property of axes objects.

`NumberTitle`

Prepend figure window number to title.

`on`	(Default.) Prepend the string `Figure No. N` to the figure window title.
`off`	Do not add figure number to the figure window title.

`PaperOrientation`	Paper orientation.	
	`portrait`	(Default.) Portrait orientation - longest dimension of page is vertical
	`land`	Landscape orientation - longest dimension of page is horizontal.
`PaperPosition`	Location of figure on printed page. A four-element vector, `[left,bottom,width,height]`, where `left` and `bottom` are the distance from the lower-left corner of the page to the lower-left corner of the printed figure window. `width` and `height` are the dimensions of the printed figure window. All measurements are in units specified by the `PaperUnits` property. Initial setting depends on `PaperSize` and `PaperOrientation`.	
`PaperSize`	Paper size. A two-element vector, `[width,height]`, containing the current `PaperSize` measured in `PaperUnits`. The default `PaperSize` is `[8.5 11]`.	
`PaperType`	Type of printed page.	
	`usletter`	(Default.) Standard US letter.
	`uslegal`	Standard US legal.
	`a4letter`	European A4.
	When the `PaperUnits` property is set to normal, MATLAB uses `PaperType` to scale printed figures so they fit on the page correctly.	
`PaperUnits`	Unit of measurement for hardcopy.	
	`normal`	Normalized coordinates, where the lower-left corner of the page maps to (0,0) and the upper-right corner to (1.0,1.0).
	`inches`	(Default.) Inches.
	`cent`	Centimeters.
	`points`	Points. Each point is equivalent to 1/72 of an inch.
	This property specifies the unit for `PaperPosition` and `PaperSize` measurements. If you change the value of `PaperUnits`, it is good practice to return it to its default value after completing your computation so as not to affect other functions that assume `PaperUnits` is set to the default value.	

`Parent`	Handle of the figure's parent object. A read-only property that contains the handle of the figure's parent object. The parent of a figure object is always the root, so this property is always 0.
`Pointer`	Symbol to indicate cursor position in the figure window.

`arrow`	(Default.) Arrow.
`crossh`	Crosshair.
`watch`	Watch.
`cross`	Double-line cross.
`topl`	Arrow pointing to upper-left corner of figure.
`topr`	Arrow pointing to upper-right corner of figure.
`botl`	Arrow pointing to lower-left corner of figure.
`botr`	Arrow pointing to lower-right corner of figure.
`circle`	Circle.
`fleur`	Four-headed arrow (compass).

`Position`	Size and position of figure. A four-element vector, `[left,bottom,width,height]`, where `left` and `bottom` are the distance from the lower-left corner of the screen to the lower-left corner of the figure window. `width` and `height` are the dimensions of the figure window. All measurements are in units specified by the `Units` property.
`Resize`	Window resize mode.

`on`	(Default.) Users can resize the figure window using the mouse.
`off`	Users cannot resize the figure window.

`SelectionType`	Mouse selection type. A read-only string providing information about the last mouse button press, indicating a type of selection. All selection types exist on all platforms, although the action required to implement the selections may differ. This list describes the X-Windows based selection system.

`normal`	(Default.) Click the mouse button (press and release) while the pointer is over the object to select.

For a two-button mouse, click and release the left mouse button.

`extended`	Hold down the **Shift** key and make a `normal` selection. Add selections by keeping the shift key depressed.

For a two-button mouse, make `extended` selections by clicking both buttons simultaneously. For a three-button mouse, use mouse button 2.

`alt`	Hold down the **Control** key and make a `normal` selection. For a two- or three-button mouse, press the right-most button.
`open`	Double-click the mouse while the pointer is over the object to select. For a two- or three-button mouse, use the same button for both clicks.

`ShareColors`

Share color table slots.

`yes`	(Default.) MATLAB minimizes color table use by reusing existing color table slots whenever possible. This allows the maximum number of windows to render their contents accurately.
`no`	Do not share color table slots with other windows.

`Type`

Type of graphics object.

A read-only string; always `'figure'` for a figure object.

`Units`

Unit of measurement for screen display.

`pixels`	(Default.) Screen pixels.
`normalized`	Normalized coordinates, where the lower-left corner of the figure window maps to (0,0) and the upper-right corner to (1.0,1.0).
`inches`	Inches.
`cent`	Centimeters.
`points`	Points. Each point is equivalent to 1/72 of an inch.

All units are measured from the lower-left corner of the window. If you change the value of `Units`, it is good practice to return it to its default value after completing your computation so as not to affect other functions that assume `Units` is set to the default value.

`UserData`	User-specified data.	
	Any matrix you want to associate with the figure object. The object does not use this data, but you can retrieve it using the `get` command.	
`Visible`	Figure visibility.	
	`on`	(Default.) Figure window is visible on the screen.
	`off`	Figure window is not visible on the screen.
`WindowButtonDownFcn`	Callback string, mouse button press.	
	Any legal MATLAB expression, including the name of an M-file or function. When you press a mouse button while the pointer is within the figure (but not over a uicontrol), the string is passed to the `eval` function to execute the specified function. Initially the empty matrix.	
`WindowButtonMotionFcn`	Callback string, pointer motion.	
	Any legal MATLAB expression, including the name of an M-file or function. When you move the pointer within the figure, the string is passed to the `eval` function to execute the specified function. Initially the empty matrix.	
`WindowButtonUpFcn`	Callback string, mouse button release.	
	Any legal MATLAB expression, including the name of an M-file or function. When you release a mouse button while the pointer is within the figure, the string is passed to the `eval` function to execute the specified function. If you move the pointer between figures, the callback executed is that associated with the figure where the button was pressed. Initially the empty matrix.	

See Also `axes`, `clf`, `close`, `gcf`, `subplot`

fill

Purpose Filled 2-D polygons.

Synopsis

```
fill(X,Y,C)
fill(X1,Y1,C1,X2,Y2,C2,...)
```

Description `fill(X,Y,C)` fills the two-dimensional polygon defined by vectors `X` and `Y` with the color specified by `C`. The vertices of the polygon are specified by pairs of components of `X` and `Y`. If necessary, the polygon is closed by connecting the last vertex to the first.

If `C` is a single character string chosen from the list `'r'`, `'g'`, `'b'`, `'c'`, `'m'`, `'y'`, `'w'`, `'k'`, or a red- green-blue row vector triple, `[r g b]`, the polygon is filled with the constant specified color.

If C is a vector the same length as X and Y, its elements are scaled by caxis and used as indices into the current colormap to specify colors at the vertices; the color within the polygon is obtained by bilinear interpolation of the vertex colors.

If X and Y are matrices of the same size, fill(X,Y,C) draws one polygon per column. In this case, C is a row vector for *flat* polygon colors, and C is a matrix for *interpolated* polygon colors.

If either X or Y is a matrix, and the other is a column vector with the same number of rows, the column vector argument is replicated to produce a matrix of the required size.

fill(X1,Y1,C1,X2,Y2,C2,...) is another way of specifying multiple filled areas.

fill sets the patch object FaceColor property to 'flat', 'interp', or a ColorSpec depending upon the value of the C matrix.

fill returns a column vector of handles to patch objects, one handle per patch. The X, Y, C triples can be followed by parameter/value pairs to specify additional properties of the patches.

Examples Create a red STOP sign (without the lettering).

```
t = (1/16:1/8:1)'*2*pi;
x = sin(t);
y = cos(t);
fill(x,y,'r')
axis square
```

See Also colormap, fill3, patch

fill3

Purpose Filled 3-D polygons in 3-space.

Synopsis

```
fill3(X,Y,Z,C)
fill3(X1,Y1,Z1,C1,X2,Y2,Z2,C2,...)
```

Description fill3(X,Y,Z,C) fills the three-dimensional polygon defined by vectors X, Y, and Z with the color specified by C. The vertices of the polygon are specified by triples of components of X, Y, and Z. If necessary, the polygon is closed by connecting the last vertex to the first.

If C is a single character string chosen from the list 'r','g','b', 'c','m','y','w','k', or a red-green-blue row vector triple, [r g b], the polygon is filled with the constant specified color.

If C is a vector the same length as X, Y, and Z, its elements are scaled by caxis and used as indices into the current colormap to specify colors at the vertices; the color within the polygon is obtained by bilinear interpolation in the vertex colors.

If X, Y, and Z are matrices the same size, fill3(X,Y,Z,C) draws one polygon per column. In this case, C is a row vector for *flat* polygon colors, and C is a matrix for *interpolated* polygon colors.

If any one of X, Y, or Z is a matrix, and the others are column vectors with the same number of rows, the column vector arguments are replicated to produce matrices of the required size.

`fill3(X1,Y1,Z1,C1,X2,Y2,Z2,C2,...)` is another way of specifying multiple filled areas.

`fill3` sets the patch object `FaceColor` property to `flat`, `interp`, or a `ColorSpec` depending upon the value of the `C` matrix.

`fill3` returns a column vector of handles to patch objects, one handle per patch. The `X`, `Y`, `Z`, `C` parameters can be followed by parameter/value pairs to specify additional properties of the patches.

Examples Fill four random triangles with color.

```
colormap(cool)
fill3(rand(3,4),rand(3,4),rand(3,4),rand(3,4))
```

See Also `colormap`, `fill`, `patch`

---

filter

Purpose Filter data with an infinite impulse response (IIR) or finite impulse response (FIR) filter.

Synopsis

```
y = filter(b,a,x)
[y,zf] = filter(b,a,x)
y = filter(b,a,x,zi)
```

Description `filter` filters a data sequence using a digital filter. The filter realization is the *transposed direct form II* structure, which can handle both FIR and IIR filters [1].

`y = filter(b,a,x)` filters the data in vector `x` with the filter described by numerator coefficient vector `b` and denominator coefficient vector `a` to create the filtered data vector `y`. If `a(1) ~= 1`, `filter` normalizes the filter coefficients by `a(1)`. If `a(1) = 0`, `filter` returns an error.

`[y,zf] = filter(b,a,x)` returns the final state values in vector `zf`.

`y = filter(b,a,x,zi)` specifies initial state conditions in vector `zi`.

The size of the initial/final condition vector is `max(size(a),size(b))`.

`filter` works for both real and complex inputs.

Algorithm `filter` is implemented as a transposed direct form II structure,

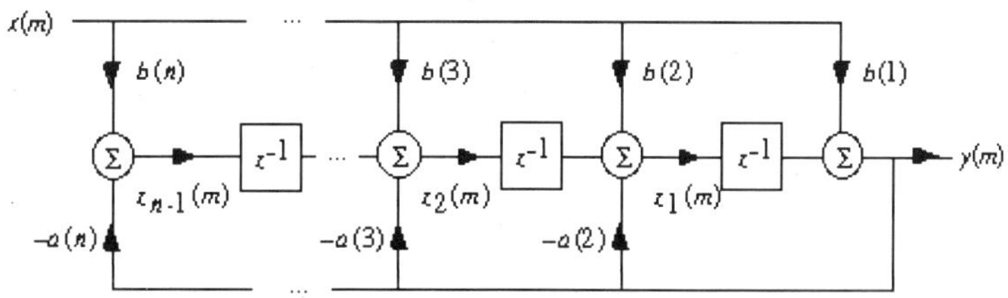

where n-1 is the filter order.

The operation of `filter` at sample m is given by the time domain difference equations

$$\begin{aligned} y(m) &= b(1)x(m) + z_1(m-1) \\ z_1(m) &= b(2)x(m) + z_2(m-1) - a(2)y(m) \\ \vdots &= \vdots \qquad\qquad \vdots \\ z_{n-2}(m) &= b(n-1)x(m) + z_{n-1}(m-1) - a(n-1)y(m) \\ z_{n-1}(m) &= b(n)x(m) - a(n)y(m) \end{aligned}$$

The input-output description of this filtering operation in the z-transform domain is a rational transfer function,

$$Y(z) = \frac{b(1) + b(2)z^{-1} + \ldots + b(nb+1)z^{-nb}}{1 + a(2)z^{-1} + \ldots + a(na+1)z^{-na}} X(z)$$

See Also `filter2`

`filtic` in the Signal Processing Toolbox

References [1] Oppenheim, A. V. and R.W. Schafer. *Discrete-Time Signal Processing*, pp. 311–312. Englewood Cliffs, NJ: Prentice Hall, 1989.

---

filter2

Purpose 2-D digital filtering.

Synopsis

```
Y = filter2(B,X)
Y = filter2(B,X,'shape')
```

Description `Y = filter2(B,X)` filters the data in `X` with the two-dimensional FIR filter in the matrix `B`. The result, `Y`, is computed using two-dimensional convolution and is the same size as `X`.

`Y = filter2(B,X,'shape')` returns `Y` computed via two-dimensional convolution with size specified by *shape*:

- `same` returns the central part of the convolution that is the same size as `X` (default).
- `full` returns the full two-dimensional convolution, `size(Y) > size(X)`.
- `valid` returns only those parts of the convolution that are computed without the zero-padded edges, `size(Y) < size(X)`.

Algorithm `filter2` uses `conv2` to compute the full two-dimensional convolution of the FIR filter with the input matrix. By default, `filter2` extracts and returns the central part of the convolution that is the same size as the input matrix. Use the *shape* parameter to specify an alternate part of the convolution for return.

See Also `conv2`, `filter`

find

Purpose Find indices and values of nonzero elements.

Synopsis

```
k = find(X)
[i,j] = find(X)
[i,j,s] = find(X)
```

Description `k = find(X)` returns the indices of the vector `X` that point to nonzero elements. If none is found, `find` returns an empty matrix. If `X` is a matrix, `find` regards `X` as `X(:)`, which is the long column vector formed by concatenating the columns of `X`.

`[i,j] = find(X)` returns the row and column indices of the nonzero entries in the matrix `X`. This is often used with sparse matrices.

`[i,j,s] = find(X)` also returns a column vector of the nonzero entries in `X`. Note that `find(X)` and `find(X ~= 0)` produce the same `i` and `j`, but the latter produces an `s` with all 1s.

Examples Some operations on a vector

```
x = [11  0  33  0  55]';
find(x) =
    1
    3
    5
find(x == 0) =
    2
    4
find(0 < x & x < 10*pi) =
    1
```

And on a matrix

```
M = magic(3)
M =
     8     1     6
     3     5     7
     4     9     2
[i,j,m] = find(M > 6)
i =
     1
     3
     2
j =
     1
     2
     3
m =
     1
     1
     1
```

See Also <, <=,>,>=,==, ~=, isempty, nonzeros, sparse

---

findobj

Purpose Find objects with specified properties.

Synopsis

```
h = findobj('PropertyName',PropertyValue,...)
h = findobj(obj_h,'PropertyName',PropertyValue,...)
h = findobj(obj_h,'flat','PropertyName',PropertyValue,...)
h = findobj
h = findobj(obj_h)
```

Description `h = findobj('PropertyName',PropertyValue,...)` finds all objects with properties matching those specified in the property name/property value list.

`h = findobj(obj_h,'PropertyName',PropertyValue,...)` restricts the search to objects whose handles are listed in the `obj_h` vector, as well as the descendants of those objects.

`h = findobj(obj_h,'flat','PropertyName',PropertyValue,...)` restricts the search to objects whose handles are listed in the `obj_h` vector. Descendants of these objects are not included.

`h = findobj` returns a vector of handles of all existing objects, including the root object.

`h = findobj(obj_h)` returns a vector of handles for the objects specified in `obj_h`, as well as their descendants.

See Also gca, gcf, get, set

findstr

Purpose Find one string within another.

Synopsis `k = findstr('s1','s2')`

Description `k = findstr('s1','s2')`, where *s1* and *s2* are both character strings, finds the starting indices of any occurrences of the shorter string within the longer.

Examples

```
s1 = 'Find the starting indices of the shorter string.';
s2 = 'the';
findstr(s1,s2)

ans =
     6    30
```

See Also `strcmp`

finite

Purpose Detect infinities.

Synopsis `Y = finite(X)`

Description `Y = finite(X)` returns 1s where the elements of `X` are finite and 0s where they are infinite or `NaN`. `X` can be a vector or a matrix.

For any `x`, exactly one of the three quantities `finite(x)`, `isinf(x)`, and `isnan(x)` is equal to one.

Examples Let

```
x = [-2  -1  0  1  2]
```

Then

```
finite(1./x) =  [1  1  0  1  1]
isinf(1./x)  =  [0  0  1  0  0]
isnan(1./x)  =  [0  0  0  0  0]
```

and

```
finite(0./x) =  [1  1  0  1  1]
isinf(0./x)  =  [0  0  0  0  0]
isnan(0./x)  =  [0  0  1  0  0]
```

See Also `all`, `any`, `find`, `isempty`, `isinf`, `isnan`

fix

Purpose Round towards zero.

Synopsis `Y = fix(X)`

Description `fix(X)` rounds the elements of `X` to integers by eliminating the fractional part.

Examples

```
X = [-1.9 -0.2 3.4 5.6 7.0]
fix(X) = [-1 0 3 5 7]
```

See Also `ceil`, `floor`, `round`

---

flag

Purpose Alternating red, white, blue, and black colormap.

Synopsis `cmap = flag(m)`

Description `flag(m)` returns an `m`-by-3 matrix containing the `flag` colormap. The default value for `m` is the length of the current colormap. Increasing the value of `m` increases the granularity emphasized by the map. This colormap show image detail by completely changing color with each index increment.

Examples The `cape` file in the `demos` directory includes a matrix `X` with altitude data over New England, and a custom colormap, `map`. Viewing the same data with the `flag` colormap shows more detail.

```
load cape
image(X)
colormap(map)
pause
colormap(flag)
```

See Also `bone`, `colormap`, `cool`, `copper`, `gray`, `hot`, `hsv`, `jet`, `pink`, `prism`, `white`

---

fliplr

Purpose Flip matrices left-right.

Synopsis `B = fliplr(A)`

Description `fliplr(A)` returns `A` with columns flipped in the left-right direction, that is, about a vertical axis.

Examples

```
A =
    1    2    3
    4    5    6
fliplr(A) =
    3    2    1
    6    5    4
```

See Also `:`, `'`, `flipud`, `reshape`, `rot90`

---

flipud

Purpose Flip matrices up-down.

Synopsis `B = flipud(A)`

Description `flipud(A)` returns A with rows flipped in the up-down direction, that is, about a vertical axis.

Examples

```
A =
      1    4
      2    5
      3    6
flipud(A) =
      3    6
      2    5
      1    4
```

See Also `:`, `'`, `fliplr`, `reshape`, `rot90`

---

floor

Purpose Round towards $-\infty$.

Synopsis `Y = floor(X)`

Description `floor(X)` rounds the elements of X to the nearest integers less than or equal to X.

Examples

```
X  =  [-1.9  -0.2  3.4  5.6  7.0]
floor(X)  =  [-2  -1  3  5  7]
```

See Also `ceil`, `fix`, `round`

---

flops

Purpose Count floating-point operations.

Synopsis

```
f = flops
flops(0)
```

Description `f = flops` returns the cumulative number of floating-point operations.

`flops(0)` resets the count to zero.

Examples If A and B are real n-by-n matrices, some typical flop counts for different operations are

Operation	Flop Count
A+B	n^2
A*B	2*n^3
A^100	99*(2*n^3)
lu(A)	(2/3)*n^3

MATLAB's version of the LINPACK benchmark is

```
n = 100;
A = rand(n,n);
b = rand(n,1);
flops(0)
tic;
x = A\b;
t = toc
megaflops = flops/t/1.e6
```

Algorithm It is not feasible to count all the floating-point operations, but most of the important ones are counted. Additions and subtractions are each one flop if real and two if complex. Multiplications and divisions count one flop each if the result is real and six flops if it is complex. Elementary functions count one if real and more if complex.

---

fmin

Purpose Minimize a function of one variable.

Synopsis

```
x = fmin('function',x1,x2)
x = fmin('function',x1,x2,options)
x = fmin('function',x1,x2,options,p1,p2, ...)
[x,options] = fmin(...)
```

Description `x = fmin('`*`function`*`',x1,x2)` returns a value of `x` which is a local minimizer of *`function`*`(x)` in the interval `x1 < x < x2`. *`function`* is a string containing the name of the objective function to be minimized.

`x = fmin('`*`function`*`',x1,x2,options)` uses a vector of control parameters.

- If `options(1)` is nonzero, intermediate steps in the solution are displayed. The default value of `options(1)` is `0`.
- `options(2)` is the termination tolerance. The default value is 1.e–4.
- `options(14)` is the maximum number of steps. The default value is `500`.

Only three of the 18 components of `options` are referenced by `fmin`. Other functions in the Optimization Toolbox reference the other options.

`x = fmin('`*`function`*`',x1,x2,options,p1,p2,...)` provides up to 10 additional arguments which are passed to the objective function, `function(x,p1,p2,...)`.

`[x,options] = fmin(...)` returns a count of the number of steps taken in `options(10)`.

Examples `fmin('cos',3,4)` computes π to a few decimal places.

`fmin('cos',3,4,[1,1.e–12])` displays the steps taken to compute π to 12 decimal places.

To find the minimum of the function

$$f(x) = x^3 - 2x - 5$$

on the interval (0,2), write an M-file called `f.m`.

```
function y = f(x)
y = x.^3-2*x-5;
```

Then invoke `fmin` with

```
x = fmin('f', 0, 2)
```

The result is

```
x =
     0.8165
```

The value of the function at the minimum is

```
y = f(x)


y =
     -6.0887
```

Algorithm The algorithm is based on golden section search and parabolic interpolation. A FORTRAN program implementing the same algorithms is given in [1].

See Also `fmins`, `fzero`

`foptions` in the Optimization Toolbox

References

[1] Forsythe G. E., M. A. Malcolm, and C. B. Moler, *Computer Methods for Mathematical Computations*, Prentice Hall, 1976.

---

fmins

Purpose Minimize a function of several variables.

Synopsis

```
x = fmins('function',x0)
x = fmins('function',x0,options)
x = fmins('function',x0,options,[],arg1,arg2, ...)
[x,options] = fmins(...)
```

Description `x = fmins('`*`function`*`',x0)` returns a vector `x` which is a local minimizer of *`function`*`(x)` near the starting vector `x0`. *`function`* is a string containing the name of the objective function to be minimized. *`function`*`(x)` is a scalar valued function of a vector variable.

`x = fmins('`*`function`*`',x0,options)` uses a vector of control parameters.

- If `options(1)` is nonzero, intermediate steps in the solution are displayed. The default value of `options(1)` is 0.
- `options(2)` is the termination tolerance for `x`. The default value is 1.e–4.

- `options(3)` is the termination tolerance for `function(x)`. The default value is 1.e–4.
- `options(14)` is the maximum number of steps. The default value is `500`.

Only four of the 18 components of options are referenced by `fmins`. Other functions in the Optimization Toolbox reference the other options.

`x = fmins('function',x0,options,[],arg1,arg2,...)` provides up to 10 additional arguments which are passed to the objective function, `function(x,arg1,arg2,...)`. The dummy argument in the fourth position is necessary to provide compatibility with `fminu` in the Optimization Toolbox.

`[x,options] = fmins(...)` returns a count of the number of steps taken in `options(10)`.

Examples A classic test example for multidimensional minimization is the Rosenbrock banana function:

$$f(x) = 100\left(x_2 - x_1^2\right)^2 + \left(1 - x_1\right)^2$$

The minimum is at (1,1) and has the value 0. The traditional starting point is (-1.2,1). The M-file `banana.m` defines the function.

```
function f = banana(x)
f = 100*(x(2)–x(1)^2)^2+(1–x(1))^2;
```

The statements

```
[x,out] = fmins('banana',[–1.2, 1]);
x
out(10)
```

produce

```
x =
     1.0000     1.0000
ans =
     165
```

This indicates that the minimizer was found to at least four decimal places in 165 steps.

The location of the minimum can be moved to the point `[a,a^2]` by adding a second parameter to `banana.m`.

```
function f = banana(x,a)
if nargin < 2, a = 1; end
f = 100*(x(2)–x(1)^2)^2+(a–x(1))^2;
```

Then the following statement sets the new parameter to `sqrt(2)` and seeks the minimum to an accuracy higher than the default.

```
[x,out] = fmins('banana', [–1.2, 1], [0, 1.e–8], [], sqrt(2));
```

Algorithm The algorithm is the Nelder-Mead simplex search described in the two references. It is a direct search method that does not require gradients or other derivative information. If `n` is the length of `x`, a simplex in `n`-dimensional space is characterized by the `n+1` distinct vectors which are its vertices. In two-space, a simplex is a triangle; in three-space, it is a pyramid.

At each step of the search, a new point in or near the current simplex is generated. The function value at the new point is compared with the function's values at the vertices of the simplex and, usually, one of the vertices is replaced by the new point, giving a new simplex. This step is repeated until the diameter of the simplex is less than the specified tolerance.

See Also `fmin`, `foptions`, and the Optimization Toolbox

References
[1] Nelder, J. A. and R. Mead, "A Simplex Method for Function Minimization," *Computer Journal*, vol. 7, p. 308-313.

[2] Dennis, J. E. Jr. and D. J. Woods, "New Computing Environments: Microcomputers in Large-Scale Computing," edited by A. Wouk, *SIAM*, pp. 116-122, 1987.

---

fopen

Purpose Open a file or obtain information about open files.

Synopsis

```
fid = fopen('filename')
fid = fopen('filename', 'permission')
[fid, message] = fopen('filename','permission', 'architecture')
fids = fopen('all')
[filename, permission, architecture] = fopen(fid)
```

Description `fid = fopen('`*`filename`*`')` opens the file in *`filename`* and returns `fid`, the file identifier.

`fopen('`*`filename`*`','`*`permission`*`')` opens the file *`filename`* in the mode specified by *`permission`*. Legal file permission strings are

`'r'`	Open the file for reading (default).
`'r+'`	Open the file for reading and writing.
`'w'`	Delete the contents of an existing file or create a new file, and open it for writing.
`'w+'`	Delete the contents of an existing file or create new file, and open it for reading and writing.
`'a'`	Create and open a new file or open an existing file for writing, appending to the end of the file.
`'a+'`	Create and open new file or open an existing file for reading and writing, appending to the end of the file.

You can also add a `'b'` to these strings, for example, `'rb'`, on systems that distinguish between text and binary files. Under DOS and VMS, for example, you cannot read a binary file unless you set the permission to `'rb'`.

If `fopen` successfully opens the file, it returns a file identifier `fid`, which is an integer greater than two, and the value of `message` is empty. The `fid` is used with other file I/O routines to identify the file on which to perform the operations.

Three `fid`s are predefined: 0 corresponds to standard input, which is always open for reading (`permission` set to `'r'`), 1 corresponds to standard output, which is always open for appending (`permission` set to `'a'`), and 2 corresponds to standard error, which is always open for appending (`permission` set to `'a'`). Standard input, output and error cannot be explicitly opened or closed.

If `fopen` does not successfully open the file, it returns a -1 value for `fid`. In that case, the value of `message` is a string that can help you determine the type of error that occurred.

`[fid, message] = fopen('`*`filename`*`','`*`permission`*`', '`*`architecture`*`')` defines the numeric format of the file, *`architecture`*, allowing you to share files between machines of different architectures. The argument can be one of these strings:

`'native'` or `'n'`	The numeric format of the machine you are currently running
`'ISIEEE–LE'` or `'l'`	IEEE Little Endian formats
`'ISIEEE–BE'` or `'b'`	IEEE Big Endian formats
`'vaxdv'` or `'d'`	VAX D-float format
`'vaxg'` or `'gv'`	VAX G-float format
`'crayv'` or `'c'`	Cray numeric format

You can omit the `'`*`architecture`*`'` argument. If you do, the numeric format of the local machine is used. Individual calls to `fread` or `fwrite` can override the numeric format specified in a call to `fopen`.

`fopen('all')` returns a row vector containing the file identifiers of all open files, including 0, 1, and 2. The number of elements in the vector is equal to the number of open files.

`[filename, permission, architecture] = fopen(fid)` returns the `filename` string, the `permission` string, and the `architecture` string associated with the specified file. An invalid `fid` returns empty strings for all output arguments. Both `permission` and `architecture` are optional.

See Also `fclose`, `ferror`, `fprintf`, `fread`, `fscanf`, `fseek`, `ftell`, `fwrite`, `sprintf`, `sscanf`

for

Purpose Repeat statements a specific number of times.

Synopsis

```
for variable = expression
    statements
end
```

Description The general format is

```
for variable = expression
    statement
    ...
    statement
end
```

The columns of the *expression* are stored one at a time in the variable while the following statements, up to the end, are executed.

In practice, the *expression* is almost always of the form `scalar : scalar`, in which case its columns are simply scalars.

The scope of the `for` statement is always terminated with a matching `end`.

Examples Assume `n` has already been assigned a value. Create the Hilbert matrix, using `zeros` to preallocate the matrix to conserve memory:

```
a = zeros(n,n)  % Preallocate matrix
for i = 1:n
    for j = 1:n
        a(i,j) = 1/(i+j -1);
    end
end
```

Step `s` with increments of `-0.1`

```
for s = 1.0: -0.1: 0.0,..., end
```

Successively set `e` to the unit `n`-vectors:

```
for e = eye(n),..., end
```

The line

```
for V = A,..., end
```

has the same effect as

```
for j = 1:n, V = A(:,j);..., end
```

except `j` is also set here.

See Also `break`, `end`, `if`, `return`, `while`

format

Purpose Control the output display format.

Synopsis See "Description" below.

Description `format` switches between different display formats:

Command	Result	Example
`format short`	5 digit scaled fixed point	`3.1416`
`format long`	15 digit scaled fixed point	`3.14159265358979`
`format short e`	5 digit floating-point	`3.1416e+00`
`format long e`	16 digit floating-point	`3.141592653589793e+00`
`format hex`	Hexadecimal	`400921fb54442d18`
`format bank`	Fixed dollars and cents	`3.14`
`format rat`	Ratio of small integers	`355/113`
`format +`	+,–, blank	`+`
`format compact`	Suppresses excess line feeds and results in a slightly more compact display. Does not affect the numeric format.	
`format loose`	Reverts to the more open display. Does not affect the numeric format.	
`format`	Returns to the default formats, `short` and `loose`.	

By default, MATLAB displays numbers in a short format with five decimal digits.

Algorithms `format +` displays +, –, and blank characters for positive, negative, and zero elements. The `spy` function uses graphics to display essentially the same information more effectively. `format hex` displays the hexadecimal representation of the binary double-precision number for each element. `format rat` uses a continued fraction algorithm to approximate floating-point values by ratios of small integers. See `rat.m` for the complete code in an M-file.

See Also `fprintf`, `num2str`, `rat`, `sprintf`, `spy`

fplot

Purpose Plot the graph of a function.

Synopsis

```
fplot('function',limits)
fplot('fstring',limits)
fplot(...,'marker')
fplot(...,'marker',tol)
[x,y] = fplot(...)
```

Description `fplot('`*`function`*`',limits)` plots the function specified by *`function`* between the limits specified by the `limits` vector. *`function`* can be any MATLAB function name, including the name of a user-created function. `limits` can be a two-element vector specifying the x-axis limits for the plot, or a four-element vector specifying both x- and y-axis limits, `[xmin xmax ymin ymax]`.

`fplot('`*`fstring`*`',limits)` evaluates and plots *`fstring`*, a function string that includes a variable `x`, such as `'sin(x)'` or `'diric(x,10)'`.

`fplot(...,'`*`marker`*`')` specifies the marker type for the plot. By default, `fplot` uses only a solid line. Valid *`marker`* strings are `'−+'`, `'−x'`, `'−o'`, and `'−*'`.

`fplot(...,'`*`marker`*`',tol)` specifies the relative error tolerance `tol`. By default, `tol` is 2e–3.

`[x,y] = fplot(...)` returns the abscissae and ordinates for the function in the column vectors `x` and `y`. No plot is drawn on the screen. The function can then be plotted with `plot(x,y)`.

Examples Plot the hyperbolic tangent function from -2 to 2.

```
fplot('tanh',[−2 2])
```

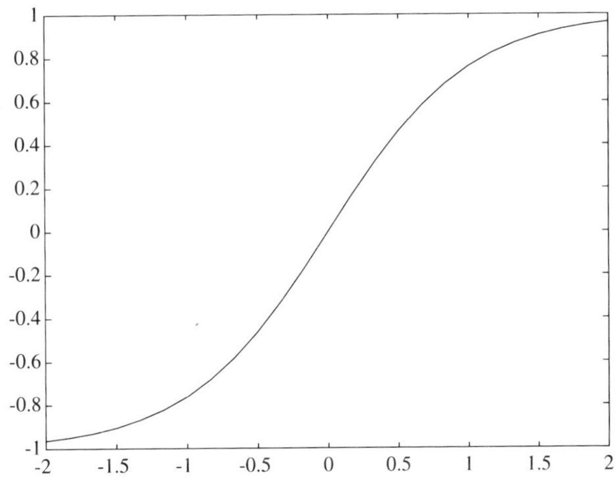

The function defined in the M-file

```
function y = myfun(x)
y(:,1) = 200*sin(x(:))./x(:);
y(:,2) = x(:).^2;
```

is graphed with the statement:

```
fplot('myfun',[-20 20])
```

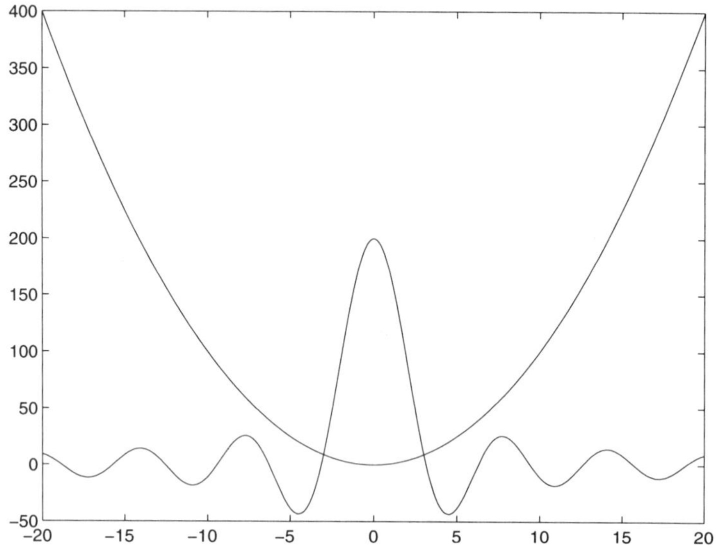

See Also `feval`, `plot`

fprintf

Purpose Write formatted data to file.

Synopsis

```
count = fprintf(fid,'format',A,...)
fprintf('format',A,...)
```

Description `count = fprintf(fid,'format',A,...)` formats the data in matrix `A` (and in any additional matrix arguments) under control of the specified *`format`* string, and writes it to the file associated with file identifier `fid`. `count` is the integer number of bytes written.

`fid` is an integer file identifier obtained from `fopen`. It can also be 1 for standard output (the screen) or 2 for standard error. On some computers, `fopen` can be used with device names. For example, in an MS-DOS environment, `COM1` might be defined as a serial port leading to a modem. In that case, use `fopen('COM1')` and `fprintf` to send characters to the modem.

Omitting `fid` from `fprintf`'s argument list causes output to appear on the screen and is the same as writing to standard output (`fid = 1`).

`format` is a string containing ordinary characters and/or conversion specifications. Ordinary characters include the normal alphanumeric characters, and escape characters. Escape characters include

`\n`	new line
`\t`	horizontal tab
`\b`	backspace
`\r`	carriage return
`\f`	form feed
`\\`	backslash

Conversion specifications involve the character `%`, optional width fields, and conversion characters. Legal conversion characters are

`%e`	exponential notation
`%f`	fixed point notation
`%g`	`%e` or `%f`, whichever is shorter; (insignificant zeros do not print)

Between the `%` and the conversion character (`e`, `f`, or `g`), you can add one or more of the following characters:

- A minus sign (–) to specify left adjustment of the converted argument in its field
- A digit string to specify a minimum field width
- A period to separate the field width from the next digit string
- A digit string specifying the precision (i.e., the number of digits to the right of the decimal point)

For more information about format strings, refer to the `printf()` and `fprintf()` routines in a C language reference manual.

`fprintf` differs from its C language `fprintf()` namesake in an important respect – it is *vectorized* for the case when input matrix `A` is nonscalar. The format string is recycled through the elements of `A` (columnwise) until all the elements are used up. It is then recycled in a similar manner, without reinitializing, through any additional matrix arguments.

Examples The statements

```
x = 0:.1:1;
y = [x; exp(x)];
fid = fopen('exp.txt','w');
fprintf(fid,'%6.2f %12.8f\n',y);
fclose(fid)
```

create a text file called `exp.txt` containing a short table of the exponential function:

```
0.00    1.00000000
0.10    1.10517092
...
1.00    2.71828183
```

The command

```
fprintf('A unit circle has circumference %g.\n',2*pi)
```

displays a line on the screen:

```
A unit circle has circumference 6.283186.
```

To insert a single quote in a string (the `format` command), use two single quotes together. For example,

```
fprint(1,'It''s Friday\n')
```

displays the following line on the screen:

```
It's Friday
```

The command

```
B = [8.8  7.7; 8800  7700]
fprintf(1,'X is %6.2f meters or %8.3f mm\n',9.9,9900,B)
```

displays the lines:

```
X is 9.90 meters or 9900.000 mm
X is 8.80 meters or 8800.000 mm
X is 7.70 meters or 7700.000 mm
```

See Also `fclose`, `ferror`, `fopen`, `fread`, `fscanf`, `fseek`, `ftell`, `fwrite`, `sprintf`

References [1] Kernighan, B.W. and D.M. Ritchie, *The C Programming Language*, Prentice Hall, Inc., 1978.

---

fread

Purpose Read binary data from file.

Synopsis

```
A = fread(fid)
[A,count] = fread(fid,size,'precision')
```

Description `A = fread(fid)` reads binary data from the specified file and writes it into matrix `A`. `fid` is an integer file identifier obtained from `fopen`.

`[A,count] = fread(fid,size,'precision')` attempts to read `size` number of elements of the specified *`precision`*. `count` is the number of elements successfully transferred.

The `size` argument can be

`n`	Reads `n` elements into a column vector.
`inf`	Reads to the end of the file, resulting in a column vector containing the same number of elements as are in the file.
`[m,n]`	Reads enough elements to fill an `m`–by–`n` matrix, filling in elements in column order, padding with zeros if the file is too small to fill the matrix.

If `size` is not specified, `Inf` is assumed.

The `'precision'` argument specifies the numeric precision of the values read. *`precision`* controls the number of bits read for each value and the interpretation of those bits as an integer, a floating-point value, or a character. The table on the next page lists the allowable *`precision`* strings and their meanings.

Note that if `fread` reaches the end of the file and the current input stream does not contain enough bits to write out a complete matrix element of the specified precision, `fread` pads the last byte or element with zero bits until the full value is obtained. If an error occurs, reading is done up to the last full value.

Numeric precisions are hardware specific; they depend on how numbers are represented in the architecture of your specific computer. Although the following table indicates certain common precision values, be aware that these values are not necessarily correct for your system except for `'float32'` and `'float64'`, which have absolute precision values and are not hardware specific, although these values may not be supported on all architectures.

This section assumes that you know the appropriate numeric precision for your hardware. If you are not familiar with the numeric precision for your hardware, refer to the hardware reference manual.

For convenience, MATLAB accepts some C and FORTRAN data type equivalents for the MATLAB precisions listed. If you are a C or FORTRAN programmer, you may find it more convenient to use the names of the data types in the language with which you are most familiar.

MATLAB	C or FORTRAN	Interpretation
'char'	'char', 'char*1'	Character; 8 bits (signed or unsigned, depending on the architecture)
'schar'	'signed char'	Signed character; 8 bits
'uchar'	'unsigned char'	Unsigned character (default); 8 bits
'short'	'short'	Integer; 16 bits
'ushort'	'unsigned short'	Unsigned integer; 16 bits
'int'	'int'	Integer; 16 or 32 bits
'uint'	'unsigned int'	Unsigned integer; 16 or 32 bits
'long'	'long'	Integer; 32 bits
'float'	'float'	Floating-point value; 32 bits
'ulong'	'unsigned long'	Unsigned integer; 32 bits
'float32'	'real*4'	32-bit floating-point value
'double'	'double'	Long floating-point value; 64 bits
'float64'	'real*8'	64-bit floating-point value
'intN'		Signed integer, N bits wide
'uintN'		Unsigned integer, N bits wide.

N represents a value between 1 and 64. In addition, the following specific equivalents prevail:

'int8'	'integer*1'	Integer; 8 bits.
'int16'	'integer*2'	Integer; 16 bits.
'int32'	'integer*4'	Integer; 32 bits.
'int64'	'integer*8'	Integer; 64 bits.

See Also fclose, ferror, fopen, fprintf, fscanf, fseek, ftell, fwrite, sprintf, sscanf

frewind

Purpose Rewind an open file.

Synopsis `frewind(fid)`

Description `frewind(fid)` sets the file pointer to the beginning of the file with identifier `fid`.

Rewinding a `fid` associated with a tape device may not work even though `frewind` does not generate an error message.

See Also fclose, ferror, fopen, fread, fscanf, fseek, ftell, fwrite, sprintf

fscanf

Purpose Read formatted data from file.

Synopsis `A = fscanf(fid,'format')`

```
[A,count] = fscanf(fid,'format',size)
```

Description `[A,count] = fscanf(fid,'format')` reads all the data from the file specified by file identifier `fid`, converts it according to the specified *`format`* string, and returns it in matrix `A`. `fid` is an integer file identifier obtained from `fopen`.

`[A,count] = fscanf(fid,'format',size)` returns `count`, the number of elements successfully read. `size` can be

`n`	Read `n` elements into a column vector.
`inf`	Read to the end of the file, resulting in a column vector containing the same number of elements as are in the file.
`[m,n]`	Read enough elements to fill an `m`-by-`n` matrix, filling the matrix in column order. `n` can be `inf`, but not `m`.

The *`format`* string specifies the format of the data to be read. When MATLAB reads the specified file, it attempts to match the data in the file to the format string. If a match occurs, the data is written into the matrix in column order. If a partial match occurs, only the matching data is written to the matrix, and the read operation stops.

The *`format`* string can consist of ordinary characters and/or conversion specifications. Conversion specifications indicate the type of data to be matched and involve the character `%`, optional width fields, and conversion characters. Legal conversion characters are

`%d`	decimal numbers
`%e, %f, %g`	floating-point numbers
`%s`	a series of non-white-space characters

Between the `%` and the conversion character (`d`, `e`, `f`, `g`, or `s`), you can add one or more of the following characters:

an asterisk (*)	Skip over the matched value, if the value is matched but not stored in the output matrix.
a digit string	Maximum field width.
a letter	The size of the receiving object; for example, `h` for short as in `%hd` for a short integer, or `l` for long as in `%ld` for a long integer or `%lg` for a double floating-point number.

For more information about format strings, refer to the `scanf()` and `fscanf()` routines in a C language reference manual.

`fscanf` differs from its C language namesakes `scanf()` and `fscanf()` in an important respect – it is *vectorized* in order to return a matrix argument. The *`format`* string is recycled through the file until an end-of-file is reached or the amount of data specified by `size` is read in.

Examples The example in `fprintf` generates an ASCII text file called `exp.txt` that looks like:

```
0.00    1.00000000
0.10    1.10517092
...
1.00    2.71828183
```

Read this ASCII file back into a two-column MATLAB matrix:

```
fid = fopen('exp.txt');
a = fscanf(fid,'%g %g',[2 inf]) % It has two rows now.
a = a';
fclose(fid)
```

See Also `fclose`, `ferror`, `fopen`, `fprintf`, `fread`, `fseek`, `ftell`, `fwrite`, `sscanf`

---

fseek

Purpose Set file position indicator.

Synopsis `status = fseek(fid, offset, 'origin')`

Description `status = fseek(fid, offset, 'origin')` repositions the file position indicator in the specified file to the specified byte `offset` for the specified *`origin`*. `fid` is a integer file identifier obtained from `fopen`.

`offset` values are interpreted as follows:

`offset > 0`	Move position indicator `offset` bytes toward the end of the file.
`offset = 0`	Do not change position.
`offset < 0`	Move position indicator `offset` bytes toward the beginning of the file.

`origin` can be

`'bof'`	–1	Beginning of file.
`'cof'`	0	Current position in file.
`'eof'`	1	End of file.

A `status` value for the operation is returned with value 0 on success and -1 on failure. If an error occurs, use the function `ferror` to get more information about the nature of the error.

See Also `fclose`, `ferror`, `fopen`, `fprint`, `fread`, `fscanf`, `ftell`, `fwrite`, `sprintf`, `sscanf`

ftell

Purpose Get file position indicator.

Synopsis `position = ftell(fid)`

Description `position = ftell(fid)` returns the location of the file position indicator in the specified file. `position` is indicated in bytes from the beginning of the file. `fid` is a integer file identifier obtained from `fopen`.

The value of `position` is a nonnegative integer if the query is successful. If -1 is returned, it indicates that the query was unsuccessful; use `ferror` to determine the nature of the error.

See Also `fopen`, `fclose`, `ferror`, `fread`, `fwrite`, `fseek`, `fprintf`, `sprintf`, `fscanf`, `sscanf`

full

Purpose Convert sparse matrices to full storage class.

Synopsis `A = full(S)`

Description `A = full(S)` converts the storage of a matrix from sparse to full. If `A` is already full, `full(A)` returns `A`.

Let `X` be an `m`-by-`n` matrix with `nz = nnz(X)` nonzero entries. Then `full(X)` requires space to store `m*n` real numbers while `sparse(X)` requires space to store `nz` real numbers and `(nz+n)` integers. On most computers, a real number requires twice as much storage as an integer. On these computers, `sparse(X)` requires less storage than `full(X)`if the density, `nnz/(m*n)`, is less than one third. Operations on sparse matrices, however, require more execution time per element than those on full matrices, so density is considerably less than two-thirds before sparse storage is used.

Examples Here is an example of a sparse matrix with a density of about two-thirds. `sparse(S)` and `full(S)` require about the same number of bytes of storage.

```
S = sparse(rand(200,200) < 2/3);
A = full(S);
whos
Name        Size     Elements      Bytes     Density  Complex
  A     200 by 200      40000     320000        Full     No
  S     200 by 200      26797     322364      0.6699     No
```

See Also `sparse`

function

Purpose Extensibility and programming.

Description New functions can be added to MATLAB's vocabulary if they are expressed in terms of other existing functions. The commands and functions that compose the new function are placed in a text file whose name, which must begin with an alphabetic character, defines the name of the new function, with a filename extension of .m appended. A line at the top of the file contains the syntax definition for the new function. The name of a function, as defined in the first line of the .m file, must be the same as the name of the file without the .m extension. For example, the existence of a file on disk called stat.m with

```
function [mean,stdev] = stat(x)
n = length(x);
mean = sum(x)/n;
stdev = sqrt(sum((x-mean).^2/n));
```

defines a new function called stat that calculates the mean and standard deviation of a vector. The variables within the body of the function are all local variables.

When a function is used that MATLAB does not recognize, it searches for a file of the same name on disk. If it is found, the function is compiled into memory for subsequent use.

If echo is enabled, the file is interpreted instead of compiled so that each input line is viewed as it is executed. Use clear to remove the function from memory.

In general, if you input the name of something to MATLAB, for example by typing foo, the MATLAB interpreter:

1. Looks for foo as a variable.
2. Checks for foo as a built-in function.
3. Looks in the current directory for files named foo.mex and foo.m.
4. Looks in the directories specified by MATLAB's search path for files named foo.mex and foo.m.

See Also echo, nargin, nargout, path, type, what

funm

Purpose Evaluate functions of a matrix.

Synopsis Y = funm(X,'*function*')

Description funm evaluates general matrix functions. For any square matrix argument X, funm(X,'*function*') evaluates the matrix function specified by *function* using Parlett's method [1].

funm(X,'sqrt') and funm(X,'log') are equivalent to sqrtm(X) and logm(X). funm(X,'exp') and expm(X) compute the same function, but by different algorithms. expm(X) is preferred.

Examples The statements

```
S = funm(X,'sin');
C = funm(X,'cos');
```

produce the same results to within roundoff error as

```
E = expm(i*X);
C = real(E);
S = imag(E);
```

In either case, the results satisfy

```
S*S+C*C = I
```

where `I = eye(size(X))`

Algorithm The matrix functions are evaluated using Partlett's algorithm, which is described in [1]. The algorithm uses the Schur factorization of the matrix and can give poor results or break down completely when the matrix has repeated eigenvalues. A warning message is printed when the results may be inaccurate.

See Also `expm`, `logm`, `sqrtm`

References

[1] Golub, G. H. and C. F. Van Loan, *Matrix Computation*, p. 384, Johns Hopkins University Press, 1983.

[2] Moler, C. B. and C. F. Van Loan, "Nineteen Dubious Ways to Compute the Exponential of a Matrix," *SIAM Review 20*, pp. 801-836, 1979.

fwrite

Purpose Write binary data from a MATLAB matrix to a file.

Synopsis `count = fwrite(fid,A,'precision')`

Description `count = fwrite(fid,A,'precision')` writes the elements of matrix `A` to the specified file, translating MATLAB values to the specified *`precision`*. The data are written to the file in column order. `fid` is a integer file identifier obtained from `fopen`. `count` is the number of elements successfully written.

The *`'precision'`* argument specifies the numeric precision of the data. See `fread` for a list of valid *`precision`* strings.

See Also `fclose`, `ferror`, `fopen`, `fprintf`, `fread`, `fscanf`, `fseek`, `ftell`, `sprintf`, `sscanf`

fzero

Purpose Zero of a function of one variable.

Synopsis

```
z = fzero('function',x0)
z = fzero('function',x0,tol)
z = fzero('function',x0,tol,trace)
```

Description `fzero('function',x0)` finds a zero of the function `function(x)` that is near `x0`. `fzero` identifies only points where the function actually crosses the x-axis. Points where the function touches the x-axis, but does not cross it, are not considered zeros.

`fzero('function',x0,tol)` returns an answer accurate to within a relative error of `tol`. The default value for `tol` is `eps`.

`z = fzero('function',x0,tol,trace)` displays information at each iteration if `trace` is nonzero.

Examples Calculate pi by finding the zero of the `sine` function near 3.

```
x = fzero('sin',3)
x =
    3.1416
```

To find a zero of the function

$$f(x) = x^3 - 2x - 5$$

write an M-file called `f.m`.

```
function y = f(x)
y = x.^3-2*x-5;
```

To find the zero near 2

```
z = fzero('f',2)
z =
    2.0946
```

Since this function is actually a polynomial, the statement

```
roots([1 0 -2 -5])
```

finds the same real zero, and a complex conjugate pair of zeros.

```
    2.0946
   -1.0473 + 1.1359i
   -1.0473 - 1.1359i
```

Algorithm `fzero` is an M-file. The algorithm, which was originated by T. Dekker, uses a combination of bisection, secant, and inverse quadratic interpolation methods. An Algol 60 version, with some improvements, is given in [1]. A FORTRAN version, upon which the `fzero` M-file is based, is in [2].

Limitations `fzero` defines a *zero* as a point where the function crosses the x-axis. Points where the function touches, but does not cross, the `x`-axis are not valid zeros. For example, `y = x.^2` is a parabola that touches the x-axis at (0,0). Since the function never crosses the x-axis, however, no zero is found. For functions with no valid zeros, `fzero` executes until you terminate it.

See Also `eps`, `fmin`, `roots`

References

[1] Brent, R., *Algorithms for Minimization Without Derivatives*, Prentice Hall, 1973.

[2] Forsythe, G. E., M. A. Malcolm, and C. B. Moler, *Computer Methods for Mathematical Computations*, Prentice Hall, 1976.

gallery

Purpose Small test matrices.

Synopsis

```
X = gallery(3)
X = gallery(5)
```

Description `X = gallery(3)` returns a badly conditioned 3-by-3 matrix.

`X = gallery(5)` returns a 5-by-5 matrix that poses an interesting eigenvalue problem. Using exact arithmetic, `gallery(5)` has a single eigenvalue at zero with algebraic multiplicity 5; and a single eigenvector. Numerical calculations, however, yield five distinct eigenvalues: two pairs of complex eigenvalues and a real eigenvalue.

For any input other than 3 or 5, `gallery` returns an empty matrix.

See Also `hadamard`, `hilb`, `invhilb`, `magic`, `rosser`, `wilkinson`

gamma, gammainc, gammaln

Purpose Gamma function.

Synopsis

```
y = gamma(a)
y = gammainc(x,a)
y = gammaln(a)
```

Description `y = gamma(a)` returns the gamma function evaluated at the elements of a. The gamma function is defined by the integral:

$$\Gamma(a) = \int_0^\infty e^{-t} t^{a-1} dt$$

The gamma function interpolates the factorial function. For integer `n`

```
gamma(n+1) = n! = prod(1:n)
```

`y = gammainc(x,a)` returns the incomplete gamma function defined by

$$P(x,a) = \frac{1}{\Gamma(a)} \int_0^x e^{-t} t^{a-1} dt$$

`y = gammaln(a)` returns the logarithm of the gamma function,

```
gammaln(a) = log(gamma(a))
```

`gammaln` avoids the underflow and overflow that may occur if it is computed directly using `log(gamma(a))`.

Algorithm The computations of `gamma` and `gammaln` are based on algorithms outlined in [1]. Several different minimax rational approximations are used depending upon the value of a. Computation of the incomplete gamma function is based on the algorithm in [2].

References
[1] Cody, J., *An Overview of Software Development for Special Functions*, Lecture Notes in Mathematics, 506, Numerical Analysis Dundee, G. A. Watson (ed.), Springer Verlag, Berlin, 1976.

[2] Abramowitz, M. and I.A. Stegun, *Handbook of Mathematical Functions*, National Bureau of Standards, Applied Math. Series #55, Dover Publications, 1965, sec. 6.5.

gca

Purpose Get current axes handle.

Synopsis

```
h = gca
```

Description `h = gca` returns the handle to the *current axes*. The current axes is the axes to which graphics commands such as `plot`, `title`, and `surf` plot their results.

Each figure has its own current axes. Changing the current figure causes `gca` to return the correct axes handle for the newly current figure.

Use the command `axes` or `subplot` to change the current axes to a different axes or to create new ones. Use `cla` to reset the current axes.

See Also `axes`, `cla`, `delete`, `hold`, `gcf`, `subplot`

gcd

Purpose Greatest common divisor.

Synopsis

```
g = gcd(a,b)
[g,c,d] = gcd(a,b)
```

Description `g = gcd(a,b)` is the greatest common divisor of the integers `a` and `b`. By convention, `gcd(0,0)` returns a value of 0; all other inputs return positive integers.

`[g,c,d] = gcd(a,b)` also returns the scalars `c` and `d` such that `g = a*c+b*d`.

Examples

```
[g,c,d] = gcd(80,120)
g =
     40
c =
     -1
d =
      1
```

See Also `lcm`

References

[1] Knuth, Donald, *The Art of Computer Programming*, Vol. 2, Addison-Wesley:Reading MA, 1973. Section 4.5.2, Algorithm X.

gcf

Purpose Get current figure handle.

Synopsis

```
h = gcf
```

Description `gcf` returns the handle to the *current figure*. The current figure is the figure (graphics window) to which graphics commands such as `plot`, `title`, and `surf` plot their results.

Each figure has a current axes.

Use the command `figure` to change the current figure to a different figure, or to create a new figure. Use `clf` to reset the current figure.

See Also `axes`, `clf`, `close`, `figure`, `gca`, `subplot`

---

gco

Purpose Handle of current object.

Synopsis

```
object = gco
object = gco(figure)
```

Description `object = gco` returns the handle of the current object in the current figure. The current object is the last object clicked on with the mouse.

`object = gco(figure)` returns the handle of the current object in the figure specified by the handle `figure`.

See Also `gca`, `gcf`

---

get

Purpose Get object properties.

Synopsis

```
V = get(h,'PropertyName')
get(h)
```

Description Use the `get` function to obtain the current values of object properties. The handle argument `h` identifies the object whose properties you are querying.

`V = get(h,'`*`PropertyName`*`')` returns the current value of the specified property of the object identified by the handle `h`.

`get(h)`, without a property name, lists all the properties that apply to the object with the handle `h`, along with their current values.

To query the value of any default property, concatenate the word `Default` with the object type and the property name. For example, to obtain the default value of the `Color` property for figure objects, use the following command:

```
get(0,'DefaultFigureColor')
```

To obtain a list of all default values currently defined by an object for its descendants, use

```
get(h,'Default')
```

where `h` is the handle of the object.

Examples Use `get` to obtain the handles of objects already on the screen. For example, the following statements create a surface and a text string:

```
surf(peaks)
text(26,50,7,'The peak of peaks')
```

If you want to change the color of the text, but did not save its handle when creating it, you can use `get` to query the properties of the objects in the current axes.

```
h = get(gca,'Children');
get(h(1),'Type')
ans =
    Text
get(h(2),'Type')
ans =
    Surface
```

In this case, `h(1)` contains the handle of the text object. You can now use this handle to set the text color:

```
set(h(1),'Color',[1 0 0])
```

To obtain a list of the surface's properties and their current values, use `get` on the surface's handle:

```
get(h(2))
CData = [ (too many rows) ]
EdgeColor = black
FaceColor = flat
LineStyle = –
MarkerSize = [6]
MeshStyle = both
XData = [ (1 by 49) ]
YData = [ (49 by 1) ]
ZData = [ (49 by 49) ]
Children = []
Clipping = on
Parent = [10.000366211]
Type = Surface
UserData = []
Visible = on
```

In cases where the property value is a matrix too large to display conveniently, the string `too many rows` is returned.

See Also `gca`, `gcf`, `set`

---

getenv

Purpose Get environment variable.

Synopsis `getenv('`*`string`*`')`

Description `getenv('`*`string`*`')` returns the text associated with the environment variable specified by *`string`*.

Examples To get MATLAB's initial search path on a UNIX system

```
s = getenv('MATLABPATH')
s =
    /matlab:/matlab/signal
```

On other systems you might get

DOS:	`\MATLAB;\MATLAB\SIGNAL`
Macintosh:	`MyDisk:MATLAB:MyDisk:MATLAB:SIGNAL`
VMS:	`DISK1:[MATLAB],DISK1:[MATLAB.SIGNAL]`

See Also `path`

References See `getenv` in any C library or UNIX reference.

getframe

Purpose Get movie frame.

Synopsis

```
M = getframe
M = getframe(h)
M = getframe(h,rect)
```

Description `getframe` returns a column vector with one movie frame. The frame is a snapshot (pixmap) of the current axes. `getframe` is usually put in a `for` loop to assemble movie matrix `M` for playback using `movie`.

`getframe(h)` gets a frame from object `h`, where `h` is a handle to the root, a figure, or an axes.

`getframe(h,rect)` specifies the rectangle to copy the bitmap from, relative to the lower-left corner of object `h`, and in the units of its `Units` property.

```
rect = [left  bottom  width  height]
```

where `width` and `height` define the dimensions of the rectangle.

To prevent excessive memory use, it is preferable to preallocate movie matrix `M` before building the movie. The function `moviein` generates a matrix of zeros of the appropriate size.

Because of the matrix limitation of *The Student Edition of MATLAB*, the size of the frame must be small.

Examples Make the peaks function vibrate:

```
z = peaks;
surf(z)
lim = axis;
M = moviein(20);
for j = 1:20
    surf(sin(2*pi*j/20)*z,z)
    axis(lim)
    M(:,j) = getframe;
end
movie(M,20)% Play the movie twenty times
```

See Also `movie`, `moviein`

---

ginput

Purpose Graphical input using the mouse in the graph window.

Synopsis

```
[x,y] = ginput(n)
[x,y] = ginput
[x,y,button] = ginput(...)
```

Description `ginput` provides the means of selecting points from the figure window using a mouse or arrow keys. Note that the window focus must be on the figure window for `ginput` to receive input.

`[x,y] = ginput(n)` gets `n` points from the current axes and returns the *x*- and *y*-coordinates in the column vectors `x` and `y`, respectively.

Use the mouse (or the arrow keys on some systems) to position the cursor. Enter data points by pressing a mouse button or a key on the keyboard. To terminate input before entering `n` points, press the **Return** key.

`[x,y] = ginput` gathers an unlimited number of points until the **Return** key is pressed.

`[x,y,button] = ginput(n)` and `[x,y,button] = ginput` return a third result, `button`, that contains a vector of integers specifying which mouse buttons were used (1, 2, 3 from left) or ASCII numbers if keys on the keyboard were used. If you do not specify an input argument, `ginput` gathers an unlimited number of points until you press the **Return** key.

See Also `gtext`

---

global

Purpose Define global variables.

Synopsis `global X Y Z`

Description `global X Y Z` defines `X`, `Y`, and `Z` as global in scope.

Ordinarily, each MATLAB function, defined by an M-file, has its own local variables, which are separate from those of other functions, and from those of the base workspace and nonfunction scripts. However, if several functions, and possibly the base workspace, *all* declare a particular name as global, then they all share a single copy of that variable. Any assignment to that variable, in any function, is available to all the other functions declaring it global.

Stylistically, global variables often have long names with all capital letters, but this is not required.

Examples Here is the code for the functions `tic` and `toc`, which manipulate a stopwatch-like timer. The global variable `TICTOC` is shared by the two functions, but it is invisible in the base workspace or in any other functions that do not declare it.

```
function tic
%    TIC Start a stopwatch timer.
%    The sequence of commands
%        TIC
%        any stuff
%        TOC
%    prints the time required.
%    See also: TOC, CLOCK.
global TICTOC
TICTOC = clock;
function t = toc
%    TOC Read the stopwatch timer.
%    TOC prints the elapsed time since TIC was used.
%    t = TOC; saves elapsed time in t, does not print.
%    See also: TIC, ETIME.
global TICTOC
if nargout < 1
    elapsed_time = etime(clock,TICTOC)
else
    t = etime(clock,TICTOC);
end
```

See Also `clear`, `isglobal`, `who`

---

gplot

Purpose Plot a graph theoretic graph.

Synopsis
```
gplot(A,xy)
gplot(A,xy,lc)
```

Description `gplot(A,xy)` plots the graph specified by `A` and `xy`. A graph, G, is a set of nodes numbered from 1 to n, and a set of connections, or edges, between them. In order to plot G, two matrices are needed. The adjacency matrix, `A`, has `A(i,j)` nonzero if node `i` is connected to node j. The coordinates array, xy, is an n-by-2 matrix with the position for node i in the i-th row, xy(i,:) = [x(i) y(i)].

gplot(A,xy,lc) uses a specified line type and color instead of the default, solid red lines. See plot for a description of valid line types.

Examples To draw the graph of half of a Bucky ball:

```
[B,xy] = bucky;
gplot(B(1:30,1:30),xy(1:30,:))
axis square
```

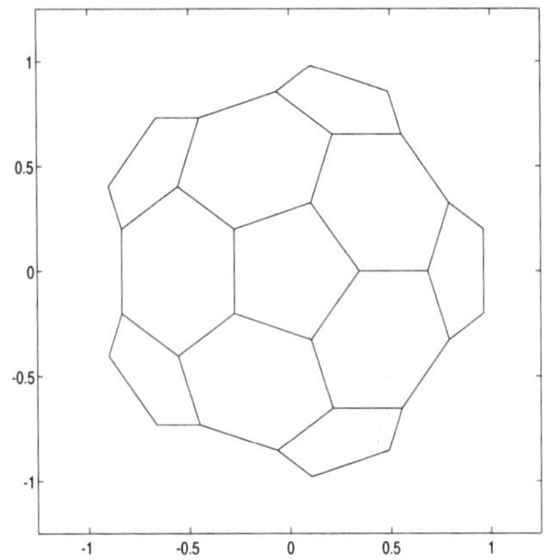

See Also plot, spy

gradient

Purpose Approximate gradient.

Synopsis

```
[PX,PY] = gradient(Z)
[PX,PY] = gradient(Z,x,y)
[PX,PY] = gradient(Z,dx,dy)
p = gradient(Z,...)

d = gradient(y)
d = gradient(y,x)
d = gradient(y,dx)
```

Description [PX,PY] = gradient(Z), with a single matrix argument, computes a numerical approximation to the gradient field of the function tabulated in Z. The result is ordinarily two matrices, the same size as Z, containing horizontal and vertical first differences. One-sided differences are used at the edges of the matrix and centered differences are used in the interior.

[PX,PY] = gradient(Z,x,y), with one matrix and two vector arguments, uses divided differences involving the vector x in the horizontal direction and the vector y in the vertical direction.

`[PX,PY] = gradient(Z,dx,dy)`, with one matrix and two scalar arguments, divides the horizontal difference by the scalar `dx` and the vertical difference by the scalar `dy`.

`P = gradient(Z,...)`, with one output argument, returns a complex result, `P = PX + i*PY`.

`d = gradient(y)`, with a single vector argument, computes a numerical approximation to the first derivative of the function tabulated in `y`. The result is a vector, the same size as `y`, containing first differences. One-sided differences are used at the ends of the vector and centered differences are used in the interior.

`d = gradient(y,x)`, with two vector arguments, uses divided differences to approximate the derivative.

`d = gradient(y,dx)`, with one vector and one scalar argument, divides the first difference by the scalar `dx`.

Examples The statements

```
x = -pi:pi/20:pi;
y = -1:.05:1;
[X,Y] = meshgrid(x,y);
Z = sin(X) + Y.^3;
[PX,PY] = gradient(Z,x,y);
```

produce a matrix `PX` approximating the partial derivative with respect to *x*, which is `cos(X)`, and a matrix `PY` approximating the partial derivative with respect to `y`, which is `3*Y.^2`.

The statements

```
x = -pi:pi/500:pi;
y = tan(sin(x)) - sin(tan(x));
d = gradient(y,x);
```

produce a vector which approximates the derivative *dy/dx*.

See Also `contour`, `del2`, `diff`, `quiver`

---

gray

Purpose Linear grayscale colormap.

Synopsis `cmap = gray(m)`

Description `gray(m)` returns an `m`-by-3 matrix containing a grayscale colormap. The default value for `m` is the length of the current colormap.

Examples To set the colormap of the current figure to gray scale.

```
colormap(gray)
```

See Also `bone`, `contrast`, `colormap`, `cool`, `copper`, `flag`, `hot`, `hsv`, `jet`, `pink`, `prism`, `white`

graymon

Purpose Set default figure properties for grayscale monitors.

Synopsis `graymon`

Description `graymon` changes the default graphics properties for optimal results on a grayscale monitor. It sets the `DefaultAxesColorOrder` for figures to shades of gray, enhancing the visibility of plots and text.

See Also `axes` (properties), `figure` (properties)

grid

Purpose Grid lines for 2-D and 3-D plots.

Synopsis

```
grid on
grid off
grid
```

Description `grid on` adds grid lines on the current axes.

`grid off` takes them off.

`grid`, by itself, toggles the grid state.

Algorithm `grid` sets the `XGrid`, `YGrid`, and `ZGrid` properties of the current `axes`.

See Also `axes`, `plot`

The `XGrid`, `YGrid`, and `ZGrid` properties of `axes` objects.

griddata

Purpose Grid scattered data.

Synopsis

```
ZI = griddata(x,y,z,XI,YI)
[XI,YI,ZI] = griddata(x,y,z,XI,YI)
```

Description `ZI = griddata(x,y,z,XI,YI)` returns matrix `ZI` containing elements corresponding to the elements of matrices `XI` and `YI` and determined by interpolation within the two-dimensional function described by the (usually) nonuniformly spaced vectors `x`, `y`, and `z`.

`XI` can be a row vector, in which case it specifies a matrix with constant columns. Similarly, `YI` can be a column vector and it specifies a matrix with constant rows. `[XI,YI,ZI] = griddata(x,y,z,XI,YI)` returns the `XI` and `YI` formed this way, which are the same as the matrices returned by `meshgrid`.

`griddata` uses an inverse distance method.

Examples Sample a function at 100 random points between ±2.0:

```
rand('seed',0)
x = rand(100,1)*4-2;
y = rand(100,1)*4-2;
z = x.*exp(-x.^2-y.^2);
```

x, y, and z are now vectors containing nonuniformly sampled data. Define a regular grid, and grid the data to it:

```
ti = -2:.25:2;
[XI,YI] = meshgrid(ti,ti);
ZI = griddata(x,y,z,XI,YI);
```

Plot the gridded data along with the nonuniform data points used to generate it:

```
mesh(XI,YI,ZI)
hold
plot3(x,y,z,'o')
hold off
```

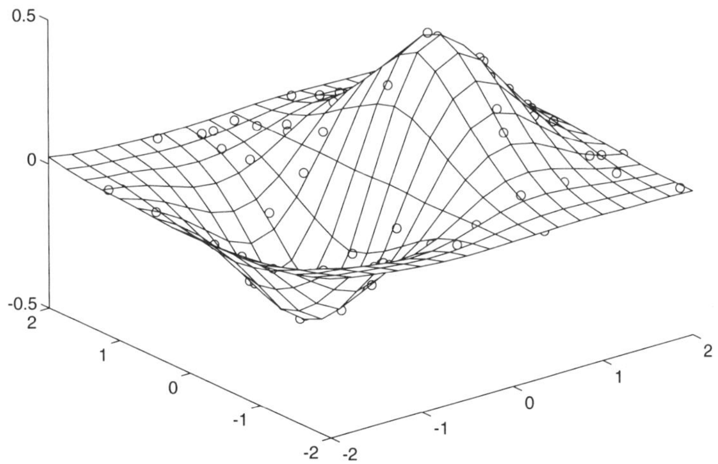

See Also interp1, interp2

gtext

Purpose Mouse placement of text on a graph.

Synopsis gtext('*string*')

Description `gtext('string')` waits for a mouse button or keyboard key to be pressed while the mouse pointer is within the graph window. Pressing a mouse button or any key writes the text string onto the graph at the selected location.

Examples Label an interesting portion of the current plot with

```
gtext('Note this peak!')
```

Algorithm `gtext` uses the functions `ginput` and `text`.

See Also `ginput`, `text`

---

hadamard

Purpose Hadamard matrix.

Synopsis `H = hadamard(n)`

Description `hadamard(n)` returns the Hadamard matrix of order `n`.

Hadamard matrices have applications in several different areas, including combinatorics, signal processing, and numerical analysis, [1], [2]. They are matrices of 1s and -1s whose columns are orthogonal,

```
H'*H = n*I
```

where `[n n] = size(H)` and `I = eye(n,n)`.

An `n`-by-`n` Hadamard matrix with `n > 2` exists only if `rem(n,4) = 0`. This function handles only the cases where `n`, `n/12`, or `n/20` is a power of 2.

Examples `hadamard(4)` is the 4-by-4 matrix

```
1   1   1   1
1  –1   1  –1
1   1  –1  –1
1  –1  –1   1
```

`contour(hadamard(32))` is a plot reminiscent of a Navajo rug.

See Also `compan`, `hankel`, `toeplitz`

References
[1] Ryser, H. J., *Combinatorial Mathematics*, John Wiley and Sons, 1963.

[2] Pratt, W. K., *Digital Signal Processing*, John Wiley and Sons, 1978.

---

hankel

Purpose Hankel matrix.

Synopsis

```
H = hankel(c)
H = hankel(c,r)
```

Description A Hankel matrix is a matrix that is symmetric and constant across the anti-diagonals, and has elements `h(i,j) = r(i+j–1)`, where vector `r` completely determines the Hankel matrix.

`hankel(c)` returns the square Hankel matrix whose first column is `c` and whose elements are zero below the first anti-diagonal.

`hankel(c,r)` returns a Hankel matrix whose first column is `c` and whose last row is `r`. If the last element of `c` differs from the first element of `r`, the last element of `c` wins the disagreement.

Examples A Hankel matrix with anti-diagonal disagreement is

```
c = 1:3; r = 7:10;
h - hankel(c,r)
h =
     1    2    3    8
     2    3    8    9
     3    8    9   10
```

See Also `hadamard`, `toeplitz`, `vander`

---

help

Purpose Online Help for MATLAB functions and M-files.

Synopsis

```
help
help topic
```

Description `help`, by itself, lists all primary help topics. Each main help topic corresponds to a directory name on MATLAB's search path.

`help` *`topic`* gives help on the specified topic. The topic can be a function name or a directory name. If it is a function name, `help` displays information on that function. If it is a directory name, `help` displays the contents file for the specified directory. It is not necessary to give the full pathname of the directory; the last component, or the last several components, is sufficient.

You can write help text for your own M-files and toolboxes; see the following sections for more information.

Algorithm MATLAB's Help system, like MATLAB itself, is highly extensible. This means that you can write help descriptions for your own M-files and toolboxes – using the same self-documenting method that MATLAB's M-files and toolboxes use.

`help`, by itself, lists all help topics by displaying the first line (the H1 line) of the contents files in each directory on MATLAB's search path. The contents files are the M-files named `Contents.m` within each directory.

`help` *`topic`*, where *`topic`* is a directory name, displays the comment lines in the `Contents.m` file located in that directory. If a contents file does not exist, `help` displays the H1 lines of all the files in the directory.

`help` *`topic`*, where *`topic`* is a function name, displays help on the function by listing the first comment lines in the M-file *`topic.m`*.

Creating Online Help for Your Own M-Files

You can create self-documenting online help for your own M-files by entering text on one or more contiguous comment lines, beginning with the second line of the file. For example, the M-file `angle.m` provided with MATLAB contains:

```
function p = angle(h)
% ANGLE Polar angle.
%    ANGLE(H) returns the phase angles, in radians, of a matrix
%    with complex elements. Use ABS for the magnitudes.
p = atan2(imag(h),real(h));
```

When you execute `help angle`, lines 2, 3, and 4 display. These lines are the first block of contiguous comment lines. The help system ignores comment lines that appear later in an M-file, after any executable statements, or after a blank line.

The first comment line in any M-file (the H1 line) is special. It contains the function name and a succinct description of the function. The `lookfor` command searches and displays this line, and `help` displays these lines in directories that do not contain a `Contents.m` file.

Creating Contents Files for Your Own M-File Directories

A `Contents.m` file is provided for each M-file directory included with the MATLAB software. If you create directories in which to store your own M-files, you can create `Contents.m` files for them too. To do so, simply follow the format used in an existing `Contents.m` file.

Examples The following command gives help on the `demos` directory:

```
help demos
```

To prevent long descriptions from scrolling off the screen before you have time to read them, enter `more on`; then enter the `help` command.

See Also `dir`, `lookfor`, `more`, `path`, `what`, `which`

---

hess

Purpose Hessenberg form of a matrix.

Synopsis

```
[P,H] = hess(A)
H = hess(A)
```

Description `hess` finds the Hessenberg form of a matrix. A Hessenberg matrix is zero below the first subdiagonal. If the matrix is symmetric or Hermitian, the form is tridiagonal.

`[P,H] = hess(A)` produces a Hessenberg matrix `H` and a unitary matrix `P` so that `A = P*H*P'` and `P'*P = eye(size(A))`.

`hess(A)` returns `H`.

Examples The matrix `gallery(3)` is a 3-by-3 eigenvalue test matrix:

```
gallery(3) =
   -149     -50    -154
    537     180     546
    -27      -9     -25
```

Its Hessenberg form introduces a single zero in the (3,1) position:

```
hess(gallery(3)) =
   -149.0000    42.2037   -156.3165
   -537.6783   152.5511   -554.9272
           0     0.0728      2.4489
```

This matrix has the same eigenvalues as the original, but more computation is needed to reveal them.

Algorithm For real matrices, `hess` uses the EISPACK routines `ORTRAN` and `ORTHES`. `ORTHES` converts a real general matrix to Hessenberg form using orthogonal similarity transformations. `ORTRAN` accumulates the transformations used by `ORTHES`.

When `hess` is used with a complex argument, the solution is computed using the QZ algorithm by the EISPACK routines `QZHES`. It has been modified for complex problems and to handle the special case $B = I$.

For detailed write-ups on these algorithms, see the *EISPACK Guide*.

See Also `eig`, `qz`, `schur`

References

[1] Smith, B. T., J. M. Boyle, J. J. Dongarra, B. S. Garbow, Y. Ikebe, V. C. Klema, and C. B. Moler, *Matrix Eigensystem Routines – EISPACK Guide*, Lecture Notes in Computer Science, volume 6, second edition, Springer-Verlag, 1976.

[2] Garbow, B. S., J. M. Boyle, J. J. Dongarra, and C. B. Moler, *Matrix Eigensystem Routines – EISPACK Guide Extension*, Lecture Notes in Computer Science, volume 51, Springer-Verlag, 1977.

[3] Moler, C.B. and G. W. Stewart, "An Algorithm for Generalized Matrix Eigenvalue Problems," *SIAM J. Numer. Anal.*, Vol. 10, No. 2, April 1973.

---

hex2dec

Purpose Hexadecimal to decimal number conversion.

Synopsis `d = hex2dec('`*`string`*`')`

Description `hex2dec('`*`string`*`')` converts the hexadecimal integer *`string`* stored in a MATLAB string to its decimal representation.

Examples `hex2dec('3ff')` is 1023.

See Also `dec2hex`, `format`, `hex2num`

---

hex2num

Purpose Hexadecimal to double number conversion.

Synopsis `f = hex2num('string')`

Description `hex2num('string')` converts the hex value in string *`string`* to the IEEE double precision floating-point number it represents. `NaN`, `Inf`, and denormalized numbers are all handled correctly. Fewer than 16 characters are padded on the right with zeros.

Examples

```
f = hex2num('400921fb54442d18')
f =
    3.14159265358979
```

Limitations `hex2num` only works for IEEE numbers; it does not work for the floating-point representation of the VAX or other non-IEEE computers.

See Also `format`, `hex2dec`

hidden

Purpose Remove hidden lines from mesh plot.

Synopsis

```
hidden on
hidden off
hidden
```

Description `hidden on` turns on hidden line removal for the current graph so that lines in the back of a mesh are hidden by those in front. This is the default.

`hidden off` turns off hidden line removal.

`hidden`, by itself, toggles the hidden line state.

Examples Try the statements:

```
mesh(peaks)
hidden off
hidden on
```

Algorithm `hidden` sets the `FaceColor` property of `surface` objects. `hidden on` corresponds to `FaceColor = BackgroundColor`, which is usually `black`. `hidden off` corresponds to `FaceColor = none`.

See Also `shading`
`surface` object properties `FaceColor` and `EdgeColor`

hilb

Purpose Hilbert matrix.

Synopsis `H = hilb(n)`

Description `hilb(n)` returns the Hilbert matrix of order `n`. The elements of the Hilbert matrices are

$$H(i,j) = \frac{1}{i+j-1}$$

The Hilbert matrix is a notable example of a poorly conditioned matrix [1].

Examples Even the fourth order Hilbert matrix shows signs of poor conditioning.

```
cond(hilb(4)) =
    1.5514e+04
```

Algorithm See the M-file for a good example of efficient MATLAB programming where conventional `for` loops are replaced by vectorized statements.

See Also `invhilb`

References
[1] G. E. Forsythe and C. B. Moler, *Computer Solution of Linear Algebraic Systems*, Chapter 19, Prentice-Hall, 1967.

---

hist

Purpose Histogram plot.

Synopsis

```
hist
hist(y)
hist(y,nb)
hist(y,x)
[n,x] = hist(y,...)
```

Description `hist` calculates or plots histograms.

`hist(y)` draws a 10-bin histogram for the data in vector `y`. The bins are equally spaced between the minimum and maximum values in `y`.

`hist(y,nb)` draws a histogram with `nb` bins.

`hist(y,x)`, if `x` is a vector, draws a histogram using the bins specified in `x`.

`[n,x] = hist(y)`, `[n,x] = hist(y,nb)`, and `[n,x] = hist(y,x)` do not draw graphs, but return vectors `n` and `x` containing the frequency counts and the bin locations such that `bar(x,n)` plots the histogram. This is useful in situations where more control is needed over the appearance of a graph, for example, to combine a histogram into a more elaborate plot.

Examples Generate bell-curve histograms from Gaussian data.

```
x = -2.9:0.1:2.9;
y = randn(10000,1);
hist(y,x)
```

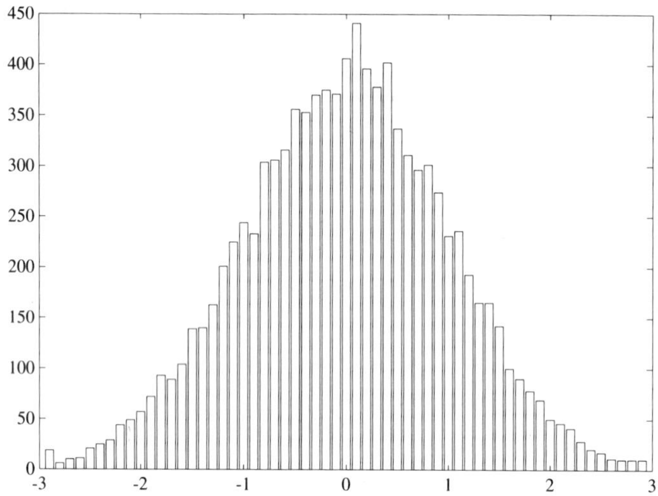

See Also `bar`, `stairs`

---

hold

Purpose Hold the current graph.

Synopsis

```
hold on
hold off
hold
```

Description `hold on` holds the current plot and all axes properties so that subsequent graphing commands add to the existing graph.

`hold off` returns to the default mode whereby plot commands erase previous plots and reset all axes properties before drawing new plots.

`hold`, by itself, toggles the `hold` state.

`hold` is a property of `axes`. If several axes exist in a figure window, each has its own `hold` state.

Algorithm `hold` manipulates the `NextPlot` property of `figure` and `axes` objects.

`hold on` sets the `NextPlot` property of the current `figure` and `axes` to `add`.

`hold off` sets the `NextPlot` property of the current `axes` to `replace`.

See Also `axis`, `cla`

The `NextPlot` property of `axes` objects.

---

home

Purpose Send the cursor home.

Synopsis `home`

Description `home` returns the cursor to the upper-left corner of the command window.

Examples Display a sequence of random matrices, on top of each other.

```
clc, for i = 1:25, home, a = rand(5), end
```

See Also `clc`

---

hostid

Purpose MATLAB server host identification number.

Synopsis `id = hostid`

Description `hostid` returns a string containing the host identification number of the MATLAB server.

See Also `matlabroot`, `version`

---

hot

Purpose Black, red, yellow, and white colormap.

Synopsis `cmap = hot(m)`

Description `hot(m)` returns an `m`-by-3 matrix containing the `hot` colormap. The default value for `m` is the length of the current colormap. The `hot` colormap varies smoothly from black, through shades of red, orange and yellow, to white.

Examples To color the `peaks` function with the `hot` colormap:

```
peaks
colormap(hot)
```

See Also `bone`, `colormap`, `cool`, `copper`, `flag`, `gray`, `hsv`, `jet`, `pink`, `prism`, `white`

---

hsv

Purpose Hue-saturation-value colormap.

Synopsis

```
map = hsv
map = hsv(m)
hsv
```

Description `hsv` is used in conjunction with `colormap` to specify a colormap which varies the hue component of the hue-saturation-value color model. The colors begin with red, pass through yellow, green, cyan, blue, magenta, and return to red. The map is particularly appropriate for displaying periodic functions.

`hsv(m)` returns an `m`-by-3 matrix containing an `hsv` colormap.

`hsv`, by itself, is the same as `hsv(64)`.

Examples The complex function demonstration

```
cplxdemo
```

shows off the `hsv` colormap.

The statements

```
format rat
M = hsv(19)
```

produce

```
M =
          1           0           0
          1        6/19           0
          1       12/19           0
          1       18/19           0
      14/19           1           0
       8/19           1           0
       2/19           1           0
          0           1        4/19
          0           1       10/19
          0           1       16/19
          0       16/19           1
          0       10/19           1
          0        4/19           1
       2/19           0           1
       8/19           0           1
      14/19           0           1
          1           0       18/19
          1           0       12/19
          1           0        6/19
```

The rows of `M` that contain just 0s and 1s represent the basic colors red, yellow, green, cyan, blue, magenta and red again. The other rows represent intermediate colors.

Algorithm `hsv(m)` is the same as `hsv2rgb([h ones(m,2)])` where `h` is the linear ramp,

```
h = (0:m–1)'/(m–1);
```

See Also `colormap`, `hsv2rgb`, `rgb2hsv`

`bone`, `cool`, `copper`, `flag`, `gray`, `hot`, `jet`, `pink`, `prism`, `white`

hsv2rgb

Purpose Hue–saturation–value to RGB conversion.

Synopsis

```
M = hsv2rgb(H)
```

Description `M = hsv2rgb(H)` converts a hue-saturation-value colormap to a red-green-blue (RGB) colormap. Each map is a matrix with any number of rows, exactly three columns, and elements in the interval from 0 to 1. The columns of the input matrix, `H`, represent hue, saturation, and value, respectively. The columns of the resulting output matrix, `M`, represent intensities of red, green, and blue, respectively.

As `H(:,1)`, the hue, varies from 0 to 1, the resulting color varies from red, through yellow, green, cyan, blue and magenta, back to red. When `H(:,2)` is 0, the colors are unsaturated; they are simply shades of gray. When `H(:,2)` is 1, the colors are fully saturated; they contain no white component. When `H(:,3)`, the color value, varies from 0 to 1, the brightness increases.

The MATLAB default colormap, `hsv`, is `hsv2rgb([h s v])` where `h` is a linear ramp from 0 to 1 and both `s` and `v` are all 1s.

See Also `brighten`, `colormap`, `hsv`, `rgb2hsv`

i

Purpose Imaginary unit.

Synopsis

```
i
a+bi
x+i*y
```

Description As the basic complex unit `sqrt(-1)`, `i` is used to enter complex numbers. Since `i` is a function, it can be overridden and used as a variable, if desired. Thus, you can still use `i` as an index in `for` loops, etc.

The character `i` can also be used without a multiplication sign as a suffix in forming a complex numerical constant.

You can also use the character `j` as the complex unit.

Examples

```
Z = 2+3i
Z = x+i*y
Z = r*exp(i*theta)
```

See Also `conj`, `imag`, `j`, `real`

if

Purpose Conditionally execute statements.

Synopsis

```
if expression
    statements
end
```

```
if expression1
    statements
elseif expression2
    statements
else
    statements
end
```

Description `if` conditionally executes statements. The simple form is

```
if expression
    statements
end
```

The *statements* execute if the *expression* has all nonzero elements. The expression is usually the result of

```
expression rop expression
```

where *rop* is one of ==, <, >, <=, >=, or ~=.

More complicated forms use `else` or `elseif`. Each `if` must be paired with a matching `end`.

Examples Here is an example showing `if`, `else`, and `elseif`:

```
for i = 1:n
    for j = 1:n
        if i == j
            a(i,j) = 2;
        elseif abs([i j]) == 1
            a(i,j) = 1;
        else
            a(i,j) = 0;
        end
    end
end
```

See Also `break`, `else`, `end`, `for`, `return`, `while`

---

ifft

Purpose Inverse 1-D fast Fourier transform.

Synopsis

```
y = ifft(x)
y = ifft(x,n)
```

Description `ifft(x)` is the inverse fast Fourier transform of vector `x`.

`ifft(x,n)` is the `n`-point inverse FFT.

Examples For any `x`, `ifft(fft(x))` equals `x` to within roundoff error. If `x` is real, `ifft(fft(x))` may have small imaginary parts.

Algorithm The algorithm for `ifft(x)` is the same as the algorithm for `fft(x)`, except for a sign change and a scale factor of `n = length(x)`. So the execution time is fastest when `n` is a power of 2 and slowest when `n` is a large prime.

See Also `fft`, `fft2`, `fftshift`, `filter`

`dftmtx`, `freqz`, `specplot`, and `spectrum` in the Signal Processing Toolbox

---

ifft2

Purpose Inverse 2-D fast Fourier transform.

Synopsis

```
Y = ifft2(X)
Y = ifft2(X,m,n)
```

Description `ifft2(X)` returns the two-dimensional inverse fast Fourier transform of matrix `X`.

`ifft(X,m,n)` returns the `m`–by–`n` inverse transform.

Examples For any `X`, `ifft2(fft2(X))` equals `X` to within roundoff error. If `X` is real, `ifft2(fft2(X))` may have small imaginary parts.

Algorithm The algorithm for `ifft2(X)` is the same as the algorithm for `fft2(X)`, except for a sign change and scale factors of `[m,n] = size(X)`. The execution time is fastest when `m` and `n` are powers of 2 and slowest when they are large primes.

See Also `fft`, `fft2`, `fftshift`, `filter`, `ifft`

`dftmtx`, `freqz`, `specplot`, and `spectrum` in the Signal Processing Toolbox

---

imag

Purpose Imaginary part.

Synopsis

```
Y = imag(Z)
```

Description `imag(Z)` is the imaginary part of the elements of `Z`.

Examples

```
imag(2+3i)
ans =
     3
```

See Also `conj`, `real`

---

image

Purpose Display image by creating image object.

Synopsis

```
image(C)
image(x,y,C)
image('PropertyName',PropertyValue,...)
h = image(...)
```

Description `image` is both a high-level command for displaying images and a low-level function for creating image objects.

Image objects are children of axes objects. MATLAB displays an image by mapping each element in a matrix to an entry in the figure's colormap. The `image` function allows you to specify property name/property value pairs as input arguments. These properties, which control various aspects of the image object, are described under "Object Properties." You can also set and query property values after creating the object using the `set` and `get` commands.

`image(C)` displays matrix `C` as an image. Each element of `C` specifies the color of a rectilinear patch in the image. The elements of `C` are used as indices into the current colormap to determine the color.

`image(x,y,C)`, where `x` and `y` are vectors, specifies the bounds of the image data on the x- and y-axes, but produces the same image as `image(C)`. The matrix arguments can be followed by property name/property value pairs to specify additional image properties. You can omit the matrix arguments entirely and specify all properties using property name/property value pairs.

`h = image('`*`PropertyName`*`',`*`PropertyValue`*`,...)` specifies property name/property value pairs as input arguments.

`h = image(...)` returns the handle of the object it creates.

`image` and `pcolor` are similar, however, there are important differences:

`image(C)`:

- Specifies the colors of patches
- Uses pixel replication to fill each rectangle
- Uses matrix coordinates if `hold` is not `on` (y-axis reversed)
- Uses indices into the colormap directly

`pcolor(C)`:

- Specifies the colors of vertices
- Uses bilinear interpolation when shading mode is `interp`
- Goes through `caxis` limits into the colormap

The number of rectangles for `image(C)` is the same as the number of vertices for `pcolor(C)`.

Object Properties

This section lists property names along with the type of values each accepts.

`ButtonDownFcn`	Callback string, object selection. Any legal MATLAB expression, including the name of an M-file or function. When you select the object, the string is passed to the `eval` function to execute the specified function. Initially the empty matrix.

`CData`	Color data. A matrix of values that specifies the color at each element of the image. `image(C)` assigns `C` to `CData`. The elements of `CData` are indices into the current colormap, determining the color at each rectilinear patch in the image. Values with a decimal portion are fixed to the nearest, lower integer. Values less than 1 are mapped to the lowest value in the colormap, and values greater than `length(colormap)` are mapped to the highest value.	
`Children`	Children of image. Always the empty matrix; image objects have no children.	
`Clipping`	Clipping mode.	
	`on`	(Default.) Any portion of the image outside the axes rectangle is not displayed.
	`off`	Image data is not clipped.
`Interruptible`	Callback interruptibility.	
	`yes`	The callback specified by `ButtonDownFcn` is interruptible by other callbacks.
	`no`	(Default.) The `ButtonDownFcn` callback is not interruptible.
`Parent`	Handle of the image's parent object. A read-only property that contains the handle of the image's parent object. The parent of an image object is the axes in which it is displayed.	
`Type`	Type of object. A read-only string; always `'image'` for an image object.	
`UserData`	User-specified data. Any matrix you want to associate with the image object. The object does not use this data, but you can retrieve it using the `get` command.	
`Visible`	Image visibility.	
	`on`	(Default.) Image is visible on the screen.
	`off`	Image is not displayed.
`XData`	Image *x*-data. Specifies the position of the image rows. If omitted, the row numbers of `CData` are used.	

`YData`	Image *y*-data. Specifies the position of the image columns. If omitted, the column numbers of `CData` are used.

Examples Some purely mathematical constructs generate interesting images. For example,

```
colormap(cool)
image(((real(fft(eye(64)))+1)/2)*64)
```

Alternatively, you can read images such as photographs and drawings from external files:

```
load earth
colormap(map)
image(X)
```

See Also `colormap`, `pcolor`

imagesc

Purpose Scale data and display as image.

Synopsis

```
h = imagesc(C)
h = imagesc(x,y,C)
```

Description `imagesc(...)` is the same as `image(...)` except the data is scaled to use the full colormap. An optional final arguement `clims = [clow chigh]` can specify the scaling.

Algorithms `imagesc` places scaling information in the displayed image's UserData property so the `colorbar` command can annotate meaningful labels.

See Also `image`, `colorbar`

Inf

Purpose Infinity.

Synopsis `Inf`

Description `Inf` is the IEEE arithmetic representation for positive infinity. Infinity results from operations like division by zero and overflow, which lead to results too large to represent as conventional floating-point values.

Examples `1/0`, `1.e1000`, `2^1000`, and `exp(1000)` all produce `Inf`.

`log(0)` produces `-Inf`.

`Inf-Inf` and `Inf/Inf` both produce `NaN`, Not-a-Number.

See Also `dbstop`, `finite`, `isinf`, `isnan`, `NaN`

info

Purpose Information about MATLAB, MATLAB toolboxes, and The MathWorks.

Synopsis `info`

`info toolboxpath`

Description `info` displays information about

- Supported platforms
- Available toolboxes
- Contacting The MathWorks
- Becoming a subscribing MATLAB user or joining the MATLAB User's Group

`info toolboxpath` displays the `README` file for the specified toolbox.

See Also `whatsnew`

---

input

Purpose Request user input.

Synopsis

```
x = input('prompt')
x = input('prompt','string')
```

Description `x = input('prompt')` displays the text string *prompt* as a prompt on the screen, waits for input from the keyboard, and returns the value entered in `x`. The response can be any MATLAB expression, including function evaluation and reference to M-files. If the function produces multiple results, they are passed as multiple results from `input`.

`x = input('prompt','string')` returns the entered string as a text variable.

If you press the **Return** key without entering input, `input` returns an empty matrix.

Examples Press **Return** to select a default value by detecting an empty matrix:

```
i = input('Do you want more? Y/N [Y]: ','s');
if isempty(i)
    i = 'Y';
end
```

See Also `ginput`, `keyboard`, `menu`, `uicontrol`

---

int2str

Purpose Integer to string conversion.

Synopsis `string = int2str(n)`

Description `string = int2str(n)` converts an integer to a string with integer format.

Examples `int2str(2+3)` is the string `'5'`.

A plot can be labeled with

```
title(['case number ' int2str(n)])
```

See Also `fprintf`, `hex2num`, `num2str`, `setstr`, `sprintf`

interp1

Purpose 1-D data interpolation (table lookup).

Synopsis

```
yi = interp1(x,y,xi)
yi = interp1(x,y,xi,'method')
```

Description `interp1` interpolates between data points. It finds values of a one-dimensional function underlying the data at intermediate points.

`yi = interp1(x,y,xi)` returns vector `yi` containing elements corresponding to the elements of `xi` and determined by interpolation within vectors `x` and `y`.

Interpolation is the same operation as *table lookup*. Described in table lookup terms, the *table* is `tab = [x,y]` and `interp1` *looks up* the elements of `xi` in `x`, and, based upon their locations, returns values `yi` interpolated within the elements of `y`.

`interp1` performs multiple output table lookup if `y` is a matrix. If `y` is a matrix with `length(x)` rows, and `n` columns, `interp1` returns a `length(xi)`-by-`n` matrix `yi` containing the multi-output table lookup results.

`yi = interp1(x,y,xi,'method')` specifies alternative interpolation methods, where *`method`* can be:

`'linear'` for linear interpolation

`'spline'` for cubic spline interpolation

`'cubic'` for cubic interpolation

All the interpolation methods require that `x` be monotonic. The `'cubic'` method also requires that `x` contain uniformly spaced points.

Examples Here are two vectors representing the census years from 1900 to 1990 and the corresponding United States population in millions of people.

```
t = 1900:10:1990;
p = [75.995   91.972  105.711  123.203  131.669 ...
    150.697  179.323  203.212  226.505  249.633];
```

The expression

```
interp1(t,p,1975)
```

interpolates within the census data to estimate the population in 1975. The result is

```
ans =
    214.8585
```

Now interpolate within the data at every year from 1900 to 2000, and plot the result.

```
x = 1900:1:2000;
y = interp1(t,p,x,'spline');
plot(t,p,'o',x,y)
```

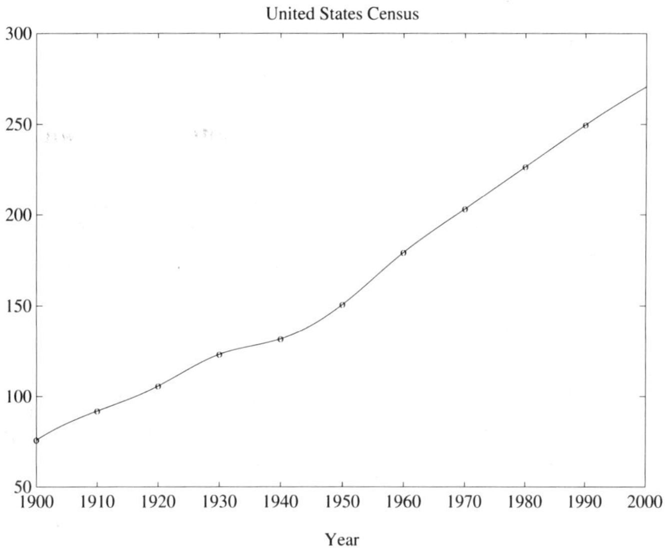

Sometimes it is more convenient to think of interpolation in table lookup terms where the data are stored in a single table. If a portion of the census data is stored in single 5-by-2 table,

```
tab =
     1950     150.697
     1960     179.323
     1970     203.212
     1980     226.505
     1990     249.633
```

then the population in 1975, obtained by table lookup within the matrix `tab`, is

```
p = interp1(tab(:,1),tab(:,2),1975)
p =
     214.8585
```

Algorithm `interp1` is a MATLAB M-file. The `linear` and `cubic` methods have fairly straightforward implementations. For the `spline` method, `interp1` calls a function `spline` that uses the M-files `ppval`, `mkpp`, and `unmkpp`. These routines form a small suite of functions for working with piecewise polynomials. `spline` uses them in a fairly simple fashion to perform cubic spline interpolation. For access to the more advanced features, see these M-files and the Spline Toolbox.

See Also `griddata`, `interp2`, `interpft`

References [1] de Boor, C. *A Practical Guide to Splines*, Springer-Verlag, 1978.

interp2

Purpose 2-D data interpolation (table lookup).

Synopsis

```
ZI = interp2(X,Y,Z,XI,YI)
ZI = interp2(X,Y,Z,XI,YI,'method')
```

Description `interp2` interpolates between data points. It finds values of a two-dimensional function underlying the data at intermediate points.

`ZI = interp2(X,Y,Z,XI,YI)` returns matrix `ZI` containing elements corresponding to the elements of `XI` and `YI` and determined by interpolation within the two-dimensional function specified by matrices `X`, `Y`, and `Z`. Matrix `Z` contains the values of a two-dimensional function at the corresponding abscissae contained in matrices `X` and `Y`.

`X` can be a row vector, in which case the elements are assumed to apply to the columns of `Z`. Similarly, `Y` can be a column vector and its elements are assumed to apply across the rows of `Z`.

Interpolation is the same operation as table lookup. Described in table lookup terms, the table is `tab = [NaN,Y; X,Z]` and `interp2` looks up the elements of `XI` in `X`, `YI` in `Y`, and, based upon their location, returns values `ZI` interpolated within the elements of `Z`.

`ZI = interp2(X,Y,Z,XI,YI,'method')` specifies alternative interpolation methods, where *`method`* can be:

`'linear'` for linear interpolation (default)

`'cubic'` for cubic interpolation

All interpolation methods require that `X` and `Y` be monotonic. The `'cubic'` method also requires that `X` and `Y` contain uniformly spaced points.

Examples Interpolate the peaks function over a finer grid:

```
[X,Y] = meshgrid(-3:.5:3);
Z = peaks(X,Y);
[XI,YI] = meshgrid(-3:.25:3);
ZI = interp2(X,Y,Z,XI,YI);
mesh(X,Y,Z), hold, mesh(XI,YI,ZI+15)
hold off
axis([-3 3 -3 3 -5 20])
```

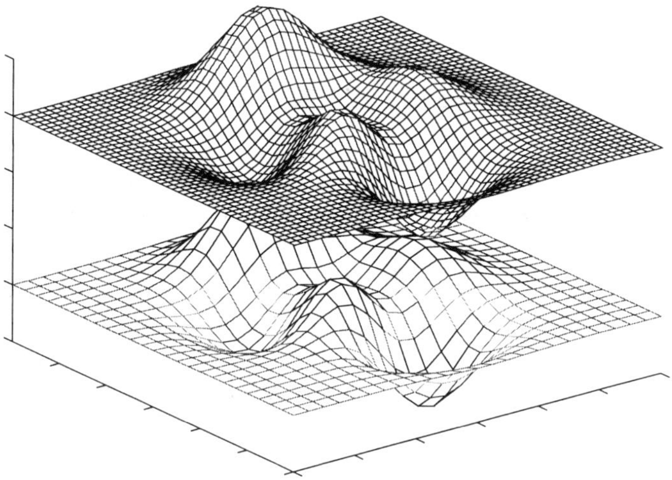

If Poptable is the 6-by-4 table:

```
Poptable =
        0.     10.000     20.000     30.000
     1950.    150.697    199.592    187.625
     1960.    179.323    195.072    250.287
     1970.    203.212    179.092    322.767
     1980.    226.505    153.706    426.730
     1990.    249.633    120.281    598.243
```

Then the statements

```
years = Poptable(2:6,1);
values = Poptable(1,2:4);
T = poptable(2:6,2:4)
```

extract a years vector (the first column of the table), a values vector (the first row of the table), and a matrix T containing all the population data in the table (everything except the first row and the first column).

Now the statement

```
p = interp2(values,years,T,15,1975)
```

returns

```
p =
    190.629
```

This is the average of the four table values.

```
    203.212    179.092
    226.505    153.706
```

See Also griddata, interp1, interpft

interpft

Purpose 1-D interpolation using the FFT method.

Synopsis `y = interpft(x,n)`

Description `y = interpft(x,n)` returns the vector `y` that contains the value of the periodic function `x` resampled to `n` equally spaced points. If `length(x) = m`, and `x` has sample interval `dx`, then the new sample interval for `y` is

```
dy = dx*n/m
```

Note that `n` cannot be smaller than `m`.

Algorithm `interpft` uses the FFT method. The original vector `x` is transformed to the Fourier domain using `fft` and then transformed back with more points.

See Also interp1, interp2

inv

Purpose Matrix inverse.

Synopsis `Y = inv(X)`

Description `inv(X)` is the inverse of the square matrix `X`. A warning message is printed if `X` is badly scaled or nearly singular.

In practice, it is seldom necessary to form the explicit inverse of a matrix. A frequent misuse of `inv` arises when solving the system of linear equations $Ax = b$. One way to solve this is with `x = inv(A)*b`. A better way, from both an execution time and numerical accuracy standpoint, is to use the matrix division operator `x = A\b`. This produces the solution using Gaussian elimination, without forming the inverse. See `\` and `/` for further information.

Examples Here is an example demonstrating the difference between solving a linear system by inverting the matrix with `inv(A)*b` and solving it directly with `A\b`. A matrix `A` of order 100 has been constructed so that its condition number, `cond(A)`, is `1.e10`, and its norm, `norm(A)`, is `1`. The exact solution `x` is a random vector of length 100 and the right-hand side is `b = A*x`. Thus the system of linear equations is badly conditioned, but consistent.

On a 20 MHz 386SX notebook computer, the statements

```
tic, y = inv(A)*b, toc
err = norm(y-x)
res = norm(A*y-b)
```

produce

```
elapsed_time =
    9.6600
err =
    2.4321e-07
res =
    1.8500e-09
```

while the statements

```
tic, z = A\b, toc
err = norm(z-x)
res = norm(A*z-b)
```

produce

```
elapsed_time =
    3.9500
err =
    6.6161e-08
res =
    9.1103e-16
```

It takes almost two and one half times as long to compute the solution with `y = inv(A)*b` as with `z = A\b`. Both produce computed solutions with about the same error, 1.e–7, reflecting the condition number of the matrix. But the size of the residuals, obtained by plugging the computed solution back into the original equations, differs by several orders of magnitude. The direct solution produces residuals on the order of the machine accuracy, even though the system is badly conditioned.

The behavior of this example is typical. Using `A\b` instead of `inv(A)*b` is two to three times as fast and produces residuals on the order of machine accuracy, relative to the magnitude of the data.

Algorithm `inv` uses the subroutines `ZGEDI` and `ZGEFA` from LINPACK. For more information, see the *LINPACK User's Guide*.

Diagnostics From `inv`, if the matrix is singular,

```
Matrix is singular to working precision.
```

On machines with IEEE arithmetic, this is only a warning message. `inv` then returns a matrix with each element set to `Inf`. On machines without IEEE arithmetic, like the VAX, this is treated as an error.

If the inverse was found, but is not reliable, this message is displayed.

```
Warning: Matrix is close to singular or badly scaled.
    Results may be inaccurate. RCOND = xxx
```

See Also `\`, `/`, `det`, `lu`, `rcond`, `rref`

References [1] Dongarra, J.J., J.R. Bunch, C.B. Moler, and G.W. Stewart, *LINPACK User's Guide*, SIAM, Philadelphia, 1979.

---

invhilb

Purpose Inverse of the Hilbert matrix.

Synopsis `H = invhilb(n)`

Description The exact inverse of the exact Hilbert matrix is a matrix whose elements are large integers. These integers can be represented as floatingpoint numbers without roundoff error as long as the order of the matrix, `n`, is less than 12 or 13. `invhilb(n)` generates this matrix. For larger `n`, `invhilb(n)` generates an approximation to the inverse Hilbert matrix.

Comparing `invhilb(n)` with `inv(hilb(n))` involves the effects of two or three sets of roundoff errors:

- The errors caused by representing `hilb(n)`
- The errors in the matrix inversion process
- The errors, if any, in representing `invhilb(n)`

It turns out that the first of these, which involves representing fractions like 1/3 and 1/5 in floating-point, is the most significant.

Examples `invhilb(4)` is

```
    16    -120     240    -140
  -120    1200   -2700    1680
   240   -2700    6480   -4200
  -140    1680   -4200    2800
```

See Also `hilb`

References [1] Forsythe, G. E. and C. B. Moler, *Computer Solution of Linear Algebraic Systems*, Chapter 19, Prentice-Hall, 1967.

---

isempty

Purpose Detect empty matrices.

Synopsis `k = isempty(X)`

Description `isempty(X)` returns 1 if `X` is an empty matrix and 0 otherwise. An empty matrix has size at least one dimension of size zero, for example, 0-by-0 or 0-by-5.

See Also `find`, `size`

isglobal

Purpose True for global variables.

Synopsis `k = isglobal(name)`

Description `isglobal(name)` is 1 if `name` has been declared to be a global variable, and is 0 if it has not been so declared.

See Also `clear`, `global`, `who`

ishold

Purpose True if `hold` is on.

Synopsis `k = ishold`

Description `ishold` returns 1 if `hold` is on, and 0 if it is not.

If `hold` is on, MATLAB preserves the current plot and all axes properties so that subsequent graphing commands add to the existing graph. In this case, the `NextPlot` property of both the current figure and axes is set to `'add'`. If `hold` is `'off'`, `NextPlot` is set to `'replace'`.

See Also `axes`, `figure`, `hold`, `newplot`

isieee

Purpose True for IEEE floating-point arithmetic.

Synopsis `k = isieee`

Description `isieee` returns 1 on machines with IEEE arithmetic and 0 on machines without IEEE arithmetic.

Machines with IEEE arithmetic include

- IBM PC
- Most UNIX workstations
- Macintosh

Machines without IEEE arithmetic include

- VAX
- Cray

See Also `computer`

isinf

Purpose Detect infinities.

Synopsis `k = isinf(X)`

Description `isinf(X)` returns 1 where the elements of `X` are `+Inf` or `–Inf` and 0 where they are not.

Examples

```
A = [pi  NaN  Inf  –Inf]
A =
    3.1416    NaN    Inf    –Inf
isinf(A)
ans =
    0    0    1    1
any(isinf(A))
ans =
    1
```

See Also `finite`, `isnan`

---

isletter

Purpose True for alphabetic character.

Synopsis `k = isletter('str')`

Description `isletter('str')` returns 1s where the elements of *str* are letters of the alphabet and 0s where they are not.

Examples

```
s = 'A1,B2,C3';
isletter(s)

ans =
    1    0    0    1    0    0    1    0
```

See Also `isstr`

---

isnan

Purpose Detect NaNs.

Synopsis `k = isnan(X)`

Description `isnan(X)` returns 1 where the elements of `X` are NaNs and 0 where they are not.

See Also `finite`, `isinf`

---

isreal

Purpose True for matrix consisting of all real elements.

Synopsis `k = isreal(X)`

Description `isreal(X)` returns 1 if all elements of `X` are real numbers, and 0 if any element has a nonzero imaginary component.

isspace

Purpose True for space character.

Synopsis `k = isspace('str')`

Description `isspace('str')` returns 1s where the elements of *str* are white spaces and 0s where they are not. White spaces are space, newline, carriage return, tab, vertical tab, or formfeed characters.

Examples

```
s = 'A B C '
isspace(s)
ans =
     0    1    0    1    0    1
```

See Also `isletter`, `isstr`

issparse

Purpose Check for sparse matrix storage class.

Synopsis `k = issparse(X)`

Description `issparse(X)` returns 1 if the storage class of `X` is sparse and 0 otherwise. Since most of MATLAB's functions work correctly and efficiently with both sparse and full matrices, `issparse(X)` is not needed very often.

Examples

```
if issparse(X)
    nrm = normest(X);
else
    nrm = norm(X);
end
```

See Also `full`, `sparse`

isstr

Purpose Detect strings.

Synopsis `k = isstr(t)`

Description The statement `t = 'Hello, World.'` creates a vector whose components are the ASCII codes for the characters. The column dimension of `t` is the number of characters. It is no different than other MATLAB vectors, except that when displayed, text is shown instead of the decimal ASCII codes.

```
t =
    Hello, World.
```

Associated with each MATLAB variable is a flag that, if set, tells the MATLAB output routines to display the variable as text.

`isstr(t)` returns 1 if the text display flag for `t` is set, and 0 otherwise.

`isstr` also works on matrices.

Examples `isstr('Hello')` is 1.

`isstr(abs('Hello'))` is 0.

See Also `setstr`, `strcmp`, `strings`

---

j

Purpose Imaginary unit.

Synopsis

```
j
x+yj
x+j*y
```

Description You can use the character `j` in place of the character `i` as the imaginary unit.

As the basic complex unit `sqrt(-1)`, `j` is used to enter complex numbers. Since `j` is a function, it can be overridden and used as a variable, if desired. Thus, you can still use `j` as an index in `for` loops, etc.

You can also use the character `j` without a multiplication sign as a suffix in forming a numerical constant.

Examples

```
Z = 2+3j
Z = x+j*y
Z = r*exp(j*theta)
```

See Also `conj`, `i`, `imag`, `real`

---

jet

Purpose Variation of HSV colormap.

Synopsis `cmap = jet(m)`

Description `jet(m)` returns an `m`-by-3 matrix containing the `jet` colormap. This colormap begins with blue, and passes through cyan, yellow, orange, and red. The default value for `m` is the length of the current colormap.

Examples The jet colormap is associated with an astrophysical fluid jet simulation from the National Center for Supercomputer Applications.

```
load flujet
colormap(jet)
```

See Also `bone`, `colormap`, `cool`, `copper`, `flag`, `gray`, `hot`, `hsv`, `pink`, `prism`, `white`

---

keyboard

Purpose Invoke the keyboard as an M-file.

Synopsis `keyboard`

Description keyboard invokes the keyboard as if it were a script M-file. When placed in an M-file, keyboard stops execution of the file and gives control to the keyboard. The special status is indicated by a K appearing before the prompt. You can examine or change variables; all MATLAB commands are valid.

Executing the command return terminates the keyboard mode. Type the six letters

```
return
```

then press the **Return** key.

See Also dbstop, debug, input, quit

kron

Purpose Kronecker tensor product.

Synopsis

```
K = kron(X,Y)
```

Description kron(X,Y) is the Kronecker tensor product of X and Y. The result is a large matrix formed by taking all possible products between the elements of X and those of Y. If X is m-by-n and Y is p-by-q, then kron(X,Y) is m*p-by-n*q.

Examples If X is 2-by-3, then kron(X,Y) is

```
[ X(1,1)*Y X(1,2)*Y X(1,3)*Y
  X(2,1)*Y X(2,2)*Y X(2,3)*Y ]
```

The matrix representation of the discrete Laplacian operator on a two-dimensional, n-by-n grid is a n^2-by-n^2 sparse matrix. There are at most five nonzero elements in each row or column. The matrix can be generated as the Kronecker product of one-dimensional difference operators with the following statements.

```
I = speye(n,n);
E = sparse(2:n,1:n-1,1,n,n);
D = E+E'-2*I;
A = kron(D,I)+kron(I,D);
```

lasterr

Purpose Last error message.

Synopsis

```
str = lasterr
lasterr('')
```

Description str = lasterr returns the last error message generated by MATLAB.

lasterr('') resets the lasterr function so it returns an empty matrix until the next error occurs.

lasterr is useful in conjunction with the two-argument form of eval:

```
eval('str1','str2')
```

where str2 examines the lasterr string to determine the cause of the error and take appropriate action.

Example Here is an example of a function that examines the `lasterr` string and displays its own message based on the error that last occurred. This example deals with two cases, each of which are errors that can result from a matrix multiply.

```
function catch
l = lasterr;
j = findstr(l,'Inner matrix dimensions');
if j~=[]
    disp('Wrong dimensions for matrix multiply')
else
    k = findstr(l,'Undefined function or variable')
    if (k~=[])
        disp('At least one operand does not exist')
    end
end
```

The two-argument form of the `eval` function,

```
eval('str1','str2')
```

evaluates the string `str1` and returns if no error occurs. If an error does occur, `eval` executes `str2`. Try using this form of `eval` with the `catch` function shown above:

```
clear
A = [1  2  3; 6  7  2; 0  -1  5];
B = [9  5  6; 0  4  9];
eval('A*B','catch')
```

MATLAB responds with

```
Wrong dimensions for matrix multiply
```

See Also `error`, `eval`

---

lcm

Purpose Least common multiple.

Synopsis `l = lcm(a,b)`

Description `l = lcm(a,b)` returns the least common multiple of the positive integers `a` and `b`.

Example

```
lcm(8,40)
ans =
    40
```

See Also `gcd`

---

legend

Purpose Add legend to plot.

Synopsis `legend('`*`string1`*`','`*`string2`*`',...)`

```
legend('linetype1','string1','linetype2','string2',...)
legend(h,...)
legend(...,tol)
legend(M)
legend(hl,M)
legend(...)
legend off
```

Description `legend('string1','string2',...)` adds a legend box to the plot in the current figure, using the specified strings as labels. It shows a sample of the linestyle used for each plot line, obtaining this information from the plot.

`legend('linetype1','string1','linetype2','string2',...)` uses the specified *linetype* for each string, instead of obtaining the *linetype* from the plot. Each *linetype* must be a valid line style string as for the `plot` function.

`legend(h,...)` associates the handle `h` with the legend box.

`legend(...,tol)` uses the specified tolerance `tol` in placing the legend box:

- If the legend box would cover `tol` or more points if placed inside the plot, `legend` places it outside the plot.
- If `tol = -1`, the legend box is outside the plot.
- If `tol = 0`, the legend box is inside the plot if there is any location where no data points are covered; otherwise, the legend box is outside the plot.

`legend(M)`, where `M` is a matrix of text strings, adds a legend box containing the rows of `M` as labels.

`legend(hl,M)`, where `hl` is a vector of line handles, associates each row of `M` with the corresponding line.

`h = legend(...)` returns a handle to the legend box axes.

`legend off` removes the legend box from the current axes.

To move a legend box, hold down the left mouse button and drag the box to a new location.

Examples Add a legend to a simple plot.

```
x1 = 1:10; y1 = 2:2:20;
x2 = 1:10; y2 = sqrt(x2);
plot(x1,y1,'+',x2,y2,'*')
legend('Function 1','Function 2')
```

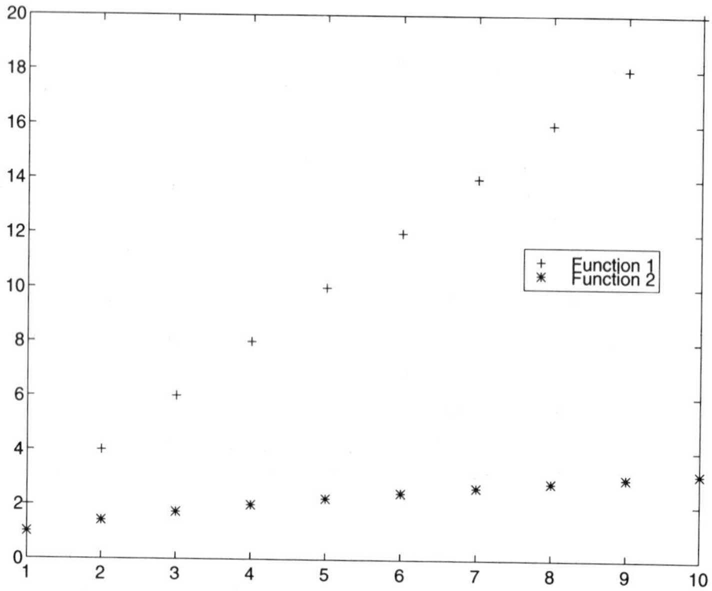

To prevent grid or plot lines from obscuring the legend, make the legend the current axes before printing:

```
h = legend(...)
axes(h)
```

When the legend box is the current axes, the figure window is not redrawn. To force a redraw, use the `refresh` command.

See Also `plot`, `refresh`

---

legendre

Purpose Associated Legendre functions.

Synopsis `P = legendre(n,x)`

Description `P = legendre(n,x)` computes the associated Legendre functions of degree `n` and order `m = 0,1,...,n`, evaluated at `x`. `n` is an integer less than or equal to 256. `x` is a vector whose elements are real and have absolute value less than or equal to one. `P` is a matrix with `n+1` rows and `length(x)` columns. Each element of the returned matrix, `P(i,j)`, corresponds to the associated Legendre function of degree `n` and order `(i–1)` evaluated at `x(j)`. The first row of `P` is the Legendre polynomial evaluated at `x`.

Examples

```
legendre(2,0:0.1:0.2)
ans =
    -0.5000    -0.4850    -0.4400
          0    -0.2985    -0.5879
     3.0000     2.9700     2.8800
```

Note that this answer is of the form

```
P(2,0;0)   P(2,0;0.1)   P(2,0;0.2)
P(2,1;0)   P(2,1;0.1)   P(2,1;0.2)
P(2,2;0)   P(2,2;0.1)   P(2,2;0.2)
```

Algorithm The mathematical definition is

$$P_n^m(x) = (-1)^m \left(1 - x^2\right)^{m/2} \frac{d^m}{dx^m} P_n(x)$$

where

$$P_n(x)$$

is the Legendre polynomial of degree n.

---

length

Purpose Length of vector.

Synopsis `n = length(x)`

Description `length(x)` is the length of vector `x`. It is equivalent to `max(size(x))`.

See Also `size`

---

lin2mu

Purpose Linear to μ-law conversion.

Synopsis `mu = lin2mu(y)`

Description `mu = lin2mu(y)` converts linear audio signal amplitudes in the range $-1 \le$ `y` ≤ 1 to μ–law encoded data in the range $0 \le$ `mu` ≤ 255.

The μ-law standard is widely used in communications, particularly in speech compression and digitization.

See Also `auwrite`, `mu2lin`

---

line

Purpose Create line object.

Synopsis

```
line(x,y)
line(x,y,z)
line('PropertyName',PropertyValue...)
```

```
line(...)
```

Description `line` is the low-level graphics function for creating line objects. Line objects are children of axes objects. They are the fundamental graphics objects used to generate plots, contours and edges of surfaces. The `line` function allows you to specify property name/property value pairs as input arguments. These properties, which control various aspects of the line object, are described under "Object Properties". You can also set and query property values after creating the object using the `set` and `get` functions.

Unlike high-level functions such as `plot`, `line` does not clear the axes, set viewing parameters, or perform any actions other than generating a line object in the current axes. `line` is a low-level function that, ordinarily, is not used directly. Use `plot` and `plot3` instead.

`line(x,y)` adds the line in vectors `x` and `y` to the current axes. If `x` and `y` are matrices of the same size, `line` draws one line per column.

`line(x,y,z)` creates lines in three-dimensional coordinates.

The `x`, `y` pair (`x`, `y`, `z` triple for three-dimensional) can be followed by property name/property value pairs to specify additional line properties. You can omit the `x`, `y`, pair (`x`, `y`, `z` triple) entirely and specify all properties using property name/property value pairs. For example, the following statements are equivalent:

```
line('XData',x,'YData',y,'ZData',z)
line(x,y,z)
```

`line('`*`PropertyName`*`',`*`PropertyValue`*`...)` specifies property name/property value pairs as input arguments.

`h = line(...)` returns a column vector of handles corresponding to each line object the function creates.

Object Properties

This section lists property names along with the type of values each accepts.

`ButtonDownFcn` — Callback string, object selection.

Any legal MATLAB expression, including the name of an M-file or function. When you select the object, the string is passed to the `eval` function to execute the specified function. Initially the empty matrix.

`Children` — Children of line.

Always the empty matrix; line objects have no children.

`Clipping` — Clipping mode.

`on`	(Default.) Any portion of the line outside the axes rectangle is not displayed.
`off`	Line data is not clipped.

`Color`	Line color. A three-element RGB vector or one of MATLAB's predefined names, specifying the line color. See the `ColorSpec` reference page for more information on specifying color.
`EraseMode`	Erase mode. This property controls the technique MATLAB uses to draw and erase line objects. This property is useful in creating animated sequences, where control of individual object redraw is necessary to improve performance and obtain the desired effect.
	`normal` — (Default.) Redraws the affected region of the display, performing the three-dimensional analysis necessary to ensure that all objects are rendered correctly. This mode produces the most accurate picture, but is the slowest. The other modes are faster, but do not perform a complete redraw and are therefore less accurate.
	`none` — The line is not erased when it is moved or destroyed.
	`xor` — The line is drawn and erased by performing an exclusive OR (XOR) with the color of the screen beneath it. When the line is erased, it does not damage the objects beneath it. Lines are dependent on the color of the screen beneath them, however, and are correctly colored only when over the figure background color.
	`background` — The line is erased by drawing it in the figure's background color. This damages objects that are behind the erased line, but lines are always properly colored.
`Interruptible`	Callback interruptibility.
	`yes` — The callback specified by `ButtonDownFcn` is interruptible by other callbacks.
	`no` — (Default.) The `ButtonDownFcn` callback is not interruptible.
`LineStyle`	Line style. This property determines the type of line drawn. Specify a linestyle to be plotted through the data points or a scalable marker type to be placed only at data points:

- Line styles: solid (–), dashed (—), dotted (:), dashdot (–.).
- Marker types: circle (o), plus (+), point (.), star (*), x-mark (x).

The default line style is a solid line.

`LineWidth`	Line width. The width, in points, of the line object. The default line width is 0.5.	
`MarkerSize`	Marker scale factor. A scalar specifying the scale factor, in points, for line markers. This applies only to the marker types circle, plus, point, star, and x-mark. The default marker size is 6 points.	
`Parent`	Handle of the line's parent object. A read-only property that contains the handle of the line's parent object. The parent of a line object is the axes in which it is displayed.	
`Type`	Type of graphics object. A read-only string; always `'line'` for a line object.	
`UserData`	User-specified data. Any matrix you want to associate with the line object. The object does not use this data, but you can retrieve it using the `get` command.	
`Visible`	Line visibility.	
	`on`	(Default.) Line is visible on the screen.
	`off`	Line is not visible on the screen.
`XData`	Line *x*-coordinates. A vector of *x*-coordinates for the line. If a matrix, each column is a separate line, and all data (`YData` and `ZData`) must have the same number of rows. `line` replicates a single column to match the number of columns in other data properties (see "Examples").	
`YData`	Line *y*-coordinates. A vector of *y*-coordinates for the line. If a matrix, each column is a separate line, and all data (`XData` and `ZData`) must have the same number of rows. `line` replicates a single column to match the number of columns in other data properties (see "Examples").	

`ZData` Line z-coordinates.

A vector of z-coordinates for the line. If a matrix, each column is a separate line, and all data (`XData` and `YData`) must have the same number of rows. `line` replicates a single column to match the number of columns in other data properties (see "Examples").

Examples This statement reuses the one column of data specified for `ZData` to produce two lines, each having four points.

```
line(rand(4,2),rand(4,2),rand(4,1))
```

If all the data has the same number of columns and one row each, MATLAB transposes the matrices to produce data for plotting. For example

```
line(rand(1,4),rand(1,4),rand(1,4))
```

is changed to

```
line(rand(4,1),rand(4,1),rand(4,1))
```

This also applies to the case when just one or two matrices have one column. For example, the statement

```
line(rand(2,4),rand(2,4),rand(1,4))
```

is equivalent to

```
line(rand(4,2),rand(4,2),rand(4,1))
```

See Also `patch`, `plot`, `plot3`, `text`

---

linspace

Purpose Generate linearly spaced vectors.

Synopsis

```
y = linspace(x1,x2)
y = linspace(x1,x2,n)
```

Description `linspace` generates linearly spaced vectors. It is similar to the colon operator ":", but gives direct control over the number of points.

`y = linspace(x1,x2)` generates a vector `y` of 100 points linearly spaced between `x1` and `x2`.

`linspace(x1,x2,n)` generates `n` points.

See Also `:`, `logspace`

---

load

Purpose Retrieve variables from disk.

Synopsis

```
load
load filename
load filename.extension
```

Description `load` and `save` are the MATLAB commands to retrieve and store variables on disk. They can also import and export ASCII data files.

MAT-files are double-precision binary MATLAB format files created by the `save` command and readable by the `load` command. They can be created on one machine and later read by MATLAB on another machine with a different floating-point format, retaining as much accuracy and range as the disparate formats allow. They can also be manipulated by other programs, external to MATLAB.

`load`, by itself, loads all the variables saved in the file `'matlab.mat'`.

`load` *`filename`* retrieves the variables from `'filename.mat'`.

`load` *`filename.extension`* reads the file `filename.extension`, which can be an ASCII file with a rectangular array of numeric data, arranged in `m` lines with `n` values in each line. The result is an `m`-by-`n` matrix with same name as the file with the extension stripped.

The ASCII data must be in matrix form, or MATLAB may be unable to use the data when you load it.

You can also load a file whose name is stored in a variable. The commands

```
str = 'fname.mat'
load (str)
```

retrieve the variables from the binary file `'filename.mat'`.

Examples If the file `sample.dat` contains four lines of numeric text,

```
1.1    1.2    1.3
2.1    2.2    2.3
3.1    3.2    3.3
4.1    4.2    4.3
```

then the command,

```
load sample.dat
```

produces a 4-by-3 matrix named `sample` in the workspace.

See Also `fprintf`, `fscanf`, `save`, `spconvert`

---

log

Purpose Natural logarithm.

Synopsis `Y = log(X)`

Description `log` is an elementary function that operates element-wise on matrices. Its domain includes complex numbers, which can lead to unexpected results if used unintentionally.

`log(X)` is the natural logarithm of the elements of `X`. For complex or negative z, where $z = x + y{*}i$, the complex logarithm is returned:

$$\log(z) = \log(abs(z)) + i \ \mathrm{atan2}(y, x)$$

Examples The statement `log(-1)` is a clever way to generate `pi`:

```
ans =
    0.0000 + 3.1416i
```

See Also `exp`, `log10`, `logm`

log10

Purpose Common logarithm.

Synopsis `Y = log10(X)`

Description `log10` is an elementary function that operates element-by-element on matrices. Its domain includes complex numbers, which can lead to unexpected results if used unintentionally.

`log10(X)` is the base 10 logarithm of the elements of `X`.

Examples On a computer with IEEE arithmetic

```
log10(realmax)
```

is

```
308.2547
```

and

```
log10(eps)
```

is

```
-15.6536
```

See Also `log`

log2

Purpose Dissect floating-point numbers.

Synopsis

```
Y = log2(X)
[F,E] = log2(X)
```

Description `log2(X)` computes the base 2 logarithm of the elements of `X`.

`[F,E] = log2(X)` for a real matrix `X`, returns a matrix `F` of real numbers, usually in the range `0.5` ≤ `abs(F) < 1`, and a matrix `E` of integers, so that `X = F.* 2.^E`. Any zeros in `X` produce `F = 0` and `E = 0`.

This function corresponds to the ANSI C function `frexp()` and the IEEE floating-point standard function `logb()`.

Examples For IEEE arithmetic, the statement `[F,E] = log2(X)` yields the following values.

X	F	E
1	1/2	1
pi	pi/4	2
–3	–3/4	2
eps	1/2	–51
realmax	1–eps/2	1024
realmin	1/2	–1021

See Also `format hex`, `log`, `pow2`, `realmax`, `realmin`

loglog

Purpose Log-log scale plot.

Synopsis

```
loglog(x,y)
loglog(x,y,'linetype')
loglog(x1,y1,'linetype1',x2,y2,'linetype2',...)
```

Description `loglog(...)` is the same as `plot(...)` except logarithmic scales are used for both the *x*- and *y*-axes (log-log scales). See `plot` for information on parameters and options.

Examples A simple `loglog` plot:

```
x = logspace(–1,2);
loglog(x,exp(x))
```

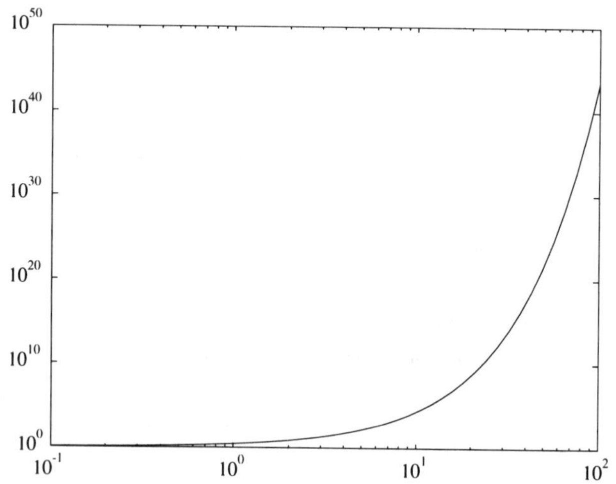

See Also `plot`, `semilogx`, `semilogy`

logm

Purpose Matrix logarithm.

Synopsis `Y = logm(X)`

Description `logm(X)` is the inverse function of `expm(X)` in that for most matrices `A`

```
logm(expm(X)) = X= expm(logm(X))
```

These identities may fail for some `X`. For example, if the computed eigenvalues of `X` include an exact zero, then `logm(X)` generates infinity. Or, if the elements of `X` are too large, `expm(X)` may overflow.

Examples Start with the matrix `X` produced by the example in `expm`.

```
X =

    2.7183    1.7183    1.0862
         0    1.0000    1.2642
         0         0    0.3679
```

Then `A = logm(X)` produces the original matrix `A` used in the `expm` example.

```
A =

    1.0000    1.0000    0.0000
         0         0    2.0000
         0         0   -1.0000
```

But `log(X)` involves taking the logarithm of zero, and so produces

```
ans =

    1.0000    0.5413    0.0826
      -Inf         0    0.2345
      -Inf      -Inf   -1.0000
```

Algorithm The matrix functions are evaluated using an algorithm due to Parlett, which is described in [1]. The algorithm uses the Schur factorization of the matrix and can give poor results or break down completely when the matrix has repeated eigenvalues. A warning message is printed when the results may be inaccurate.

See Also `expm`, `funm`, `sqrtm`

References

[1] Golub, G. H. and C. F. Van Loan, *Matrix Computation*, Johns Hopkins University Press, 1983, p. 384.

[2] Moler, C. B. and C. F. Van Loan, "Nineteen Dubious Ways to Compute the Exponential of a Matrix," *SIAM Review* 20, pp. 801-836, 1979.

logspace

Purpose Generate logarithmically spaced vectors.

Synopsis `y = logspace(a,b)`

```
y = logspace(a,b,n)
y = logspace(a,pi)
```

Description `logspace` generates logarithmically spaced vectors. Especially useful for creating frequency vectors, it is a logarithmic equivalent of `linspace` and the ":" or colon operator.

`y = logspace(a,b)` generates a row vector `y` of 50 logarithmically spaced points between decades `10^a` and `10^b`.

`logspace(a,b,n)` generates `n` points.

`y = logspace(a,pi)` generates the points between `10^a` and `pi`, which is useful for digital signal processing where frequencies over this interval go around the unit circle.

See Also `:`, `linspace`

---

lookfor

Purpose Keyword search through all help entries.

Synopsis `lookfor` *`topic`*

Description The command

```
lookfor xyz
```

looks for the string *`xyz`* in the first comment line (the H1 line) of the help text in all M-files found on MATLAB's search path. For all files in which a match occurs, `lookfor` displays the H1 line.

Examples For example

```
lookfor inverse
```

finds at least a dozen matches, including H1 lines containing "inverse hyperbolic cosine," "two-dimensional inverse FFT," and "pseudoinverse." Contrast this with

```
which inverse
```

or

```
what inverse
```

These commands run more quickly, but probably fail to find anything because MATLAB does not ordinarily have a function `inverse`.

In summary, `what` lists the functions in a given directory, `which` finds the directory containing a given function, and `lookfor` finds all functions in all directories that might have something to do with a given keyword.

See Also `dir`, `help`, `what`, `which`, `who`

---

lower

Purpose Convert string to lowercase.

Synopsis `t = lower('`*`str`*`')`

Description `t = lower('str')` returns the string formed by converting any uppercase characters in the string *str* to the corresponding lowercase characters and leaving all other characters unchanged.

Examples `lower('MathWorks')` is `'mathworks'`.

Algorithm Any values in the range `'A':'Z'` are decremented by `'A'-'a'`.

See Also `isstr`, `strcmp`, `upper`

---

ls

Purpose Directory listing.

Synopsis `ls`

Description `ls` lists the contents of the current working directory.

See Also `matlabroot`, `pwd`

---

lscov

Purpose Least squares solution in the presence of known covariance.

Synopsis `x = lscov(A,b,V)`

Description `x = lscov(A,b,V)` returns the vector `x` that minimizes `(A*x-b)'*inv(V)*(A*x-b)` for the case in which `length(b) > length(x)`. This is the over-determined least squares problem with covariance `V`. `V` is a square symmetric matrix with dimensions equal to `length(b)`. The solution is found without inverting `V`.

Algorithm The classical linear algebra solution to this problem is

```
x = inv(A'*inv(V)*A)*A'*inv(V)*b
```

but this function computes the QR decomposition of `A` and then modifies `Q` by `V`.

See Also `\`, `nnls`, `qr`

Reference Strang, G., *Introduction to Applied Mathematics*, Wellesley-Cambridge, p. 398, 1986.

---

lu

Purpose LU factorization of a matrix.

Synopsis

```
[L,U] = lu(X)
[L,U,P] = lu(X)
lu(X)
```

Description `lu(X)` expresses any square matrix `X` as the product of two essentially triangular matrices, one of them a permutation of a lower triangular matrix and the other an upper triangular matrix. The factorization is often called the *LU*, or sometimes the *LR*, factorization. Most of the algorithms for computing it are variants of Gaussian elimination. The factorization is a key step in obtaining the inverse with `inv` and the determinant with `det`. It is also the basis for the linear equation solution or matrix division obtained with \ and / .

`[L,U] = lu(X)` stores an upper triangular matrix in `U` and a psychologically lower triangular matrix (i.e., a product of lower triangular and permutation matrices) in `L`, so that `X = L*U`.

`[L,U,P] = lu(X)` stores an upper triangular matrix in `U`, a lower triangular matrix in `L`, and a permutation matrix in `P`, so that `L*U = P*X`.

By itself, `lu(X)` returns the output from the LINPACK routine ZGEFA.

Examples Start with

```
A =
     1     2     3
     4     5     6
     7     8     0
```

To see the LU factorization, call `lu` with two output arguments:

```
[L,U] = lu(A)
L =
     0.1429    1.0000         0
     0.5714    0.5000    1.0000
     1.0000         0         0
U =
     7.0000    8.0000    0.0000
          0    0.8571    3.0000
          0         0    4.5000
```

Notice that `L` is a permutation of a lower triangular matrix that has 1s on the permuted diagonal, and that `U` is upper triangular. To check that the factorization does its job, compute the product:

```
L*U
```

which returns the original `A`. Using three arguments on the left-hand side to get the permutation matrix as well:

```
[L,U,P] = lu(A)
```

returns the same value of U, but L is reordered:

```
L =

    1.0000         0         0
    0.1429    1.0000         0
    0.5714    0.5000    1.0000

U =

    7.0000    8.0000         0
         0    0.8571    3.0000
         0         0    4.5000

P =

     0     0     1
     1     0     0
     0     1     0
```

To verify that L*U is a permuted version of A, compute L*U and subtract it from P*A:

```
P*A – L*U
```

The inverse of the example matrix, X = inv(A), is actually computed from the inverses of the triangular factors:

```
X = inv(U)*inv(L)
```

The determinant of the example matrix is

```
d = det(A)
```

which gives

```
d =
    27
```

It is computed from the determinants of the triangular factors:

```
d = det(L)*det(U)
```

The solution to $Ax = b$ is obtained with matrix division:

```
x = A\b
```

The solution is actually computed by solving two triangular systems:

```
y = L\b, x = U\y
```

Triangular factorization is also used by a specialized function, rcond. This quantity is produced by several of the LINPACK subroutines as an estimate of the reciprocal condition number of the input matrix.

Algorithm lu uses the subroutines ZGEDI and ZGEFA from LINPACK. For more information, see the *LINPACK User's Guide*.

See Also \, /, det, inv, qr, rcond, rref

References [1] J.J. Dongarra, J.R. Bunch, C.B. Moler, and G.W. Stewart, *LINPACK User's Guide*, SIAM, Philadelphia, 1979.

---

magic

Purpose Magic square.

Synopsis `M = magic(n)`

Description `M = magic(n)`, for $n \geq 3$, is an n-by-n matrix constructed from the integers 1 through `n^2` with equal row and column sums. A magic square, scaled by its magic sum, is doubly stochastic.

Examples The magic square of order 3 is

```
M = magic(3)
M =
     8     1     6
     3     5     7
     4     9     2
```

This is called a magic square because the sum of the elements in each column is the same.

```
sum(M) =
     15     15     15
```

And the sum of the elements in each row, which can be obtained by transposing twice, is the same.

```
sum(M')' =
     15
     15
     15
```

This is also a special magic square because the diagonal elements have the same sum.

```
sum(diag(M)) =
     15
```

The value of the characteristic sum for a magic square of order `n` is

```
sum(1:n^2)/n
```

which, when `n = 3`, is `15`.

Algorithm There are three different algorithms: one for odd `n`, one for even `n` not divisible by four, and one for even `n` divisible by four.

The following demonstration makes this apparent:

```
for n = 3:20
    A = magic(n);
    plot(A,'-')
    r(n) = rank(A);
end
r
```

See Also ones, rand

---

matlabrc

Purpose MATLAB startup M-file.

Synopsis

```
matlabrc
startup
```

Description At startup time, MATLAB automatically executes the master M-file `matlabrc.m` and, if it exists, `startup.m`. On multiuser or networked systems, `matlabrc.m` is reserved for use by the system manager. The file `matlbrc.m` invokes the file `startup.m` if it exists on MATLAB's search path.

As an individual user, you can create a startup file in your own MATLAB directory. Use these files to include physical constants, engineering conversion factors, or anything else you want predefined in your workspace.

Examples In the file `startup.m`

```
% Add my test directory as the first entry
% of MATLABPATH
path('/my_home/my_tests',path)
```

Algorithm Only `matlabrc` is actually invoked by MATLAB at startup. However, `matlabrc.m` contains the statements:

```
if exist('startup') == 2
    startup
end
```

that invoke `startup.m`. You can extend this process to create additional startup M-files, if required.

See Also !, exist, quit, path

---

matlabroot

Purpose Root directory of MATLAB installation.

Synopsis `rd = matlabroot`

Description `rd = matlabroot` returns the name of the directory in which the MATLAB software is installed.

See Also ls, pwd

max

Purpose Maximum elements of a matrix.

Synopsis

```
y = max(X)
[y,i] = max(X)
C = max(A,B)
```

Description For vectors, `max(X)` is the largest element in `X`. For matrices, `max(X)` is a row vector containing the maximum element from each column and `max(max(X))` is the largest element in the entire matrix.

`[y,i] = max(X)` stores the indices of the maximum values in vector `i`. If there are several identical maximum values, the index of the first one found is returned.

`C = max(A,B)` returns a matrix the same size as `A` and `B` with the largest elements taken from `A` or `B`.

For complex input `X`, `max` returns the complex number with the largest modulus, computed with `max(abs(X))`.

If any element of the matrix is a `NaN`, `max` returns `NaN`. To test for this occurrence, use the function `isnan`.

Examples Start with `A = magic(3)`:

```
A =
     8     1     6
     3     5     7
     4     9     2
```

Then `max(A)` is the vector

```
     8     9     7
```

and `max(max(A))` is scalar

```
     9
```

Furthermore, `[m,i] = max(A)` produces

```
m =
     8     9     7
i =
     1     3     2
```

See Also `isnan`, `min`, `sort`

mean

Purpose Average or mean value of vectors and matrices.

Synopsis

```
m = mean(X)
```

Description `mean(X)`, where `X` is a vector, is the mean value of the elements in vector `X`. For matrices, `mean(X)` is a row vector containing the mean value of each column.

See Also corrcoef, cov, median, std

median

Purpose Median value of vectors and matrices.

Synopsis `m = median(X)`

Description `m = median(X)`, where `X` is a vector, is the median value of the elements in vector `X`. For matrices, `median(X)` is a row vector containing the median value of each column. Since `median` is implemented using `sort`, it can require long execution times for large matrices.

See Also corrcoef, cov, mean, std

menu

Purpose Generate a menu of choices for user input.

Synopsis `k = menu('mtitle','opt1','opt2',...,'optn')`

Description `menu('mtitle','opt1','opt2',...,'optn')` displays the menu whose title is in the string variable `'mtitle'` and whose choices are string variables `'opt1'`, `'opt2'`, and so on. `menu` returns the value you entered.

Examples `k = menu('Choose a color','Red','Green','Blue')` displays

```
——— Choose a color ———
    1) Red
    2) Green
    3) Blue
Select a menu number:
```

After you input a number, use `k` to control the color of a graph.

```
color = ['r','g','b']
plot(t,s,color(k))
```

Computers that employ a window system use a bitmapped menu.

See Also demo, input, uicontrol

mesh, meshc, meshz

Purpose 3-D mesh surface plot.

Synopsis

```
mesh(X,Y,Z,C)
mesh(X,Y,Z)
mesh(x,y,Z,C)
mesh(x,y,Z)
mesh(Z,C)
mesh(Z)
h = mesh(...)
h = meshc(...)
```

```
h = meshz(...)
```

Description For a complete discussion of parametric surfaces, please refer to `surf`.

In its most general invocation, `mesh` has four matrix input arguments. `mesh(X,Y,Z,C)` plots the colored grid lines on the parametric surface specified by `X`, `Y`, and `Z`, with color specified by `C`. In simpler uses, `X` and `Y` can be vectors, or can be omitted, and `C` can be omitted.

The view point is specified by `view`. Axis labels are determined by the range of `X`, `Y`, and `Z`, or by the current setting of `axis`. The color scaling is determined by the range of `C`, or by the current setting of `caxis`. The scaled color values are used as indices into the current colormap.

`mesh(X,Y,Z)` uses `C = Z`, so color is proportional to surface height.

`mesh(x,y,Z,C)` and `mesh(x,y,Z)`, with two vector arguments replacing the first two matrix arguments, must have `length(x) = n` and `length(y) = m` where `[m,n] = size(Z)`. In this case, the intersections of the grid lines are the triples `(x(j),y(i),Z(i,j))`. Note that `x` corresponds to the columns of `Z` and `y` corresponds to the rows.

`mesh(Z,C)` and `mesh(Z)` use `x = 1:n` and `y = 1:m`.

`h = mesh(...)` returns a handle to a `surface` object. `surface` objects are children of `axes` objects.

`meshc(...)` is the same as `mesh(...)` except that a contour plot is drawn beneath the mesh.

`meshz(...)` is the same as `mesh(...)` except that a curtain plot, or reference plane, is drawn beneath the mesh.

Examples Produce a combination mesh and contour plot of the peaks surface.

```
[x,y] = meshgrid(-3:0.25:3);
z = peaks(x,y);
meshc(x,y,z);
axis([-3 3 -3 3 -10 5])
```

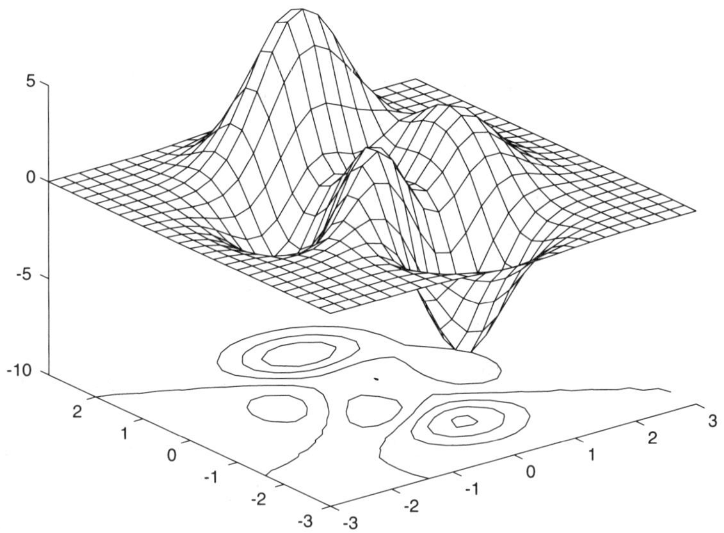

Generate the curtain plot for the peaks function.

```
[x,y] = meshgrid(-3:0.25:3);
z = peaks(x,y);
meshz(x,y,z)
```

Algorithm meshc calls mesh, turns hold on, and then calls contour with an offset. If you need additional control over the appearance of the contours, you can issue these commands directly. You can also combine other types of graphs in this manner, for example surf and pcolor plots.

Limitations meshc assumes that vectors x and y are monotonically increasing. If x and y are irregularly spaced, the contours are simply stretched apart, which is not the same as contouring the irregularly spaced data.

See Also axis, caxis, colormap, contour, hold, image, meshc, pcolor, shading, surf, surfc, surfl, view

---

meshgrid

Purpose Generate X and Y matrices for 3-D plots.

Synopsis

```
[X,Y] = meshgrid(x,y)
[X,Y] = meshgrid(x)
```

Description `[X,Y] = meshgrid(x,y)` transforms the domain specified by vectors `x` and `y` into matrices `X` and `Y` that can be used for the evaluation of functions of two variables and three-dimensional mesh/surface plots. The rows of the output matrix `X` are copies of the vector `x` and the columns of the output matrix `Y` are copies of the vector `y`.

`[X,Y] = meshgrid(x)` is an abbreviation of `[X,Y] = meshgrid(x,x)`.

Examples To evaluate and plot the function

$$z = xe^{\left(-x^2-y^2\right)}$$

over the range $-2 \le x \le 2$, $-2 \le y \le 2$,

```
[X,Y] = meshgrid(-2:.2:2);
Z = X.*exp(-X.^2 - Y.^2);
mesh(Z)
axis([0 20 0 20 min(min(Z)) max(max(Z))])
```

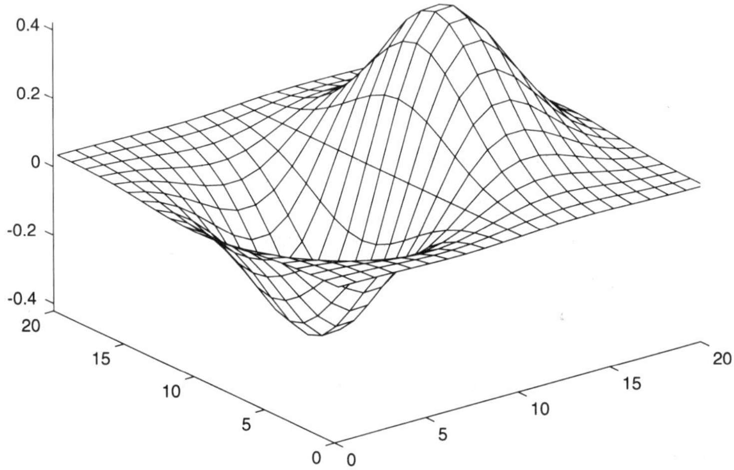

See Also `griddata`, `mesh`, `surf`

---

min

Purpose Minimum elements of a matrix.

Synopsis

```
y = min(X)
[y,i] = min(X)
C = min(A,B)
```

Description `min(X)`, where `X` is a vector, is the smallest element in `X`. For matrices, `min(X)` is a row vector containing the minimum element from each column and `min(min(X))` is the smallest element in the entire matrix.

`[y,i] = min(X)` stores the indices of the minimum values in vector `i`. If there are several identical minimum values, the index of the first one found is returned.

`C = min(A,B)` returns a matrix the same size as `A` and `B` with the smallest elements taken from `A` or `B`.

For complex input `X`, `min` returns the complex number with the smallest modulus, computed with `min(abs(X))`.

For matrices, if a column of `X` contains a `NaN`, the element of the answer corresponding to that column will be `NaN`.

For vectors, if any element of `X` is `NaN`, the answer will be `NaN`.

Examples Start with `A = magic(3)`

```
A =
      8     1     6
      3     5     7
      4     9     2
```

Then `min(A)` is the vector

```
      3     1     2
```

and `min(min(A))` is the scalar

```
       1
```

Furthermore, `[m,i] = min(A)` produces

```
   m =
      3     1     2
   i =
      2     1     3
```

See Also `isnan`, `max`, `sort`

---

more

Purpose Control paged output for the command window.

Synopsis

```
more off
more on
more(n)
```

Description `more off` disables paging of the output in the MATLAB command window.

`more on` enables paging of the output in the MATLAB command window.

`more(n)` specifies the size of the page as `n` lines.

The defaults are `off` and `n = 23`.

When `more` is enabled and output is being paged, press the **Return** key to advance to the next line of output. Press the space bar to advance to the next page of output. Press the **q** (for quit) key to terminate display of the text.

See Also `diary`

---

movie

Purpose Play recorded movie frames.

Synopsis

```
movie(M)
movie(M,n)
movie(M,n,fps)
movie(h,...)
movie(h,m,n,fps,loc)
```

Description `movie(M)` plays the movie in matrix `M` once. `M` must be a matrix whose columns are movie frames (usually from `getframe`).

`movie(M,n)` plays the movie `n` times. If `n` is negative, each *play* is once forward and once backward. If `n` is a vector, elements 2 and above specify the order in which to play the frames. For example, if `M` has three columns, `n = [10 3 2 1]` plays the movie 10 times backwards.

`movie(M,n,fps)` plays the movie at `fps` frames per second. The default if `fps` is omitted is 12 frames per second. Machines that can't achieve the specified `fps` play as fast as they can.

`movie(h,...)` plays the movie in object `h`, where `h` is a handle to a figure or an axes.

`movie(h,m,n,fps,loc)` specifies the location `loc` to play the movie, relative to the lower-left corner of object `h` and in the units of its `Units` property. `loc` is a four-element vector, `loc = [x y 0 0]`. Note that the first two elements of `loc` specify the location of the lower-left corner of the movie; the last two elements are not used, but must be present.

Because of the matrix limitation in *The Student Edition of MATLAB,* recorded movie frames must be small in size.

Examples Make the `peaks` function vibrate:

```
z = peaks;
surf(z);
lim = axis;
M = moviein(20);
for j = 1:20 % Record the movie
    surf(sin(2*pi*j/20)*z,z)
    axis(lim)
    M(:,j) = getframe;
end
movie(M,20) % Play the movie twenty times
```

See Also `getframe`, `moviein`

moviein

Purpose Allocate matrix for movie frames.

Synopsis

```
M = moviein(n)
M = moviein(n,h)
M = moviein(n,h,rect)
```

Description `M = moviein(n)` creates a matrix `M` large enough to hold `n` frames of a movie based on the current axis. The matrix has enough rows to store `n` copies of the output from `getframe`, one in each column.

`M = moviein(n,h)` specifies a graphics object other than the current axis. `h` is a handle to a figure or an axes.

`M = moviein(n,h,rect)` specifies the rectangle to copy the bitmap from, relative to the lower-left corner of object `h` and in the units of its `Units` property,

```
rect = [left  bottom  width  height]
```

where `width` and `height` define the dimensions of the rectangle.

In each case, `moviein` allocates an appropriate matrix for the `getframe` function. Use `getframe` with the same parameters to actually create the movie; for example,

```
M = moviein(n)
for i = 1:n
    plot_command
    M(:,i) = getframe
end
```

Because of the matrix limitation in *The Student Edition of MATLAB,* recorded movie frames must be small in size.

See Also `getframe`, `movie`

mu2lin

Purpose μ-law to linear conversion.

Synopsis

```
y = mu2lin(mu)
```

Description `y = mu2lin(mu)` converts μ-law encoded 8-bit audio signals, stored as *flints* (floating-point integers) in the range 0 ≤ `mu` ≤ 255, to linear audio signal amplitudes in the range –`s` ≤ `y` ≤ `s`. The value of `s` is 32124/32768, or approximately 0.9803.

The μ-law encoded input is often obtained using

```
fread(...,'uchar')
```

to read byte-encoded audio files.

See Also `auread`, `lin2mu`

NaN

Purpose Not-a-Number.

Synopsis `NaN`

Description `NaN` is the IEEE arithmetic representation for Not-a-Number (`NaN`). `NaN`s result from operations which have undefined numerical results.

Examples The following operations produce `NaN`:

- Any arithmetic operation on a `NaN`, such as `sqrt(NaN)`
- Addition or subtraction, such as magnitude subtraction of infinities as `(+Inf)+(-Inf)`
- Multiplication, such as `0*Inf`
- Division, such as `0/0` and `Inf/Inf`
- Remainder, such as `rem(x,y)` where `y` is zero or `x` is infinity
- Comparison by way of predicates involving < or >, such as `min([Inf NaN])` and `max([Inf NaN])`

Logical operations involving `NaN`s always return false, except ~= (not equal). Consequently, the statement `NaN ~= NaN` is true while the statement `NaN == NaN` is false.

See Also `dbstop`, `finite`, `Inf`, `isinf`, `isnan`

nargchk

Purpose Check number of input arguments.

Synopsis `msg = nargchk(low,high,number)`

Description `msg = nargchk(low,high,number)` returns an error message if `number` is less than `low` or greater than `high`. If `number` is between `low` and `high` (inclusive), `nargchk` returns an empty matrix.

Use the function `nargin` to determine the number of input arguments passed to a function.

See Also `nargin`, `nargout`

nargin, nargout

Purpose Number of function arguments.

Synopsis

```
n = nargin
n = nargout
```

Description In the body of a function M-file, `nargin` and `nargout` indicate how many input or output arguments, respectively, a user has supplied.

`nargin` returns the number of input arguments specified for a function.

`nargout` returns the number of output arguments specified for a function.

Examples This example shows portions of the code for the function `fplot`, which has an optional number of input and output arguments:

```
function [x0,y0] = fplot(fname,lims,npts,angl,subdiv)
%      FPLOT  Plot a function.
%      FPLOT(fname,lims,npts,angl,subdiv)
%      The first two input arguments are
%      required; the other three have default values.
...
%      [x,y] = fplot(...) returns x and y instead
%      of plotting them.
...
if nargin < 5, subdiv = 20; end
if nargin < 4, angl = 10; end
if nargin < 3, npts = 25; end
...
if nargout == 0
    plot(x,y)
else
    x0 = x;
    y0 = y;
end
```

newplot

Purpose Graphics M-file preamble to complement the `NextPlot` property.

Synopsis `h = newplot`

Description `h = newplot` is a standard command for the beginning of any graphics M-file function that uses low-level object creation commands to draw graphs. `newplot` uses the `NextPlot` property of figures and axes to determine where to place the graph. For the figure `NextPlot` property:

- If `NextPlot` is `'new'`, open a new figure.
- If `NextPlot` is `'replace'`, clear and reset the current figure.

For the axes `NextPlot` property:

- If `NextPlot` is `'new'`, open a new axes.
- If `NextPlot` is `'replace'`, clear and reset the current axes.

`newplot` returns a handle to the appropriate axes.

See Also `axes`, `figure`, `hold`, `ishold`

nextpow2

Purpose Next power of two.

Synopsis
```
p = nextpow2(n)
p = nextpow2(x)
```

Description `p = nextpow2(n)` returns the first `p` such that `2^p ≥ abs(n)`. It is often useful for finding the nearest power of two sequence length for FFT operations.

`p = nextpow2(x)`, where `x` is a vector, is `nextpow2(length(x))`.

Examples For any integer `n` in the range from 513 to 1024, `nextpow2(n)` is 10.

See Also `fft`, `log2`, `pow2`

---

nnls

Purpose Nonnegative least squares.

Synopsis
```
x = nnls(A,b)
x = nnls(A,b,tol)
[x,w] = nnls(A,b)
[x,w] = nnls(A,b,tol)
```

Description `x = nnls(A,b)` solves the system of equations

$$Ax = b$$

in a least squares sense, subject to the constraint that the solution vector `x` have non-negative elements

$$x_j \geq 0, \quad j = 1, 2, \ldots n$$

`x = nnls(A,b,tol)` specifies a tolerance `tol`, which determines when an element is less than 0. By default, `tol` is

```
max(size(A))*norm(A,1)*eps
```

`[x,w] = nnls(...)` returns the dual vector `w`. The elements of `x` and `w` are related by

$$w_i < 0, \ \left(i \,\middle|\, x_i = 0\right)$$

$$w_i = 0, \ \left(i \,\middle|\, x_i > 0\right)$$

Examples Compare the unconstrained least squares solution to the nnls solution for a 4-by-2 problem:

```
A =
      0.0372    0.2869
      0.6861    0.7071
      0.6233    0.6245
      0.6344    0.6170
b =
      0.8587
      0.1781
      0.0747
      0.8405
[A\b nnls(A,b)] =
     -2.5625         0
      3.1106    0.6929
[norm(A*(a\b)-b) norm(A*nnls(a,b)-b)] =
       0.6677    0.9119
```

The solution from nnls does not fit as well, but has no negative components.

Algorithm nnls uses the algorithm described in [1], Chapter 23. The algorithm starts with a set of possible basis vectors, computes the associated dual vector w, and selects the basis vector corresponding to the maximum value in w to swap out of the basis in exchange for another possible candidate, until w ≤ 0.

See Also \

References [1] Lawson, C. L. and R. J. Hanson, *Solving Least Squares Problems,* Chapter 23, Prentice Hall, 1974.

---

nnz

Purpose Number of nonzero entries.

Synopsis `nz = nnz(X)`

Description `nz = nnz(X)` is the number of nonzeros in matrix X.

The density of a sparse matrix, which is printed by the whos command, is `nnz(X)/prod(size(X))`.

Examples The matrix

```
w = sparse(wilkinson(21));
```

is a tridiagonal matrix with 20 nonzeros on each of three diagonals, so `nnz(w) = 60`.

See Also find, issparse, nonzeros, nzmax, size, whos

nonzeros

Purpose The nonzero entries in a matrix.

Synopsis `s = nonzeros(X)`

Description `s = nonzeros(X)` is a full column vector of the nonzero elements in `X`, ordered by columns.

This gives the `s`, but not the `i` and `j`, from `[i,j,s] = find(X)`. Generally,

`length(s) = nnz(X) ≤ nzmax(X) ≤ prod(size(X))`

See Also `find`, `issparse`, `nnz`, `nzmax`, `size`, `whos`

norm

Purpose Vector and matrix norms.

Synopsis

```
n = norm(X)
n = norm(X,p)
n = norm(X,'fro')
```

Description The *norm* of a matrix is a scalar that gives some measure of the magnitude of the elements of the matrix. Several different types of norms can be calculated:

`n = norm(X)`, where `X` is a matrix, is the largest singular value of `X`.

`n = norm(X,p)`, lets you specify a value to indicate largest singular value, largest column sum, or largest row sum of matrix `X`:

- `norm(X,1)` is the 1-norm, or largest column sum of `X`, `max(sum(abs((X))))`.
- `norm(X,2)` is the same as `norm(X)`.
- `norm(X,inf)` is the infinity norm, or largest row sum of `X`, `max(sum(abs(X')))`.

`n = norm(X,'fro')` is the F-norm of matrix `X`, `sqrt(sum(diag(X'*X)))`.

When the `X` is a vector, slightly different rules apply:

- `norm(x,p) = sum(abs(x).^p)^(1/p)`.
- `norm(x) = norm(x,2)`.
- `norm(x)/sqrt(n)` is the root-mean-square (RMS) value.
- `norm(x,inf) = max(abs(x))`.
- `norm(x, -inf) = min(abs(x))`.

Examples For a matrix

```
A =
      1    2    3
      4    5    6
      7    8    9
norm(A)          = 16.8481
norm(A,1)        = 18
norm(A,2)        = 16.8481
norm(A,inf)      = 24
norm(A,'fro')    = 16.8819
```

For a vector

```
v =
      1    2    3
norm(v)          = 3.7417
norm(v,1)        = 6
norm(v,2)        = 3.7417
norm(v,Inf)      = 3
norm(v,pi)       = 3.2704
```

See Also `cond`, `max`, `min`, `rcond`, `svd`

---

normest

Purpose Estimate matrix z–norm.

Synopsis

```
nrm = normest(S)
nrm = normest(S,tol)
[nrm,cnt] = normest(S)
```

Description This function is intended primarily for sparse matrices, although it works correctly and may be useful for large, full matrices as well.

`nrm = normest(S)` is an estimate of the 2-norm of the matrix S.

`nrm = normest(S,tol)` uses relative error `tol` instead of the default tolerance 1.e-6. The value of `tol` determines when the estimate is considered acceptable.

`[nrm,cnt] = normest(S)` also gives the number of power iterations used.

Examples The matrix `W = wilkinson(101)` is a tridiagonal matrix. Its order, 101, is small enough that `norm(full(W))`, which involves `svd(full(W))`, is feasible. The computation takes 4.13 seconds (on a SPARC 1) and produces the exact norm, 50.7462. On the other hand, `normest(sparse(W))` requires only 1.56 seconds and produces the estimated norm, 50.7458.

Algorithm The power iteration involves repeated multiplication by the matrix `S` and its transpose, `S'`. The iteration is carried out until two successive estimates agree to within the specified relative tolerance.

See Also `condest`, `norm`, `rcond`, `svd`

null

Purpose Null space of a matrix.

Synopsis `Q = null(X)`

Description `Q = null(X)` returns an orthonormal basis for the null space of `X`. If `Q` is not equal to the empty matrix, the number of columns of `Q` is the nullity of `X`.

See Also `orth`, `qr`

num2str

Purpose Number to string conversion.

Synopsis

```
s = num2str(x)
s = num2str(x,precision)
```

Description `num2str` converts numbers to their string representations. This function is useful for labeling and titling plots with numeric values.

`s = num2str(x)` converts the scalar number `x` into a string representation `s` with roughly four digits of precision and an exponent if required.

`s = num2str(x,precision)` converts the scalar number `x` into a string representation `s` with maximum precision specified by `precision`.

Examples `num2str(pi)` is `'3.142'`.

`num2str(eps)` is `'2.22e-16'`.

Algorithm `num2str` uses `sprintf`. If desired, the file can be modified to change the number of digits.

See Also `fprintf`, `hex2num`, `int2str`, `setstr`, `sprintf`

nzmax

Purpose Amount of storage allocated for nonzeros in matrix.

Synopsis `n = nzmax(S)`

Description `n = nzmax(S)`, where `S` is a sparse matrix, is the number of storage locations allocated for the nonzero elements in `S`. For a full matrix, `nzmax(A) = prod(size(A))`.

Often, `nnz(S)` and `nzmax(S)` are the same. But if `S` has been created by an operation, which produces fill-in matrix elements, such as sparse matrix multiplication or sparse LU factorization, more storage may be allocated than is actually required, and `nzmax(S)` reflects this. Alternatively,

```
sparse(i,j,s,m,n,nzmax)
```

or its simpler form, `spalloc(m,n,nzmax)`, can set `nzmax` in anticipation of later fill-in.

See Also `find`, `issparse`, `nnz`, `nonzeros`, `size`, `whos`

ode23, ode45

Purpose Solve ordinary differential equations.

Synopsis

```
[t,x] = ode23('xprime',t0,tf,x0)
[t,x] = ode23('xprime',t0,tf,x0,tol,trace)
[t,x] = ode45('xprime',t0,tf,x0)
[t,x] = ode45('xprime',t0,tf,x0,tol,trace)
```

Description `ode23` and `ode45` are functions for the numerical solution of ordinary differential equations. They can solve simple differential equations or simulate complex dynamical systems.

A system of nonlinear differential equations can always be expressed as a set of first order differential equations:

$$\frac{dx}{dt} = f(t, x)$$

where t is (usually) time, x is the state vector, and f is a function that returns the state derivatives as a function of t and x.

`ode23` integrates a system of ordinary differential equations using second and third order Runge-Kutta formulas. `ode45` uses fourth and fifth order formulas.

The input arguments to `ode23` and `ode45` are

`xprime`	A string variable with the name of the M-file that defines the differential equations to be integrated. The function needs to compute the state derivative vector, given the current time and state vector. It must take two input arguments, scalar `t` (time) and column vector `x` (state), and return output argument `xdot`, a column vector of state derivatives: $\dot{x}_i = \frac{dx_i}{dt}$
`t0`	The starting time for the integration (initial value of t).
`tf`	The ending time for the integration (final value of t).
`x0`	A column vector of initial conditions.
`tol`	Optional - the desired accuracy of the solution. The default value is 1.e–3 for `ode23`, 1.e–6 for `ode45`.
`trace`	Optional - flag to print intermediate results. The default value is 0 (don't trace).

Examples Consider the second order differential equation known as the Van der Pol equation:

$$\ddot{x} + \left(x^2 - 1\right)\dot{x} + x = 0$$

You can rewrite this as a system of coupled first order differential equations:

$$\dot{x}_1 = x_1\left(1 - x_2^2\right) - x_2$$

$$\dot{x}_2 = x_1$$

The first step towards simulating this system is to create a function M-file containing these differential equations. Call it `vdpol.m`:

```
function xdot = vdpol(t,x)
xdot = [x(1).*(1-x(2).^2)-x(2); x(1)]
```

Note that `ode23` requires this function to accept two inputs, `t` and `x`, although the function does not use the `t` input in this case.

To simulate the differential equation defined in `vdpol` over the interval $0 \le t \le 20$, invoke `ode23`:

```
t0 = 0; tf = 20;
x0 = [0 0.25]';  % Initial conditions
[t,x] = ode23('vdpol',t0,tf,x0);
plot(t,x)
```

The general equations for a dynamical system are sometimes written as

$$\frac{dx}{dt} = f(x,t) + u(t)$$

$$y = g(x,t)$$

where u is a time dependent vector of external inputs and y is a vector of final measured outputs of the system. Simulation of equations of this form are accomplished in the current framework by recognizing that $u(t)$ can be formed within the `xdot` function, possibly using a global variable to hold a table of input values, and that $y = g(x,t)$ is simply a final function application to the simulation results.

Algorithm `ode23` and `ode45` are M-files that implement algorithms from [1]. On many systems, MEX-file versions are provided for speed.

`ode23` and `ode45` are automatic step-size Runge-Kutta-Fehlberg integration methods. `ode23` uses a simple second and third order pair of formulas for medium accuracy and `ode45` uses a fourth and fifth order pair for higher accuracy.

Automatic step-size Runge-Kutta algorithms take larger steps where the solution is more slowly changing. Since `ode45` uses higher order formulas, it usually takes fewer integration steps and gives a solution more rapidly. Interpolation (using `interp1`) can be used to give a smoother time history plot.

Diagnostics If `ode23` or `ode45` cannot perform the integration over the full time range requested, it displays the message

```
Singularity likely.
```

See Also `odedemo`, `SIMULINK`

References

[1] Forsythe, G.E., M.A. Malcolm and C.B. Moler, *Computer Methods for Mathematical Computations*, PrenticeHall, 1977.

ones

Purpose Create vector or matrix of all ones.

Synopsis

```
Y = ones(n)
Y = ones(m,n)
Y = ones(size(A))
```

Description `Y = ones(n)` is an n-by-n matrix of 1s.

`Y = ones(m,n)` is an m-by-n matrix of 1s.

`Y = ones(size(A))` is the same size as `A` and consists of all 1s.

See Also `eye`, `rand`, `randn`, `zeros`

orient

Purpose Hardcopy paper orientation.

Synopsis

```
orient
orient portrait
orient landscape
orient tall
```

Description `orient` sets the `PaperOrientation` and `PaperPosition` properties of the current figure window. `orient`, by itself, returns a string with the current paper orientation, either `portrait`, `landscape`, or `tall`:

- `portrait` orientation means that the paper is held with the largest dimension up and down (default).
- `landscape` orientation means that the largest dimension is left to right.
- `tall` orientation means that the figure maps to the whole page in portrait orientation.

`orient portrait` returns to the default portrait orientation with the figure window occupying a rectangle with aspect ratio 4/3 in the middle of the page.

`orient landscape` generates the output for subsequent print operations from the current figure window in full page landscape orientation.

`orient tall` causes the figure window to map to the whole page in portrait orientation.

See Also `print`

`PaperOrientation` and `PaperPosition` properties of `figure` objects

orth

Purpose Range space of a matrix.

Synopsis

```
Q = orth(X)
```

Description `Q = orth(X)` returns an orthonormal basis for the range of `X`. The columns of `Q` span the same space as the columns of `X`, and the columns of `Q` are orthogonal, so that

```
Q'*Q = eye(min(size(X)))
```

The number of columns of `Q` is the rank of `X`.

See Also `null`, `qr`

---

pack

Purpose Consolidate workspace memory.

Synopsis

```
pack
pack filename
```

Description If you get the `Out of memory` message from MATLAB, the `pack` command may find you some free memory without forcing you to delete variables.

`pack` frees up needed space by compressing information into the minimum memory required. Since MATLAB uses a heap method of memory management, extended MATLAB sessions may cause memory to become fragmented. When memory is fragmented, there may be plenty of free space, but not enough contiguous memory to store a new large variable.

The `pack` command

- Saves all variables on disk in a temporary file called `pack.tmp`.
- Clears all variables and functions from memory.
- Reloads the variables back from `pack.tmp`.
- Deletes the temporary file `pack.tmp`.

This results in your workspace having the variables packed or compressed into the minimum memory required, with no wasted space.

`pack` *`filename`* accepts an optional *`filename`* for the temporary file used to hold the variables. Otherwise it uses the file named `pack.tmp`.

If you use `pack` and there is still not enough free memory to proceed, you must clear some variables.

If you run out of memory often, here are some system-specific tips:

- **MS-DOS:** Add more memory.
- **Macintosh**: Under MultiFinder, change the application memory size using **Get Info** on the program icon. Under Single Finder, install more memory.
- **VAX/VMS**: Ask your system manager to increase your working set.
- **UNIX**: Ask your system manager to increase your swap space.

See Also `clear`

pascal

Purpose Pascal matrix.

Synopsis

```
X = pascal(n)
X = pascal(n,1)
X = pascal(n,2)
```

Description `X = pascal(n)` returns the Pascal matrix of order `n`. `X` is a symmetric, positive, definite matrix with integer entries made up from Pascal's triangle. The inverse of `X` has integer entries.

`X = pascal(n,1)` is the lower triangular Cholesky factor (up to the signs of the columns) of the Pascal matrix. It is *involuntary*, that is, it is its own inverse.

`X = pascal(n,2)` is a transposed and permuted version of `pascal(n,1)`. `X` is a cube root of the identity matrix.

Examples

```
(pascal(3,2))^3
ans =
     1     0     0
     0     1     0
     0     0     1
```

See Also `chol`

patch

Purpose Create a patch object.

Synopsis

```
patch(x,y,c)
patch(x,y,z,c)
patch('PropertyName',PropertyValue,...)
h = patch(...)
```

Description `patch` is the low-level graphics function for creating patch objects. Patch objects are children of axes objects. A patch is a filled polygonal area that optionally accepts color data to use for shading. The `patch` function allows you to specify property name/property value pairs as input arguments. These properties, which control various aspects of the patch object, are described under "Object Properties." You can also set and query property values after creating the object using the `set` and `get` functions.

Unlike high-level area creating functions such as `fill`, `patch` does not clear the axes, set viewing parameters, nor perform any actions other than to generate a patch object in the current axes. If points `(x,y)` do not define a closed polygon, `patch` closes the polygon. The points in `x` and `y` can define a concave or self-intersecting polygon.

`patch(x,y,c)` adds the filled two-dimensional polygon defined by vectors `x` and `y` to the current axes. `c` specifies the color used to fill the polygon. The vertices of the polygon are specified by pairs of components of `x` and `y`.

If `c` is a scalar, it simply specifies the color of the polygon ("flat" coloring). If it is a vector the same length as `x` and `y`, its elements are scaled by `caxis` and used as indices into the current colormap to specify colors at the vertices; the color within the polygon is obtained by bilinear interpolation in the vertex colors.

If `c` is a string, the polygon(s) are filled with the specified color. The string can be either the letter: `r`, `g`, `b`, `c`, `m`, `y`, `w`, or `k`, or the names of the colors: `red`, `green`, `blue`, `cyan`, `magenta`, `yellow`, `white`, or `black`. In either case, the string must be enclosed in single quotes.

If `x` and `y` are matrices of the same size, `patch` draws one polygon per column. In this case, `c` can be one of the following:

- A row vector whose size equals `size(x,2)` (i.e., equal to the number of columns in `x` or `y`) for flat shading.
- A matrix the same size as `x` and `y` for interpolated shading.
- A column vector having the same number of rows as `x` and `y`, in which case each patch uses this column to obtain vertex colors for interpolated shading.

`patch` sets its `FaceColor` property to `flat`, `interp`, or a `ColorSpec` depending on the value of `c`.

`patch(x,y,z,c)` creates patches in three-dimensional coordinates.

`h = patch(...)` returns a column vector of handles corresponding to each patch object it creates.

The `x`, `y` pair (`x`, `y`, `z` triple for three-dimensional space) can be followed by property name/property value pairs to specify additional patch properties. You can omit the `x`, `y`, pair (`x`, `y`, `z` triple for three-dimensional space) entirely and specify all properties using property name/property value pairs.

`patch` is a low-level function that, ordinarily, is not used directly. Use `fill` and `fill3` instead.

Object Properties

This section lists property names along with the type of values each accepts.

`ButtonDownFcn`	Callback string, object selection. Any legal MATLAB expression, including the name of an M-file or function. When you select the object, the string is passed to the `eval` function to execute the specified function. Initially the empty matrix.
`CData`	Color data. A vector of values that specifies the color at every point along the edge of the patch. MATLAB uses these values only if `EdgeColor` or `FaceColor` is set to `interp` or `flat`.
`Children`	Children of patch. Always the empty matrix; patch objects have no children.

`Clipping`	Clipping mode.	
	`on`	(Default.) Any portion of the patch outside the axes rectangle is not displayed.
	`off`	Patch data is not clipped.
`EdgeColor`	Patch edge color.	
	`ColorSpec`	A three-element RGB vector or one of MATLAB's predefined names, specifying a single color for edges. The default edge color is black. See the `ColorSpec` reference page for more information on specifying color.
	`none`	Edges are not drawn.
	`flat`	Edges are a single color determined by the average of the color data for that patch.
	`interp`	Edge color is determined by linear interpolation through the values at the patch vertices.
`EraseMode`	Erase mode. This property controls the technique MATLAB uses to draw and erase patch objects. This property is useful in creating animated sequences, where control of individual object redraw is necessary to improve performance and obtain the desired effect.	
	`normal`	(Default.) Redraws the affected region of the display, performing the three-dimensional analysis necessary to ensure that all objects are rendered correctly. This mode produces the most accurate picture, but is the slowest. The other modes are faster, but do not perform a complete redraw and are therefore less accurate.
	`none`	The patch is not erased when it is moved or destroyed.
	`xor`	The patch is drawn and erased by performing an exclusive OR (XOR) with the color of the screen beneath it. When the patch is erased, it does not damage the objects beneath it. Patch objects are dependent on the color of the screen beneath them, however, and are correctly colored only when over the figure background color.

	`background`	The patch is erased by drawing it in the figure's background color. This damages objects that are behind the erased patch, but patch objects are always properly colored.
`FaceColor`	Patch face color.	
	`ColorSpec`	A three-element RGB vector or one of MATLAB's predefined names, specifying a single color for faces. See the `ColorSpec` reference page for more information on specifying color.
	`none`	Faces are not drawn. You can still draw edges, however.
	`flat`	(Default.) The values in `c` determine the face color for each patch.
	`interp`	Face color is determined by linear interpolation through the values specified in the `CData` property.
`Interruptible`	Callback interruptibility.	
	`yes`	The callback specified by `ButtonDownFcn` is interruptible by other callbacks.
	`no`	(Default.) The `ButtonDownFcn` callback is not interruptible.
`LineWidth`	Line width for patch edges.	
	A scalar specifying the width, in points, of the patch edges. The default line width is 0.5.	
`Parent`	Handle of the patch's parent object.	
	A read-only property that contains the handle of the patch's parent object. The parent of a patch object is the axes in which it is displayed.	
`Type`	Type of object.	
	A read-only string; always `'patch'` for a patch object.	
`UserData`	User-specified data.	
	Any matrix you want to associate with the patch object. The object does not use this data, but you can retrieve it using the `get` command.	
`Visible`	Patch visibility.	
	`on`	(Default.) Patch is visible on the screen.
	`off`	Patch is not drawn.

`XData`	Patch vertices *x*-coordinates. *X*-coordinates of the points along the edge of the patch. If a matrix, each column represents the *x*-coordinates for a separate patch. In this case, `XData`, `YData`, and `ZData` must all have the same number of rows.
`YData`	Patch vertices *y*-coordinates. *Y*-coordinates of the points along the edge of the patch. If a matrix, each column represents the *y*-coordinates for a separate patch. In this case, `XData`, `YData`, and `ZData` must all have the same number of rows.
`ZData`	Patch vertices *z*-coordinates. *Z*-coordinates of the points along the edge of the patch. If a matrix, each column represents the *z*-coordinates for a separate patch. In this case, `XData`, `YData`, and `ZData` must all have the same number of rows.

Examples `patch` replicates a single column of vertex data to match the number of columns specified in other arguments. For example, this statement reuses the one column of data specified for `ZData` to produce two patches, each having four vertices.

```
patch(rand(4,2),rand(4,2),rand(4,1))
```

If all the data has the same number of columns and one row each, MATLAB transposes the matrices to produce plotable data. For example,

```
patch(rand(1,4),rand(1,4),rand(1,4))
```

is changed to

```
patch(rand(4,1),rand(4,1),rand(4,1))
```

This also applies to the case when just one or two matrices have one column. For example, this statement,

```
patch(rand(2,4),rand(2,4),rand(1,4))
```

is equivalent to

```
patch(rand(4,2),rand(4,2),rand(4,1))
```

See Also `fill`, `fill3`, `line`, `text`

---

path

Purpose Control MATLAB's directory search path.

Synopsis

```
path
p = path
path(p)
path(p1,p2)
```

Description MATLAB has a *search path*. If you enter a name, such as `fox`, the MATLAB interpreter

1. Looks for `fox` as a variable.
2. Checks for `fox` as a built-in function.
3. Looks in the current directory for `fox.mex` and `fox.m`.
4. Searches the directories specified by `path` for `fox.mex` and `fox.m`.

`path`, by itself, prints out the current setting of MATLAB's search path. The search path list originates from the environment variable `MATLABPATH` in the underlying operating system, which is set in MATLAB's startup script, or is set by `matlabrc.m`, and is perhaps individualized by `startup.m`.

`p = path` returns the current search path in string variable `p`.

`path(p)` changes the path to the string in `p`.

`path(p1,p2)` changes the path to the concatenation of the two path strings `p1` and `p2`. Thus `path(path,p)` appends a new directory to the current path and `path(p,path)` prepends a new path.

Examples Add a new directory to the search path on various operating systems:

UNIX:	`path(path,'/home/myfriend/goodstuff')`
VMS:	`path(path,'DISKS1:[MYFRIEND.GOODSTUFF]')`
MS-DOS:	`path(path,'TOOLS\GOODSTUFF')`
Macintosh:	`path(path,'Tools:GoodStuff')`

See Also `cd`, `dir`, `getenv`, `what`

---

pause

Purpose Halt execution temporarily.

Synopsis

```
pause
pause(n)
pause on
pause off
```

Description `pause` causes M-files to stop and wait for you to press any key before continuing.

`pause(n)` pauses for `n` seconds before continuing.

`pause on` indicates that subsequent pause commands do pause.

`pause off` indicates that any subsequent `pause` or `pause(n)` commands do not actually pause. This allows normally interactive scripts to run unattended.

Examples An important use of `pause` is to halt M-files temporarily when graphics commands are encountered. If `pause` is not used, the graphics may not be visible. An example of this is

```
for n = 3:22
    mesh(magic(n))
    pause
end
```

pcolor

Purpose Pseudocolor plot.

Synopsis

```
pcolor(C)
pcolor(x,y,C)
pcolor(X,Y,C)
h = pcolor(...)
```

Description `pcolor(C)` draws a pseudocolor plot, a rectangular array of cells with colors determined by the elements of `C`. A pseudocolor plot is actually a surface plot seen from the top. MATLAB creates this type of plot by using each set of four adjacent points in `C` to define a patch object.

With the default shading mode, `faceted`, each cell has a constant color determined by the corner with the smallest xy coordinates, and the last row and column of `C` are not used. With `interp` shading, each cell has color resulting from bilinear interpolation of the color at its four vertices and all elements of `C` are used. The mapping from `C` to `color` is defined by `colormap` and `caxis`.

`pcolor(x,y,C)` with two vector arguments must have `length(x) = n` and `length(y) = m` where `[m,n] = size(C)`. The spacing of the grid lines is set by `x` and `y`, so the grid lines are straight, but not necessarily evenly spaced. Note that `x` corresponds to the columns of `C` and `y` corresponds to the rows.

`pcolor(X,Y,C)` has three matrix arguments, all the same size. A logically rectangular, two-dimensional grid is drawn with vertices at the points `[X(i,j), Y(i,j)]`. The color in the `(i,j)`th cell is determined by `C(i,j)` if shading is `faceted` or `flat`, or by interpolation between the colors at the four vertices if shading is `interp`.

`pcolor` is closely related to `surf`; in fact, `pcolor(X,Y,C)` is the same as viewing `surf(X,Y,0*Z,C)` from *above*, that is, `view([0 90])`.

`pcolor` and `image` are similar, but `pcolor(C)` specifies the colors of vertices whereas `image(C)` specifies the colors of cells and directly indexes into the colormap with no scaling. Consequently, the number of vertices for `pcolor(C)` is the same as the number of cells for `image(C)`. With three arguments, `pcolor(X,Y,C)` can produce parametric grids, which are not possible with `image`.

`h = pcolor(...)` returns a handle to a `surface` object. `surface` objects are children of `axes`.

Examples A Hadamard matrix has elements which are +1 and -1, so a colormap with only two entries is appropriate.

```
pcolor(hadamard(20))
colormap(gray(2))
axis('ij')
axis('square')
```

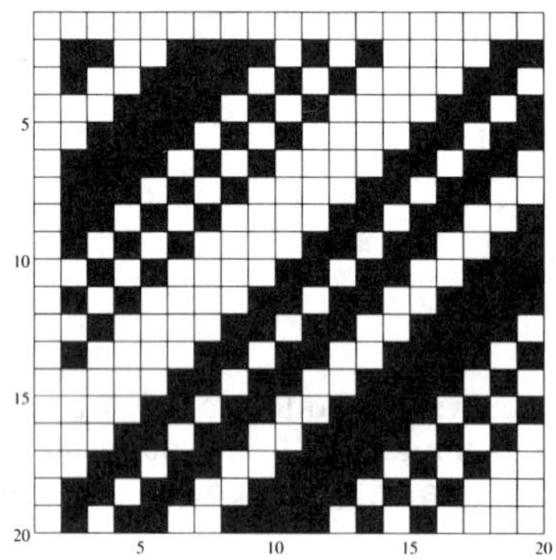

A simple color wheel illustrates a polar coordinate system.

```
n = 6; r = (0:n)'/n;
theta = pi*(-n:n)/n;
X = r*cos(theta);
Y = r*sin(theta);
C = r*cos(2*theta);
pcolor(X,Y,C)
axis('square')
```

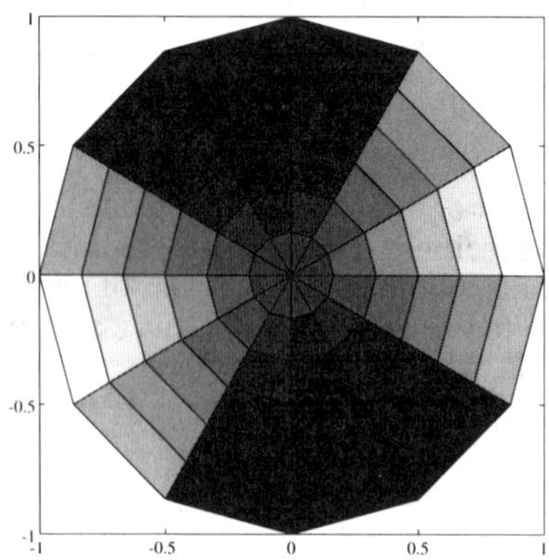

See Also image, surf, view

pi

Purpose Provide π

Synopsis pi

Description pi is a function that returns the floating-point number nearest the value of π.

4*atan(1) and imag(log(-1)) provide the same value.

Examples sin(pi) is not exactly zero because pi is not exactly π.

See Also ans, eps, finite, i, Inf, isnan, j, NaN

pink

Purpose Pink shades colormap.

Synopsis cmap = pink(m)

Description pink(m) returns an m-by-3 matrix containing the pink colormap. The default value for m is the length of the current colormap.

Examples The pink colormap can provide sepia tone colorization of gray scale photographs. One example is MATLAB's default image

```
image
colormap(flipud(pink(15)))
axis image
```

See Also bone, colormap, cool, copper, flag, gray, hot, hsv, jet, prism, white

pinv

Purpose Moore-Penrose pseudoinverse of a matrix.

Synopsis
```
X = pinv(A)
X = pinv(A,tol)
```

Description X = pinv(A) produces the Moore-Penrose pseudoinverse, which is a matrix X of the same dimensions as A' satisfying four conditions:

```
A*X*A = A
X*A*X = X
A*X is Hermitian
X*A is Hermitian
```

The computation is based on svd(A) and any singular values less than a tolerance are treated as zero. The default tolerance is

```
tol = max(size(A))*norm(A)*eps
```

This tolerance can be overridden with X = pinv(A,tol).

Examples If A is square and not singular, then pinv(A) is an expensive way to compute inv(A). If A is not square, or is square and singular, then inv(A) does not exist. In these cases, pinv(A) has some of, but not all, the properties of inv(A).

If A has more rows than columns and is not of full rank, then the overdetermined least squares problem

```
minimize norm(A*x-b)
```

does not have a unique solution. Two of the infinitely many solutions are

```
x = pinv(A)*b
```

and

```
y = A\b
```

These two are distinguished by the facts that norm(x) is smaller than the norm of any other solution and that y has the fewest possible nonzero components.

For example, the matrix generated by

```
A = magic(8); A = A(:,1:6)
```

is an 8-by-6 matrix which happens to have rank(A) = 3.

```
A =
     64     2     3    61    60     6
      9    55    54    12    13    51
     17    47    46    20    21    43
     40    26    27    37    36    30
     32    34    35    29    28    38
     41    23    22    44    45    19
     49    15    14    52    53    11
      8    58    59     5     4    62
```

The right-hand side is b = 260*ones(8,1),

```
b =
      260
      260
      260
      260
      260
      260
      260
      260
```

The scale factor 260 is the 8-by-8 magic sum. With all eight columns, one solution to A*x = b would be a vector of all 1s. With only six columns, the equations are still consistent, so a solution exists, but it is not all 1s. Since the matrix is rank deficient, there are infinitely many solutions. Two of them are

```
x = pinv(A)*b
```

which is

```
x =
     1.1538
     1.4615
     1.3846
     1.3846
     1.4615
     1.1538
```

and

```
y = A\b
```

which is

```
y =
     3.0000
     4.0000
          0
          0
     1.0000
          0
```

Both of these are exact solutions in the sense that `norm(A*x-b)` and `norm(A*y-b)` are on the order of roundoff error. The solution `x` is special because

```
norm(x) = 3.2817
```

is smaller than the norm of any other solution, including

```
norm(y) = 6.4807
```

On the other hand, the solution `y` is special because it has only three nonzero components.

See Also `inv`, `qr`, `rank`, `svd`

---

plot

Purpose Linear 2–D plot.

Synopsis

```
plot(Y)
plot(X,Y)
plot(X,Y,'linetype')
plot(X1,Y1,'linetype1',X2,Y2,'linetype2',...)
h = plot(...)
```

Description `plot(Y)` plots the columns of `Y` versus their index. If `Y` is complex, `plot(Y)` is equivalent to `plot(real(Y),imag(Y))`. In all other uses of `plot`, the imaginary part is ignored.

`plot(X,Y)` plots vector `X` versus vector `Y`. If `X` or `Y` is a matrix, then the vector is plotted versus the rows or columns of the matrix, whichever line up.

Various line types, plot symbols and colors can be obtained with `plot(X,Y,'linetype')` where *linetype* is a 1-, 2-, or 3-character string made from the following characters:

.	point	y	yellow
o	circle	m	magenta
x	x-mark	c	cyan
+	plus	r	red
*	star	g	green
–	solid line	b	blue
:	dotted line	w	white
–.	dashdot line	k	black
—	dashed line		

For example, `plot(X,Y,'c+')` plots a cyan plus at each data point.

`plot(X1,Y1,'linetype1',X2,Y2,'linetype2',...)` combines the plots defined by the `(X,Y,linetype1')` triples, where the X's and Y's are vectors or matrices and the *linetypes* are strings.

For example, `plot(X,Y,'–',X,Y,'go')` plots the data twice, with a solid yellow line interpolating green circles at the data points.

The `plot` command, if no color is specified, makes automatic use of the colors in the above table. The default is yellow for one line, and for multiple lines, to cycle through the first six colors in the table.

`h = plot(...)` returns a column vector of handles to `line` objects, one handle per line. The `line` objects that `plot` creates are children of the current `axes`.

The X,Y pairs, or X, Y, *linetype* triples, can be followed by parameter value pairs to specify additional properties of the lines. See `line` for more information.

Examples The statements

```
x = –pi:pi/500:pi;
y = tan(sin(x)) – sin(tan(x));
plot(x,y)
```

produce

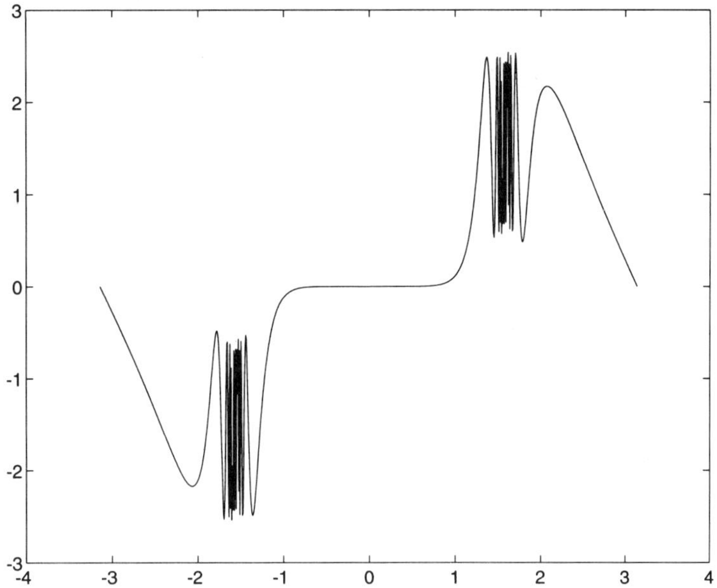

See Also axis, grid, hold, line, loglog, polar, semilogx, semilogy, subplot, title, xlabel, ylabel

---

plot3

Purpose Plot points and lines in 3-D space.

Synopsis

```
plot3(X,Y,Z)
plot3(X,Y,Z,'linetype')
plot3(X1,Y1,Z1,'linetype1',X2,Y2,Z2,'linetype2',...)
h = plot3(...)
```

Description `plot3` is a three-dimensional analog of `plot`.

`plot3(X,Y,Z)`, where `X`, `Y`, and `Z` are three vectors of the same length, plots a line in 3-D space through the points whose coordinates are the elements of `X`, `Y` and `Z`. If `X`, `Y`, and `Z` are three matrices of the same size, `plot3` plots several lines obtained from the columns of `X`, `Y`, and `Z`.

Various line types, plot symbols and colors can be obtained with `plot3(X,Y,Z,s)` where `s` is a 1, 2, or 3 character string made from the characters listed under the `plot` command.

`plot3(x1,y1,z1,s1,x2,y2,z2,s2,x3,y3,z3,s3,...)` combines the plots defined by the `(x,y,z,s)` quads, where the x's, y's, and z's are vectors or matrices and the s's are strings.

h = plot3(...) returns a column vector of handles to line objects, one handle per line. The line objects created by plot3 are children of the current axes.

The x,y,z triples, or x,y,z,s quads can be followed by parameter value pairs to specify additional properties of the lines.

Examples Plot a three-dimensional helix:

```
t = 0:pi/50:10*pi;
plot3(sin(t),cos(t),t)
```

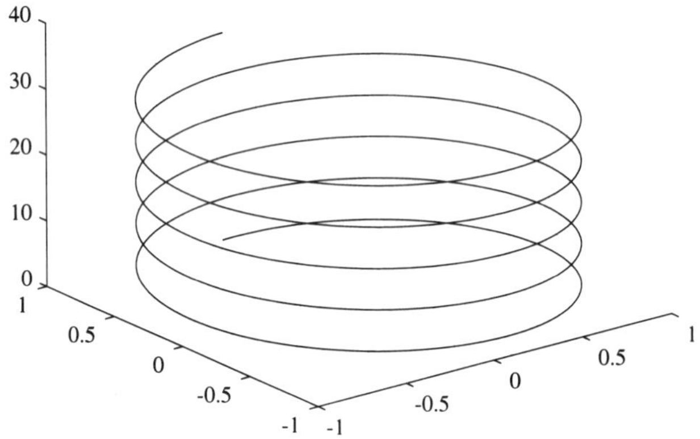

pol2cart

Purpose Transform polar coordinates to Cartesian.

Synopsis

```
[x,y] = pol2cart(theta,rho)
[x,y,z] = pol2cart(theta,rho,z)
```

Description [x,y] = pol2cart(theta,rho) transforms data stored in polar coordinates to Cartesian, or *xy*, coordinates. theta and rho must be the same size. theta is in radians.

[x,y,z] = pol2cart(theta,rho,z) transforms data stored in cylindrical coordinates to Cartesian coordinates. theta, rho, and z must all be the same size.

Example

```
theta = 0:.01:2*pi;
rho = sin(2*theta).*cos(2*theta);
polar(theta,rho)
```

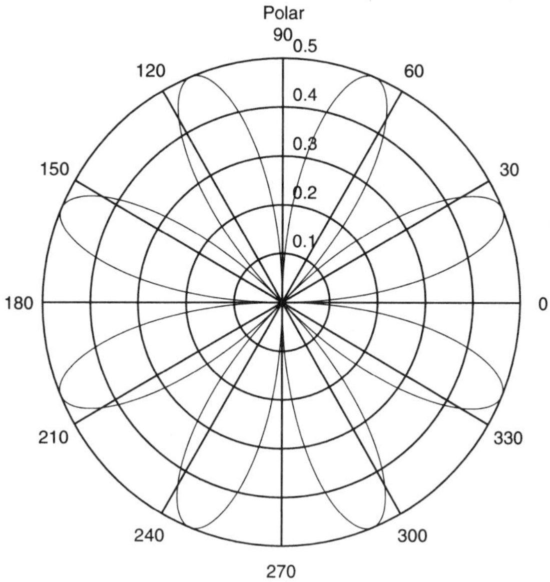


```
[x,y] = pol2cart(theta,rho);
plot(x,y)
```

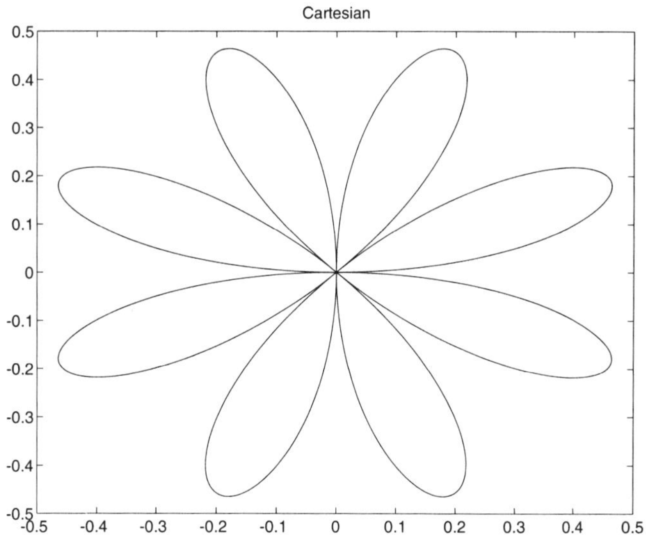

See Also `cart2pol`, `cart2sph`, `sph2cart`

polar

Purpose Polar coordinate plot.

Synopsis
```
polar(theta,rho)
polar(theta,rho,'linetype')
```

Description `polar(theta,rho)` makes a polar coordinate plot of the angle `theta`, in radians, versus the radius `rho`.

`polar(theta,rho,'linetype')` specifies a one-, two-, or three-character line style string made from the characters listed under the `plot` command.

Examples A simple polar plot:

```
t = 0:.01:2*pi;
polar(t,sin(2*t).*cos(2*t))
```

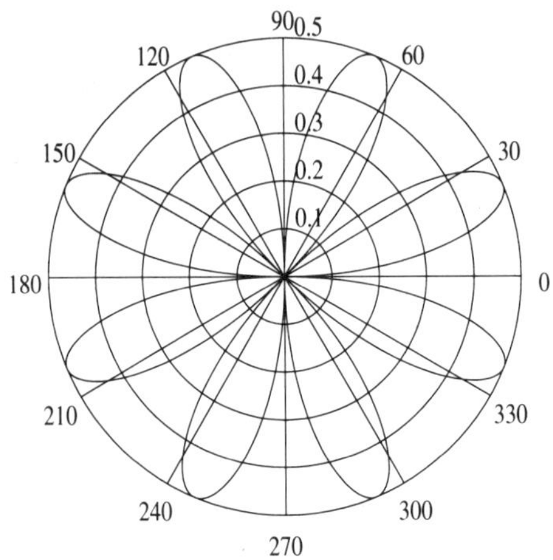

See Also `compass`, `plot`, `rose`

poly

Purpose Polynomial with specified roots.

Synopsis
```
p = poly(A)
p = poly(r)
```

Description If `A` is an `n`-by-`n` matrix, `poly(A)` is an `n+1` element row vector whose elements are the coefficients of the characteristic polynomial, $det(sI - A)$. The coefficients are ordered in descending powers: if a vector `c` has `n+1` components, the polynomial it represents is

$$c_1 s^n + \ldots + c_n s + c_{n+1}$$

If `r` is a column vector containing the roots of a polynomial, `poly(r)` returns a row vector whose elements are the coefficients of the polynomial.

If `c` is a row vector containing the coefficients of a polynomial, `roots(c)` is a column vector whose elements are the roots of the polynomial.

For vectors, `roots` and `poly` are inverse functions of each other, up to ordering, scaling, and roundoff error.

Examples MATLAB displays polynomials as row vectors containing the coefficients ordered by descending powers. The characteristic equation of the matrix

```
A =
    1    2    3
    4    5    6
    7    8    0
```

is returned in a row vector by `poly`:

```
p = poly(A)
p =
    1    –6    –72    –27
```

The roots of this polynomial (eigenvalues of matrix `A`) are returned in a column vector by `roots`:

```
r = roots(p)
r =
    12.1229
    –5.7345
    –0.3884
```

`poly` reassembles these into the coefficients of a polynomial:

```
p2 = poly(r)
p2 =
    1    –6    –72    –27
```

Algorithm The algorithms employed for `poly` and `roots` illustrate an interesting aspect of the modern approach to eigenvalue computation. `poly(A)` generates the characteristic polynomial of `A` and `roots(poly(A))` finds the roots of that polynomial, which are the eigenvalues of `A`. But both `poly` and `roots` use EISPACK eigenvalue subroutines, which are based on similarity transformations. The classical approach, which characterizes eigenvalues as roots of the characteristic polynomial, is actually reversed.

If `A` is an `n`-by-`n` matrix, `poly(A)` produces the coefficients `c(1)` through `c(n+1)`, with `c(1) = 1`, in

$$\det(\lambda I - A) = c_1\lambda^n + \ldots + c_n\lambda + c_{n+1}$$

The algorithm is expressed in an M-file:

```
z = eig(A);
c = zeros(n+1,1); c(1) = 1;
for j = 1:n
    c(2:j+1) = c(2:j+1)-z(j)*c(1:j);
end
```

This recursion is easily derived by expanding the product.

$$(\lambda - \lambda_1)(\lambda - \lambda_2)\ldots(\lambda - \lambda_n)$$

It is possible to prove that `poly(A)` produces the coefficients in the characteristic polynomial of a matrix within roundoff error of `A`. This is true even if the eigenvalues of `A` are badly conditioned. The traditional algorithms for obtaining the characteristic polynomial, which do not use the eigenvalues, do not have such satisfactory numerical properties.

See Also `conv`, `polyval`, `residue`, `roots`

---

polyder

Purpose Polynomial derivative.

Synopsis

```
k = polyder(p)
k = polyder(a,b)
[q,d] = polyder(b,a)
```

Description `k = polyder(p)`, where `p` is a vector whose elements are the coefficients of a polynomial in descending powers, returns the derivative of that polynomial. `k` is a vector containing the coefficients of the derivative in descending powers.

`k = polyder(a,b)` returns the derivative of the product of the polynomials `a` and `b`.

`[q,d] = polyder(b,a)` returns the derivative of the polynomial quotient `b/a`, where the derivative is represented by `q/d`.

Examples The derivative of the product

$$(3x^2 + 6x + 9)(x^2 + 2x)$$

is obtained with

```
a = [3 6 9];
b = [1 2 0];
k = polyder(a,b)
k =
    12    36    42    18
```

This result represents the polynomial

$$12x^3 + 36x^2 + 42x + 18$$

polyeig

Purpose Solve polynomial eigenvalue problem.

Synopsis `[X,e] = polyeig(A0,A1,...Ap)`

Description `[X,e] = polyeig(A0,A1,...Ap)` solves the polynomial eigenvalue problem of degree `p`

$$\left(A_0 + \lambda A_1 + \ldots + \lambda^p A_p\right)x = 0$$

where `p` is an integer from 1 to 10. The input matrices `A0`, `A1`,...`Ap` are size `p+1`-by-`p+1` matrices, all of order `n`. The output matrix `X`, of size `n`-by-`n*p`, contains the eigenvectors in its columns. The output vector `e`, of length `n*p`, contains the eigenvalues.

Based on the values of `p` and `n`, `polyeig` handles several special cases:

`p = 0`, or `polyeig(A)`, is the standard eigenvalue problem `eig(A)`.

`p = 1`, or `polyeig(A,B)` is the generalized eigenvalue problem `eig(A,-b)`.

`n = 1`, or `polyeig(a0,a1,...ap)` for scalars `a0`, `a1` ..., `ap` is the standard polynomial problem `roots([ap ... a1 a0])`.

Algorithm If both `A0` and `Ap` are singular, the problem is potentially *ill posed*; solutions might not exist or they might not be unique. In this case, the computed solutions may be inaccurate. `polyeig` attempts to detect this situation and display an appropriate warning message. If either one, but not both, of `A0` and `Ap` is singular, the problem is well posed but some of the eigenvalues may be zero or infinite (`Inf`).

`polyeig` uses the QZ factorization method to find intermediate results in the computation of generalized eigenvalues. It uses these intermediate results to determine if the eigenvalues are well-determined. See the descriptions of `eig` and `qz` for more on this, as well as the *EISPACK Guide*.

See Also `eig`, `qz`

polyfit

Purpose Polynomial curve fitting.

Synopsis `p = polyfit(x,y,n)`

Description `p = polyfit(x,y,n)` finds the coefficients of a polynomial `p(x)` of degree `n` that fits the data, `p(x(i))` to `y(i)`, in a least squares sense. The result `p` is a row vector of length `n+1` containing the polynomial coefficients in descending powers.

$$p(x) = p_1 x^n + p_2 x^{n-1} + \cdots p_n x + p_{n+1}$$

Examples This example involves fitting the error function, `erf(x)`, by a polynomial in `x`. This is a risky project because `erf(x)` is a bounded function, while polynomials are unbounded, so the fit might not be very good.

First generate a vector of `x`-points, equally spaced in the interval `[0, 2.5]`; then evaluate `erf(x)` at those points.

```
x = (0: 0.1: 2.5)';
y = erf(x);
```

The coefficients in the approximating polynomial of degree 6 are

```
p = polyfit(x,y,6)
p =
 0.0084  -0.0983   0.4217   -0.7435  0.1471   1.1064  0.0004
```

There are seven coefficients and the polynomial is

$$p(x) = .0084x^6 - .0983x^5 + \cdots + 1.1064x + .0004$$

To see how good the fit is, evaluate the polynomial at the data points with

```
f = polyval(p,x);
```

A table showing the data, fit, and error is

```
table = [x y f y-f]
table =
   0          0          0.0004    -0.0004
   0.1000     0.1125     0.1119     0.0006
   0.2000     0.2227     0.2223     0.0004
   0.3000     0.3286     0.3287    -0.0001
   0.4000     0.4284     0.4288    -0.0004
   ...
   2.1000     0.9970     0.9969     0.0001
   2.2000     0.9981     0.9982    -0.0001
   2.3000     0.9989     0.9991    -0.0003
   2.4000     0.9993     0.9995    -0.0002
   2.5000     0.9996     0.9994     0.0002
```

So, on this interval, the fit is good to between three and four digits. Beyond this interval the graph shows that the polynomial behavior takes over and the approximation quickly deteriorates.

```
x = (0: 0.1: 5)';
y = erf(x);
f = polyval(p,x);
plot(x,y,'o',x,f,'-')
axis([0  5  0  2])
```

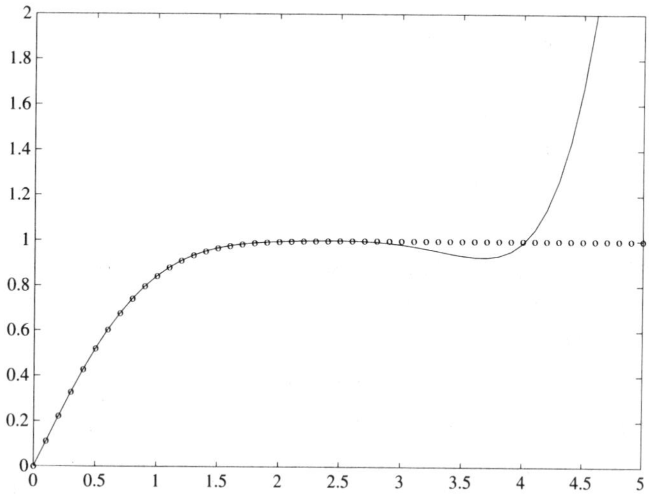

Algorithm The M-file forms the Vandermonde matrix, V, whose elements are powers of x.

$$v_{i,j} = x_i^{n-j}$$

It then uses the backslash operator, \, to solve the least squares problem

$$Vp \cong y$$

The M-file can be modified to use other functions of x as the basis functions.

See Also `conv`, `poly`, `polyval`, `roots`, `vander`

---

polyval

Purpose Polynomial evaluation.

Synopsis `y = polyval(p,S)`

Description `y = polyval(p,S)`, where `p` is a vector whose elements are the coefficients of a polynomial in descending powers, is the value of the polynomial evaluated at `S`. If `S` is a matrix or vector, the polynomial is evaluated at each of the elements.

`polyvalm(p,S)`, with `S` a matrix, evaluates the polynomial in a matrix sense. See `polyvalm` for more information.

Examples The polynomial

$$p(s) = 3s^2 + 2s + 1$$

is evaluated at $s = 5$ with

```
p = [3 2 1];
polyval(p,5)
```

which results in

```
ans =
      86
```

For another example, see `polyfit`.

See Also `conv`, `poly`, `polyfit`, `polyvalm`, `residue`, `roots`

---

polyvalm

Purpose Matrix polynomial evaluation.

Synopsis `Y = polyvalm(p,S)`

Description `polyvalm(p,S)` evaluates a polynomial in a matrix sense. `p` is a vector whose elements are the coefficients of a polynomial in descending powers, and `S` is a square matrix

Examples The Pascal matrices are formed from Pascal's triangle of binomial coefficients. Here is the Pascal matrix of order 4.

```
S = pascal(4)
S =
     1     1     1     1
     1     2     3     4
     1     3     6    10
     1     4    10    20
```

Its characteristic polynomial can be generated with the `poly` function.

```
p = poly(S)
p =
     1   –29    72   –29     1
```

This represents the polynomial

$$s^4 - 29s^3 + 72s^2 - 29s + 1$$

Pascal matrices have the curious property that the vector of coefficients of the characteristic polynomial is palindromic – it is the same forward and backward.

Evaluating this polynomial at each element of S is not very interesting.

```
polyval(p,S)
ans =
    16      16      16      16
    16      15    -140    -563
    16    -140   -2549  -12089
    16    -563  -12089  -43779
```

But evaluating it in a matrix sense is interesting.

```
polyvalm(p,S)
ans =
    0    0    0    0
    0    0    0    0
    0    0    0    0
    0    0    0    0
```

The result is the zero matrix. This is an instance of the Cayley-Hamilton theorem: a matrix that satisfies its own characteristic equation.

See Also `conv`, `poly`, `polyval`, `residue`, `roots`

---

pow2

Purpose Scale floating-point numbers.

Synopsis

```
X = pow2(Y)
X = pow2(F,E)
```

Description `X = pow2(Y)` returns a matrix `X` whose elements are the corresponding elements of `Y` raised to the power 2. `Y` can also be a vector or a scalar.

`X = pow2(F,E)`, for a real matrix `F` and a matrix of integers `E`, computes `X(i,j) = F(i,j).*(2.^E(i,j))`. The result is computed quickly by simply adding `E` to the floating-point exponent of `F`.

This function corresponds to the ANSI C function `ldexp()` and the IEEE floating-point standard function `scalbn()`.

Examples For IEEE arithmetic, the statement `X = pow2(F,E)` yields the following values:

F	E	X
1/2	1	1
pi/4	2	pi
–3/4	2	–3
1/2	–51	eps
1–eps/2	1024	realmax
1/2	–1021	realmin

See Also `^`, `.^`, `exp`, `hex2num`, `log2`, `realmax`, `realmin`

---

print, printopt

Purpose Create hardcopy output of current figure window.

Synopsis

```
print
print [-ddevicetype] [-options] [filename]
[pcmd,dev] = printopt
```

Description `print` with no arguments sends the contents of the current figure window to the default printer.

`print filename` saves the current figure window to the designated filename in the default device format. If `filename` does not include an extension, then an appropriate extension like `.ps` or `.eps` is appended to it, depending on the default device.

Directly supported device types include:

Device	Description
`–dps`	PostScript for black and white printers
`–dpsc`	PostScript for color printers
`–dps2`	Level 2 PostScript for black and white printers
`–dpsc2`	Level 2 PostScript for color printers
`–deps`	Encapsulated PostScript (EPSF)
`–depsc`	Encapsulated color PostScript (EPSF)
`–deps2`	Encapsulated Level 2 PostScript (EPSF)
`–depsc2`	Encapsulated Level 2 color PostScript (EPSF)

In general, Level 2 PostScript files are smaller and render more quickly when printing. However, not all PostScript printers support Level 2, so try to determine the capabilities of your printer before using the `print` command.

Additional devices supported via the Ghostscript post processor, which converts PostScript files into other formats, are listed below. This feature is available only on UNIX and PC systems, not including *The Student Edition of MATLAB*.

Ghostscript Device	Description
`-dlaserjet`	HP LaserJet
`-dljetplus`	HP LaserJet+
`-dljet2p`	HP LaserJet IIP
`-dljet3`	HP LaserJet III
`-dcdeskjet`	HP DeskJet 500C with 1 bit/pixel color
`-dcdjcolor`	HP DeskJet 500C with 24 bit/pixel color and high-quality color (Floyd-Steinberg) dithering
`-dcdjmono`	HP DeskJet 500C printing black only

Ghostscript Device	Description
`-ddeskjet`	HP DeskJet and DeskJet Plus
`-dpaintjet`	HP PaintJet color printer
`-dpjetxl`	HP PaintJet XL color printer
`-dbj10e`	Canon BubbleJet BJ10e
`-dln03`	DEC LN03 printer
`-depson`	Epson-compatible dot matrix printer (9- or 24-pin)
`-deps9high`	Epson-compatible 9-pin, interleaved lines (triple resolution)
`-depsonc`	Epson LQ-2550 and Fujitsu 3400/2400/1200 color printers
`-dgif8`	8-bit color GIF file format
`-dpcx16`	Older color PCX file format (EGA/VGA, 16-color)
`-dpcx256`	Newer color PCX file format (256-color)

`print` *`-options`* `filename` specifies an additional print option. Options supported on all devices are:

Option	Description
`-append`	Append to specified file, do not overwrite
`-epsi`	Add 1-bit deep EPSI preview
`-Pprinter`	Specify printer to use
`-fhandle`	Handle of figure to print
`-sname`	Name of SIMULINK system window to print

MS-Windows-only devices and options are:

Windows Option	Description
`-dwin`	Use Windows printing services
`-dwinc`	Use color Windows printing services
`-dmeta`	Put on clipboard in Windows metafile format
`-dbitmap`	Put on clipboard in Windows bitmap format
`-dsetup`	Bring up **Print Setup** dialog box, but do not print
`-v`	Verbose mode - bring up **Print** dialog box (normally suppressed)

Macintosh-only devices and options are:

Mac Option	Description
`-dmac`	Send figure, in monochrome, to currently installed printer

By default, MATLAB changes the normally black figure background color to white and changes the normally white axis lines and labels to black. You can maintain the white on black figure colors by setting the `InvertHardCopy` figure property to `off`:

```
set(gcf,'InvertHardCopy','off')
```

To obtain hardcopy output that matches what you see on the screen, set `InvertHardCopy` to `off` and select a color device type. You must send the file to a color printer unless the figure colormap is a grayscale (such as that created with `colormap(gray)`). In addition, to obtain true WYSIWYG output, you must adjust the width and height of the figure's `PaperPostion` property so that it has the same aspect ratio as the width and height of the figure's `Position` property. See `figure` for details.

MATLAB always produces graduated output for surfaces and patches, even for black and white output devices. However lines and text are printed in black or white.

Note that uicontrols and uimenus are not printed. See `capture` (UNIX systems only) to print a figure with these special widgets.

`[pcmd,dev] = printopt` returns strings describing the current settings for `printopt`, an M-file used by `print` that you or your system manager can edit to indicate your default printer type and destination. `pcmd` is a string containing the `print` command that `print` uses to spool a file to the printer. Its default is platform-dependent:

UNIX	`lpr -r`
Windows	`PRINT`
Macintosh	`unused`
VMS	`PRINT/DELETE`
SGI	`lp`

`dev` is a string that contains the device options for the `print` command. Its default is also platform-dependent:

UNIX & VMS	`-dps`
Windows	`-dwin`
Macintosh	`-dwin`

Example The statement

```
print meshdata -depsc2
```

saves the contents of the current figure as Level 2 color Encapsulated PostScript in the file called `meshdata.eps`.

See Also `orient`

All the properties of `figure` objects.

prism

Purpose Colormap of prism colors.

Synopsis
```
cmap = prism(m)
prism
```

Description `prism(m)` returns an `m`-by-3 colormap matrix that repeats the colors red, orange, yellow, green, blue, violet. The default value for `m` is the length of the current colormap.

`prism`, with no input or output arguments, changes the colors of any line objects in the current axes to the prism colors, repeating the colors as needed.

Examples The `prism` colormap is especially useful with contour plots.

```
[x,y,z] = peaks;
contour(x,y,z,32)
prism
```

See Also `bone`, `colormap`, `contour`, `cool`, `copper`, `flag`, `gray`, `hot`, `hsv`, `jet`, `pink`, `white`

prod

Purpose Product of matrix elements.

Synopsis
```
p = prod(X)
```

Description `prod(X)`, where `X` is a vector, is the product of the elements of `X`. For matrices, `prod(X)` is a row vector with the product over each column.

Examples The magic square of order 3 is

```
M = magic(3)
M =
     8    1    6
     3    5    7
     4    9    2
```

The product of the elements in each column is

```
prod(M) =
     96     45     84
```

The product of the elements in each row can be obtained by transposing twice.

```
prod(M')' =
      48
     105
      72
```

See Also `cumprod`, `sum`

pwd

Purpose Show current working directory.

Synopsis `dir = pwd`

Description `dir = pwd` returns a string containing the name of the current working directory.

See Also `matlabroot`

qr

Purpose Orthogonal-triangular decomposition.

Synopsis

```
[Q,R] = qr(X)
[Q,R,E] = qr(X)
A = qr(X)
```

Description `qr` performs the orthogonal-triangular decomposition of a matrix. This factorization is useful for both square and rectangular matrices. It expresses the matrix as the product of a real orthonormal or complex unitary matrix and an upper triangular matrix.

`[Q,R] = qr(X)` produces an upper triangular matrix `R` of the same dimension as `X` and a unitary matrix `Q` so that `X = Q*R`.

`[Q,R,E] = qr(X)` produces a permutation matrix `E`, an upper triangular matrix `R` with decreasing diagonal elements, and a unitary matrix `Q` so that `X*E = Q*R`.

`A = qr(X)` returns the output of the LINPACK subroutine `ZQRDC`. `triu(qr(X))` is `R`.

Examples Start with

```
A =
     1     2     3
     4     5     6
     7     8     9
    10    11    12
```

This is a rank-deficient matrix; the middle column is the average of the other two columns. The rank deficiency is revealed by the factorization:

```
[Q,R] = qr(A)
Q =
     -0.0776   -0.8331    0.5444    0.0605
     -0.3105   -0.4512   -0.7709    0.3251
     -0.5433   -0.0694   -0.0913   -0.8317
     -0.7762    0.3124    0.3178    0.4461
R =
    -12.8841   -14.5916   -16.2992
           0    -1.0413    -2.0826
           0          0     0.0000
           0          0          0
```

The triangular structure of R gives it zeros below the diagonal; the zero on the diagonal in R(3,3) implies that R, and consequently A, does not have full rank.

The QR factorization is used to solve linear systems with more equations than unknowns. For example

```
b =
     1
     3
     5
     7
```

The linear system $Ax = b$ represents four equations in only three unknowns. The best solution in a least squares sense is computed by

```
x = A\b
```

which produces

```
Warning: Rank deficient, rank = 2, tol = 1.4594E-014
x =
     0.5000
     0.0000
     0.1667
```

The quantity `tol` is a tolerance used in deciding that a diagonal element of R is negligible. If `[Q,R,E] = qr(A)`, then

```
tol = max(size(A))*eps*abs(R(1,1))
```

The solution x was computed using the factorization and the two steps

```
y = Q'*b;
x = R\y
```

The computed solution can be checked by forming Ax. This equals b to within roundoff error, which indicates that even though the simultaneous equations $Ax = b$ are overdetermined and rank deficient, they happen to be consistent. There are infinitely many solution vectors `x`; the QR factorization has found just one of them.

Algorithm `qr` uses the LINPACK routines `ZQRDC` and `ZQRSL`. `ZQRDC` computes the QR decomposition, while `ZQRSL` applies the decomposition.

See Also `\`, `/`, `lu`, `null`, `orth`, `qrdelete`, `qrinsert`

References Dongarra, J.J., J.R. Bunch, C.B. Moler, and G.W. Stewart, *LINPACK User's Guide*, SIAM, Philadelphia, 1979.

qrdelete

Purpose Delete a column from the QR factorization.

Synopsis `[Q,R] = qrdelete(Q,R,j)`

Description `[Q,R] = qrdelete(Q,R,j)`, where the inputs `Q` and `R` are the original QR factorization of a matrix `A`, returns `Q` and `R` as the QR factorization of the matrix `A` with columns `A(:,j)` removed.

Algorithm `qrdelete` uses a series of Givens rotations to zero out the appropriate elements of the factorization.

See Also `qr`, `qrinsert`

qrinsert

Purpose Insert a column in the QR factorization.

Synopsis `[Q,R] = qrinsert(Q,R,j,x)`

Description `[Q,R] = qrinsert(Q,R,j,x)`, where the inputs `Q` and `R` are the original QR factorization of a matrix `A`, returns `Q` and `R` as the QR factorization of the matrix `A` with an additional column, `x`, inserted before `A(:,j)`. If `A` has `n` columns and `j = n+1`, then `qrinsert` inserts `x` after the last column of `A`.

Algorithm `qrinsert` inserts the values of `x` into the `j`-th column of `R`. It then uses a series of Givens rotations to zero out the nonzero elements of `R` on and below the diagonal in the `j`-th column.

See Also `qr`, `qrdelete`

quad, quad8

Purpose Numerical evaluation of integrals.

Synopsis

```
a = quad('function',a,b)
a = quad('function',a,b,tol)
a = quad('function',a,b,tol,trace)

a = quad8('function',a,b)
a = quad8('function',a,b,tol)
a = quad8('function',a,b,tol,trace)
```

Description *Quadrature* is a numerical method of finding the area under the graph of a function, that is, computing a definite integral.

$$q = \int_a^b f(x)dx$$

`quad` and `quad8` implement two different quadrature algorithms. `quad` implements a low order method using an adaptive recursive Simpson's rule. `quad8` implements a higher order method using an adaptive recursive Newton-Cotes 8 panel rule.

`q = quad('`*`function`*`',a,b)` returns the result of numerically integrating the function `fun(x)` between the limits `a` and `b`. The function *`function`* must return a vector of output values when given a vector of input values.

`q = quad('`*`function`*`',a,b,tol)` iterates until the relative error is less than `tol`. The default value for `tol` is `1.e–3`.

If the final argument `trace` is nonzero, `quad` plots a graph showing the progress of the integration.

`quad8` has the same calling sequence as `quad`.

Examples Integrate the sine function from 0 to π:

```
a = quad('sin',0,pi)
a =
    2.0000
```

Algorithm `quad` uses an adaptive recursive Simpson's rule. `quad8` uses an adaptive recursive Newton-Cotes 8 panel rule. `quad8` is better than `quad` at handling functions with soft singularities:

$$\int_0^1 \sqrt{x}dx$$

Diagnostics `quad` and `quad8` have recursion level limits of 10 to prevent infinite recursion for a singular integral. Reaching this limit in one of the integration intervals produces the warning message:

```
Recursion level limit reached in quad. Singularity likely.
```

The computation continues using the best value available in that interval.

Limitations Neither `quad` nor `quad8` is set up to handle integrable singularities:

$$\int_0^1 \frac{1}{\sqrt{x}}dx$$

If you need to evaluate an integral with such a singularity, recast the problem by transforming the problem into one in which you can explicitly evaluate the integrable singularities and let `quad` or `quad8` take care of the remainder.

See Also `quaddemo` demonstration program

References
[1] Forsythe, G.E., M.A. Malcolm and C.B. Moler, *Computer Methods for Mathematical Computations*, Prentice-Hall, 1977.

quit

Purpose Terminate MATLAB.

Synopsis `quit`

Description `quit` terminates MATLAB without saving the workspace. To save your workspace variables, use the `save` command before quitting.

See Also `save`, `startup`

quiver

Purpose Quiver or needle plot.

Synopsis

```
quiver(X,Y,DX,DY)
quiver(DX,DY)
quiver(x,y,dx,dy,s)
quiver(dx,dy,s)
quiver(....'linetype')
```

Description `quiver(X,Y,DX,DY)` draws arrows at every pair of elements in matrices `X` and `Y`. The pairs of elements in matrices `DX` and `DY` determine the direction and relative magnitude of the arrows. If `X` and `Y` are vectors, they must have `length(x) = n` and `length(y) = m` where `[m,n] = size(DX) = size(DY)`. In this case, the arrows represent `(x(j),y(i),DX(i,j), DY(i,j))`. Note that vector `X` corresponds to the columns of `DX` and `DY` and `Y` corresponds to the rows.

`quiver(DX,DY)` uses `x = 1:n` and `y = 1:m`. In this case `DX` and `DY` are defined over a geometrically rectangular grid.

`quiver(x,y,dx,dy,s)` and `quiver(dx,dy,s)` apply scalar `s` as a scale factor to the lengths of the arrow. For example, `s = 2` doubles their relative length and `s = 0.5` halves them.

`quiver(....'`*`linetype`*`')` specifies line type and color using any legal line specification as described under the `plot` command.

Examples Plot the gradient field of the function

$$z = xe^{\left(-x^2 - y^2\right)}$$

```
[x,y] = meshgrid(-2:.2:2);
z = x.*exp(-x.^2-y.^2);
[dx,dy] = gradient(z,.2,.2);
contour(x,y,z)
hold on
quiver(x,y,dx,dy)
hold off
```

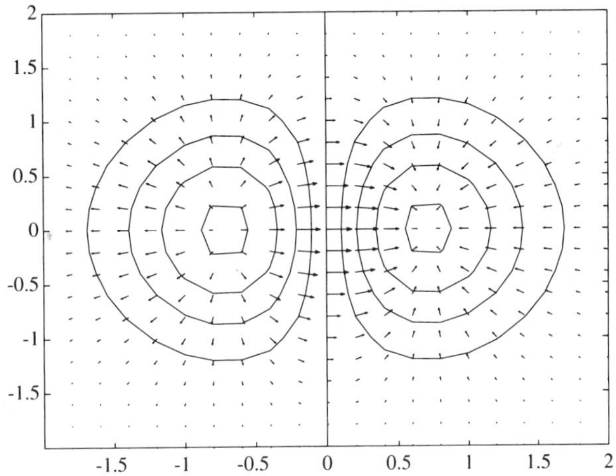

See Also `contour`, `plot`

---

qz

Purpose QZ factorization for generalized eigenvalues.

Synopsis `[AA,BB,Q,Z,V] = qz(A,B)`

Description The `qz` function gives access to what are normally only intermediate results in the computation of generalized eigenvalues. For square matrices `A` and `B`, the statement

```
[AA,BB,Q,Z,V] = qz(A,B)
```

produces upper triangular matrices AA and BB, and matrices Q and Z containing the products of the left and right transformations, such that

```
Q*A*Z = AA
Q*B*Z = BB
```

qz also returns the generalized eigenvector matrix V.

The generalized eigenvalues are the diagonal elements of AA and BB so that

```
A*V*diag(BB) = B*V*diag(AA)
```

Algorithm Complex generalizations of the EISPACK routines QZHES, QZIT, QZVAL, and QZVEC implement the QZ algorithm.

See Also eig

References
[1] Moler, C. B. and G.W. Stewart, "An Algorithm for Generalized Matrix Eigenvalue Problems", *SIAM J. Numer. Anal.*, Vol. 10, No. 2, April 1973.

---

rand

Purpose Uniformly distributed random numbers and matrices.

Synopsis

```
Y = rand(n)
Y = rand(m,n)
Y = rand(size(A))
rand('seed')
rand('seed',n)
rand
```

Description rand generates random numbers and matrices whose elements are uniformly distributed in the interval (0,1).

rand(n) is an n-by-n matrix with random entries.

rand(m,n) is an m-by-n matrix with random entries.

rand(size(A)) is the same size as A and has random entries.

rand('seed') returns the current value of the seed used by the generator.

rand('seed',n) sets the seed to n. rand('seed',0) sets the seed to its startup value.

rand with no arguments is a scalar whose value changes each time it is referenced.

The functions rand and randn maintain separate generators with separate seeds.

Examples If rand has not yet been used during a particular MATLAB session, then R = rand(3,4) produces

```
R =

    0.2190    0.6793    0.5194    0.0535
    0.0470    0.9347    0.8310    0.5297
    0.6789    0.3835    0.0346    0.6711
```

A random choice between two equally probable alternatives can be made with

```
if rand < .5
   'heads'
else
   'tails'
end
```

Algorithm rand is a uniform random number generator based on the linear congruential method described in [1]. The basic algorithm is

$$seed = \left(7^5 seed\right) \mathrm{mod}\left(2^{31} - 1\right)$$

See Also randn, randperm

Reference

[1] Park, S.K. and K.W. Miller, "Random Number Generators: Good ones are hard to find," *Comm. A.C.M.*, vol. 31, n. 10, pgs. 1192-1201, Oct. 1988.

---

randn

Purpose Normally distributed random numbers and matrices.

Synopsis

```
Y = randn(n)
Y = randn(m,n)
Y = randn(size(A))
randn('seed')
randn('seed',n)
randn
```

Description randn generates random numbers and matrices whose elements are normally distributed with mean 0 and variance 1.

randn(n) is an n-by-n matrix with random entries.

randn(m,n) is an m-by-n matrix with random entries.

randn(size(A)) is the same size as A and has random entries.

randn('seed') returns the current value of the seed used by the generator.

randn('seed',n) sets the seed to n. randn('seed',0) sets the seed to its startup value.

randn with no arguments is a scalar whose value changes each time it is referenced.

The functions rand and randn maintain separate generators with separate seeds.

Examples If randn has not yet been used during a particular MATLAB session, then R = randn(3,4) produces

```
R =
    1.1650    0.3516    0.0591    0.8717
    0.6268   -0.6965    1.7971   -1.4462
    0.0751    1.6961    0.2641   -0.7012
```

For a histogram of the `randn` distribution, see `hist`.

Algorithm `randn` uses a second copy of the uniform generator used by `rand` and transforms the resulting values according to the algorithm in [1].

See Also `rand`, `randperm`

References
[1] Forsythe, G.E., M.A. Malcolm and C.B. Moler, *Computer Methods for Mathematical Computations*, Prentice Hall, 1977.

---

randperm

Purpose Random permutation.

Synopsis `p = randperm(n)`

Description `p = randperm(n)` is a random permutation of the integers `1:n`.

Examples `randperm(6)` might be the vector

```
[3  2  6  4  1  5]
```

or it might be some other permutation of `1:6`.

See Also `rand`, `sprandn`

---

rank

Purpose Rank of a matrix.

Synopsis

```
k = rank(X)
k = rank(X,tol)
```

Description `rank(X)` is the number of singular values of `X` that are larger than `max(size(X))*norm(X)*eps`.

`rank(X,tol)` is the number of singular values of `X` that are larger than `tol`.

Algorithm There are a number of ways to compute the rank of a matrix. MATLAB uses the method based on the singular value decomposition, or SVD, described in Chapter 11 of the *LINPACK User's Guide*. The SVD algorithm is the most time consuming, but also the most reliable.

The `rank` algorithm is

```
s = svd(X);
tol = max(size(X))*s(1)*eps;
r = sum(s > tol);
```

References
[1] Dongarra, J.J., J.R. Bunch, C.B. Moler, and G.W. Stewart, *LINPACK User's Guide*, SIAM, Philadelphia, 1979.

---

rat, rats

Purpose Rational fraction approximation.

Synopsis `[N,D] = rat(X,tol)`

```
[N,D] = rat(X)
rat(...)
s = rats(X,strlen)
s = rats(X)
```

Description Even though all floating-point numbers are rational numbers, it is sometimes desirable to approximate them by simple rational numbers, which are fractions whose numerator and denominator are small integers. `rat(X)` and `rats(X)` attempt to do this. The rational approximations are generated by truncating continued fraction expansions.

`[N,D] = rat(X,tol)` returns two integer matrices so that `N./D` is close to `X` in the sense that `abs(N./D - X) <= tol*abs(X)`.

`[N,D] = rat(X)` uses a default tolerance, `1.e-6*norm(X(:),1)`.

`rat(...)` displays the continued fraction.

`s = rats(X,strlen)` uses `rat` to display simple rational approximations to the elements of `X`. The string length for each element is `strlen`. The default is `strlen = 13`, which allows 6 elements in 78 spaces. Asterisks are used for elements which cannot be printed in the allotted space, but are not negligible compared to the other elements in `X`.

`s = rats(X)` is an M-file which produces the same results as those printed by MATLAB with

```
format rat
```

Examples Ordinarily, the statement

```
s = 1 - 1/2 + 1/3 - 1/4 + 1/5 - 1/6 + 1/7
```

produces

```
s =
    0.7595
```

However, with

```
format rat
```

or with

```
rats(s)
```

the printed result is

```
s =
    319/420
```

This is a simple rational number. Its denominator is 420, the least common multiple of the denominators of the terms involved in the original expression. Even though the quantity `s` is stored internally as a binary floating-point number, the desired rational form can be reconstructed.

To see how the rational approximation is generated, the statement `rat(s)` produces

```
1 + 1/(-4 + 1/(-6 + 1/(-3 + 1/(-5))))
```

And the statement

```
[n,d] = rat(s)
```

produces

```
n = 319, d = 420
```

The mathematical quantity π is certainly not a rational number, but the MATLAB quantity `pi` which approximates it is a rational number. With IEEE floating-point arithmetic, `pi` is the ratio of a large integer and 2^{52}:

```
14148475504056880/4503599627370496
```

However, this is not a simple rational number. The value printed for `pi` with `format rat`, or with `rats(pi)`, is

```
355/113
```

This approximation was known in Euclid's time. Its decimal representation is

```
3.14159292035398
```

and so it agrees with `pi` to seven significant figures. The statement

```
rat(pi)
```

produces

```
3 + 1/(7 + 1/(16))
```

This shows how the `355/113` was obtained. The less accurate, but more familiar approximation `22/7` is obtained from the first two terms of this continued fraction.

Algorithm `rat(X)` approximates each element of `X` by a continued fraction of the form:

$$\frac{n}{d} = d_1 + \frac{1}{\left(d_2 + \frac{1}{\left(d_3 + \ldots + \frac{1}{d_k}\right)}\right)}$$

The d's are obtained by repeatedly picking off the integer part and then taking the reciprocal of the fractional part. The accuracy of the approximation increases exponentially with the number of terms and is worst when `X = sqrt(2)`. For `x = sqrt(2)`, the error with `k` terms is about `2.68*(.173)^k`, so each additional term increases the accuracy by less than one decimal digit. It takes 21 terms to get full floating-point accuracy.

rbbox

Purpose Rubberband box for region selection.

Synopsis `rbbox(rect,xy)`

Description `rbbox(rect,xy)` initializes and tracks a rubberband box in the current figure. It initially displays a rubberband box described by `rect`, which has the form

```
rect = ([x1  y1  width  height])
```

where `(x1,y1)` is the lower-left corner of the rectangle and `width` and `height` define the rectangle size.

`rbbox` starts tracking at the point `xy`, where `xy` has the form

```
xy = ([xstart  ystart])
```

All measurements are in pixels, with the origin at the lower-left corner of the figure window.

The mouse button must be down when `rbbox` is called. The corner of the rectangle closest to the `xy` point follows the cursor when the user drags the mouse. The function returns when the mouse button is released.

To determine the extent of the rectangle, use the figure's `CurrentPoint` property. If `p` is the figure's `CurrentPoint` prior to calling `rbbox`, and `q` is the figure's `CurrentPoint` when `rbbox` returns, then

```
[min(p,q), abs(p–q)]
```

is the extent of the user-defined rectangle.

`rbbox` is useful in M-files, along with the command `waitforbuttonpress`, to control dynamic behavior. Typical uses for `rbbox` include:

- Box resizing. In this case, `rect` defines the initial rectangle which the user must resize, and `xy` corresponds to one of the corners of the rectangle.
- Box definition. In this case, `rect` is defined as

  ```
  [x y 0 0]
  ```

  where `(x,y)` is equal to the initial point `xy`.

---

rcond

Purpose Matrix condition number estimate.

Synopsis `c = rcond(X)`

Description `rcond(X)` is an estimate for the reciprocal of the condition of `X` in 1-norm using the LINPACK condition estimator. If `X` is well conditioned, `rcond(X)` is near 1.0. If `X` is badly conditioned, `rcond(X)` is near 0.0. Compared to `cond`, `rcond` is a more efficient, but less reliable, method of estimating the condition of a matrix.

Algorithm `rcond` uses the condition estimator from the LINPACK routine `ZGECO`.

See Also `cond`, `condest`, `norm`, `normest`, `rank`, `svd`

References

[1] Dongarra, J.J., J.R. Bunch, C.B. Moler, and G.W. Stewart, *LINPACK User's Guide*, SIAM, Philadelphia, 1979.

real

Purpose Real part.

Synopsis `X = real(Z)`

Description `real(Z)` is the real part of the elements of `Z`.

Examples `real(2+3*i)` is 2.

See Also `conj`, `imag`

realmax

Purpose Largest positive floating-point numbers.

Synopsis `n = realmax`

Description `realmax` is the largest floating-point number representable on a particular computer. Anything larger overflows.

Examples On machines with IEEE floating-point format, `realmax` is one bit less than 2^{1024} or about 1.7977e+308.

Algorithm `realmax` is `pow2(2–eps,maxexp)` where `maxexp` is the largest possible floating-point exponent.

Execute `type realmax` to see `maxexp` for various computers.

See Also `eps`, `log2`, `pow2`, `realmin`

realmin

Purpose Smallest positive floating-point numbers.

Synopsis `n = realmin`

Description `realmin` is the smallest positive normalized floating-point number on a particular computer. Anything smaller underflows or is an IEEE denormal.

Examples On machines with IEEE floating-point format, `realmin` is `2^(–1022)` or about 2.2251e–308.

Algorithm `realmin` is `pow2(1,minexp)` where `minexp` is the smallest possible floating-point exponent.

Execute `type realmin` to see `minexp` for various computers.

See Also `eps`, `log2`, `pow2`, `realmax`

refresh

Purpose Redraw current figure window.

Synopsis

```
refresh
refresh(h)
```

Description `refresh` redraws the current figure window.

`refresh(h)` redraws the figure window specified by the handle `h`.

---

rem

Purpose Remainder after division.

Synopsis `r = rem(x,y)`

Description `rem(x,y)` is `x–n.*y`, where `n = fix(x./y)` is the integer part of the quotient, `x./y`. For matrices, this function operates element-by-element.

See Also `ceil`, `exp`, `fix`, `floor`, `log`, `round`, `sign`

---

reset

Purpose Reset `axes` or `figure` properties to their defaults.

Synopsis `reset(h)`

Description `reset(h)` resets all properties, except `Position`, of the object with handle `h` to the properties' default values.

Examples `reset(gca)` resets the properties of the current axes.

`reset(gcf)` resets the properties of the current figure.

See Also `cla`, `clf`, `gca`, `gcf`, `hold`

---

reshape

Purpose Reshape matrix.

Synopsis `B = reshape(A,m,n)`

Description `B = reshape(A,m,n)` returns the `m`-by-`n` matrix `B` whose elements are taken column-wise from `A`. An error results if `A` does not have `m*n` elements.

Examples Reshape a 3-by-4 matrix into a 2-by-6 matrix:

```
A =

     1     4     7    10
     2     5     8    11
     3     6     9    12

B = reshape(A,2,6)

B =

     1     3     5     7     9    11
     2     4     6     8    10    12
```

Algorithm MATLAB's colon notation can achieve the same effect, but `reshape` is less cryptic. `reshape` uses the equivalent colon notation:

```
B = zeros(m,n);
B(:) = A;
```

See Also `:`, `fliplr`, `flipud`, `rot90`

---

residue

Purpose Convert between partial fraction expansion and polynomial coefficients.

Synopsis

```
[r,p,k] = residue(b,a)
[b,a] = residue(r,p,k)
```

Description `[r,p,k] = residue(b,a)` find the residues, poles, and direct term of a partial fraction expansion of the ratio of two polynomials, $b(s)$ and $a(s)$. If there are no multiple roots

$$\frac{b(s)}{a(s)} = \frac{r_1}{s-p_1} + \frac{r_2}{s-p_2} + \ldots + \frac{r_n}{s-p_n} + k(s)$$

Vectors `b` and `a` specify the coefficients of the polynomials in descending powers of s. The residues are returned in the column vector `r`, the pole locations in column vector `p`, and the direct terms in row vector `k`. The number of poles `n` is

```
n = length(a)-1 = length(r) = length(p)
```

The direct term coefficient vector is empty if `length(b) < length(a)`; otherwise

```
length(k) = length(b)-length(a)+1
```

If `p(j) = ... = p(j+m-1)` is a pole of multiplicity `m`, then the expansion includes terms of the form

$$\frac{r_j}{s-p_j} + \frac{r_{j+1}}{\left(s-p_j\right)^2} + \cdots + \frac{r_{j+m-1}}{\left(s-p_j\right)^m}$$

`[b,a] = residue(r,p,k)`, with three input arguments and two output arguments, converts the partial fraction expansion back to the polynomials with coefficients in `b` and `a`.

Algorithm `residue` is an M-file. It first obtains the poles with `roots`. Next, if the fraction is nonproper, the direct term `k` is found using `deconv`, which performs polynomial long division. Finally, the residues are determined by evaluating the polynomial with individual roots removed. For repeated roots, the M-file `resi2` computes the residues at the repeated root locations.

Limitations Numerically, the partial fraction expansion of a ratio of polynomials represents an ill-posed problem. If the denominator polynomial, $a(s)$, is near a polynomial with multiple roots, then small changes in the data, including roundoff errors, can make arbitrarily large changes in the resulting poles and residues. Problem formulations making use of state-space or zero-pole representations are preferable.

See Also `deconv`, `poly`, `roots`

References [1] Oppenheim, A.V. and R.W. Schafer, *Digital Signal Processing*, p. 56, Prentice Hall, 1975.

---

return

Purpose Return to the invoking function.

Synopsis `return`

Description `return` causes a normal return to the invoking function or to the keyboard.

Examples If the determinant function were an M-file, it might use a `return` statement in handling the special case of an empty matrix as follows:

```
function d = det(A)
%DET det(A) is the determinant of A.
if isempty(A)
    d = 1;
    return
else
    ...
end
```

See Also `break`, `disp`, `end`, `error`, `for`, `if`, `while`

---

rgb2hsv

Purpose Red-green-blue to hue-saturation-value conversion.

Synopsis `H = rgb2hsv(M)`

Description `H = rgb2hsv(M)` converts an RGB colormap to an HSV colormap. Each map is a matrix with any number of rows, exactly three columns, and elements in the interval from 0 to 1. The columns of the input matrix, `M`, represent intensities of red, green, and blue respectively. The columns of the resulting output matrix, `H`, represent hue, saturation, and value, respectively.

See Also `brighten`, `colormap`, `hsv`, `hsv2rgb`

---

rgbplot

Purpose Plot colormap.

Synopsis `rgbplot(map)`

Description `rgbplot(map)` plots a colormap, an `m`-by-3 matrix which is appropriate input for `colormap`. The three columns of `map` are plotted with red, green and blue lines.

Examples `rgbplot(hsv)` plots the default colormap.

See Also `colormap`, `spinmap`

root object

Purpose Root object properties.

Description The *root* is a handle graphics object that corresponds to the computer screen. There is only one root object and it has no parent. The children of the root object are figure windows.

The root object exists when you start MATLAB; you never have to create it. Use the `set` function to alter the root's properties, listed under "Object Properties" below. Use `get` to query the root's properties.

Object Properties

This section lists property names along with the type of values each accepts.

`BlackAndWhite` — Automatic hardware checking flag.

- `on` — Assume the display is monochrome. This is useful if MATLAB is running on color hardware, but is displaying on a monochrome terminal. Prevents MATLAB from determining erroneously that the display is color.
- `off` — (Default.) Differentiate between color and monochrome.

`ButtonDownFcn` — Callback string.

Any legal MATLAB expression, including the name of an M-file or function. When you select the object, the string is passed to the `eval` function to execute the `Callback`. Initially the empty matrix.

`CaptureMatrix` — Matrix of screen data.

A read-only matrix that contains the image data of the region enclosed by the `CaptureRect` rectangle. Use the `get` function to obtain this matrix. Use `image` to display the captured matrix.

`CaptureRect` — Size and position of rectangle to capture.

A four-element vector, `[left,bottom,width,height]`, that defines the region captured by `CaptureMatrix`. `left` and `bottom` define the location of the lower-left corner of the rectangle and `width` and `height` define the dimensions of the rectangle. The `Units` property determines what units are used to specify these dimensions. Initial value is `[0 0 0 0]`.

Property	Description	
`Children`	Handles of child objects.	
	A read-only vector containing the handles of all figure objects.	
`Clipping`	Data clipping.	
	`on`	(Default.) No effect for root objects.
	`off`	No effect for root objects.
`CurrentFigure`	Current figure window.	
	The handle of the current figure.	
`Diary`	Diary file mode (see also `diary` command).	
	`on`	MATLAB maintains a file (with name `DiaryFile`) that saves a copy of all keyboard input and most of the resulting output.
	`off`	(Default.) MATLAB does not save input and output to a file.
`DiaryFile`	Diary file name.	
	A string containing the name of the diary file. The default name is `diary`.	
`Echo`	Script echoing mode (see also `echo` command).	
	`on`	MATLAB displays each line of a script file as it executes.
	`off`	(Default.) MATLAB does not echo during script execution.
`Format`	Output format mode (see also `format` command).	
	`short`	(Default.) Fixed-point format with 5 digits.
	`shortE`	Floating-point format with 5 digits.
	`long`	Scaled fixed-point format with 15 digits.
	`longE`	Floating-point format with 15 digits.
	`bank`	Fixed-format of dollars and cents.
	`hex`	Hexadecimal format.
	`+`	Displays + and – symbols.
`FormatSpacing`	Output format spacing (see also `format` command).	
	`compact`	Suppress extra line feeds for more compact display.
	`loose`	(Default.) Display extra line feeds for looser display.

`Interruptible`	Callback interruptibility.	
	`yes`	The callback specified by `ButtonDownFcn` is interruptible by other callbacks.
	`no`	(Default.) The `ButtonDownFcn` callback is not interruptible.
`Parent`	Handle of parent object. A read-only property that always contains the empty matrix, as the root object has no parent.	
`PointerLocation`	Current location of pointer. A read-only vector containing the x- and y-coordinates of the pointer position, measured from the lower-left corner of the screen. The `Units` property determines the units of this measurement. This property always contains the instantaneous pointer location, even if the pointer is not in a MATLAB window. This location may change by the time the value is returned.	
`PointerWindow`	Handle of window containing the pointer. A read-only property containing the handle of the figure window that contains the pointer. If the pointer is not in a MATLAB window, the value of the property is 0.	
`ScreenDepth`	Screen depth. An integer specifying the screen depth in bits.	
`ScreenSize`	Screen size. A four-element read-only vector, `[left,bottom,width,height]`, that defines the display size. `left` and `bottom` are always 0. `width` and `height` are the screen dimensions in units specified by the `Units` property.	
`TerminalOneWindow`	Indicates if there is only one window on your terminal.	
	`yes`	There is only one window on your terminal.
	`no`	Multiple windows.
	If the terminal uses only one window, MATLAB waits for you to press a key before it switches from graph mode back to command mode. This property is only used by the terminal graphics driver.	
`TerminalProtocol`	Terminal type.	
	`x`	X display server. If you are using X Windows and MATLAB can connect to your X display server, this property is automatically set to `x`.

	`tek401x`	Terminals that emulate Tektronix 4010/4014.
	`tek410x`	Terminals that emulate Tektronix 4100/4105.
	`none`	Not in terminal mode, not connected to an X server.

Once this property is set, it cannot be changed unless you quit and restart MATLAB.

`Type` — Type of graphics object.

A read-only string; always `'root'` for the root object.

`Units` — Unit of measurement.

`pixels`	(Default.) Screen pixels.
`normalized`	Normalized coordinates, where the lower-left corner of the screen maps to (0,0) and the upper-right corner to (1.0,1.0).
`inches`	Inches.
`cent`	Centimeters.
`points`	Points. Each point is equivalent to 1/72 of an inch.

This property affects `CaptureRect`, `PointerLocation`, and `ScreenSize`. If you change the value of `Units`, it is good practice to return it to its default value after completing your computation so as not to affect other functions that assume `Units` is set to the default value.

`UserData` — User-specified data.

Any matrix you want to associate with the root object. The object does not use this data, but you can retrieve it using the `get` command.

`Visible` — Object visibility.

`on`	(Default.) No effect for root objects.
`off`	No effect for root objects.

See Also `diary`, `echo`, `figure`, `format`, `gcf`

---

roots

Purpose Polynomial roots.

Synopsis `r = roots(p)`

Description Polynomial coefficients are ordered in descending powers: if a vector c has n+1 components, the polynomial it represents is

$$c_1 s^n + \ldots + c_n s + c_{n+1}$$

If c is a row vector containing the coefficients of a polynomial, roots(c) is a column vector whose elements are the roots of the polynomial.

If r is a column vector containing the roots of a polynomial, poly(r) returns a row vector whose elements are the coefficients of the polynomial.

For vectors, roots and poly are inverse functions of each other, up to ordering, scaling, and roundoff error.

Examples The polynomial

$$s^3 - 6s^2 - 72s - 27$$

is represented in MATLAB as

```
p = [1 -6 -72 -27]
```

The roots of this polynomial are returned in a column vector by

```
r = roots(p)
r =
     12.1229
     -5.7345
     -0.3884
```

Algorithm The algorithm simply involves computing the eigenvalues of the companion matrix:

```
A = diag(ones(n-1,1)),-1);
A(1,:) = -c(2:n-1)/c(1);
eig(A)
```

It is possible to prove that the results produced are the exact eigenvalues of a matrix within roundoff error of the companion matrix A, but this does not mean that they are the exact roots of a polynomial with coefficients within roundoff error of those in c.

See Also conv, fzero, poly, polyval, residue

---

rose

Purpose Angle histogram.

Synopsis

```
rose(theta)
rose(theta,n)
rose(theta,x)
[t,r] = rose(...)
```

Description rose(theta) plots an angle histogram for the angles in theta, which must be in radians. The angle histogram is a plot in polar coordinates of the number of theta points or samples within each of 20 angle bins.

`rose(theta,n)` where `n` is a scalar, uses `n` equally spaced bins in the range `[0, 2*pi]`.

`rose(theta,x)` where `x` is a vector, uses the bins specified in `x`. The values in `x` specify the center angle of each bin.

`[t,r] = rose(...)` returns the vectors `t` and `r` such that `polar(t,r)` is the histogram for the data. No plot is drawn.

Examples Create a rose plot for one year of simulated wind direction measurements:

```
wind = 360*rand(365,1)
rose(wind*pi/180,24)  %histogram every 15 degrees
```

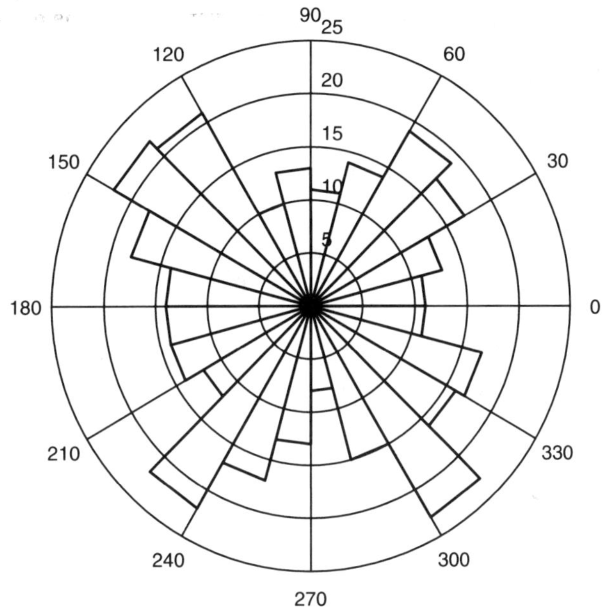

See Also `compass`, `feather`, `hist`, `polar`

---

rosser

Purpose A classic symmetric eigenvalue test problem.

Synopsis `A = rosser`

Description This matrix was a challenge for many matrix eigenvalue algorithms. But the Francis QR algorithm, as perfected by Wilkinson and implemented in EISPACK and MATLAB, has no trouble with it. The matrix is 8-by-8 with integer elements. It has

- A double eigenvalue
- Three nearly equal eigenvalues
- Dominant eigenvalues of opposite sign
- A zero eigenvalue

- A small, nonzero eigenvalue

The Rosser matrix is:

611	196	−192	407	−8	−52	−49	29
196	899	113	−192	−71	−43	−8	−44
−192	113	899	196	61	49	8	52
407	−192	196	611	8	44	59	−23
−8	−71	61	8	411	−599	208	208
−52	−43	49	44	−599	411	208	208
−49	−8	8	59	208	208	99	−911
29	−44	52	−23	208	208	−911	99

Its exact eigenvalues are

```
10*sqrt(10405)
1020
510 + 100*sqrt(26)
1000
1000
510 − 100*sqrt(26)
0
−10*sqrt(10405)
```

See Also `eig`, `gallery`, `wilkinson`

---

rot90

Purpose Rotate matrix 90°.

Synopsis

```
B = rot90(A)
B = rot90(A,k)
```

Description `B = rot90(A)` rotates the matrix `A` counterclockwise by 90 degrees.

`B = rot90(A,k)` rotates the matrix `A` counterclockwise by `k*90` degrees. `k` must be an integer.

Examples The matrix

```
X =
     1     2     3
     4     5     6
     7     8     9
```

rotated by 90 degrees is

```
Y = rot90(X)
Y =
     3     6     9
     2     5     8
     1     4     7
```

See Also fliplr, flipud, reshape

---

rotate

Purpose Rotate an object.

Synopsis

```
rotate(h,azel,alpha)
rotate(h,azel,alpha,origin)
```

Description rotate(h,azel,alpha) rotates the object specified by handle h through alpha degrees. The axis of rotation is specified by azel, a two-element or three-element vector. A two-element vector specifies *azimuth*, or horizontal rotation, and vertical elevation,

```
[azimuth  elevation]
```

A three-element vector specifies the desired direction vector in Cartesian coordinates,

```
[x y z]
```

rotate(h,azel,alpha,origin) specifies the origin of rotation, origin, as a three-element vector

```
[x y z]
```

Examples

```
[x,y,z] = cylinder(10,25);
h = surf(x,y,z)
```

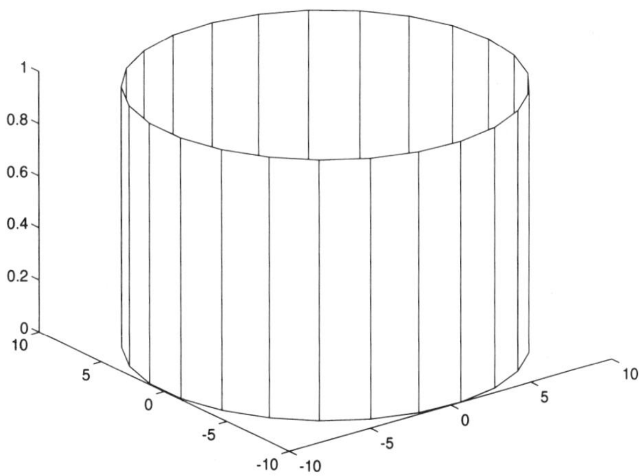

```
rotate(h,[90 0],45)
```

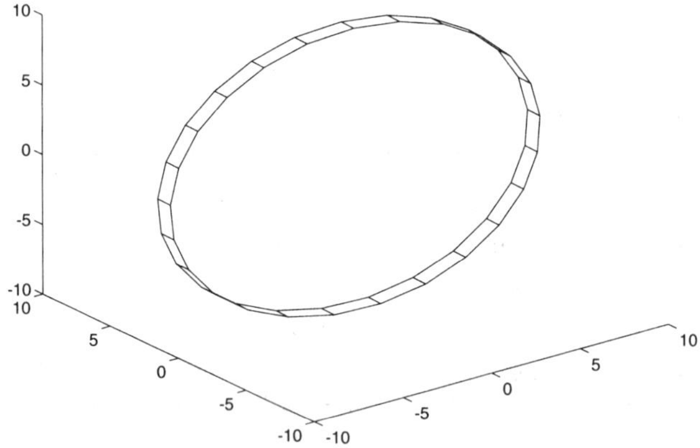

See Also `view`, `viewmtx`

---

round

Purpose Round to nearest integers.

Synopsis `Y = round(X)`

Description `round(X)` rounds the elements of `X` to the nearest integers.

Examples

```
X =          [-1.9  -0.2  3.4  5.6  7.0]
round(X) =   [-2  0  3  6  7]
```

See Also `ceil`, `fix`, `floor`

---

rref, rrefmovie

Purpose Reduced row echelon form.

Synopsis

```
R = rref(A)
R = rref(A,tol)
R = rrefmovie(A)
```

Description `R = rref(A)` produces the reduced row echelon form of `A` using Gauss Jordan elimination with partial pivoting. A default tolerance of

```
tol = max(size(A))*eps*norm(A,inf)
```

is used to test for negligible column elements.

`R = rref(A,tol)` uses `tol` for the tolerance.

`R = rrefmovie(A)` shows a movie of the algorithm working.

Examples Use `rref` on a rank-deficient magic square:

```
A = magic(4), R = rref(A)
A =
    16     2     3    13
     5    11    10     8
     9     7     6    12
     4    14    15     1
R =
     1     0     0     1
     0     1     0     3
     0     0     1    -3
     0     0     0     0
```

See Also `inv`, `lu`, `rank`

---

rsf2csf

Purpose Convert real Schur form to complex Schur form.

Synopsis `[U,T] = rsf2csf(U,T)`

Description The *complex Schur form* of a matrix is upper triangular with the eigenvalues of the matrix on the diagonal. The *real Schur form* has the real eigenvalues on the diagonal and the complex eigenvalues in 2-by-2 blocks on the diagonal.

`[U,T] = rsf2csf(U,T)` converts the real Schur form to the complex form. See `schur` for details.

See Also `eig`, `hess`, `schur`

---

save

Purpose Save workspace variables on disk.

Synopsis

```
save
save filename
save filename variables
save filename keywords
save filename variables keywords
```

Description `save` and `load` are the MATLAB commands to store and retrieve variables on disk. They can also import and export ASCII data files. MAT-files are full-precision, binary, MATLAB format files created by the `save` command and readable by the `load` command. They can be created on one machine and later read by MATLAB on another machine with a different floating-point format, retaining as much accuracy and range as the disparate formats allow. They can also be manipulated by other programs, external to MATLAB.

`save` by itself, stores all workspace variables in a binary format in the file named `matlab.mat`. The data can be retrieved with `load`.

`save filename` uses file `filename.mat` instead of the default `matlab.mat`.

`save filename variables` saves a selection of the current workspace variables by listing the variables after the filename.

`save filename keywords` specifies characteristics for data to be saved in ASCII format, instead of in the binary MAT-file format. Valid keywords are `–ascii`, `–double`, and `–tabs`.The command

```
save  filename  variables  –ascii
```

uses 8-digit ASCII form.

The command

```
save  filename  variables  –ascii  –double
```

uses 16-digit ASCII form.

The data can be separated with tabs by appending the keyword `tabs`.

```
save  filename  variables  –ascii  –tabs
save  filename  variables  –ascii  –double  –tabs
```

Variables saved in ASCII format merge into a single variable that takes the name of the ASCII file. Therefore, loading the file *filename* shown above results in a single workspace variable named *filename*. Use the colon operator to access individual variables.

The ASCII data must be in matrix form, or MATLAB may be unable to use the data when you load it.

Saving complex data with the `–ascii` keyword causes the imaginary part of the data to be lost, as MATLAB cannot load nonnumeric data (`'i'`).

If the *filename* is the special string `stdio`, `save` sends the data as standard output.

Algorithm The binary formats used by `save` depend on the size and type of each matrix. Matrices with any noninteger entries and matrices with 10,000 or fewer elements are saved in floating-point formats requiring eight bytes per real element. Matrices with all integer entries and more than 10,000 elements are saved in the following formats, requiring fewer bytes per element.

Element Range	Bytes per Element
[0:255]	1
[0:65535]	2
[–32767:32767]	2
[–2^31+1:2^31–1]	4
other	8

See Also `fprintf`, `fwrite`, `load`

---

saxis

Purpose Sound axis scaling.

Synopsis `saxis([smin smax])`

```
saxis('auto')
v = saxis
```

Description The `sound` function ordinarily scales its input vector `y` so that the maximum and minimum values in `y` correspond to the maximum and minimum input ranges allowed by the sound hardware. Usually this is in the range ±1.0. `saxis` is analogous to `caxis` and `axis`, only its scaling applies to sound limits rather than graphical limits.

`saxis([smin smax])` disables `sound`'s automatic scaling and sets the scaling so that `smin` and `smax` correspond to the minimum and maximum input ranges allowed by the sound hardware.

`saxis('auto')` sets sound axis scaling back to automatic.

`v = saxis` returns the two-element row vector `v` containing the `[smin smax]` currently in effect.

Examples Set the sound limits to ±1.0 and play a 1000 Hz tone using an 8192 Hz sample rate:

```
saxis([-1 1])
t = (0:10000)/8192;
y = sin(2*pi*1000*t);
sound(y,8192)
saxis([-2  2])       %increase the volume
saxis([-.5  .5])     %decrease the volume
```

See Also `axis`, `caxis`, `sound`

---

schur

Purpose Schur decomposition.

Synopsis

```
[U,T] = schur(A)
T = schur(A)
```

Description `schur` computes the Schur form of a matrix. The *complex Schur form* of a matrix is upper triangular with the eigenvalues of the matrix on the diagonal. The *real Schur form* has the real eigenvalues on the diagonal and the complex eigenvalues in 2-by-2 blocks on the diagonal.

`[U,T] = schur(A)` produces a Schur matrix `T`, and a unitary matrix `U` so that `A = U*T*U'` and `U'*U = eye(size(A))`.

`T = schur(A)` returns just the Schur matrix `T`.

If the matrix `A` is real, `schur` returns the real Schur form. If `imag(A)` is nonzero, `schur` returns the complex Schur form. The function `rsf2csf` converts the real form to the complex form.

Examples The matrix `gallery(3)` is a 3-by-3 eigenvalue test matrix:

```
gallery(3) =
    -149    -50   -154
     537    180    546
     -27     -9    -25
```

Its Schur form is

```
schur(gallery(3)) =
    1.0000    7.1119   815.8706
         0    2.0000   -55.0236
         0         0     3.0000
```

The eigenvalues, which in this case are 1, 2, and 3, are on the diagonal. The fact that the off-diagonal elements are so large indicates that this matrix has poorly conditioned eigenvalues; small changes in the matrix elements produce relatively large changes in its eigenvalues.

Algorithm For real matrices, `schur` uses the EISPACK routines `ORTRAN`, `ORTHES`, and `HQR2`. `ORTHES` converts a real general matrix to Hessenberg form using orthogonal similarity transformations. `ORTRAN` accumulates the transformations used by `ORTHES`. `HQR2` finds the eigenvalues of a real upper Hessenberg matrix by the QR method.

The EISPACK subroutine `HQR2` has been modified to allow access to the Schur form, ordinarily just an intermediate result, and to make the computation of eigenvectors optional.

When `schur` is used with a complex argument, the solution is computed using the `QZ` algorithm by the EISPACK routines `QZHES`, `QZIT`, `QZVAL`, and `QZVEC`. They have been modified for complex problems and to handle the special case $B = I$.

For detailed descriptions of these algorithms, see the *EISPACK Guide*.

See Also `eig`, `hess`, `qz`, `rsf2csf`

References

[1] Smith, B. T., J. M. Boyle, J. J. Dongarra, B. S. Garbow, Y. Ikebe, V. C. Klema, and C. B. Moler, *Matrix Eigensystem Routines – EISPACK Guide*, Lecture Notes in Computer Science, volume 6, second edition, Springer-Verlag, 1976.

[2] Garbow, B. S., J. M. Boyle, J. J. Dongarra, and C. B. Moler, *Matrix Eigensystem Routines – EISPACK Guide Extension*, Lecture Notes in Computer Science, volume 51, Springer-Verlag, 1977.

[3] Moler, C.B. and G. W. Stewart, "An Algorithm for Generalized Matrix Eigenvalue Problems," *SIAM J. Numer. Anal.*, Vol. 10, No. 2, April 1973.

sec, sech

Purpose Secant and hyperbolic secant.

Synopsis

```
Y = sec(X)
Y = sech(X)
```

Description `sec` and `sech` operate element-wise on matrices. Their domains and ranges include complex values. All angles are in radians.

`sec(X)` is a matrix the same size as `X` containing the secant of the elements of `X`.

`sech(X)` is a matrix the same size as `X` containing the hyperbolic secant of the elements of `X`.

Examples `sec(pi/2) = 1.6332e+16` is not infinite, but rather a value the size of the reciprocal of the floating point accuracy `eps`, because `pi` is only a floating-point approximation to the exact value.

Algorithm

$$\sec(z) = \frac{1}{\cos(z)}$$

$$\text{sech}(z) = \frac{1}{\cosh(z)}$$

See Also `asec`, `asech`

---

semilogx, semilogy

Purpose Semi-logarithmic plots.

Synopsis

```
semilogx(x,y)
semilogy(x,y)
```

Description `semilogx` and `semilogy`, as well as `loglog`, are used exactly the same as `plot`, but they result in graphs on different scales.

`semilogx(x,y)` makes a plot using a base 10 logarithmic scale for the x-axis and a linear scale for the y-axis.

`semilogy(x,y)` makes a plot using a base 10 logarithmic scale for the y-axis and a linear scale for the x-axis.

Examples A simple semilogy plot is:

```
x = 0:.1:10;
semilogy(x,10.^x)
```

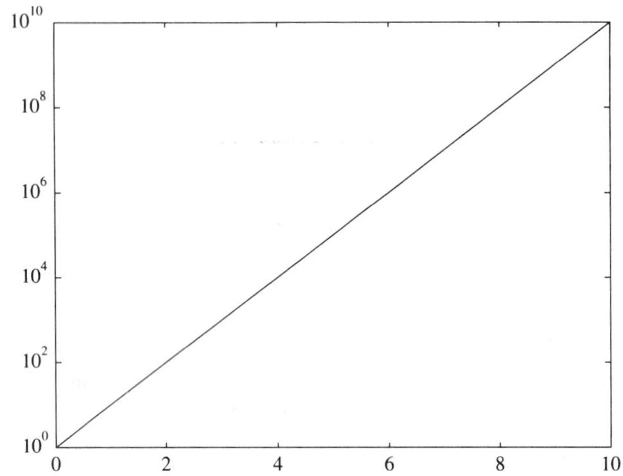

See Also `loglog`, `plot`

---

set

Purpose Set object properties.

Synopsis

```
set(h,'PropertyName',PropertyValue,...)
set(h)
set(h,'PropertyName')
```

Description Use the `set` function to specify the value of object properties (except those that are read only).

`set(h,'PropertyName',PropertyValue,...)` sets properties for the object with handle `h` (or objects if `h` is a vector). A single call to `set` can specify any number of property name/property value pairs. MATLAB processes a single long call faster than several short ones.

To set default object properties values, concatenate the words `Default`, the object name, and the property name. For example, to set the default axes color for new axes in the current figure, use the statement

```
set(gcf,'DefaultAxesColor',[1 1 1])
```

Since objects check for default property values on parent objects, set the default axes properties in the figure in which you want the value used. Similarly, set default figure properties at the root level.

`set(h)`, where `h` is an object handle, displays a list of all the properties you can set for that object along with the possible values for each property.

`set(h,'`*`PropertyName`*`')` lists the possible values for the specified property.

`set` also accepts the strings `default`, `factory`, and `remove` as property values.

Specifying a value of `default` sets the property to the first encountered default value defined for that property. For example, these statements set the surface's `EdgeColor` property to green:

```
h = surf(peaks)
set(0,'DefaultSurfaceEdgeColor','g')
set(h,'EdgeColor','default')
```

If a default value for surface `EdgeColor` existed on the figure or axes level, these statements would instead set the `EdgeColor` of the surface `h` to that default value.

Specifying a value of `factory` sets the property to its factory setting. For example, these statements

```
h = surf(peaks)
set(0,'DefaultSurfaceEdgeColor','g')
set(h,'EdgeColor','factory')
```

set the `EdgeColor` of surface `h` to black (its factory setting) regardless of what default values are defined.

The string `remove` gets rid of a default value. The statement

```
set(0,'DefaultSurfaceEdgeColor','remove')
```

removes the definition of the default surface `EdgeColor` on the root. The value for surface `EdgeColor` then reverts to the factory setting.

To use `default`, `factory`, or `remove` as strings in an axis label, for example, as opposed to property values as discussed above, place a backslash before the word:

```
set(gca,'XLabel','\default')
```

Examples The following statement sets the current axes limits to 0 to 10 for the x-axis, -25 to 25 for the y-axis, and -8 to 10 for the z-axis:

```
set(gca,'Xlim',[0 10],'Ylim',[-25 25],'Zlim',[-8 10])
```

Use set to list the properties of a particular object that you can specify. For example, the following statement lists the properties that apply to a surface object. Default values are indicated by curly brackets:

```
h = surf(peaks);
set(h)
CData
EdgeColor: [ none | {flat} | interp ] -or- a ColorSpec.
FaceColor: [ none | {flat} | interp | texturemap ] -or- a
     ColorSpec.
LineStyle: [ {-} | — | : | -. | + | o | * | . | x ]
MarkerSize
MeshStyle: [ {both} | row | column ]
XData
YData
ZData
Children
Clipping: [ on | off ]
UserData
Visible: [ on | off ]
```

See Also gca, gcf, get

---

setstr

Purpose Set string flag.

Synopsis STR = setstr(T)

Description STR = setstr(T), where T is a vector or matrix of numeric values, does not alter the numeric values in T, but it does cause the result to be interpreted as ASCII characters when it is printed.

Ordinarily the elements of T are integers in the range 32:127, which are the printable ASCII characters, or in the range 0:255, which are all 8-bit values. For noninteger values, or values outside the range 0:255, the characters printed are determined by fix(rem(t,256)).

Examples The statement

```
ascii = setstr(reshape(32:127,32,3)')
```

prints a 3-by-32 display of the printable ASCII characters

```
ascii =
 ! " # $ % & ' ( ) *+ , - . / 0 1 2 3 4 5 6 7 8 9 : ; < = > ?
@ A B C D E F G H I J K L M N O P Q R S T U V W X Y Z [ \ ] ^ _
' a b c d e f g h i j k l m n o p q r s t u v w x y z { | } ~
```

The infinite loop

```
while 1
    disp(setstr(7))
    pause(2)
end
```

attracts attention.

shading

Purpose Set color shading properties.

Synopsis

```
shading flat
shading faceted
shading interp
```

Description `shading` controls the color shading of `surface` and `patch` objects. `surface` and `patch` objects are created by the functions `surf`, `mesh`, `pcolor`, `fill`, and `fill3`.

`shading flat` sets the shading of the current graph to `flat`. Flat shading is piecewise constant; each mesh line segment or surface patch has a constant color determined by the color values at the end points of the segment or the corners of the patch.

`shading faceted` sets the shading to `faceted`, which is the default. Faceted shading is flat shading with superimposed black mesh lines. This is often the most effective.

`shading interp` sets the shading to `interpolated`. Interpolated shading, which is also known as Gouraud shading, is piecewise bilinear; the color in each segment or patch varies linearly and interpolates the end or corner values.

Algorithm `shading` sets the `EdgeColor` and `FaceColor` properties of all surface objects in the current axes. It sets them to the correct values, depending on whether the `surface` objects represent meshes or surfaces.

See Also `fill`, `fill3`, `hidden`, `mesh`, `patch`, `pcolor`, `surf`, `surface`

`surface` object properties `EdgeColor` and `FaceColor`

sign

Purpose Signum function.

Synopsis `Y = sign(X)`

Description `Y = sign(X)` returns a matrix `Y` the same size as `X`, where each element of `Y` is

- 1 if the element is greater than zero
- 0 if it equals zero
- -1 if it is less than zero

For nonzero complex `X`, `sign(X) = X./abs(X)`.

See Also `abs`, `conj`, `imag`, `real`

sin, sinh

Purpose Sine and hyperbolic sine.

Synopsis

```
Y = sin(X)
Y = sinh(X)
```

Description The trigonometric functions operate element-wise on matrices. Their domains and ranges include complex values. All angles are measured in radians.

`sin(X)` is the circular sine of the elements of `X`.

`sinh(X)` is the hyperbolic sine of the elements of `X`.

Examples `sin(pi)` might not be exactly zero, but rather a value the size of the floating-point accuracy `eps`, because `pi` is only a floating-point approximation to the exact value.

Algorithm

$$\sin(x+iy) = \sin(x)\cosh(y) + i\cos(x)\sinh(y)$$

$$\sin(z) = \frac{e^{iz} - e^{-iz}}{2i}$$

$$\sinh(z) = \frac{e^{-z} - e^{z}}{2}$$

See Also `acos`, `asin`, `atan`, `cos`, `exp`, `expm`, `funm`, `tan`

size

Purpose Matrix dimensions.

Synopsis

```
d = size(X)
[m n] = size(X)
m = size(X,1)
n = size(X,2)
```

Description `d = size(X)`, where `X` is an `m`-by-`n` matrix, returns a vector with two integer components, `d = [m n]`.

`[m n] = size(X) size` returns the two integers in two separate variables.

With a second input argument

`m = size(X,1)` returns the row size of `X`.

`n = size(X,2)` returns the column size of `X`.

See Also `exist`, `length`, `whos`

slice

Purpose Volumetric slice plot.

Synopsis

```
h = slice(V,sx,sy,sz,nx)
```

```
h = slice(X,Y,Z,V,sx,sy,sz,nx)
```

Description `h = slice(V,sx,sy,sz,nx)` draws slices of the volume matrix `V` for the locations specified in the index vectors `sx`, `sy`, and `sz` along the *x*, *y*, and *z* directions respectively. `nx` is the number of rows in `V`. `h` is a vector of handles to the created surface objects.

`h = slice(X,Y,Z,V,sx,sy,sz,nx)` draws the slices for the volume locations specified by the triples `(X(i),Y(i),Z(i))`.

Examples

```
[x,y,z] = meshgrid(-2:.2:2, -2:.2:2, -2:.2:2);
v = x.*exp(-x.^2-y.^2-z.^2);
slice(v,[5 15 21],21,[1 10],21)
```

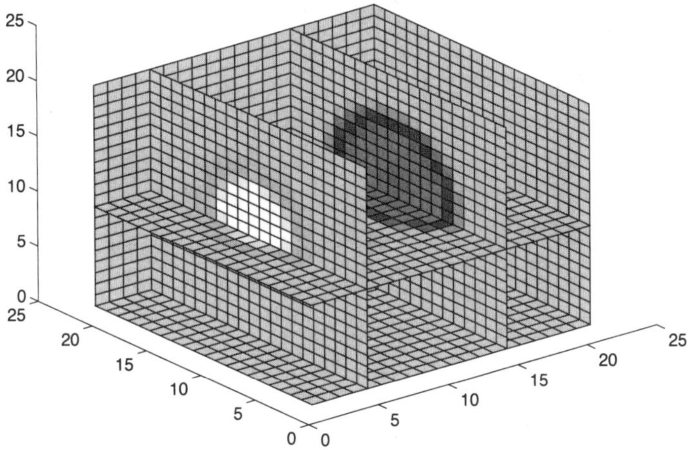

---

sort

Purpose Sort elements in ascending order.

Synopsis

```
Y = sort(X)
[Y,I] = sort(X)
```

Description `sort(X)`, where `X` is a matrix of real values, sorts each column of `X` in ascending order. If `X` is a vector of real values, `sort` sorts the elements of `X` in ascending order. `NaN`s appear at the end of the sorted list, after the highest integers. For identical values in `X`, the location in the input vector or matrix determines location in the sorted list.

`sort` orders complex input according to magnitude. Again, `NaN`s are appear at the end of the sorted list. For identical values, the phase of the input value on the interval [-`pi pi`] determines the final location. (A value that occurs directly on the negative real axis has phase `-pi`.) If the phase of the values is equal, `sort` uses location in the input vector or matrix to determine location in the sorted list.

`[Y,I] = sort(X)`, where `X` is a matrix, also returns an `m`-by-`n` index matrix `I`, with each column a permutation of `1:m`, so that for `j = 1:n`, `Y(:,j) = X(I(:,j),j)`. If `X` is a vector, `I` is an index vector, a permutation of `1:n`, so that `Y = X(I)`.

If `X` has repeated elements of equal value, indices are returned that preserve the original relative ordering.

When `X` is complex, the elements are sorted by `abs(X)`.

Examples Let `X = magic(4)`

```
X =
      16     2     3    13
       5    11    10     8
       9     7     6    12
       4    14    15     1
```

Then

```
[Y,I] = sort(X)
```

produces

```
Y =
       4     2     3     1
       5     7     6     8
       9    11    10    12
      16    14    15    13
 I =
       4     1     1     4
       2     3     3     2
       3     2     2     3
       1     4     4     1
```

See Also `find`, `max`, `mean`, `median`, `min`

---

sound

Purpose Convert vector into sound.

Synopsis

```
sound(y)
sound(y,Fs)
```

Description `sound(y)` sends the signal in vector `y` to the speaker on the following computers:

- SPARC
- Hewlett-Packard
- SGI
- Macintosh
- IBM-compatible

The vector is auto-scaled to provide maximum amplitude. If your system does not include the hardware needed to play sound files, the `sound` function displays an error message.

The sound is played at the default sample rate. On the SPARC, for example, the sample rate is fixed at 8192 Hz. On the Macintosh, the default sample rate is 22.255K Hz.

`sound(y,Fs)`, on the PC, Macintosh, and SGI, plays the sound at the specified sample frequency `Fs`, where `Fs` is measured in Hz.

The sound input vector `y` is automatically scaled so that the maximum and minimum values in `y` correspond to the maximum and minimum input ranges allowed by the sound hardware. On the Macintosh a,d SGI, the volume control on the Control Panel determines the final sound level.

See the function `saxis` to override the automatic scaling and control the sound range explicitly.

Examples There are a number of MAT-files in the `sounds` directory that contain example sounds, mostly sampled at 8192 Hz. List a directory of them with the command `what sounds`. Each MAT-file contains a vector `y` with the sound, and a scalar `Fs` with its associated sample frequency. Load one of them into the workspace, play it, and plot it with

```
load handel
sound(y,Fs)
t = (0:length(y)-1)/Fs;
plot(t,y)
```

See Also `auread`, `auwrite`, `saxis`

---

spalloc

Purpose Allocate space for sparse matrices.

Synopsis `S = spalloc(m,n,nzmax)`

Description `S = spalloc(m,n,nzmax)` creates an all zero sparse matrix `S` of size `m`-by-`n` with room to hold `nzmax` nonzeros. The matrix can then be generated column by column without requiring repeated storage allocation as the number of nonzeros grows.

`spalloc(m,n,nzmax)` is shorthand for

```
sparse([],[],[],m,n,nzmax)
```

Examples To generate a sparse matrix efficiently, which has an average of at most three nonzero elements per column

```
S = spalloc(n,n,3*n);
for j = 1:n
    S(:,j) = { the j-th column }
end
```

sparse

Purpose Create sparse matrices.

Synopsis

```
S = sparse(...)
S = sparse(A)
S = sparse(i,j,s,m,n,nzmax)
S = sparse(i,j,s,m,n)
S = sparse(i,j,s)
S = sparse(m,n)
```

Description `S = sparse(...)` is the built-in function, which generates matrices in MATLAB's sparse storage organization. It can be called with one, two, three, five, or six arguments.

`S = sparse(A)` converts a full matrix to sparse form by squeezing out any zero elements. If `S` is already sparse, `sparse(S)` returns `S`.

`S = sparse(i,j,s,m,n,nzmax)` uses vectors `i`, `j`, and `s` to generate an `m`-by-`n` sparse matrix with space allocated for `nzmax` nonzeros. The two vectors `i` and `j` contain the row and column coordinates, respectively, for the non-zero elements of the matrix. `s` contains the values for each non-zero element. Any elements of `s` that are zero are ignored, along with the corresponding values of `i` and `j`. `i`, `j`, and `s` are all the same length.

There are several simplifications of this six-argument call.

The argument `s` and one of the arguments `i` or `j` can be scalars, in which case they are expanded so that the first three arguments all have the same length.

`S = sparse(i,j,s,m,n)` uses `nzmax = length(s)`.

`S = sparse(i,j,s)` uses `m = max(i)` and `n = max(j)`. The maxima are computed before any zeros in `s` are removed, so one of the rows of `[i j s]` might be `[m n 0]`.

`S = sparse(m,n)` abbreviates `sparse([],[],[],m,n,0)`. This generates the ultimate sparse matrix, an `m`-by-`n` all zero matrix.

All of MATLAB's built-in arithmetic, logical, and indexing operations can be applied to sparse matrices, or to mixtures of sparse and full matrices. Operations on sparse matrices return sparse matrices and operations on full matrices return full matrices.

In most cases, operations on mixtures of sparse and full matrices return full matrices. The exceptions include situations where the result of a mixed operation is structurally sparse, for example, `A.*S` is at least as sparse as `S`.

Some operations, such as `S >= 0`, generate *Big Sparse,* or *BS*, matrices – matrices with sparse storage organization but few zero elements.

Examples `S = sparse(1:n,1:n,1)` generates a sparse representation of the `n`-by-`n` identity matrix. The same `S` results from `S = sparse(eye(n,n))`, but this would also temporarily generate a full `n`-by-`n` matrix with most of its elements equal to zero.

`B = sparse(10000,10000,pi)` is probably not very useful, but is legal and works; it sets up a `10000`-by-`10000` matrix with only one nonzero element. Don't try `full(B)`; it requires 800 megabytes of storage.

This dissects and then reassembles a sparse matrix:

```
[i,j,s] = find(S);
[m,n] = size(S);
S = sparse(i,j,s,m,n);
```

So does this, if the last row and column have nonzero entries:

```
[i,j,s] = find(S);
S = sparse(i,j,s);
```

See Also `diag`, `find`, `full`, `nnz`, `nonzeros`, `nzmax`, `sparse`, `spones`, `sprandn`, `sprandsym`, `spy`

The `sparfun` directory

---

spaugment

Purpose Create an augmented matrix associated with least squares problems.

Synopsis

```
S = spaugment(A,c)
S = spaugment(A)
```

Description `S = spaugment(A,c)` creates the sparse, square, symmetric, and indefinite matrix.

```
S = [c*I  A; A'  0]
```

The augmented matrix can compute the solution to the least squares problem.

$$\text{minimize}\|b - Ax\|$$

If $r = b - Ax$ is the residual and c is a residual scaling factor, the least squares problem is equivalent to the system of linear equations

$$\begin{pmatrix} cI & A \\ A' & 0 \end{pmatrix}\begin{pmatrix} r/c \\ x \end{pmatrix} = \begin{pmatrix} b \\ 0 \end{pmatrix}$$

If `A` is `m`-by-`n`, the augmented matrix `S` is `(m+n)`-by-`(m+n)`. The augmented linear system can be written

```
S * [r/c; x] = [b; z]
```

where `z = zeros(n,1)`.

`S = spaugment(A)` uses a value of `max(max(abs(A)))/1000` for `c`. The optimum value of the residual scaling factor, `c`, involves `min(svd(A))` and `norm(r)`, which are usually too expensive to compute. You can use the command `spparms` to change the default.

The sparse backslash operation `x = A\b` automatically forms the same augmented matrix, with the default value of `c`. `spparms` can set other values of `c` for use with the backslash operation.

Examples Here is a simple system of three equations in two unknowns.

$$x_1 + 2x_2 = 5$$
$$3x_1 = 6$$
$$4x_2 = 7$$

Since it is such a small system, it is ordinarily solved in MATLAB with

```
A = [1  2; 3  0; 0  4];
b = [5  6  7]';
x = A\b
```

This produces the least squares solution

```
x =
        1.9592
        1.7041
```

The sparse augmented systems approach uses

```
S = spaugment(A,1)
```

which gives

```
full(S) =
        1     0     0     1     2
        0     1     0     3     0
        0     0     1     0     4
        1     3     0     0     0
        2     0     4     0     0
```

This example used `c = 1` and printed `full(S)` to improve readability. The default value of `c` for this example is 0.0040. Even in this small example, more than half of the elements of `S` are zero.

The augmented systems also use an extended right-hand side

```
y = [b; 0; 0];
```

Now

```
z = S\y
```

involves a square matrix, so MATLAB factors that matrix by Gaussian elimination instead of orthogonalization. For sparse matrices, this usually involves less fill-in and requires less storage.

The computed solution is

```
z =
       -0.3673
        0.1224
        0.1837
        1.9592
        1.7041
```

The first three components are the residual `r` and the last two components are the same solution `x`, as was computed by the first approach.

See Also `\`, `diag`, `spparms`

---

spconvert

Purpose Convert external form of a sparse matrix.

Synopsis

```
S = spconvert(D)
```

Description The `load` and `save` commands handle sparse matrices, so there is no need for a special sparse `load` or `save` command. However, conversion between the external form and internal form of a sparse matrix is needed after loading from an ASCII text file.

`S = spconvert(D)` converts a matrix `D` with rows containing `[i,j,s]` or `[i,j,r,s]`to the corresponding sparse matrix. `D` must have an `nnz` or `nnz+1` row and three or four columns. Three elements per row generate a real matrix and four elements per row generate a complex matrix. A row of the form `[m n 0]` or `[m n 0 0]` anywhere in `D` can be used to specify `size(S)`. If `D` is already sparse, no conversion is done, so `spconvert` can be used after `D` is loaded from either a MAT-file or an ASCII file.

Examples Suppose the ASCII file `uphill.dat` contains

```
1    1    1.000000000000000
1    2    0.500000000000000
2    2    0.333333333333333
1    3    0.333333333333333
2    3    0.250000000000000
3    3    0.200000000000000
1    4    0.250000000000000
2    4    0.200000000000000
3    4    0.166666666666667
4    4    0.142857142857143
4    4    0.000000000000000
```

Then the statements

```
load uphill.dat
H = spconvert(uphill)
```

recreate `sparse(triu(hilb(4)))`, possibly with roundoff errors. In this case, the last line of the input file is not necessary because the earlier lines already specify that the matrix is at least 4-by-4.

---

spdiags

Purpose Extract and create sparse band and diagonal matrices.

Synopsis

```
[B,d] = spdiags(A)
B = spdiags(A,d)
A = spdiags(B,d,A)
```

```
A = spdiags(B,d,m,n)
```

Description The `spdiags` function, which generalizes the built-in function `diag`, deals with three matrices, in various combinations, as both input and output:

- `A` is an `m`-by-`n` matrix, usually (but not necessarily) sparse, with its nonzero or specified elements located on `p` diagonals.
- `B` is a `min(m,n)`-by-`p` matrix, usually (but not necessarily) full, whose columns are the diagonals of `A`.
- `d` is a vector of length `p` whose integer components specify the diagonals in `A`.

Roughly, `A`, `B`, and `d` are related by

```
for k = 1:p
    B(:,k) = diag(A,d(k))
end
```

Four different operations, distinguished by the number of input arguments, are possible with `spdiags`:

`[B,d] = spdiags(A)` extracts all nonzero diagonals.

`B = spdiags(A,d)` extracts the diagonals specified by d.

`A = spdiags(B,d,A)` replaces the specified diagonals.

`A = spdiags(B,d,m,n)` creates a sparse matrix from its diagonals.

The precise relationship among `A`, `B`, and `d` is

```
if m >= n
    for k = 1:p
        for j = max(1,1+d(k)):min(n,m+d(k))
            B(j,k) = A(j-d(k),j);
        end
    end
if m < n
    for k = 1:p
        for i = max(1,1-d(k)):min(m,n-d(k))
            B(i,k) = A(i,i+d(k));
        end
    end
end
```

Some elements of `B`, corresponding to positions outside of `A`, are not defined by these loops. They are not referenced when `B` is input and are set to zero when `B` is output.

Examples This example generates a sparse tridiagonal representation of the classic second difference operator on `n` points.

```
e = ones(n,1);
A = spdiags([e -2*e e], -1:1, n, n)
```

Turn it into Wilkinson's test matrix (see `wilkinson`):

```
A = spdiags(abs(-(n-1)/2:(n-1)/2)',0,A)
```

Finally recover the three diagonals:

```
B = spdiags(A)
```

The second example is not square.

```
A = [11   0  13   0
      0  22   0  24
      0   0  33   0
     41   0   0  44
      0  52   0   0
      0   0  63   0
      0   0   0  74]
```

Here `m = 7, n = 4,` and `p = 3.`

The statement `[B,d] = spdiags(A)` produces `d = [-3 0 2]'` and

```
B = [41  11   0
     52  22   0
     63  33  13
     74  44  24]
```

Conversely, with the above `B` and `d`, the expression `spdiags(B,d,7,4)` reproduces the original `A`.

See Also `diag`

---

specular

Purpose Specular reflectance.

Synopsis

```
r = specular(Nx,Ny,Nz,S,V)
r = specular(Nx,Ny,Nz,S,V,spread)
```

Description `specular` returns the reflectance `r` of a specular surface from the normal vector components. The reflectance is the fraction of light that reflects from the surface toward the viewer. `r` is a scalar that varies from 0.0 (no light reflected) to 1.0 (all the light reflected).

`r = specular(Nx,Ny,Nz,S,V)` returns the reflectance of the surface with normal vector components `[Nx,Ny,Nz]`. The normal vector components can be matrices so that the normal is

```
n(i,j) = [Nx(i,j), Ny(i,j), Nz(i,j)]
```

These normal components can be calculated using `surfnorm`. The light source vector `S = [Sx,Sy,Sz]` is a three-element vector that specifies the direction from the surface to the light source. `V` is three-element vector that specifies the viewpoint. Both can be two-element vectors specifying azimuth and elevation.

`r = specular(Nx,Ny,Nz,S,V,spread)` specifies the surface spread exponent.

`specular` is used as part of the function `surfl` to produce shaded surfaces.

Algorithm `specular` implements a common approximation for the reflectance of shiny metal-like surfaces. The specular highlight is strongest when the normal vector is in the direction of $(s+v)/2$, where s is the source direction and v is the view direction.

See Also `diffuse`, `surfl`, `surfnorm`

---

speye

Purpose Sparse identity matrix.

Synopsis

```
S = speye(m,n)
S = speye(n)
```

Description `S = speye(m,n)` forms an `m`-by-`n` sparse matrix with 1s on the main diagonal.

`S = speye(n)` abbreviates `speye(n,n)`.

Examples `I = speye(1000)` forms the sparse representation of the 1000-by-1000 identity matrix, which requires only about 16 kilobytes of storage. This is the same final result as `I = sparse(eye(1000,1000))`, but the latter requires eight megabytes for temporary storage for the full representation.

See Also `spalloc`, `spones`

---

spfun

Purpose Apply function to nonzero entries.

Synopsis

```
f = spfun('function',S)
```

Description `f = spfun('`*`function`*`',S)` evaluates *`function`*`(S)` on the nonzero elements of `S`. *`function`* must be the name of a function, usually defined in an M-file, which can accept a matrix argument, `S`, and evaluate the function at each element of `S`.

Examples `f = spfun('exp',S)` has the same sparsity pattern as `S` (except for underflow), whereas `exp(S)` has 1s where `S` has 0s.

---

sph2cart

Purpose Transform spherical coordinates to Cartesian.

Synopsis

```
[x,y,z] = sph2cart(az,el,r)
```

Description `[x,y,z] = sph2cart(az,el,r)` transforms data stored as spherical coordinates to Cartesian, or *xy*, coordinates. `az`, `el`, and `r` must all be the same size. `az` and `el` are in radians.

See Also `cart2pol`, `cart2sph`, `pol2cart`

---

sphere

Purpose Generate sphere.

Synopsis

```
[X,Y,Z] = sphere(n)
```

```
sphere(n)
sphere
```

Description `sphere` generates the *x*-, *y*-, and *z*-coordinates of a unit sphere for use with `surf` and `mesh`.

`[X,Y,Z] = sphere(n)` returns the coordinates of a sphere in three `(n+1)`-by-`(n+1)` matrices. You can graph the sphere with

```
surf(X,Y,Z)
```

`sphere(n)`, with no output arguments, draws a `surf` plot of the sphere on the screen. `sphere` with no input arguments defaults to `n = 20`.

Examples Generate and plot a sphere:

```
[X,Y,Z] = sphere(10);
mesh(X,Y,Z)
```

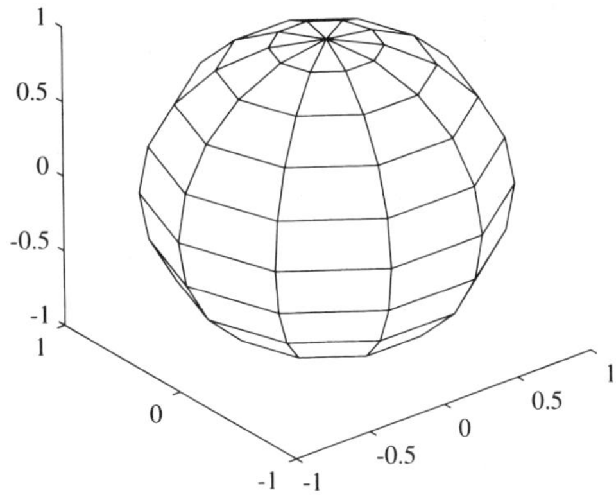

See Also `cylinder`

spinmap

Purpose Spin colormap.

Synopsis

```
spinmap
spinmap(t)
spinmap(t,inc)
spinmap(inf)
```

Description `spinmap` cyclically rotates the colormap for about five seconds.

`spinmap(t)` rotates it for about `t` seconds.

`spinmap(t,inc)` uses the specified increment. The default is `inc` = 2, so `inc` = 1 is a slower rotation, `inc` = 3 is faster, `inc` = –2 is the other direction, and so on.

`spinmap(inf)` is an infinite loop. To break the loop, press **Ctrl-C**.

See Also `colormap`

---

spline

Purpose Cubic spline interpolation.

Synopsis

```
yi = spline(x,y,xi)
pp = spline(x,y)
```

Description `spline` interpolates between data points using cubic spline fits.

`yi = spline(x,y,xi)` accepts vectors `x` and `y` that contain coarsely spaced data, and vector `xi` that specifies a new, more finely spaced abscissa. The function uses cubic spline interpolation to find a vector `yi` corresponding to `xi`.

`pp = spline(x,y)` returns the pp-form of the cubic spline interpolant, for later use with `ppval` and other spline functions.

Examples The following two vectors represent the census years from 1900 to 1990 and the corresponding United States population in millions of people.

```
t = 1900:10:1990;
p = [ 75.995  91.972  105.711  123.203  131.669 ...
     150.697 179.323  203.212  226.505  249.633 ]';
```

The expression

```
spline(t,p,2000)
```

uses the cubic spline to extrapolate and predict the population in the year 2000. The result is

```
ans =
    270.6060
```

The following statements interpolate the data with a cubic spline, evaluate that spline for each year from 1900 to 2000, and plot the result.

```
x = 1900:1:2000;
y = spline(t,p,x);
plot(t,p,'o',x,y)
title('United States Census')
xlabel('year')
```

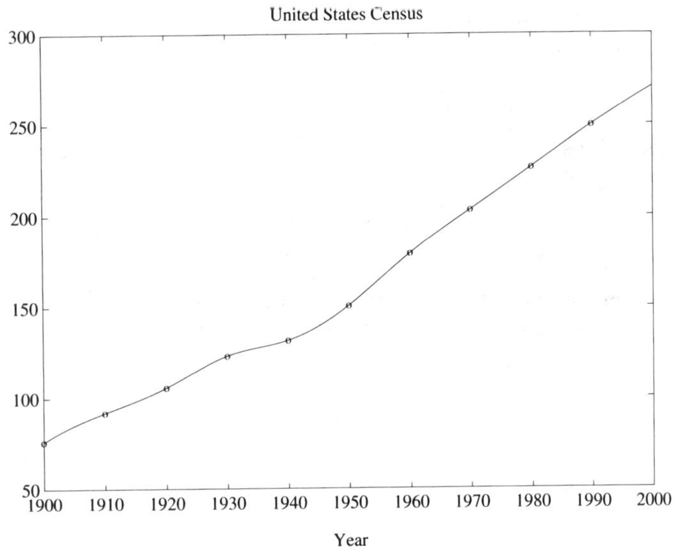

Algorithm `spline` is a MATLAB M-file. It uses the M-files `ppval`, `mkpp`, and `unmkpp`. These routines form a small suite of functions for working with piecewise polynomials. `spline` uses these functions in a fairly simple fashion to perform cubic spline interpolation. For access to the more advanced features, see the M-files and the Spline Toolbox.

See Also `interp1`, `polyfit`

References

[1] de Boor, C., *A Practical Guide to Splines*, Springer-Verlag, 1978.

---

spones

Purpose Replace nonzero elements with ones.

Synopsis `R = spones(S)`

Description `R = spones(S)` generates a matrix `R` with the same sparsity structure as `S`, but with 1s in the nonzero positions.

Examples `c = sum(spones(S))` is the number of nonzeros in each column.

`r = sum(spones(S'))'` is the number of nonzeros in each row.

`sum(c)` and `sum(r)` are equal, and are equal to `nnz(S)`.

spparms

Purpose Set parameters for sparse matrix ordering algorithms and direct linear equation solvers.

Synopsis

```
spparms('key',value)
spparms
values = spparms
[keys,values] = spparms
spparms(values)
value = spparms('key')
spparms('default')
spparms('tight')
```

Description Most users of the sparse matrix facilities in MATLAB need not use this function. It is intended for users wanting to see and possibly modify the detailed behavior of the minimum degree ordering algorithms and the built-in direct sparse linear equations solvers.

`spparms('key',value)` sets one or more of the *tunable* parameters used in the sparse linear equation operators, \ and /, and the minimum degree orderings, `colmmd` and `symmmd`.

The meanings of the `key` parameters are

`'spumoni'`	Sparse Monitor flag. 0 produces no diagnostic output, the default. 1 produces information about choice of algorithm based on matrix structure, and about storage allocation. 2 also produces very detailed information about the minimum degree algorithms.
`'thr_rel'`,	
`'thr_abs'`	Minimum degree threshold is `thr_rel*mindegree+thr_abs`

`'exact_d'`	Nonzero to use exact degrees in minimum degree. Zero to use approximate degrees.
`'supernd'`	If positive, minimum degree amalgamates the supernodes every `supernd` stages.
`'rreduce'`	If positive, minimum degree does row reduction every `rreduce` stages.
`'wh_frac'`	Rows with `density > wh_frac` are ignored in `colmmd`.
`'autommd'`	Nonzero to use minimum degree orderings with `\` and `/`.
`'aug_rel'`, `'aug_abs'`	Residual scaling parameter for augmented equations is `aug_rel*max(max(abs(A))) + aug_abs` For example, `aug_rel = 0`, `aug_abs = 1` puts an unscaled identity matrix in the (1,1) block of the augmented matrix.

`spparms`, by itself, prints a description of the current settings.

`values = spparms` returns a vector whose components give the current settings.

`[keys,values] = spparms` returns that vector, and also returns a character matrix whose rows are the keywords for the parameters.

`spparms(values)`, with no output argument, sets all the parameters to the values specified by the argument vector.

`value = spparms('`*`key`*`')` returns the current setting of one parameter.

`spparms('default')` sets all the parameters to their default settings.

`spparms('tight')` sets the minimum degree ordering parameters to their *tight* settings, which can lead to orderings with less fill-in, but which make the ordering functions themselves use more execution time.

The key parameters with the `default` and `tight` values are

	Keyword	Default	Tight
`values(1)`	`'spumoni'`	0.0	
`values(2)`	`'thr_rel'`	1.1	1.0
`values(3)`	`'thr_abs'`	1.0	0.0
`values(4)`	`'exact_d'`	0.0	1.0
`values(5)`	`'supernd'`	3.0	1.0
`values(6)`	`'rreduce'`	3.0	1.0
`values(7)`	`'wh_frac'`	0.5	0.5
`values(8)`	`'autommd'`	1.0	
`values(9)`	`'aug_rel'`	0.001	
`values(10)`	`'aug_abs'`	0.0	

See Also `\`, `colmmd`, `spaugment`, `symmmd`

References [1] Gilbert, John R., Cleve Moler and Robert Schreiber, "Sparse Matrices in MATLAB: Design and Implementation," *SIAM Journal on Matrix Analysis and Applications 13*, pp. 333-356, 1992.

---

sprandn

Purpose Random sparse matrices.

Synopsis

```
R = sprandn(S)
R = sprandn(m,n,density)
R = sprandn(m,n,density,rc)
```

Description `R = sprandn(S)` has the same sparsity structure as `S`, but normally distributed random entries with mean 0 and variance 1.

`R = sprandn(m,n,density)` is a random, `m`-by-`n`, sparse matrix with approximately `density*m*n` normally distributed nonzero entries `(0 ≤ density ≤ 1)`.

`R = sprandn(m,n,density,rc)` also has reciprocal condition number approximately equal to `rc`. `R` is constructed from a sum of matrices of rank one.

If `rc` is a vector of length `lr`, where `lr ≤ min(m,n)`, then `R` has `rc` as its first `lr` singular values, all others are zero. In this case, `R` is generated by random plane rotations applied to a diagonal matrix with the given singular values. It has a great deal of topological and algebraic structure.

See Also `sprandsym`

---

sprandsym

Purpose Random sparse symmetric matrices.

Synopsis

```
R = sprandsym(S)
R = sprandsym(n,density)
R = sprandsym(n,density,rc)
R = sprandsym(n,density,rc,kind)
```

Description `R = sprandsym(S)` is a symmetric random matrix whose lower triangle and diagonal have the same structure as `S`. Its elements are normally distributed, with mean 0 and variance 1.

`R = sprandsym(n,density)` is a symmetric random, `n`-by-`n`, sparse matrix with approximately `density*n*n` nonzeros; each entry is the sum of one or more normally distributed random samples, and `(0 ≤ density ≤ 1)`.

`R = sprandsym(n,density,rc)` has a reciprocal condition number equal to `rc`. The distribution of entries is nonuniform; it is roughly symmetric about 0; all are in [-1,1].

If `rc` is a vector of length `n`, then `R` has eigenvalues `rc`. Thus, if `rc` is a positive (nonnegative) vector then `R` is a positive definite matrix. In either case, `R` is generated by random Jacobi rotations applied to a diagonal matrix with the given eigenvalues or condition number. It has a great deal of topological and algebraic structure.

`R = sprandsym(n, density, rc, kind)` is positive definite, where `kind` can be:

- `1` to generate `R` by random Jacobi rotation of a positive definite diagonal matrix. `R` has the desired condition number exactly.
- `2` to generate an `R` that is a shifted sum of outer products. `R` has the desired condition number only approximately, but has less structure.
- `3` to generate an `R` that has the same structure as the matrix `S` and approximate condition number `1/rc`. `density` is ignored.

See Also `sprandn`

sprank

Purpose Sparse matrix structural rank.

Synopsis `r = sprank(A)`

Description `r = sprank(A)` is the structural rank of the matrix `A`. This is also known as maximum traversal, maximum assignment, and size of a maximum matching in the graph of `A`.

It is always true that `sprank(A) >= rank(A)`.

Furthermore, in exact arithmetic

```
sprank(A) == rank(sprand(A))
```

with probability one.

Examples The 3-by-3 matrix

```
A = [2   0   1
     0   0   0
     3   0   x]
```

has `sprank(A) = 2` for any value of `x`. It also has `rank(A) = 2`, unless `x = 3/2` in which case, `rank(A) = 1`.

See Also `dmperm`, `rank`

sprintf

Purpose Write formatted data to a string.

Synopsis `s = sprintf('format',A,...)`

Description `num2str`, `int2str`, and `sprintf` all convert numbers to their MATLAB string representations and can be used for labeling and titling plots with numeric values.

`s = sprintf('format',A,...)` formats the data in matrix `A` (and in any additional matrix arguments) under control of the specified *`format`* string and returns it in the MATLAB string variable `s`.

`sprintf` is the same as `fprintf` except that it returns the data in a MATLAB string variable rather than writing it to a file. See `fprintf` for full information on creating format strings.

Examples The statement

```
S = sprintf('rho is %6.3f',(1+sqrt(5))/2)
```

produces the string

```
S = 'rho is 1.618'
```

See Also `fprintf`, `int2str`, `num2str`, `sscanf`

---

spy

Purpose Visualize matrix sparsity patterns.

Synopsis

```
spy(S)
spy(S,marksize)
spy(S,'color')
spy(S,'color',marksize)
```

Description `spy(S)` plots the sparsity pattern of any matrix `S`.

`spy(S,marksize)`, where `marksize` is an integer, uses the specified marker size.

`spy(S,'color')`, where *`color`* is a string, uses the specified color for the plot markers.

`spy(S,'color',marksize)` uses the specified color and size for the plot markers.

`S` is usually a sparse matrix, but full matrices are acceptable, in which case the locations of the nonzero elements are plotted.

`spy` replaces `format +`, which takes much more space to display essentially the same information.

See Also `find`, `gplot`, `symmmd`, `symrcm`

---

sqrt

Purpose Square root.

Synopsis `Y = sqrt(X)`

Description `sqrt(X)` is the square root of the elements of `X`. For the elements of `X` that are negative or complex, `sqrt(X)` produces complex results.

See `sqrtm` for the matrix square root.

Examples

```
sqrt((-2:2)')
ans =
          0 + 1.4142i
          0 + 1.0000i
          0
     1.0000
     1.4142
```

See Also `exp`, `log`, `sqrtm`

sqrtm

Purpose Matrix square root.

Synopsis `Y = sqrtm(X)`

Description `Y = sqrtm(X)` is one of the many matrices that satisfy

```
Y*Y = X
```

If `X` is symmetric and positive definite, then `Y` is the unique positive definite square root.

Examples A matrix representation of the fourth difference operator is

```
X =

     5    -4     1     0     0
    -4     6    -4     1     0
     1    -4     6    -4     1
     0     1    -4     6    -4
     0     0     1    -4     5
```

This matrix is symmetric and positive definite. Its unique positive definite square root, `Y = sqrtm(X)`, is a representation of the second difference operator.

```
Y =

     2    -1     0     0     0
    -1     2    -1     0     0
     0    -1     2    -1    -0
     0     0    -1     2    -1
     0     0    -0    -1     2
```

The matrix

```
X =

     7    10
    15    22
```

has four square roots. Two of them are

```
Y1 =

    1.5667    1.7408
    2.6112    4.1779
```

and

```
Y2 =

     1     2
     3     4
```

The other two are –Y1 and –Y2. All four can be obtained from the eigenvalues and vectors of X.

```
[V,D] = eig(X);
D =
     0.1386          0
          0    28.8614
```

The four square roots of the diagonal matrix D result from the four choices of sign in

```
S =
     ±0.3723          0
          0     ±5.3723
```

All four Ys are of the form

```
Y = V*S/V
```

The sqrtm function chooses the two plus signs and produces Y1, even though Y2 is more natural because its entries are integers.

Finally, the matrix

```
X =
     0     1
     0     0
```

does not have any square roots. There is no matrix Y, real or complex, for which Y*Y = X. The statement

```
Y = sqrtm(X)
```

produces

```
WARNING: Result from FUNM is probably inaccurate.
Y =
     0     0
     0     0
```

Algorithm The function sqrtm(X) is an abbreviation for funm(X,'sqrt'). The algorithm used by funm is based on a Schur decomposition. It can fail in certain situations where X has repeated eigenvalues. See funm for details.

See Also expm, funm, logm

---

sscanf

Purpose Read string under format control.

Synopsis

```
A = sscanf(s,'format')
A = sscanf(s,'format',size)
[A,count] = sscanf(...)
```

Description A = sscanf(s,'*format*') reads data from the MATLAB string variable s, converts it according to the specified *format* string, and returns it in matrix A.

`A = sscanf(s,'format',size)` specifies the number of elements to generate.

`[A,count] = sscanf(...)` also returns the number of elements successfully read.

`sscanf` is the same as `fscanf` except that it reads the data from a MATLAB string variable rather than reading it from a file. See `fscanf` for full information on how to use `sscanf`.

Example The statements

```
s = '2.7183  3.1416';
A = sscanf(s,'%f')
```

create a two-element vector containing poor approximations to e and `pi`.

See Also `eval`, `fscanf`, `sprintf`

---

stairs

Purpose Stairstep plot.

Synopsis

```
stairs(y);
stairs(x,y);
[xb,yb] = stairs(y)
[xb,yb] = stairs(x,y)
```

Description `stairs(y)` draws a stairstep graph of the elements of vector `y`. A stairstep graph is similar to a bar graph, but the vertical lines dropping to the *x*-axis are omitted. Stairstep plots are useful for drawing time history plots of digital sampled data systems.

`stairs(x,y)` draws a stairstep graph of the elements in vector `y` at the locations specified in vector `x`. The values in `x` must be evenly spaced in ascending order.

`[xb,yb] = stairs(y)` and `[xb,yb] = stairs(x,y)` do not draw graphs, but return vectors `xb` and `yb` such that `plot(xb,yb)` plots the stairstep graph.

Examples Create a stairstep plot of a sine wave:

```
x = 0:.25:10;
stairs(x,sin(x))
```

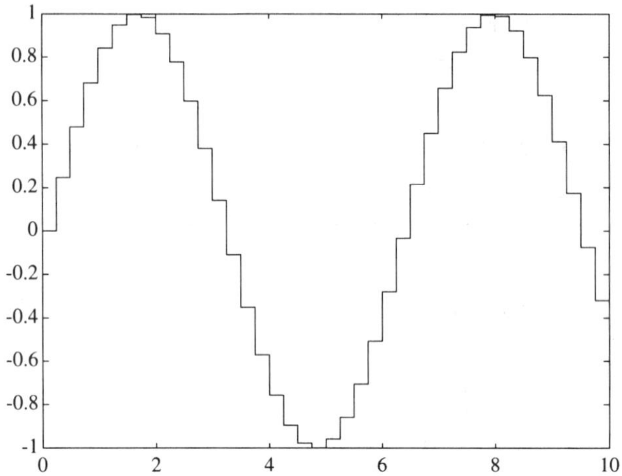

See Also `bar`, `hist`

---

startup

Purpose MATLAB startup M-file.

Synopsis `startup`

Description At startup time, MATLAB automatically executes the master M-file `matlabrc.m` and, if it exists, `startup.m`. On multiuser or networked systems, `matlabrc.m` is reserved for use by the system manager. The file `matlbrc.m` invokes the file `startup.m` if it exists on MATLAB's search path.

You can create a startup file in your own MATLAB directory. The file can include physical constants, engineering conversion factors, or anything else you want predefined in your workspace.

Examples In the file `startup.m`

```
% Add my test directory as the first entry
% on MATLAB's search path
path('/my_home/my_tests',path)
```

Algorithm Only `matlabrc.m` is actually invoked by MATLAB at startup. However, `matlabrc.m` contains the statements

```
if exist('startup')==2
    startup
end
```

that invoke `startup.m`. You can extend this process to create additional startup M-files, if required.

See Also `!`, `exist`, `matlabrc`, `path`, `quit`

---

std

Purpose Standard deviation.

Synopsis `s = std(X)`

Description `s = std(X)`, where `X` is a matrix, returns a row vector `s` containing the standard deviation of each column.If `X` is a vector, `s` is the standard deviation of the elements of `X`.

The standard deviation `s` of a sample of data is

$$s = \left(\frac{1}{n-1} \sum_{i=1}^{n} \left(x_i - \bar{x} \right)^2 \right)^{\frac{1}{2}}$$

where

$$\bar{x} = \frac{1}{n} \sum_{i=1}^{n} x_i$$

and n is the number of elements in the sample.

Using this definition, s^2 is the minimum variance unbiased estimate of sigma squared (the scale parameter of the normal distribution).

See Also `corrcoef`, `cov`, `mean`, `median`

---

stem

Purpose Stem plot for discrete sequence data.

Synopsis

```
stem(y)
stem(x,y)
stem(y,'linetype')
stem(x,y,'linetype')
```

Description `stem(y)` plots the data sequence `y` as stems from the *x*-axis. Each stem is terminated with a circle whose y-position represents the data value.

`stem(x,y)` plots the data sequence `y` at the values specified in `x`.

`stem(y,'`*`linetype`*`')` and `stem(x,y,'`*`linetype`*`')` specify a line type for the stems of the plot. See *`plot`* for valid *`linetypes`*.

Examples

```
y = randn(50,1);
stem(y)
```

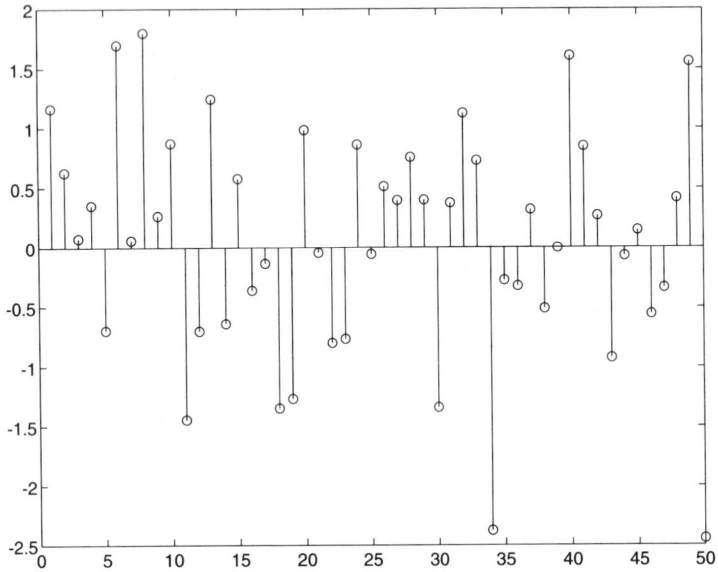

See Also `bar`, `plot`, `stairs`

---

str2mat

Purpose Form string matrix from individual strings.

Synopsis `S = str2mat('t1','t2','t3',...)`

Description `S = str2mat('t1','t2','t3',...)` forms the matrix `S` containing the text strings *t1*, *t2*, *t3*, etc. as rows. `str2mat` automatically pads each string with spaces in order to form a valid matrix. Up to 11 text strings can be used to form `S`. In addition, each text parameter, *tn*, can itself be a string matrix. This allows the creation of arbitrarily large string matrices.

Examples Form a string matrix from the strings `'One'`, `'Twenty'`, and `'Thirty-six'`:

```
s = str2mat('One','Twenty','Thirty-six')
s =
    One
    Twenty
    Thirty-six
```

See Also `int2str`, `isstr`, `num2str`, `setstr`

str2num

Purpose String to number conversion.

Synopsis `x = str2num('str')`

Description `x = str2num('str')` converts the string *str*, which is an ASCII character representation of a numeric value, to MATLAB's numeric representation. The string can contain digits, a decimal point, a leading + or – sign, an `e` preceding a power of 10 scale factor, and/or an `i` for a complex unit.

Examples `str2num('3.14159e0')` is approximately π.

Algorithm The M-file for `x = str2num(s)` is a single line:

```
eval(['x = ' s ';']);
```

See Also `hex2num`, `num2str`, `sscanf`

strcmp

Purpose Compare strings.

Synopsis `l = strcmp('str1','str2')`

Description `l = strcmp('str1','str2')` compares the strings *str1* and `str2` and returns 1 if the two are identical, and 0 otherwise. Note that the value returned by `strcmp` is not the same as the C language convention.

`strcmp` is case sensitive; any leading and trailing blanks in either of the strings are explicitly included in the comparison.

Examples

```
strcmp('Yes','No') =
   0
strcmp('Yes','Yes') =
   1
```

See Also `isstr`, `lower`, `setstr`, `upper`

strings

Purpose MATLAB string handling.

Description The statement `t = 'Hello, World.'` creates a vector whose components are the ASCII codes for the characters. The column dimension of `t` is the number of characters. It is no different than other MATLAB vectors, except that it displays as text instead of the decimal ASCII codes:

```
t =
     Hello, World.
```

Associated with each MATLAB variable is a flag that, if set, tells the MATLAB output routines to display the variable as text. `t = abs(t)` is one way of clearing the flag so the vector is displayed in its decimal ASCII representation. `t = setstr(t)` sets the flag back to text display.

`isstr(t)` returns 1 if `t` is a string, and 0 otherwise.

Two single quotes indicate a quote within a string.

To concatenate strings, treat them as vectors:

```
x = 'Hello, ';
y = 'World';
z = [x y]
z =
    Hello, World.
```

Examples The string

```
s = ['It is 1 o''clock', 7, 13, 'It is 2']
```

uses two quotes to indicate a single quote, and includes a bell (ASCII 7) on all platforms except Macintosh, and a carriage return (ASCII 13) on all platforms.

See Also `abs`, `isstr`, `setstr`, `strcmp`

---

strrep

Purpose String search and replace.

Synopsis `str = strrep('str1','old_str','new_str')`

Description `str = strrep('str1','old_str','new_str')` replaces all occurrences of a substring with another string, where

- *str1* is the string in which to search and replace.
- *old_str* is the substring to replace within *str1*.
- *new_str* is the string that replaces *old_str*.

The function returns the modified string `str`.

Examples

```
s1 = 'This is a good example.';
str = strrep(s1,'good','great')
str =
This is a great example.
```

See Also `findstr`

---

strtok

Purpose First token in string.

Synopsis

```
token = strtok('str',del)
token = strtok('str')
[token,rem] = strtok('str',del)
```

Description `token = strtok('str',del)` returns the first token in the text string *str*, that is, the first set of characters before a delimiter is encountered. `del` is a vector containing valid delimiter characters.

`token = strtok('str')` uses the default delimiter, a white space.

`[token,rem] = strtok('str',del)` returns the remainder `rem` of the original string. The remainder consists of all characters from the first delimiter on.

Examples

```
s = 'This is a good example.';
[token,rem] = strtok(s)
token =
This
rem =
 is a good example.
```

See Also `isspace`

subplot

Purpose Create and control tiled axes.

Synopsis

```
h = subplot(m,n,p)
subplot(m,n,p)
subplot(h)
```

Description `h = subplot(m,n,p)` breaks the figure window into an `m`-by-`n` matrix of small rectangular panes, creates an axes in the `p`-th pane, makes it current, and returns the handle to this new axes. The axes are counted along the top row of the figure window, then the second row, etc.

`subplot(m,n,p)`, if an axes already exists at a specified location, makes it current.

`subplot(h)`, where `h` is an axes handle, is another way of making an axes current for subsequent plotting commands.

If a `subplot` specification causes an new axes to overlap an existing axes, the existing axes is deleted. For example, the statement `subplot(1,1,1)` deletes all existing smaller axes in the figure window and creates a new full-figure axes.

Use `clf` or `subplot(1,1,1)` to delete all axes and return to the default `subplot(1,1,1)` configuration.

Examples To plot `income` on the top half of the screen and `outgo` on the bottom half,

```
income = [3.2  4.1  5.0]
outgo = [3.0  3.2  3.35]
subplot(2,1,1), plot(income)
subplot(2,1,2), plot(outgo)
```

See Also `axes`, `cla`, `clf`, `figure`, `gca`

subscribe

Purpose Become a subscribing user of MATLAB.

Synopsis `subscribe`

Description `subscribe` helps you become a subscribing user of MATLAB. Subscribing users:

- Receive a subscription to the *MathWorks Newsletter* by regular mail.
- Receive an email subscription to the MATLAB *News Digest*.
- Obtain technical support.

Any MATLAB user can become a subscribing user. It is not necessary to be a registered user.

`subscribe` prompts for a name and address. For UNIX systems, the function emails this information directly to The MathWorks. Otherwise, it creates a file you can mail or fax to The MathWorks.

See Also `info`

subspace

Purpose Angle between two subspaces.

Synopsis `t = subspace(A,B)`

Description `t = subspace(A,B)` finds the angle between two subspaces specified by the columns of `A` and `B`. If `A` and `B` are vectors of unit length, this is the same as `acos(A'*B)`.

Examples If the angle between the two subspaces is small, the two spaces are nearly linearly dependent. In a physical experiment described by some observations `A`, and a second realization of the experiment described by `B`, `subspace(A,B)` gives a measure of the amount of new information afforded by the second experiment not associated with statistical errors of fluctuations.

For example, consider a study that requires two groups of medical patients, a control group and a test group. Assume that data exists for several groups of patients, and this data is organized into one matrix per group. Each row of a matrix represents various test results for a single patient. The two groups with the smallest returned `angle` are the most similar with regard to these test results.

sum

Purpose Sum of elements.

Synopsis `y = sum(X)`

Description `y = sum(X)`, where `X` is a matrix, returns a row `y` vector with the sum over each column. If `X` is a vector, `y` is the sum of the elements of `X`.

Examples The magic square of order 3 is

```
M = magic(3)
M =
      8     1     6
      3     5     7
      4     9     2
```

This is called a magic square because the sum of the elements in each column is the same.

```
sum(M) =
        15    15    15
```

And the sum of the elements in each row, which can be obtained by transposing twice, is the same.

```
sum(M')' =
        15
        15
        15
```

See Also cumsum, prod, trace

surf, surfc

Purpose 3-D shaded surface plot.

Synopsis

```
surf(X,Y,Z,C)
surf(X,Y,Z)
surf(x,y,Z,C)
surf(x,y,Z)
surf(Z,C)
surf(Z)
h = surf(...)

h = surfc(...)
```

Description In its most general invocation, `surf` takes four matrix input arguments. `surf(X,Y,Z,C)` plots the colored parametric surface specified by `X`, `Y`, and `Z`, with color specified by `C`. In simpler uses, `X` and `Y` may be vectors, or may be omitted, and `C` may be omitted.

The viewpoint is specified by `view`. The axis labels are determined by the range of `X`, `Y`, and `Z`, or by the current setting of `axis`. The color scaling is determined by the range of `C`, or by the current setting of `caxis`. The scaled color values are used as indices into the current colormap.

`surf(X,Y,Z)` uses `C = Z`, so color is proportional to surface height.

`surf(x,y,Z,C)` and `surf(x,y,Z)` with two vector arguments replacing the first two matrix arguments, must have `length(x) = n` and `length(y) = m` where `[m,n] = size(Z)`. In this case, the vertices of the surface patches are the triples `(x(j),y(i),Z(i,j))`. Note that `x` corresponds to the columns of `Z` and `y` corresponds to the rows.

`surf(Z,C)` and `surf(Z)` use `x = 1:n` and `y = 1:m`. In this case, the height, `Z`, is a single-valued function, defined over a geometrically rectangular grid.

`h = surf(...)` returns a handle to a `surface` object. `surface` objects are children of `axes` objects.

`h = surfc(...)` is the same as `surf(...)` except that a contour plot is drawn beneath the surface.

Algorithm Abstractly, a parametric surface is parametrized by two independent variables, `i` and `j`, which vary continuously over a rectangle, for example, `1 ≤ i ≤ m` and `1 ≤ j ≤ n`. The surfaces are specified by three functions, `x(i,j)`, `y(i,j)` and `z(i,j)`. When `i` and `j` are restricted to integer values, they define a rectangular grid with integer grid points. The functions `x(i,j)`, `y(i,j)` and `z(i,j)` become three `m`-by-`n` matrices, `X`, `Y` and `Z`. Surface color is a fourth function, `c(i,j)`, which leads to a fourth matrix, `C`.

Each point in the rectangular grid can be thought of as connected to its four nearest neighbors:

```
         i-1,j
           |
i,j-1 - i,j - i,j+1
           |
         i+1,j
```

This underlying rectangular grid induces four-sided patches on the surface. To express this another way, `[X(:) Y(:) Z(:)]` returns a list of triples specifying points in 3-space. Each interior point is connected to the four neighbors inherited from the matrix indexing. Points on the edge of the surface have three neighbors and the four points at the corners of the grid have only two neighbors. This defines a mesh of quadrilaterals or *quad-mesh* for short.

Surface color can be specified in two different ways: at the vertices or at the centers of each patch. In this general setting, the surface need not be a single valued function of `x` and `y`. Moreover, the four-sided surface patches need not be planar. For example, surfaces defined in polar, cylindrical, and spherical coordinates systems can be represented.

`shading` sets the shading. If the shading is `interp`, then `C` must be the same size as `X`, `Y`, and `Z`; it specifies the colors at the vertices. The color within a patch is a bilinear function of the local coordinates. If the shading is `faceted` (the default) or `flat`, then `C(i,j)` specifies the constant color in the patch:

```
  (i,j)    -    (i,j+1)
    |      C(i,j)    |
 (i+1,j)   -   (i+1,j+1)
```

In this case, `C` can have the same size as `X`, `Y`, and `Z` and its last row and column are ignored, or its row and column dimensions can be one less than those of `X`, `Y`, and `Z`.

Examples Produce a combination surface and contour plot of the peaks surface.

```
[X,Y] = meshgrid(-3:.125:3);
Z = peaks(X,Y);
surfc(X,Y,Z)
axis([-3 3 -3 3 -10 5])
```

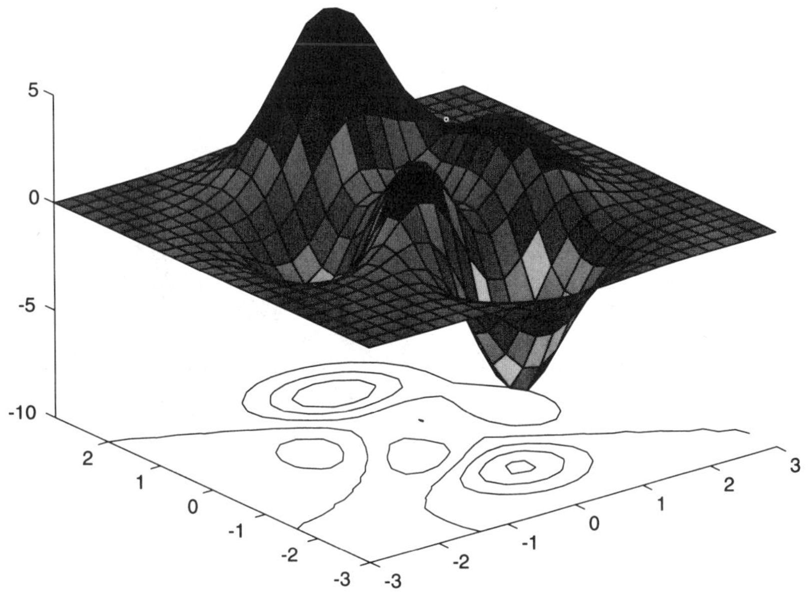

A more complex example colors a sphere with the pattern of +1 and -1s in a Hadamard matrix.

```
k = 5;
n = 2^k-1;
[x,y,z] = sphere(n);
c = hadamard(2^k);
surf(x,y,z,c);
colormap([1  1  0; 0  1  1])
```

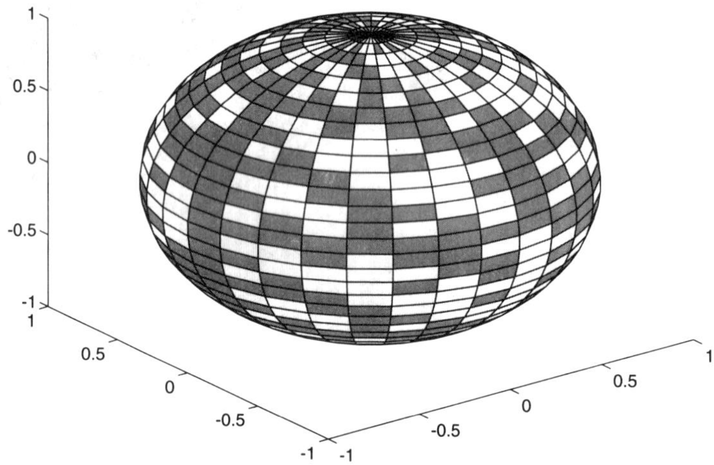

See Also axis, caxis, colormap, contour, mesh, pcolor, shading, view

axes, figure, and surface object properties

---

surface

Purpose Create surface object.

Synopsis

```
surface(X,Y,Z,C)
surface(X,Y,Z)
surface(Z,C)
surface(Z)
surface('PropertyName',PropertyValue,....)
h = surface(...)
```

Description `surface` is a low-level graphics function for creating surface objects. Surface objects are children of axes objects. A surface is a plot of a matrix of data with the element index representing the *x*- and *y*- coordinates and the value of each element representing either a height above a plane or an index into the colormap. `surface` returns the handle of the surface object.

`surface` accepts property name/property value pairs as input arguments. These properties, which control various aspects of the surface object, are described under "Object Properties." You can also set and query property values after the object is created using the `set` and `get` commands.

Unlike high level surface generation functions such as `surf` or `mesh`, `surface` does not clear the axes, set viewing parameters, or perform any actions other than to generate a surface object in the current axes. This is useful if you want to add a surface to existing axes or to tailor the way surfaces are drawn by controlling its properties at creation time. The `axis`, `caxis`, `colormap`, `hold`, `shading`, and `view` commands set graphics properties that affect `surface`.

`surface` also provides convenient forms that allow you to omit the property name for certain properties. For example, the following statements are equivalent:

```
surface('XData',X,'YData',Y,'ZData',Z,'CData',C)
```

```
surface(X,Y,Z,C)
```

`surface(X,Y,Z,C)` plots the parametric surface specified by `X`, `Y` and `Z`, with color specified by `C`. In simpler uses, `X` and `Y` can be vectors or can be omitted, and `C` can be omitted. The `X` and `Y` arguments can be followed by property name/property value pairs to specify additional surface properties.

`surface(X,Y,Z)` uses `C = Z`, so color is proportional to surface height.

`surface(x,y,Z)` and `h = surface(x,y,Z,C)`, with two vector arguments replacing the first two matrix arguments, must have `length(x) = n` and `length(y) = m` where `[m,n] = size(Z)`. In this case, the vertices of the surface patches are the triples `(x(j),y(i),Z(i,j))`. Note that `x` corresponds to the columns of `Z` and `y` corresponds to the rows of `Z`.

`surface(Z)` and `surf(Z,C)` use `x = 1:n` and `y = 1:m`. In this case, the height, `Z`, is a single-valued function, defined over a geometrically rectangular grid.

`surface('`*`PropertyName`*`',PropertyValue,....)` omits the matrix arguments entirely and specifies all values using property name/property value pairs.

`h = surface(...)` returns a handle to a `surface` object.

For a complete discussion of parametric surfaces, see the `surf` reference page.

Object Properties This section lists property names along with the type of values each accepts.

`ButtonDownFcn` Callback string, object selection.

Any legal MATLAB expression, including the name of an M-file or function. When you select the object, the string is passed to the `eval` function to execute the specified function. Initially the empty matrix.

`CData` Color data.

A matrix of values that specifies the color at every point in `ZData`. However, `CData` does not need to be the same size as `ZData`. If it is not, MATLAB treats it as a texture map. In this case, the image contained in `CData` is made to conform the surface defined by `ZData`.

`Children`	Children of surface. Always the empty matrix; surface objects have no children.	
`Clipping`	Clipping mode.	
	`on`	(Default.) Any portion of the surface outside the axes rectangle is not displayed.
	`off`	Surface data is not clipped.
`EdgeColor`	Surface edge color.	
	`ColorSpec`	A three-element RGB vector or one of MATLAB's predefined names, specifying a single color for edges. The default edge color is black. See the `ColorSpec` reference page for more information on specifying color.
	`none`	Edges are not drawn.
	`flat`	Edges are a single color determined by the first `CData` entry for that face.
	`interp`	Edge color is determined by linear interpolation through the values at the vertices.
`EraseMode`	Erase mode. This property controls the technique MATLAB uses to draw and erase surface objects. This property is useful in creating animated sequences, where control of individual object redraw is necessary to improve performance and obtain the desired effect.	
	`normal`	(Default.) Redraws the affected region of the display, performing the three-dimensional analysis necessary to ensure that all objects are rendered correctly. This mode produces the most accurate picture, but is the slowest. The other modes are faster, but do not perform a complete redraw and are therefore less accurate.
	`none`	The surface is not erased when it is moved or destroyed.
	`xor`	The surface is drawn and erased by performing an exclusive OR (XOR) with the color of the screen beneath it. When the surface is erased, it does not damage the objects beneath it. Surface objects are de-

		pendent on the color of the screen beneath them, however, and are correctly colored only when over the figure background color.
	background	The surface is erased by drawing it in the figure's background color. This damages objects that are behind the erased surface, but surface objects are always properly colored.
FaceColor	Surface face color.	
	ColorSpec	A three-element RGB vector or one of MATLAB's predefined names, specifying a single color for faces. See the ColorSpec reference page for more information on specifying color.
	none	Faces are not drawn. You can still draw edges, however.
	flat	(Default.) The first value in CData determines face color.
	interp	Face color is determined by linear interpolation through the mesh points on the surface.
Interruptible	Callback interruptibility.	
	yes	The callback specified by ButtonDownFcn is interruptible by other callbacks.
	no	(Default.) The ButtonDownFcn callback is not interruptible.

LineStyle

Edge line style.

This property determines the type of line for edges. Specify a linestyle to be plotted through the mesh points or a scalable marker type to be placed only at mesh points:

- Line styles: solid (–), dashed (– –), dotted (:), dashdot (–.).
- Marker types: circle (o), plus (+), point (.), star (*), x-mark (x).

The default line style is solid.

LineWidth

Edge line width.

The line width for edges. The default width is 0.5.

`MarkerSize`	Marker scale factor. A scalar specifying the scale factor, in points, for edge line markers. This applies only to the marker types circle, plus, point, star, and x-mark. The default marker size is 6 points.
`MeshStyle`	Draw row/column lines. `both` (Default.) Draw all edges (lines for both rows and columns). `row` Draw row edges only. `column` Draw column edges only.
`Parent`	Handle of the surface's parent object. A read-only property that contains the handle of the surface's parent object. The parent of a surface object is the axes in which it is displayed.
`Type`	Type of object. A read-only string; always `'surface'` for a surface object.
`UserData`	User-specified data. Any matrix you want to associate with the surface object. The object does not use this data, but you can retrieve it using the `get` command.
`Visible`	Surface visibility. `on` (Default.) Surface is visible on the screen. `off` Surface is not drawn.
`XData`	*X*-coordinates of surface points. *X*-position of the surface points. If a row vector, `surface` replicates the row until `XData` has the same number of rows as `ZData`.
`YData`	*Y*-coordinates of surface points. *Y*-position of the surface points. If a column vector, `surface` replicates the column until `YData` has the same number of columns as `ZData`.
`ZData`	*Z*-coordinates of surface points. *Z*-position of the surface points. See the "Description" section for more information.

See Also `mesh`, `pcolor`, `surf`

surfl

Purpose 3-D shaded surface with lighting.

Synopsis

```
surfl(Z)
surfl(x,y,Z)
surfl(Z,s)
h = surfl(x,y,Z,s)
h = surfl(...)
```

Description `surfl` produces a shaded surface plot based on a combination of diffuse, specular and ambient lighting models. View the surfaces with a grayscale or similar colormap (such as `gray`, `copper`, `bone`, `pink`) and with interpolated shading.

`surfl(Z)`, `surfl(x,y,Z)`, `surfl(Z,s)`, and `surfl(x,y,Z,s)` are used the same way as `surf(...)`, but accept an optional trailing argument `s`. Argument `s`, if specified, is a three-element vector `s = [Sx Sy Sz]` that specifies the direction of the light source. `s` can also be specified in spherical coordinates, `s = [azimuth,elevation]`.

`h = surfl(...)` returns a handle to a surface object.

The default value for `s` is 45° counterclockwise from the current view direction. `s` points from the object to the light source.

Examples A view of the `peaks` function, with a light source, is

```
[x,y] = meshgrid(-3:1/8:3);
z = peaks(x,y);
surfl(x,y,z);
shading interp
colormap(gray);
axis([-3  3  -3  3  -8  8])
```

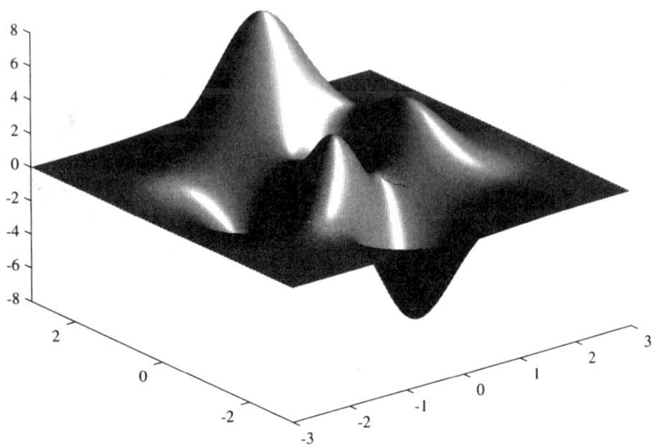

See Also `colormap`, `diffuse`, `shading`, `specular`

surfnorm

Purpose Compute and display 3-D surface normals.

Synopsis

```
[Nx,Ny,Nz] = surfnorm(X,Y,Z)
[Nx,Ny,Nz] = surfnorm(Z)
surfnorm(Z)
surfnorm(X,Y,Z)
```

Description `[Nx,Ny,Nz] = surfnorm(X,Y,Z)` returns the components of the three-dimensional surface normals for the surface defined by the matrices `X`, `Y`, and `Z`. The surface normals are unnormalized and are valid at each vertex. `surfnorm` is used by `surfl` to compute surface normals.

`[Nx,Ny,Nz] = surfnorm(Z)` returns the normal vector components for the surface defined by `Z`.

`surfnorm(Z)` and `surfnorm(X,Y,Z)` plot the surface with the normals emanating from it. Normals are not shown for surface elements that face away from the viewer.

The direction of the normals is reversed by calling `surfnorm` with transposed arguments.

```
surfnorm(X',Y',Z')
```

Algorithm The surface normals are based on a bicubic fit of the data in the `X`, `Y`, and `Z` matrices. For each vertex, diagonal vectors are computed and crossed to form the normal.

Examples Plot the normal vectors for a truncated cone.

```
[x,y,z] = cylinder(1:10);
surfnorm(x,y,z)
```

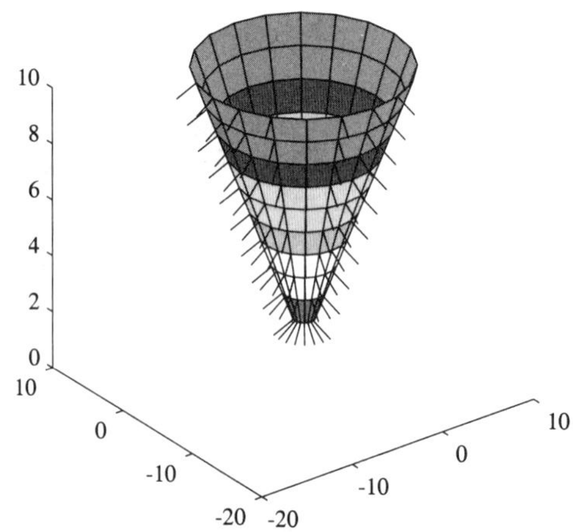

See Also `diffuse`, `specular`, `surfl`

---

svd

Purpose Singular value decomposition.

Synopsis

```
[U,S,V] = svd(X)
s = svd(X)
[U,S,V] = svd(X,0)
```

Description `svd` computes the matrix singular value decomposition.

`[U,S,V] = svd(X)` produces a diagonal matrix `S` of the same dimension as `X`, with nonnegative diagonal elements in decreasing order, and unitary matrices `U` and `V` so that `X = U*S*V'`.

`s = svd(X)` returns a vector containing the singular values.

`[U,S,V] = svd(X,0)` produces the economy size decomposition. If `X` is `m`-by-`n` with `m > n`, then `svd` computes only the first `n` columns of `U` and `S` is `n`-by-`n`.

Examples For the matrix

```
X =
     1     2
     3     4
     5     6
     7     8
```

the statement

```
[U,S,V] = svd(X)
```

produces

```
U =
    0.1525    0.8226   -0.3945   -0.3800
    0.3499    0.4214    0.2428    0.8007
    0.5474    0.0201    0.6979   -0.4614
    0.7448   -0.3812   -0.5462    0.0407
S =
    14.2691         0
          0    0.6268
          0         0
          0         0
V =
    0.6414   -0.7672
    0.7672    0.6414
```

The economy size decomposition generated by

```
[U,S,V] = svd(X,0)
```

produces

```
U =
     0.1525    0.8226
     0.3499    0.4214
     0.5474    0.0201
     0.7448   -0.3812
S =
    14.2691         0
          0    0.6268
V =
     0.6414   -0.7672
     0.7672    0.6414
```

Algorithm svd uses the LINPACK routine ZSVDC.

Diagnostics If the limit of 75 QR step iterations is exhausted while seeking a singular value, this message appears:

```
Solution will not converge.
```

References

[1] Dongarra, J.J., J.R. Bunch, C.B. Moler, and G.W. Stewart, *LINPACK User's Guide*, SIAM, Philadelphia, 1979.

---

symbfact

Purpose Symbolic analysis of sparse Cholesky and LU factorization.

Synopsis

```
count = symbfact(A)
count = symbfact(A,'f')
[count,h,parent,post,R] = symbfact(...)
```

Description `count = symbfact(A)` returns the vector of row counts for the upper triangular Cholesky factor of a symmetric matrix whose upper triangle is that of `A`, assuming no cancellation during the factorization. This function is much faster than `chol(A)`.

`count = symbfact(A,'f')` lets you specify an option, where `f` can be

- `col` to analyze `A'*A` (without forming it explicitly).
- `sym` to obtain the same result as `symbfact(A)`.

`[count,h,parent,post,R] = symbfact(...)` also returns

- the height of the elimination tree, `h`
- the elimination tree itself, `parent`
- a postordering permutation of the elimination tree, `post`
- a 0-1 matrix `R` whose structure is that of `chol(A)`

See Also `chol`, `etree`

symmmd

Purpose Symmetric minimum degree ordering for elimination sparsity.

Synopsis `p = symmmd(S)`

Description `p = symmmd(S)` returns a symmetric minimum degree ordering of `S`. For a symmetric positive definite matrix `S`, this is a permutation `p` such that `S(p,p)` tends to have a sparser Cholesky factor than `S`. Sometimes `symmmd` works well for symmetric indefinite matrices too.

The minimum degree ordering is automatically used by `\` and `/` for the solution of symmetric, positive definite, sparse linear systems.

Some options and parameters associated with heuristics in the algorithm can be changed with `spparms`.

Algorithm The symmetric minimum degree algorithm is based on the column minimum degree algorithm. In fact, `symmmd(A)` just creates a nonzero structure `K` such that `K'*K` has the same nonzero structure as `A` and then calls the column minimum degree code for `K`.

Examples Here is a comparison of reverse Cuthill-McKee and minimum degree on the Bucky ball example mentioned in the `symrcm` reference page.

```
B = bucky+4*speye(60);
r = symrcm(B);
p = symmmd(B);
R = B(r,r);
S = B(p,p);
subplot(2,2,1), spy(R), title('B(r,r)')
subplot(2,2,2), spy(S), title('B(s,s)')
subplot(2,2,3), spy(chol(R)), title('chol(B(r,r))')
subplot(2,2,4), spy(chol(S)), title('chol(B(s,s))')
```

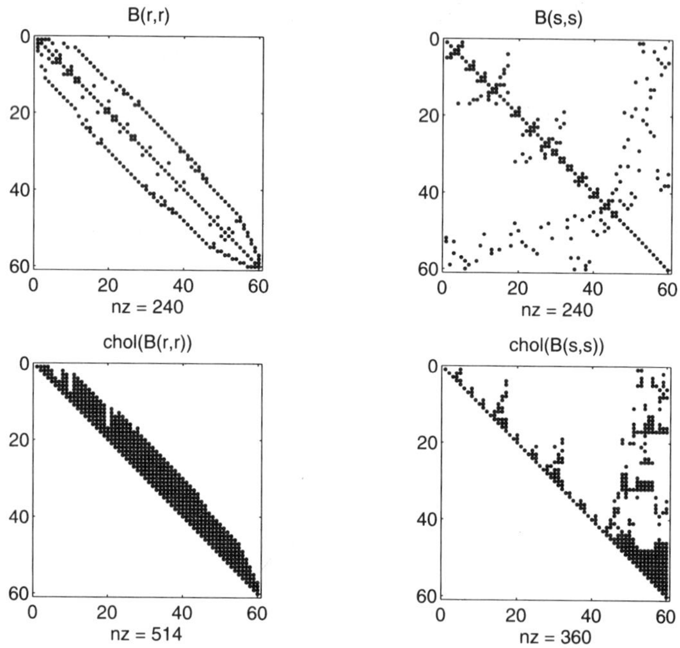

Even though this is a very small problem, the behavior of both orderings is typical. RCM produces a matrix with a narrow bandwidth which fills in almost completely during the Cholesky factorization. Minimum degree produces a structure with large blocks of contiguous zeros which do not fill in during the factorization. Consequently, the minimum degree ordering requires less time and storage for the factorization.

See Also `\`, `chol`, `colmmd`, `colperm`, `spparms`, `symrcm`

References
[1] Gilbert, John R., Cleve Moler, and Robert Schreiber, "Sparse Matrices in MATLAB: Design and Implementation," *SIAM Journal on Matrix Analysis and Applications 13*, pp. 333-356, 1992.

symrcm

Purpose Reverse Cuthill-McKee ordering to reduce profile or bandwidth.

Synopsis

```
r = symrcm(S)
```

Description `r = symrcm(S)` returns the symmetric reverse Cuthill-McKee ordering of `S`. This is a permutation `r` such that `S(r,r)` tends to have its nonzero elements closer to the diagonal. This is a good preordering for LU or Cholesky factorization of matrices that come from long, skinny problems. The ordering works for both symmetric and asymmetric `S`.

For a real, symmetric sparse matrix, `S`, the eigenvalues of `S(r,r)` are the same as those of `S`, but `eig(S(r,r))` probably takes less time to compute than `eig(S)`.

Algorithm The algorithm first finds a *pseudo-peripheral* vertex of the graph of the matrix, and then generates a level structure by breadth-first search and orders the vertices by decreasing distance from the pseudo-peripheral vertex. The implementation is based closely on the SPARSPAK implementation described by George and Liu.

Examples The statement

```
B = bucky
```

uses an M-file in the `demos` toolbox to generate the adjacency graph of a truncated icosahedron. This is better known as a soccer ball, a Buckminster Fuller geodesic dome (hence the name `bucky`), or, more recently, as a 60-atom Carbon molecule. There are 60 vertices. The vertices have been ordered by numbering half of them from one hemisphere, pentagon by pentagon, then reflecting into the other hemisphere and gluing the two halves together. With this numbering, the matrix does not have a particularly narrow bandwidth, as the first spy plot shows

```
subplot(1,2,1), spy(B), title('B')
```

The reverse Cuthill-McKee ordering is obtained with

```
p = symrcm(B);
R = B(p,p);
```

The spy plot shows a much narrower bandwidth:

```
subplot(1,2,2), spy(R), title('B(p,p)')
```

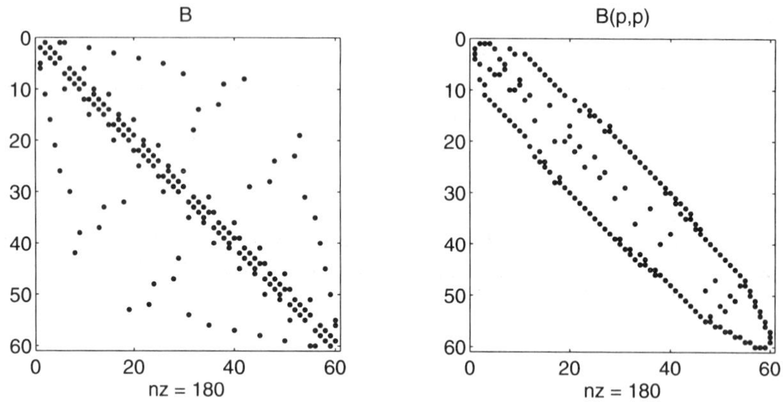

This example is continued in the reference pages for symmmd.

The bandwidth can also be computed with

```
[i,j] = find(B);
bw = max(i-j) + 1
```

The bandwidths of B and R are 35 and 12, respectively.

See Also \, chol, colmmd, colperm, eig, lu

References

[1] George, Alan and Joseph Liu, *Computer Solution of Large Sparse Positive Definite Systems*, Prentice-Hall, 1981.

[2] Gilbert, John R., Cleve Moler, and Robert Schreiber, "Sparse Matrices in MATLAB: Design and Implementation," to appear in *SIAM Journal on Matrix Analysis*, 1992. A slightly expanded version is also available as a technical report from the Xerox Palo Alto Research Center.

tan, tanh

Purpose Tangent and hyperbolic tangent.

Synopsis

```
Y = tan(X)
Y = tanh(X)
```

Description The trigonometric functions operate element-wise on matrices. Their domains and ranges include complex values. All angles are measured in radians.

`tan(X)` is the circular tangent of each element of `X`.

`tanh(X)` is the hyperbolic tangent of each element of `X`.

Examples `tan(pi/2)= 1.6332e+16` is not infinite, but rather a value the size of the reciprocal of the floating-point accuracy `eps`, because `pi` is only a floating-point approximation to the exact value.

Algorithm

$$\tan(z) = \frac{\sin(z)}{\cos(z)}$$

$$\tanh(z) = \frac{\sinh(z)}{\cosh(z)}$$

See Also `acos`, `asin`, `atan`, `cos`, `exp`, `expm`, `funm`, `sin`

tempdir

Purpose Return the name of the system's temporary directory.

Synopsis `tmp_dir = tempdir`

Description `tmp_dir = tempdir` returns the name of the system's temporary directory, if one exists. This function does not create a new directory.

See Also `tempname`

tempname

Purpose Unique name for temporary file.

Synopsis `tempname`

Description `tempname` returns a unique string beginning with the characters `tp`. This string is useful as a name for a temporary file.

terminal

Purpose Set graphics terminal type.

Synopsis

```
terminal
terminal('type')
```

Description `terminal` displays a menu of graphics terminal types and prompts for a choice. It then configures MATLAB appropriately for your terminal type.

`terminal('type')` accepts a terminal type string. Valid *type* strings are

Keystroke	Description
`tek401x`	Tektronix 4010/4014
`tek4100`	Tektronix 4100
`tek4105`	Tektronix 4105
`retro`	Retrographics card
`sg100`	Selanar Graphics 100
`sg200`	Selanar Graphics 200
`vt240tek`	VT240 & VT340 Tektronix mode
`ergo`	Ergo terminal
`graphon`	Graphon terminal
`citoh`	C.Itoh terminal
`xtermtek`	xterm, Tektronix graphics
`wyse`	Wyse WY-99GT
`kermit`	MS-DOS Kermit 2.23
`hp2647`	Hewlett-Packard 2647
`versa`	Macintosh with VersaTerm (Tektronix 4010/4014)
`versa4100`	Macintosh with VersaTerm (Tektronix 4100)
`versa4105`	Color/grayscale Macintosh with VersaTerm (Tektronix 4105)
`hde`	Human designed systems

If necessary, edit `terminal.m` to add any terminal-specific settings (escape characters, line length, and so on).

text

Purpose Add text to current plot by creating text object.

Synopsis

```
text(x,y,'text')
text(x,y,z,'text')
text('PropertyName','PropertyValue'....)
h = text(...)
```

Description `text` is both a high-level graphics command for adding text to a plot, and a low-level graphics function for creating text objects.

Text objects are children of axes. `text` is an object creation function that accepts property name/property value pairs as input arguments. These properties, which control various aspects of the text object, are described under "Object Properties." You can also set and query property values after creating the object using the `set` and `get` commands.

`text(x,y,'text')` adds the string in the quotes to the location identified by the point `(x,y)` to the current axes. Specify the point `(x,y)` in units of the current plot's data.

If x and y are vectors, `text` writes the string at all locations defined by the list of points. h is the handle of the created text object. The x, y pair can be followed by property name/property value (*'PropertyName'*, *'PropertyValue'*) pairs to specify additional text properties.

If the text string is an array that is the same length as x and y, `text` writes the corresponding row of the text array at each point specified by x and y.

`text(x,y,z,'text')` adds text in three-dimensional coordinates. The x, y, z triple for three-dimensional can be followed by property name/property value (*'PropertyName'*, 'PropertyValue') pairs to specify additional text properties.

`text('PropertyName','PropertyValue'....)` omits the x, y, pair (x, y, z triple for three-dimensional) entirely and specifies all properties using property name/property value pairs.

`h = text(..)` returns a handle to a text object.

Object Properties This section lists property names along with the type of values each accepts.

`ButtonDownFcn` — Callback string, object selection.

Any legal MATLAB expression, including the name of an M-file or function. When you select the object, the string is passed to the `eval` function to execute the specified function. Initially the empty matrix.

`Children` — Children of text.

Always the empty matrix; text objects have no children.

`Clipping` — Clipping mode.

`on`	(Default.) Any portion of the text outside the axes rectangle is not displayed.
`off`	Text object is not clipped.

`Color` — Text color.

A three-element RGB vector or one of MATLAB's predefined names, specifying the text color. The default text color is white. See the `ColorSpec` reference page for more information on specifying color.

`EraseMode` — Erase mode.

This property controls the technique MATLAB uses to draw and erase text objects. This property is useful in creating animated sequences, where control of individual object redraw is necessary to improve performance and obtain the desired effect.

`normal`	(Default.) Redraws the affected region of the display, performing the three-dimensional analysis necessary to ensure that all objects are rendered correctly. This mode produces the most accurate picture, but is

	the slowest. The other modes are faster, but do not perform a complete redraw and are therefore less accurate.
`none`	The text is not erased when it is moved or destroyed.
`xor`	The text is drawn and erased by performing an exclusive OR (XOR) with the color of the screen beneath it. When the text is erased, it does not damage the objects beneath it. Text objects are dependent on the color of the screen beneath them, however, and are correctly colored only when over the figure background color.
`background`	Text is erased by drawing it in the figure's background color. This damages objects that are behind the erased text, but text objects are always properly colored.

`Extent` — Text extent rectangle.

A four-element read-only vector, `[left,bottom,width,height]`, that defines the size and position of the text string. `left` and `bottom` are the distance from the lower-left corner of the figure window to the lower-left corner of the rectangle. `width` and `height` are the dimensions of the text rectangle. All measurements are in units specified by the `Units` property.

`FontAngle` — Italics for text object.

`normal`	(Default.) Regular font angle.
`italic`	Italics.
`oblique`	Italics on some systems.

See the *MATLAB Release Notes* for more details on the `FontAngle` property.

`FontName` — Font family.

A string specifying the name of the font to use for the text object. To display and print properly, this must be a font that your system supports. The default font is Helvetica.

`FontSize` — Font point size.

An integer specifying the font size, in points, to use for text. The default point size is 12.

`FontWeight` — Bolding for text object.

`normal`	(Default.) Regular font weight.
`bold`	Bold weight.

HorizontalAlignment	Horizontal text alignment.	
	left	(Default.) Text is left-justified with respect to its Position.
	center	Text is centered with respect to its Position.
	right	Text is right-justified with respect to its Position.
Interruptible	Callback interruptibility.	
	yes	The callback specified by ButtonDownFcn is interruptible by other callbacks.
	no	(Default.) The ButtonDownFcn callback is not interruptible.
Parent	Handle of the text's parent object.	
	A read-only property that contains the handle of the text's parent object. The parent of a text object is the axes in which it is displayed.	
Position	Location of text.	
	A two- or three-element vector, [x y (z)], that specifies the location of the text in three dimensions. If you omit the z value, it defaults to 0. All measurements are in units specified by the Units property. Initial value is [0 0 0].	
Rotation	Text orientation.	
	One of seven predefined orientations: 0, ±90, ±180, or ±270. These seven possibilities define only four different orientations. For example, –90 and +270 produce the same result. On some systems, MATLAB supports arbitrary rotation angles. The default rotation is 0.	
String	Text string.	
	The text string that is displayed.	
Type	Type of graphics object.	
	A read-only string; always 'text' for a text object.	
Units	Unit of measurement for screen display.	
	data	(Default.) Data units of the parent axes.
	pixels	Screen pixels.
	normalized	Normalized coordinates, where the lower-left corner of the axes maps to (0,0) and the upper-right corner to (1.0,1.0).
	inches	Inches.
	cent	Centimeters.

`points`	Points. Each point is equivalent to 1/72 of an inch.

This property affects the `Extent` and `Position` properties. All units are measured from the lower-left corner of the window. If you change the value of `Units`, it is good practice to return it to its default value after completing your computation so as not to affect other functions that assume `Units` is set to the default value.

`UserData` — User-specified data.

Any matrix you want to associate with the text object. The object does not use this data, but you can retrieve it using the `get` command.

`VerticalAlignment` — Vertical text alignment.

`top`	String is placed at the top of the specified *y*-position.
`cap`	Font's capital letter height is placed at the specified *y*-position.
`middle`	(Default.) String is placed at the middle of the specified *y*-position.
`baseline`	Font's baseline is placed at the specified *y*-position.
`bottom`	String is placed at the bottom of the specified *y*-position.

`Visible` — Text visibility.

`on`	(Default.) Text is visible on the screen.
`off`	Text is not visible on the screen.

Examples The statements

```
plot([1 5 10],[1 10 20],'x')
text(5,10,' Action point')
```

annotate the point at `(5,10)` with the string `Action point`, while

```
plot(x1,y1,x2,y2)
text(x1,y1,'1'), text(x2,y2,'2')
```

marks two curves so you can easily distinguish them.

See Also `gtext`, `int2str`, `num2str`, `plot`, `title`, `xlabel`

---

tic, toc

Purpose Stopwatch timer.

Synopsis `tic`

```
    any statements
toc
t = toc
```

Description `tic` starts a stopwatch timer.

`toc`, by itself, prints the elapsed time since `tic` was used.

`t = toc` returns the elapsed time in `t`.

Examples This example measures how the time required to solve a linear system varies with the order of a matrix.

```
for n = 1:100
    A = rand(n,n);
    b = rand(n,1);
    tic
    x = A\b;
    t(n) = toc;
end
plot(t)
```

See Also `clock`, `cputime`, `etime`

---

title

Purpose Graph title.

Synopsis `title('text')`

Description `title('text')` writes the text as a title at the top of the current plot.

Examples The statement

```
title(date)
```

puts today's date on the top of the current plot.

Two examples of including a variable's value in a title are

```
f = 70; c = (f-32)/1.8;
title(['Temperature is ',num2str(c),' degrees C'])
n = 3
title(['Case number #',int2str(n)])
```

Algorithm `title` sets the `Title` property of the current `axes` object to a new `text` object.

See Also `gtext`, `int2str`, `num2str`, `plot`, `text`, `xlabel`, `ylabel`, `zlabel`

---

toeplitz

Purpose Toeplitz matrix.

Synopsis

```
T = toeplitz(c,r)
T = toeplitz(r)
```

Description A *Toeplitz* matrix is defined by one row and one column. A *symmetric Toeplitz* matrix is defined by just one row. `toeplitz` generates Toeplitz matrices given just the row or row and column description.

`T = toeplitz(c,r)` returns a nonsymmetric Toeplitz matrix `T` having `c` as its first column and `r` as its first row. If the first elements of `c` and `r` are different, a message is printed and the column wins the disagreement.

`T = toeplitz(r)` returns the symmetric or Hermitian Toeplitz matrix formed from vector `r`, where `r` defines the first row of the matrix.

Examples A Toeplitz matrix with diagonal disagreement is

```
c = [1  2  3  4  5];
r = [1.5  2.5  3.5  4.5  5.5];
toeplitz(c,r)
Column wins diagonal conflict:
ans =
     1.000     2.500     3.500     4.500     5.500
     2.000     1.000     2.500     3.500     4.500
     3.000     2.000     1.000     2.500     3.500
     4.000     3.000     2.000     1.000     2.500
     5.000     4.000     3.000     2.000     1.000
```

See Also `hankel`, `vander`

---

trace

Purpose Trace of a matrix.

Synopsis `t = trace(A)`

Description `trace(A)` is the sum of the diagonal elements of the matrix `A`.

Algorithm `trace` is a single-statement M-file.

```
t = sum(diag(A));
```

See Also `det`, `eig`

---

trapz

Purpose Trapezoidal numerical integration.

Synopsis

```
z = trapz(x,y)
z = trapz(y)
```

Description `z = trapz(x,y)` computes the integral of `y` with respect to `x` using trapezoidal integration. `x` and `y` must be vectors of the same length, or `x` must be a column vector and `y` a matrix with as many rows as `x`. `trapz` computes the integral of each column of `y` separately. The resulting `z` is a scalar or a row vector.

`z = trapz(y)` computes the trapezoidal integral of `y` assuming unit spacing between the data points. To compute the integral for spacing other than one, multiply `z` by the spacing increment.

Examples The exact value of

$$\int_0^{\pi} \sin(x)\ dx$$

is 2. To approximate this numerically on a uniformly spaced grid, use

```
x = 0:pi/100:pi;
y = sin(x);
```

Then both

```
z = trapz(x,y)
```

and

```
z = pi/100*trapz(y)
```

produce

```
z =
    1.9998
```

A nonuniformly spaced example is generated by

```
x = sort(rand(1,101)*pi);
y = sin(x);
z = trapz(x,y);
```

The result is not as accurate as the uniformly spaced grid. One random sample produced

```
z =
    1.9984
```

See Also `cum`, `quad`, `quad8`, `sum`

---

tril

Purpose Lower triangular part of a matrix.

Synopsis

```
L = tril(X)
L = tril(X,k)
```

Description `tril(X)` is the lower triangular part of `X`.

`tril(X,k)` is the elements on and below the `k`-th diagonal of `X`. `k = 0` is the main diagonal, `k > 0` is above the main diagonal, and `k < 0` is below the main diagonal.

Examples `tril(ones(4,4),-1)` is

```
0  0  0  0
1  0  0  0
1  1  0  0
1  1  1  0
```

See Also `diag`, `triu`

triu

Purpose Upper triangular part of a matrix.

Synopsis

```
u = triu(X)
u = triu(X,k)
```

Description `triu(X)` is the upper triangular part of `X`.

`triu(X,k)` is the element on and above the `k`-th diagonal of `X`. `k = 0` is the main diagonal, `k > 0` is above the main diagonal, and `k < 0` is below the main diagonal.

Examples `triu(ones(4,4),-1)` is

```
1  1  1  1
1  1  1  1
0  1  1  1
0  0  1  1
```

See Also `diag`, `tril`

type

Purpose List file.

Synopsis `type` *`filename`*

Description `type` *`filename`* lists the specified file. Use pathnames and drive designators in the usual way for your operating system.

The `type` command differs from the `type` commands provided by operating systems in an important way: if a filename extension is not given, `.m` is added by default. `type` also checks the directories specified in MATLAB's search path. This makes it convenient for the most frequent use of `type`, which is to list the contents of M-files on the screen.

Examples `type foo.bar` lists the file `foo.bar`.

`type foo` lists the file `foo.m`.

See Also `!`, `cd`, `dbtype`, `delete`, `dir`, `path`, `what`, `who`

uicontrol

Purpose Create user interface control object.

Synopsis

```
uicontrol('PropertyName',PropertyValue,...)
h = uicontrol(...)
```

Description

`uicontrol('PropertyName',PropertyValue,...)` is a uicontrol object creation function that accepts property name/property value pairs as input arguments. These properties are described under "Object Properties." You can also set and query property values after creating the object by using the `set` and `get` functions.

uicontrols, or user interface controls, are useful in implementing graphical user interfaces. When selected, most uicontrol objects perform a predefined action. MATLAB supports seven styles of uicontrols, each of which is suited for a different purpose:

- Push buttons
- Check boxes
- Pop-up menus
- Radio buttons
- Sliders
- Editable text
- Frames

Push buttons are analogous to the buttons on a telephone – they generate an action with each press, but do not remain in a pressed state. To activate a push button, press and release the mouse button on the object. Push buttons are useful when the action you want to perform is not related to any other action executable by the user interface (for example, a **Quit** button).

Check boxes also generate an action when pressed, but remain in a pressed state until pressed a second time. These objects are useful when providing the user with a number of independent choices, each toggling between two states. To activate a check box, press and release the mouse button on the object. The state of the objects is indicated on the display.

Popup menus display a list of choices when you press them. When not activated, they display a single button with text indicating their current setting. Popup menus are useful when you want to provide users with a number of mutually exclusive choices, but do not want to take up the amount of space that a series of radio buttons require.

Radio buttons are similar to check boxes, but are intended to be mutually exclusive (i.e., only one is in a pressed state at any given time). To activate a radio button, press and release the mouse button on the object. The state of the objects is indicated on the display. Note that your code must implement the mutually exclusive behavior of radio buttons.

Sliders are intended to accept numeric input by allowing the user to move a sliding bar (by pressing the mouse button and dragging the mouse over the bar) within a rectangle. The location of the bar indicates a numeric value that is selected by releasing the mouse button. You can set the minimum, maximum, and initial values of the slider.

Editable text are boxes containing editable text. After typing in the desired text, press **Control-Return** or move the pointer off the object to execute its `Callback`. Use editable text when you want to input text.

Frames are boxes that enclose regions of a figure window. Frames can make a user interface easier to understand by grouping related controls.

These objects are children of figures and are therefore independent of axes.

`h = uicontrol(...)` returns the handle of a uicontrol object.

Object Properties This section lists property names along with the type of values each accepts.

`BackGroundColor` — Object color.

A three-element RGB vector or one of MATLAB's predefined names, specifying the color used to fill the rectangle defined by the control. The default color is light gray. See the `ColorSpec` reference page for more information on specifying color.

`ButtonDownFcn` — Callback string, object selection.

Any legal MATLAB expression, including the name of an M-file or function. When you select the object, the string is passed to the `eval` function to execute the specified function. Initially the empty matrix.

`CallBack` — Control action.

Any legal MATLAB expression, including the name of an M-file or function. When you activate the uicontrol object, the string is passed to the `eval` function to execute the `Callback`.

`Children` — Children of uicontrol.

Always the empty matrix; uicontrol objects have no children.

`Clipping` — Clipping mode.

`on`	(Default.) Any portion of the control outside the axes rectangle is not displayed.
`off`	The control is not clipped.

`ForeGroundColor` — Text color.

A three-element RGB vector or one of MATLAB's predefined names, specifying the color of the text on the uicontrol. The default text color is black. See the `ColorSpec` reference page for more information on specifying color.

`HorizontalAlignment` — Horizontal alignment of label string.

`left`	Text is left-justified with respect to the uicontrol.

	`center`	(Default.) Text is centered with respect to the uicontrol.
	`right`	Text is right-justified with respect to the uicontrol.
`Interruptible`	Callback interruptibility.	
	`yes`	The control callbacks (`ButtonDownFcn` and `Callback`) are interruptible by other callbacks.
	`no`	(Default.) The control callbacks are not interruptible.

`Max` — Maximum value.

The largest value allowed for the `Value` property. For radio buttons and check boxes, which operate as on/off switches, this represents the setting of `Value` while the uicontrol is in the `on` position. For popup menus, this property specifies the maximum number of choices you can define on the menu. For sliders, this property is the largest value you can select. The default maximum is 1.

`Min` — Minimum value.

The smallest value allowed for the `Value` property. For radio buttons and check boxes, which operate as on/off switches, this represents the setting of `Value` while the uicontrol is in the `off` position. For sliders, this property is the minimum value that you can select. The default minimum is 0.

`Parent` — Handle of the uicontrol's parent object.

A read-only property that contains the handle of the uicontrol's parent object. The parent of a uicontrol object is the figure in which it is displayed.

`Position` — Size and location of uicontrol.

A four-element read-only vector, `[left,bottom,width,height]`, that defines the size and position of the uicontrol. `left` and `bottom` are the distance from the lower-left corner of the figure window to the lower-left corner of the uicontrol object. `width` and `height` are the dimensions of the control rectangle. All measurements are in units specified by the `Units` property.

`String` — uicontrol label.

A string specifying the label on push buttons, radio buttons, check boxes, and popup menus. For multiple items on a popup menu or multiple lines on an editable text object, enter a string, then `|`, then the next string, and so on. Place quotes around the entire list of strings (not each individual string). For editable text, this property is set to the string typed in by the user.

`Style` Type of uicontrol object.

`pushbutton`	(Default.) Push button.
`radio`	Radio button.
`checkbox`	Check box.
`slider`	Slider.
`edit`	Editable text.
`popup`	Popup menu.

`Type` Type of graphics object.

A read-only string; always `'uicontrol'` for a uicontrol object.

`Units` Unit of measurement for screen display.

`pixels`	(Default.) Screen pixels.
`normalized`	Normalized coordinates, where the lower-left corner of the figure maps to (0,0) and the upper-right corner to (1.0,1.0).
`inches`	Inches.
`cent`	Centimeters.
`points`	Points. Each point is equivalent to 1/72 of an inch.

This property affects the `Position` properties. All units are measured from the lower-left corner of the figure window. If you change the value of `Units`, it is good practice to return it to its default value after completing your computation so as not to affect other functions that assume `Units` is set to the default value.

`UserData` User-specified data.

Any matrix you want to associate with the uicontrol object. The object does not use this data, but you can retrieve it using the `get` command.

`Value` Current value of control.

Scalar. Legal values depend on the type of control:

- Radio buttons and check boxes set `Value` to `Max` (usually 1) when they are on (when the button is pressed) and `Min` (usually 0) when off (not pressed)
- Sliders set `Value` to the number indicated by the slider bar, relative to the range established by `Min` and `Max`.

- Popup menus set `Value` to the index of the item selected. The "Examples" section shows how to use the `Value` property to determine which item has been selected.
- Push buttons and editable text objects do not set this property.

Set the `Value` property either interactively with the mouse or through a call to the `set` function. The display reflects changes made to `Value`.

`Visible` uicontrol visibility.

`on`	(Default.) uicontrol is visible on the screen.
`off`	uicontrol is not visible on the screen.

Examples The following statement creates a push button that clears the current axes when pressed:

```
h = uicontrol('Style','Pushbutton','Position',...
    [20 150 100 70], 'Callback','cla','String','Clear');
```

You can create a uicontrol object that changes figure colormaps by specifying a popup menu and supplying an M-file as the object's `Callback`:

```
hpop = uicontrol('Style','Popup','String',...
       'hsv| hot|cool|gray','Position',[20 320 100 50],...
       'Callback','setmap')
```

This call to `uicontrol` defines four individual choices in the menu: `hsv`, `hot`, `cool`, and `gray`. You specify these choices with the `String` property, separating each with the "`|`" character.

The `Callback`, in this case `setmap`, is the name of an M-file that defines a more complicated set of instructions than a single MATLAB command:

```
val = get(hpop,'Value');
if val == 1
    colormap(hsv)
elseif val == 2
    colormap(hot)
elseif val == 3
    colormap(cool)
elseif val == 4
    colormap(gray)
end
```

The `Value` property contains a number that indicates which choice you selected. The choices are numbered sequentially from one to four. The `setmap` M-file can get and then test the contents of the `Value` property to determine what action to take.

See Also `uimenu`

uigetfile

Purpose Retrieve name of a file to open through a dialog box.

Synopsis `[filename,pathname] = uigetfile('filterSpec','boxTitle',x,y)`

Description `[filename,pathname] = uigetfile('filterSpec','boxTitle',x,y)` displays a dialog box in which you can fill in or select the filename and path strings, where

- *`filterSpec`* filters the types of files initially displayed in the dialog box. For example, `'*.m'` lists all MATLAB M-files within a selected directory.
- *`boxTitle`* is a string containing the title of the dialog box.
- `x` is the *x*-position, in pixels, of the lower-left corner of the dialog box (not supported on all systems).
- `y` is the *y*-position, in pixels, of the lower-left corner of the dialog box (not supported on all systems).

All parameters are optional, but if one is present, all preceding parameters must also be present. A successful return occurs only if the file exists. If you select a file that does not exist, the function displays an error message and control returns to the dialog box. You can then enter another filename, or press the **Cancel** button.

`filename` is a string containing the name of the file selected in the dialog box. If you press the **Cancel** button or if any error occurs, `filename` is set to 0.

`pathname` is a string containing the path of the file selected in the dialog box. If you press the **Cancel** button or if any error occurs, `pathname` is set to 0.

Examples An example of `uigetfile`'s operation on one platform is

```
[fname,pname] = uigetfile('*.m','Example Dialog Box')
```

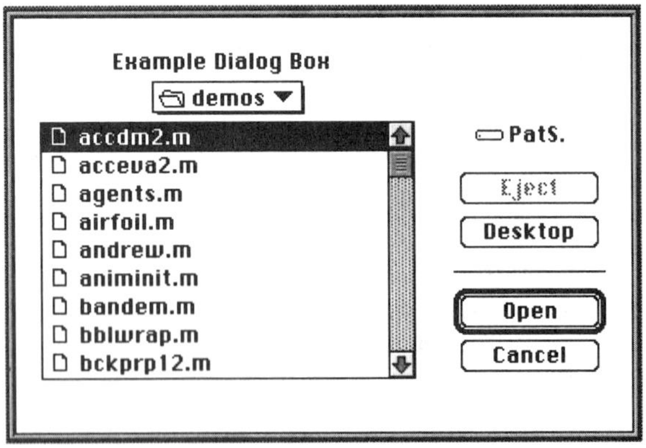


```
fname =
accdm2.m
pname =
Pats:Student MATLAB:Toolbox:matlab:demos:
```

The exact appearance of the dialog box depends on your windowing system.

See Also `uiputfile`

uimenu

Purpose Create menu objects.

Synopsis

```
h = uimenu('PropertyName',PropertyValue,...)
hsub = uimenu(h,'PropertyName',PropertyValue,...)
```

Description `uimenu` is a menu object creation function that accepts property name/property value pairs as input arguments. These properties are described under "Object Properties." You can also set and query property values after creating the object by using the `set` and `get` functions.

`h = uimenu('PropertyName',PropertyValue,...)` creates menus that can perform a predefined action. These menus display along the top of the figure on a menu bar. Each menu choice can itself be a menu that displays its submenu when selected. `h` is the handle of the created menu object.

`hsub = uimenu(h,'PropertyName',PropertyValue,...)` creates a submenu of the parent menu specified by the handle `h`. If you do not specify the handle of an existing menu, the current figure becomes the parent and the menu is a top-level menu.

Object Properties This section lists property names along with the type of values each accepts.

`Accelerator` Keyboard equivalent.

A character specifying the keyboard equivalent for the menu item. This allows users to select a particular menu choice by pressing the specified character in conjunction with another key, instead of selecting the menu item with the mouse. The key sequence is platform-specific:

- For X-Windows and PC based systems, the sequence is **Control**-`Accelerator`.
- For Macintosh systems, the sequence is **Command**-`Accelerator`.

The window focus must be in the figure window when the key sequence is entered.

`BackGroundColor` Object color.

A three-element RGB vector or one of MATLAB's predefined names, specifying the color used to fill the rectangle defined by the menu. The default color is light gray. See the `ColorSpec` reference page for more information on specifying color.

`ButtonDownFcn`	Callback string, object selection.	
	Any legal MATLAB expression, including the name of an M-file or function. When you select the object, the string is passed to the `eval` function to execute the specified function. Initially the empty matrix.	
`CallBack`	Menu action.	
	Any legal MATLAB expression, including the name of an M-file or function. When you click on the menu, the string is passed to the `eval` function to execute the `Callback`.	
`Checked`	Check mark for selected items.	
	`on`	Check mark appears next to item, indicating it is selected.
	`off`	(Default.) Check mark is not displayed.
`Children`	Children of uimenu.	
	A read-only vector containing the handles of all children of the uimenu object. The children objects of uimenus are other uimenus, which function as submenus.	
`Clipping`	Clipping mode.	
	`on`	(Default.) Any portion of the menu outside the axes rectangle is not displayed.
	`off`	The menu is not clipped.
`Enable`	Menu enable mode.	
	`on`	(Default.) Menu item label appears dimmed, indicating you cannot select it.
	`off`	Menu label does not appear dimmed.
`ForeGroundColor`	Text color.	
	A three-element RGB vector or one of MATLAB's predefined names, specifying the color of the text on the menu. The default text color is black. See the `ColorSpec` reference page for more information on specifying color.	
`Interruptible`	Callback interruptibility.	
	`yes`	The menu callbacks (`ButtonDownFcn` and `Callback`) are interruptible by other callbacks.
	`no`	(Default.) The menu callbacks are not interruptible.
`Label`	Menu label.	
	A string specifying the text label on the menu items.	

`Parent`	Handle of the uimenu's parent object. A read-only property that contains the handle of the uimenu's parent object. The parent of a uimenu is the figure in which it is displayed, or the uimenu of which it is a submenu.
`Position`	Relative menu position. Scalar indicating placement on the menu bar or within a menu. Top-level menus are placed from left to right on the menu bar according to the value of their `Position` property, with 1 representing the left-most position. The individual items within a given menu are placed from top to bottom according to the value of their `Position` property, with 1 representing the top-most position.
`Separator`	Separator line mode. `on` — A dividing line is drawn above the menu item. `off` — (Default.) No dividing line is drawn.
`Type`	Type of graphics object. A read-only string; always `'uimenu'` for a uimenu object.
`UserData`	User-specified data. Any matrix you want to associate with the uimenu object. The object does not use this data, but you can retrieve it using the `get` command.
`Visible`	uimenu visibility. `on` — (Default.) uimenu is visible on the screen. `off` — uimenu is not visible on the screen.

Examples This example creates a menu labeled **Workspace** whose choices allow users to create a new figure window, save the variables in the workspace, and exit out of MATLAB.

```
f = uimenu('Label','Workspace');
    uimenu(f,'Label','Save','Callback','save');
    uimenu(f,'Label','New Figure','Callback','figure');
    uimenu(f,'Label','Save','Callback','save');
    uimenu(f,'Label','Save','Callback','save');
           'Separator','on', 'Accelerator', 'Q');
```

See Also `uicontrol`

uiputfile

Purpose Retrieve name of a file to write to through a dialog box.

Synopsis `[filename,pathname] = uiputfile('`*`initFile`*`','`*`boxTitle`*`',x,y)`

Description `[filename,pathname] = uiputfile('initFile','boxTitle',x,y)` displays a dialog box in which you can fill in or select the filename and path strings, where

- *`initFile`* determines the initial display of files in the dialog box. This string can be a full filename, or it may include wildcards. For example, `'newfile.m'` initializes the text box to that filename, providing a default. A wildcard specification such as `'*.m'` does not provide a default, but the scroll box lists all files with the "`.m`" extension.
- *`boxTitle`* is a string containing the title of the dialog box.
- `x` is the *x*-position, in pixels, of the lower-left corner of the dialog box (not supported on all systems).
- `y` is the *y*-position, in pixels, of the lower-left corner of the dialog box (not supported on all systems).

All parameters are optional, but if one is present, all preceding parameters must also be present. If you select a file that exists, a prompt is issued asking whether you want to delete the file. A **Yes** response successfully returns but does not delete the file (which is the responsibility of the calling routines). A **No** response returns control back to the dialog box. You can then enter another filename or press the **Cancel** button.

`filename` is a string containing the name of the file selected in the dialog box. If you press the **Cancel** button or if any error occurs, `filename` is set to 0.

`pathname` is a string containing the path of the file selected in the dialog box. If you press the **Cancel** button or if any error occurs, `pathname` is set to 0.

Examples An example of `uiputfile`'s operation on one platform is

```
[fname,pname] = uiputfile('animinit.m','Example Dialog Box')
```

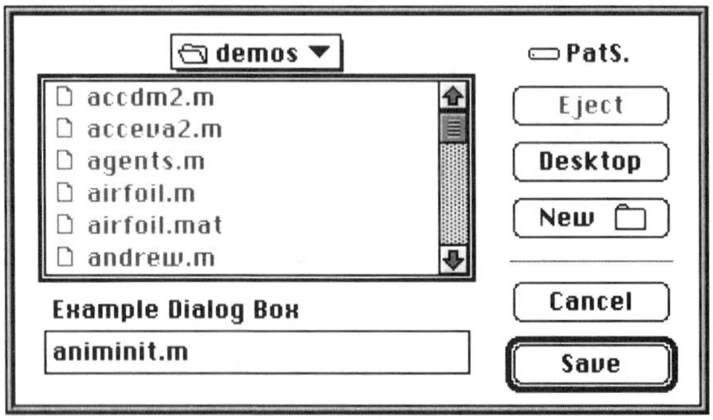


```
fname =
animinit.m
pname =
/amnt/u/home2/pat
```

The exact appearance of the dialog box depends on your windowing system.

See Also `uigetfile`

unix

Purpose Execute a UNIX command from MATLAB.

Synopsis

```
unix('command')
[status,output] = unix('command')
```

Description `unix('command')` executes a UNIX operating system command. The statement returns the resulting status as an integer and the standard output in a string variable.

Examples `[s,w] = unix('who')` returns `s = 0` and, in `w`, a MATLAB string containing a list of users currently logged in.

```
s =
    0
w =
gulley   console Apr 13 09:32
joe      ttyp0   Apr 13 09:02
sully    ttyp2   Apr 13 09:37
gulley   ttyp2   Apr 13 09:37
gulley   ttyp3   Apr 13 09:34
marianne ttyp4   Apr 13 10:43
```

If the UNIX command fails, MATLAB returns a nonzero status value. For example,

```
[s,w] = unix('why')
why: Command not found.
s =
    256
w =
    []
```

See Also `!`

unwrap

Purpose Correct phase angles.

Synopsis

```
Q = unwrap(P,cutoff)
Q = unwrap(P)
```

Description `Q = unwrap(P,cutoff)` corrects the phase angles in vector or matrix `P` by adding multiples of $\pm 2\pi$, when absolute jumps are greater than `cutoff`, to smooth the transitions across branch cuts. When `P` is a matrix, the phase angles are corrected down each column. The phase must be in radians.

`Q = unwrap(P)` uses the default value for `cutoff`, π.

Limitations `unwrap` does the best it can to detect branch cut crossings, but it can be fooled by sparse, rapidly changing phase values.

See Also `abs`, `angle`

upper

Purpose Convert string to uppercase.

Synopsis `t = upper('str')`

Description `t = upper('str')` converts any lowercase characters in the string *str* to the corresponding uppercase characters and leaves all other characters unchanged.

Examples `upper('attention!')` is `'ATTENTION!'`.

Algorithm Any values in the range `'a':'z'` are incremented by `'A'-'a'`.

See Also `isstr`, `lower`, `strcmp`

vander

Purpose Vandermonde matrix.

Synopsis `V = vander(x)`

Description `V = vander(x)` returns the Vandermonde matrix whose columns are powers of the elements of `x`. If `n = length(x)`, then the matrix is `n`-by-`n`. The elements of a Vandermonde matrix are

$$v_{i,j} = x_i^{n-j}$$

Vandermonde matrices arise in connection with fitting polynomials to data.

Examples Create a small Vandermonde matrix from a vector `x`:

```
x =
     2     3     4     5
V = vander(x)
V =
     8     4     2     1
    27     9     3     1
    64    16     4     1
   125    25     5     1
```

See Also `hankel`, `polyfit`, `toeplitz`

ver

Purpose MATLAB and toolbox versions.

Synopsis

```
ver
ver toolboxpath
```

Description `ver` displays the current versions numbers for MATLAB and all installed toolboxes.

`ver` *`toolboxpath`* displays the current version number for the specified toolbox.

See Also `version`

version

Purpose MATLAB version number.

Synopsis

```
v = version
[v,d] = version
```

Description `v = version` returns a string `v` containing the MATLAB version number.

`[v,d] = version` also returns a string `d` containing the date of the version.

See Also help, ver,Readme

view

Purpose 3-D graph viewpoint specification.

Synopsis

```
view(az,el)
view([az,el])
view([x,y,z])
view(2)
view(3)
view(T)

[az,el] = view;
T = view
```

Description `view(az,el)` and `view([az,el])` set the viewing angle for a three-dimensional plot. `az` is the azimuth or horizontal rotation and `el` is the vertical elevation (both in degrees). Azimuth revolves about the z-axis, with positive values indicating counter-clockwise rotation of the viewpoint. Positive values of elevation correspond to moving above the object; negative values move below.

`view([x,y,z])` sets the viewing angle in cartesian coordinates. The magnitude of `(x,y,z)` is ignored.

`view(2)` sets the default two-dimensional view, `az = 0, el = 90`.

`view(3)` sets the default three-dimensional view, `az = -37.5, el = 30`.

`view(T)` accepts a 4-by-4 transformation matrix, such as the perspective transformations generated by `viewmtx`.

`[az,el] = view` returns the current azimuth and elevation.

`T = view` returns the current 4-by-4 transformation matrix.

Examples `el = 90` is directly overhead.

`az = el = 0` looks directly up the first column of the matrix.

`az = 180` is behind the matrix.

`az = -37.5, el = 30` is the default for three-dimensional.

See Also `viewmtx`

axes object properties `View` and `XForm`

viewmtx

Purpose View transformation matrices.

Synopsis

```
T = viewmtx(az,el)
T = viewmtx(az,el,phi)
T = viewmtx(az,el,phi,xc)
```

Description `viewmtx` computes 4-by-4 orthographic or perspective transformation matrices that project four-dimensional homogeneous vectors onto a two-dimensional view surface (for example, your computer screen). A four-dimensional homogenous vector is formed by appending a 1 to the corresponding three-dimensional vector. For instance the four-dimensional vector corresponding to the three-dimensional point `[x,y,z]'` is `[x,y,z,1]'`. The `x` and `y` components of the projected vector are the desired two-dimensional components (see "Examples" below).

`T = viewmtx(az,el)` returns an *orthographic* transformation matrix corresponding to azimuth `az` and elevation `el`. `viewmtx` uses the same definition for azimuth and elevation as `view`, in particular, `az` and `el` are specified in degrees. `T = viewmtx(az,el)` returns the same matrix as the commands

```
view(az,el)
T = view
```

but doesn't change the current view.

`T = viewmtx(az,el,phi)` returns a *perspective* transformation matrix. `phi` is the subtended view angle of the normalized plot cube (in degrees) and controls the amount of perspective distortion:

Phi	Description
0 degrees	Orthographic projection
10 degrees	Similar to telephoto lens
25 degrees	Similar to normal lens
60 degrees	Similar to wide angle lens

The matrix returned can be used to set the view transformation with `view(T)`. The 4-by-4 perspective transformation matrix transforms four-dimensional homogeneous vectors into unnormalized vectors of the form (x,y,z,w) where w is not equal to 1. The x- and y-components of the normalized vector $(x/w, y/w, z/w, 1)$ are the desired two-dimensional components (see example below).

`T = viewmtx(az,el,phi,xc)` returns the perspective transformation matrix using `xc` as the target point within the normalized plot cube, (that is the camera is looking at the point `xc`). The coordinate center is specified as a three-element vector, `xc = [xc,yc,zc]`. Valid values for the components of `xc` are in the interval [0,1]. The default value is `xc = [0,0,0]`.

Examples Determine the projected two-dimensional vector corresponding to the three-dimensional point (.5,0,-3) using the default view direction.

```
A = viewmtx(-37.5,30);
x4d = [.5  0  -3  1]';
x2d = A*x4d;
x2d = x2d(1:2)
x2d =
      0.3967
     -2.4459
```

Vectors that trace the edges of a unit cube are

```
x = [0  1  1  0  0  0  1  1  0  0  1  1  1  1  0  0];
y = [0  0  1  1  0  0  0  1  1  0  0  0  1  1  1  1];
z = [0  0  0  0  0  1  1  1  1  1  1  0  0  1  1  0];
```

Transform the points in these vectors to the screen, and then plot the object.

```
A = viewmtx(-37.5,30);
[m,n] = size(x);
x4d = [x(:),y(:),z(:),ones(m*n,1)]';
x2d = A*x4d;
x2 = zeros(m,n); y2 = zeros(m,n);
x2(:) = x2d(1,:);
y2(:) = x2d(2,:);
plot(x2,y2)
```

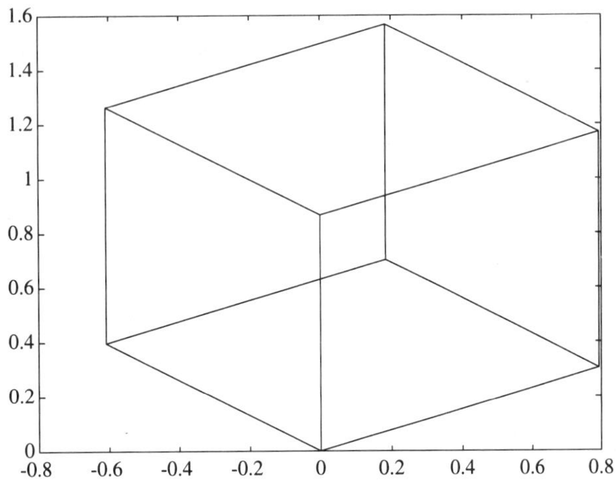

Now repeat the above, but use a perspective transformation.

```
A = viewmtx(-37.5,30,25);
x4d = [.5  0  -3  1]';
x2d = A*x4d;
x2d = x2d(1:2)/x2d(4);    % Normalize
x2d =
     0.1777
    -1.8858
```

And again transform the cube vectors to the screen, and plot the object.

```
A = viewmtx(-37.5,30,2);
[m,n] = size(x);
x4d = [x(:),y(:),z(:),ones(m*n,1)]';
x2d = A*x4d;
x2 = zeros(m,n); y2 = zeros(m,n);
x2(:) = x2d(1,:)./x2d(4,:);
y2(:) = x2d(2,:)./x2d(4,:);
plot(x2,y2)
```

See Also `view`

---

waitforbuttonpress

Purpose Wait for key or mouse button press over a figure.

Synopsis `k = waitforbuttonpress`

Description `k = waitforbuttonpress` stops execution until it detects a mouse button or key press over a figure window. It returns 1 if it detects a key press, or 0 if it detects a mouse button press. Additional information about the terminating event is available through the figure's `CurrentCharacter`, `SelectionType`, and `CurrentPoint` properties.

See Also `figure`, `gcf`, `ginput`

---

waterfall

Purpose Waterfall plot.

Synopsis

```
waterfall(X,Y,Z,C)
waterfall(X,Y,Z)
waterfall(x,y,Z,C)
waterfall(x,y,Z)
waterfall(Z,C)
waterfall(Z)

h = waterfall(...)
```

Description `waterfall` is similar to the `mesh` function but it omits the column lines of the plot to produce a "waterfall" effect. For a complete discussion of parametric surfaces and related color properties, see `surf`.

`waterfall(X,Y,Z,C)` plots the colored grid lines on the parametric surface specified by `X`, `Y`, and `Z`, with color specified by `C`. In simpler uses, `X` and `Y` may be vectors, and `Z` and `C` may be omitted.

The range of `X`, `Y`, and `Z`, or the current setting of `axis`, determines the axis labels. The range of `C`, or the current setting of `caxis`, determines the color scaling. The scaled color values are used as indices into the current color map.

`waterfall(X,Y,Z)` uses `C = Z`, so color is proportional to surface height.

`waterfall(x,y,Z,C)` and `waterfall(x,y,Z)`, with two vector arguments replacing the first two matrix arguments, must have `length(x) = n` and `length(y) = m` where `[m,n] = size(Z)`. Here, the intersections of the grid lines are the triples `(x(j),y(i),Z(i,j))`. Note that `x` corresponds to the columns of `Z` and `y` corresponds to the rows.

`waterfall(Z,C)` and `waterfall(Z)` use `x = 1:n` and `y = 1:m`.

`h = waterfall(...)` returns a vector of handles of the patch objects that comprise the plot.

Examples Produce a waterfall plot of the `peaks` surface.

```
[x,y,z] = peaks(30);
waterfall(x,y,z)
```

See Also `mesh`

---

wavread

Purpose Load Microsoft Windows 3.1 .WAV format sound files.

Synopsis

```
y = wavread('wavefile')
[y,Fs] = wavread('wavefile')
[y,Fs,format] = wavread('wavefile')
```

Description `y = wavread('wavefile')` loads the .WAV file *`wavefile`*, returning the sampled data in `y`. If *`wavefile`* does not include a file extension, the function assumes .WAV.

`[y,Fs] = wavread('wavefile')` returns the sample rate `Fs`.

`[y,Fs,format] = wavread('wavefile')` returns the six-element vector `format`, where

- `format(1)` is the data format. This is always the string `'PCM'`.
- `format(2)` is the number of channels.
- `format(3)` is the sample rate (`Fs`).
- `format(4)` is the average bytes per second (sampled).
- `format(5)` is the block alignment of the data.
- `format(6)` is the number of bits per sample.

This function currently supports only 8-bit, single channel data.

See Also `wavwrite`

---

wavwrite

Purpose Save Microsoft Windows 3.1 .WAV format sound files.

Synopsis `wavwrite('wavefile',y,Fs)`

Description `wavwrite('wavefile',y,Fs)` writes the 8-bit sampled data `y`, with sample rate `Fs`, to the file *wavefile*.

`wavwrite` creates an 8-bit, single channel file. The function truncates sample data of more than eight bits.

See Also `wavread`

---

what

Purpose Directory listing of M-files, MAT-files, and MEX-files.

Synopsis

```
what
what dirname
```

Description `what`, by itself, lists the M-files, MAT-files, and MEX-files in the current directory.

`what dirname` lists the files in directory *dirname* on MATLAB's search path. It is not necessary to enter the full pathname of the directory; the last component, or last couple of components, is sufficient.

Examples The commands

```
what general
```

and

```
what matlab/general
```

both list the M-files in the `general` directory. The syntax of the path depends on your operating system:

UNIX:	`matlab/general`
VMS:	`MATLAB.GENERAL`
MS-DOS:	`MATLAB\GENERAL`
Macintosh:	`MATLAB:General`

See Also `dir`, `lookfor`, `path`, `which`, `who`

---

whatsnew

Purpose Display `README` files for MATLAB and toolboxes.

Synopsis `whatsnew matlab`

`whatsnew` *`toolboxpath`*

Description `whatsnew` displays the `README` file for the MATLAB product or a specified toolbox. If present, the `README` file summarizes new functionality that is not described in the documentation.

`whatsnew matlab` displays the `README` file for MATLAB.

`whatsnew` *`toolboxpath`* displays the `README` file for the toolbox specified by the string *`toolboxpath`*.

Examples

```
whatsnew matlab % MATLAB README file
whatsnew signal % Signal Processing Toolbox README file
```

See Also `help`, `lookfor`, `path`, `ver`, `version`, `which`

---

which

Purpose Locate functions and files.

Synopsis `which` *`funname`*

Description `which` *`funname`* displays the full pathname of the specified function. The function can be an M-file, MEX-file, built-in function, or SIMULINK graphical function. Built-in functions and SIMULINK graphical functions display a message indicating that they are built in to MATLAB or are part of SIMULINK.

Examples For example,

```
which inv
```

reveals that `inv` is a built-in function, and

```
which pinv
```

indicates that `pinv` is in the `matfun` directory of the MATLAB Toolbox.

The statement

```
which inverse
```

probably says

```
inverse not found
```

because there is no file `inverse.m` on MATLAB's search path. Contrast this with `lookfor inverse`, which takes longer to run, but finds several matches to the keyword `inverse` in its search through all the help entries. (If `inverse.m` does exist in the current directory, or in some private directory that has been added to MATLAB's search path, `which inverse` finds it.)

See Also `dir`, `exist`, `help`, `lookfor`, `path`, `what`, `who`

---

while

Purpose Repeat statements an indefinite number of times.

Synopsis `while` *`expression`*

```
    statements
end
```

Description `while` repeats `statements` an indefinite number of times. The `statements` are executed while the *`expression`* has all nonzero elements. The *`expression`* is usually of the form

```
expression rop expression
```

where *rop* is ==, <, >, <=, >=, or ~=.

The scope of a `while` statement is always terminated with a matching `end`.

Examples The variable `eps` is a tolerance used to determine such things as near singularity and rank. Its initial value is the *machine epsilon*, the distance from 1.0 to the next largest floating-point number on your machine. Its calculation demonstrates `while` loops:

```
eps = 1;
while (1+eps) > 1
    eps = eps/2;
end
eps = eps*2
```

See Also `all`, `any`, `break`, `end`, `for`, `if`, `return`

---

white

Purpose All white monochrome colormap.

Synopsis

```
cmap = white(m)
white
```

Description `white(m)` returns an `m`-by-3 matrix containing a white colormap. For a mesh or surface plot, the white colormap ensures that all lines are visible.

`white`, with no input or output arguments, is the same length as the current colormap.

See Also `colormap`, `cool`, `copper`, `flag`, `gray`, `hot`, `hsv`, `pink`, `rgbplot`

---

whitebg

Purpose Change figure background color.

Synopsis

```
whitebg
whitebg(fig)
whitebg(c)
whitbg(fig,c)
```

Description `whitebg` toggles the background color of the current figure between white and black. The command also changes other figure properties, including text and axes color, to ensure that all objects are visible.

`whitebg(fig)`, where `fig` is a row vector of figure handles, toggles all the specified figures. The function sets the default properties of the root window such that all subsequent figure plots use the new background color.

`whitebg(c)` and `whitebg(fig,c)` change the default background color to `c`, where `c` can be an RGB triple like [1 0 0] or a color name string like `'white'` (see the `ColorSpec` reference page for details on specifying colors).

This command also affects other figure properties so text and graphics are visible against the new background color. See the `figure` reference description (`InvertHardCopy` property) for details on printing the contents of a figure window.

See Also `figure`

---

who,whos

Purpose List directory of variables in memory.

Synopsis

```
who
whos
who global
whos global
```

Description `who` lists the variables currently in memory.

`whos` lists the current variables, their sizes, and whether they have nonzero imaginary parts.

`who global` and `whos global` list the variables in the global workspace.

See Also `dir`, `exist`, `help`, `what`

---

wilkinson

Purpose Wilkinson's eigenvalue test matrix.

Synopsis `W = wilkinson(n)`

Description `wilkinson(n)` is one of J. H. Wilkinson's eigenvalue test matrices. It is a symmetric, tridiagonal matrix with pairs of nearly, but not exactly, equal eigenvalues.

Examples `wilkinson(7)` is

```
3  1  0  0  0  0  0
1  2  1  0  0  0  0
0  1  1  1  0  0  0
0  0  1  0  1  0  0
0  0  0  1  1  1  0
0  0  0  0  1  2  1
0  0  0  0  0  1  3
```

The most frequently used case is `wilkinson(21)`. Its two largest eigenvalues are both about 10.746; they agree to 14, but not to 15, decimal places.

See Also `eig`, `pascal`, `rosser`

wk1read

Purpose Read a Lotus123 WK1 spreadsheet file into a matrix.

Synopsis

```
M = wk1read('filename')
M = wk1read('filename',r,c)
M = wk1read('filename',range)
```

Description `M = wk1read('filename')` reads a Lotus123 WK1 spreadsheet file into the matrix `M`.

`M = wk1read('filename',r,c)` starts reading at the row-column cell offset specified by `(r,c)`.

`M = wk1read('filename',range)` reads the range of values specified by the parameter `range`, where `range` can be

- A four-element vector specifying the cell range in the format

  `[upper_left_x upper_left_y lower_right_x lower_right_y]`

- A cell range specified as a string; for example, `'A1...C5'`.
- A named range specified as a string; for example, `'Sales'`.

See Also `wk1write`

wk1write

Purpose Write a matrix to a Lotus123 WK1 spreadsheet file.

Synopsis

```
wk1write('filename',M)
wk1write('filename',M,r,c)
```

Description `wk1write('filename',M)` writes the matrix `M` into a Lotus123 WK1 spreadsheet file named *filename*.

`wk1write('filename',M,r,c)` starts at the row-column location of `M` specified by `(r,c)`.

See Also `wk1read`

xlabel

Purpose *x*-axis labels.

Synopsis

```
xlabel('text')
```

Description `xlabel('text')` adds text to the current two-dimensional plot *beneath* the *x*-axis, or to the side or bottom of a current three-dimensional plot.

Reissuing an `xlabel` command causes the new text to replace the old text.

Algorithm For three-dimensional graphics, MATLAB puts the label in the front or side, so that it is never hidden by the plot itself.

See Also `text`, `title`, `ylabel`, `zlabel`

xor

Purpose Exclusive or.

Synopsis `C = xor(A,B)`

Description `C = xor(A,B)` is the symmetric difference of the nonzero structures of `A` and `B`, that is, `C(i,j)` is one if `A(i,j)` or `B(i,j)`, but not both, is nonzero.

Examples To see where either `S` or `T` has a nonzero element and the other matrix does not,

```
spy(xor(S,T))
```

See Also `&`, `|`, `all`, `any`, `find`

ylabel

Purpose *y*-axis labels.

Synopsis `ylabel('`*`text`*`')`

Description `ylabel('`*`text`*`')` adds text to the current two-dimensional plot *beside* the *y*-axis, or to the side of a current three-dimensional plot.

Reissuing a `ylabel` command causes the new text to replace the old text.

Algorithm For three-dimensional graphics, MATLAB puts the label in the front or side, so that it is never hidden by the plot itself.

See Also `text`, `title`, `xlabel`, `zlabel`

zeros

Purpose All zeros.

Synopsis

```
Y = zeros(n)
Y = zeros(m,n)
Y = zeros(size(A))
```

Description `Y = zeros(n)` is an `n`-by-`n` matrix of zeros.

`Y = zeros(m,n)` is an `m`-by-`n` matrix of zeros.

`Y = zeros(size(A))` is the same size as `A` and consists of all zeros.

The MATLAB language does not have a dimension statement and MATLAB automatically allocates storage for matrices. Nevertheless, most MATLAB programs execute faster if the `zeros` function is used to set aside storage for a matrix whose elements are to be generated one at a time, or a row or column at a time.

Examples With `n = 1000`, the `for` loop

```
for i = 1:n, x(i) = i; end
```

takes about 1.2 seconds to execute on a Sun SPARC-1. If the loop is preceded by the statement

```
x = zeros(1,n);
```

the computations require less than 0.2 seconds.

See Also `eye`, `ones`, `rand`, `randn`

---

zlabel

Purpose *z*-axis labels.

Synopsis

```
zlabel('text')
```

Description `zlabel('text')` adds text to the *z*-axis of a current three-dimensional plot.

Reissuing a `zlabel` command causes the new text to replace the old text.

Algorithm For three-dimensional graphics, MATLAB puts the label in the front or side, so that it is never hidden by the plot itself.

See Also `text`, `title`, `xlabel`, `ylabel`

---

zoom

Purpose Zoom in and out of a 2-D plot.

Synopsis

```
zoom on
zoom off
zoom out
zoom
```

Description `zoom on` turns zoom on for the current figure and prompts you to define the zoom center with the mouse:

- For a single-button mouse, zoom in by pressing the mouse button and zoom out by pressing **Shift** and clicking at the same time.
- For a two- or three-button mouse, zoom in by pressing the left mouse button and zoom out by pressing the right mouse button.

Each time you click, `zoom` changes the axes limits by a factor of two (in or out). You can also click and drag to zoom into an area. Press the **Return** key to stop zooming.

`zoom off` turns zoom off.

`zoom out` zooms out fully.

`zoom` toggles zoom mode.

9

Symbolic Math Toolbox Reference

This chapter provides detailed descriptions of all *Symbolic Math Toolbox* functions. It begins with tables of these functions and continues with the reference entries in alphabetical order.

9.1 Reference Tables

Calculus	
`diff`	Differentiate or difference.
`int`	Integrate.
`jacobian`	Jacobian matrix.
`taylor`	Taylor series expansion.

Linear Algebra	
`charpoly`	Symbolic characteristic polynomial.
`colspace`	Basis for column space.
`determ`	Symbolic matrix determinant.
`eigensys`	Symbolic eigenvalues and eigenvectors.
`inverse`	Symbolic matrix inverse.
`jordan`	Jordan canonical form.
`linsolve`	Solution of simultaneous linear equations.
`nullspace`	Basis for nullspace.
`singvals`	Symbolic matrix singular values and singular vectors.
`svdvpa`	Variable-precision singular values.
`transpose`	Symbolic matrix transpose.

Simplification	
`allvalues`	Find all values for `RootOf` expression.
`collect`	Collect.
`expand`	Expand.
`factor`	Factor.
`simple`	Search for shortest form.
`simplify`	Simplify.
`symsum`	Sum symbolic series.

Solution of Equations	
`compose`	Functional composition.
`dsolve`	Symbolic solution of differential equations.
`finverse`	Functional inverse.
`solve`	Symbolic solution of algebraic equations.

Variable Precision Arithmetic	
`digits`	Set variable precision accuracy.
`vpa`	Variable precision arithmetic.

Operations on Symbolic Expressions and Matrices	
`latex`	LaTeX represention of symbolic output.
`horner`	Nested polynomial representation.
`numden`	Numerator and denominator.
`numeric`	Convert symbolic matrix to numeric form.
`poly2sym`	Coefficient vector to symbolic polynomial.
`pretty`	Pretty print symbolic output.
`subs`	Substitute for a subexpression.
`sym`	Create or modify symbolic matrix.
`sym2poly`	Symbolic polynomial to coefficient vector.
`symadd`	Add symbolic expressions.
`symdiv`	Divide symbolic expressions.
`symmul`	Multiply symbolic expressions.
`symop`	Symbolic operations.
`sympow`	Power of symbolic expression.
`symrat`	Symbolic rational approximation.
`symsize`	Size of symbolic matrix.
`symsub`	Subtract symbolic expressions.
`symvar`	Determine symbolic variables.

Pedagogical and Graphical Applications	
`ezplot`	Easy to use function plotter.
`funtool`	Function calculator.
`rsums`	Riemann sums.

Transforms and Inverse Transforms	
fourier	Fourier integral transform.
invfourier	Inverse Fourier integral transform.
invlaplace	Inverse Laplace transform.
invztrans	Inverse z-transform.
laplace	Laplace transform
ztrans	z-transform.

9.2 Symbolic Math Toolbox Commands and Functions

allvalues

Purpose Find all values for `RootOf` expression.

Synopsis

```
R = allvalues(S)
```

Description For each subexpression `RootOf(expr)` in `S`, `allvalues(S)` finds the roots of `expr`. `expr` is an algebraic expression containing exactly one symbolic variable. `S` can be a symbolic expression or a symbolic column vector. After evaluating all `RootOf` subexpressions, `allvalues` evaluates all the possible values of `S` and returns the result as a symbolic column vector.

Examples `allvalues('2*RootOf(t^3-5)')` returns

```
[                  2*5^(1/3)]
[-5^(1/3)+i*3^(1/2)*5^(1/3)]
[-5^(1/3)-i*3^(1/2)*5^(1/3)]
```

Suppose

```
v =
[RootOf(3*_Z^2+4*_Z+2)]
[RootOf(4*_Z^2+6*_Z+3)]
```

Then `allvalues(v)` returns

```
[-2/3+1/3*i*2^(1/2)]
[-2/3-1/3*i*2^(1/2)]
[-3/4+1/4*i*3^(1/2)]
[-3/4-1/4*i*3^(1/2)]
```

See Also `solve`, `eigensys`, `numeric`

---

charpoly

Purpose Symbolic characteristic polynomial.

Synopsis

```
r = charpoly(A)
r = charpoly(A,'v')
```

Description `charpoly(A)` returns the characteristic polynomial of `A`, where `A` is a numeric or symbolic matrix. The result is a symbolic polynomial in `x`.

`charpoly(A,'v')` returns the characteristic polynomial as a polynomial in `v` instead of the default variable `x`.

Note that `charpoly(A)` approximately equals `poly2sym(poly(A))`. Similarly, `poly(A)` approximately equals `sym2poly(charpoly(A))`. The approximations are due to roundoff error.

Examples The statements

```
A = gallery(3)
p = charpoly(A)
```

return

```
A =
   -149    -50   -154
    537    180    546
    -27     -9    -25

p =
x^3-6*x^2+11*x-6
```

See Also `poly2sym`, `sym2poly`, `jordan`, `eigensys`, `solve`
`poly` in *MATLAB Reference*

---

collect

Purpose Collect coefficients.

Synopsis

```
R = collect(S)
R = collect(S,'v')
```

Description `collect(S)` regards each element of `S` as a polynomial in the symbolic variable of `S`. If this symbolic variable is *x*, `collect(S)` collects all the coefficients with the same power of *x*.

`collect(S,'v')` takes *v* to be the symbolic variable in each element of `S`.

Examples `collect('(exp(x)+x)*(x+2)')` returns

```
x^2+(exp(x)+2)*x+2*exp(x)
```

`collect('(x+y)*(x^2+y^2+1)','y')` returns

```
   3      2      2                2
[ y  + x y  + (x  + 1) y + x (x  + 1) ]
```

`collect('(x+y)*(x^2+y^2+1)','x')` returns

```
   3      2      2                2
[ x  + y x  + (y  + 1) x + y (y  + 1) ]
```

The following statements

```
S = sym('(x+1)*(y+1),x+y');
R = collect(S,'y')
```

return

```
R =
[(x+1)*y+x+1, x+y]
```

See Also `symvar`, `simplify`, `simple`, `expand`, `factor`

---

colspace

Purpose Basis for column space.

Synopsis

```
B = colspace(A)
```

Description `colspace(A)` returns a matrix whose columns form a basis for the column space of `A`. `A` is a symbolic or numeric matrix. Note that `symsize(colspace(A),2)` is the rank of `A`.

Example The statements

```
A = sym('2,0;3,4;0,5')
B = colspace(A)
```

return

```
A =
[2,0]
[3,4]
[0,5]

B =
[    1,    0]
[    0,    1]
[-15/8, 5/4]
```

See Also `nullspace`
`orth` in *MATLAB Reference*

---

compose

Purpose Functional composition.

Synopsis

```
h = compose(f,g)
h = compose(f,g,'u')
h = compose(f,g,'u','w')
```

Description `h = compose(f,g)` performs functional composition. `f` is a symbolic expression containing exactly one symbolic variable, say *y*. That is, `f` represents the function *f(y)*. Similarly, `g` is a symbolic expression representing a function of exactly one symbolic variable, say `x`. That is, `g` represents the function *g(x)*. Then `h` is a symbolic expression that represents the composite function *f(g(x))*.

`h = compose(f,g,'u')` uses the variable *u* as the domain variable for both `f` and `g`. Use this form when `f` has the form *f(u,a1,a2,...)*, and `g` has the form *g(u,b1,b2,...)*. `h` represents *f(g(u))*.

`h = compose(f,g,'u','w')` uses *u* as the domain variable for `f` and *w* as the domain variable for `g`. Use this form when `f` has the form *f(u,a1,a2,...)*, and `g` has the form *g(w,b1,b2,...)*. `h` represents *f(g(w))*.

Examples `compose 1/(1+x^2) sin(x)` returns `1/(1+sin(x)^2)`.

`compose('a*cos(b)','a*sin(b)','a')` returns `a*sin(b)*cos(b)`.

`compose('3*x+y','2*x+5*y','x','y')` returns `6*x+16*y`.

`compose('y+1','2*s','s','s')` returns `y+1`.

`compose('y+1','2*s','y','s')` returns `2*s+1`.

See Also `finverse`, `symvar`, `subs`

---

determ

Purpose Symbolic matrix determinant.

Synopsis

```
r = determ(A)
r = determ(vpa(A))
```

Description `determ(A)` computes the symbolic determinant of `A`, where `A` is a symbolic or numeric matrix.

`determ(vpa(A))` uses variable precision arithmetic in calculating the determinant of the matrix A.

Example

`determ(sym('a, b; c, d'))` returns

```
a*d – b*c
```

The statements

```
digits(20)
A = [2/3 1/3;1 1]
r = determ(vpa(A))
```

return

```
A =
     0.6667     0.3333
     1.0000     1.0000

r =
.33333333333333333334
```

See Also

`vpa`

diff

Purpose

Symbolic derivatives for symbolic arguments.

Synopsis

```
R = diff(S)
R = diff(S,'v')
R = diff(S,n)
R = diff(S,'v',n)
diff
```

Description

`diff(S)` differentiates each element of S with respect to S's symbolic variable, as determined by `symvar`.

`diff(S,'v')` differentiates each element of S with respect to *v*.

`diff(S,n)` and `diff(S,'v',n)` differentiate each element of S n times.

`diff`, by itself, differentiates the previous symbolic matrix.

See Also

`diff` in *MATLAB Reference*

digits

Purpose

Set variable precision accuracy.

Synopsis

```
digits(d)
d = digits
digits
```

Description

The accuracy of variable precision arithmetic is determined by `Digits`, a global parameter in the Maple Kernel, which specifies the number of significant decimal digits to be used. The initial, default value of `Digits` is set to 16.

`digits(d)` sets `Digits` to `d` for subsequent calculations.

`d = digits` returns the current setting of `Digits`.

`digits`, by itself, displays the current setting of `Digits`.

Examples

```
z = 1.0.e-16
x = vpa([1 z -1])
y = vpa([1 1 1]')
```

Then

```
digits(15)
z = symmul(x,y)
```

uses 15-digit decimal arithmetic and so returns `z = 0`.

However

```
digits(17)
z = symmul(x,y)
```

used 17-digit decimal arithmetic and so returns `z = 1.0e-16`.

See Also

`vpa`

dsolve

Purpose

Symbolic solution of ordinary differential equations.

Synopsis

```
r = dsolve('eq1,eq2,...', 'cond1,cond2,...', 'v')
r = dsolve('eq1','eq2',...,'cond1','cond2',...,'v')
[r1,r2,...] = dsolve(...)
```

Description

`dsolve('eq1,eq2,...', 'cond1,cond2,...', 'v')` symbolically solves the ordinary differential equation(s) specified by *eq1,eq2,...* using *v* as the independent variable and the boundary and/or initial condition(s) specified by *cond1,cond2,...*.

The default independent variable is `x`. A secondary default is `t`. `t` becomes the independent variable when `x` is a dependent variable in the equation, or when `t` is a free variable in an equation.

The letter `D` denotes differentiation with respect to the independent variable; with the primary default this is d/dx. A `D` followed by a digit denotes repeated differentiation. For example, `D2` is d^2/dx^2. Any character immediately following a differentiation operator is a dependent variable. For example, `D3y` denotes the third derivative of $y(x)$ or $y(t)$.

Initial/boundary conditions are specified with equations like `y(a) = b` or `Dy(a) = b`, where `y` is a dependent variable and `a` and `b` are constants. If the number of initial conditions is less than the number of dependent variables, the resulting solutions will contain the arbitrary constants `C1`, `C2`,. . . .

You can also input each equation and/or initial condition as a separate symbolic equation. `dsolve` accepts up to 12 input arguments.

With no output arguments, `dsolve` returns a list of solutions.

`[r1,r2,...] = dsolve(...)` returns the solutions, in alphabetic order, as `r1,r2,...`. For nonlinear equations, if the results are not unique, `r1,r2,...` may be symbolic vectors.

Examples

`dsolve('Dy = a*y')` returns

```
y(x) = exp(a*x)*C1
```

`dsolve('Df = f + sin(t)')` returns

```
f(t) = -1/2*cos(t)-1/2*sin(t)+exp(t)*C1
```

`y = dsolve('(Dy)^2 + y^2 = 1','s')` returns

```
y =
[ sin(s-C1)]
[-sin(s-C1)]
```

`dsolve('Dy = a*y', 'y(0) = b')` returns

```
y(x) = exp(a*x)*b
```

`dsolve('D2y = -a^2*y', 'y(0) = 1', 'Dy(pi/a) = 0')` returns

```
y(x) = cos(a*x)
```

`[x,y] = dsolve('Dx = y', 'Dy = -x')` returns

```
x =
C2*sin(t)-C1*cos(t)

y =
C1*sin(t)+C2*cos(t)
```

See Also

`symvar`

eigensys

Purpose Symbolic matrix eigenvalues and eigenvectors.

Synopsis

```
E = eigensys(A)
[V,E] = eigensys(A)
E = eigensys(vpa(A))
[V,E] = eigensys(vpa(A))
```

Description `eigensys(A)` computes the symbolic eigenvalues of the square symbolic or numeric matrix `A`.

`[V,E] = eigensys(A)` computes the symbolic eigenvalues and eigenvectors of `A`, which is a square symbolic or numeric matrix. `E` is the vector of eigenvalues; `V` is the matrix of eigenvectors. The eigenvectors may be expressed in terms of `E(n)`, where `n` is a positive integer that indexes the eigenvalue vector `E`.

`eigensys(vpa(A))` computes the eigenvalues of the square matrix `A` using variable precision arithmetic.

`[V,E] = eigensys(vpa(A))` computes the eigenvalues and eigenvectors of `A` using variable precision arithmetic. `A` is a square matrix. `E` is the vector of eigenvalues, and `V` is the matrix of eigenvectors.

Limitations For a symbolic matrix, closed form expressions for the eigenvalues are possible only if the characteristic polynomial is factorable over the rational numbers. This is always true for matrices of dimension 4 or lower. If the characteristic polynomial is not factorable, the eigenvalues are expressed in terms of the `RootOf` notation.

Examples The statements

```
digits(20);
T = eigensys(pascal(3))
U = eigensys(vpa(pascal(3)))
```

return

```
T =
[          1]
[4+15^(1/2)]
[4-15^(1/2)]

U =
[.1270166537925840]
[1.000000000000001]
[7.872983346207417]
```

The statements

```
A = sym('a, b, c; b, c, a; c, a, b');
[V,E] = eigensys(A)
```

return

```
V =
[-(a+E(1)-b)/(a-c), -(a+E(2)-b)/(a-c), 1]
[-(b-c-E(1))/(a-c), -(b-c-E(2))/(a-c), 1]
[                1,                 1, 1]

E =
[-(a^2-c*a-b*a-c*b+c^2+b^2)^(1/2)]
[ (a^2-c*a-b*a-c*b+c^2+b^2)^(1/2)]
[                           a+c+b]
```

See Also svdvpa, vpa, allvalues, subs, charpoly, jordan

---

expand

Purpose Symbolic expansion.

Synopsis

```
R = expand(S)
```

Description `expand(S)` expands each element of `S`, where `S` is a symbolic matrix. `expand` primarily expands polynomials. It also expands several mathematical functions, such as the trigonometric functions and the exponential and logarithmic functions.

Examples `expand('(x-2)*(x-4)')` returns

```
x^2-6*x+8
```

`expand('(r+s)/(r+3)')` returns

```
     r         s
[  -----  +  -----  ]
   r + 3     r + 3
```

`expand('cos(x+y)')` returns

```
cos(x)*cos(y)-sin(x)*sin(y)
```

`expand('exp((a+b)^2)')` returns

```
exp(a^2)*exp(a*b)^2*exp(b^2)
```

`expand(sym('sin(2*t),cos(2*t)'))` returns

```
[2*sin(t)*cos(t), 2*cos(t)^2-1]
```

See Also simplify, simple, factor, collect, horner

ezplot

Purpose Easy to use function plotter.

Synopsis

```
ezplot(f)
ezplot(f,[xmin xmax])
ezplot(f,[xmin xmax],fig)
```

Description `ezplot(f)` plots a graph of *f(x)* where `f` is a symbolic expression representing a mathematical expression that involves a single symbolic variable, say `x`. The domain on the *x*-axis is usually `[-2*pi, 2*pi]`.

`ezplot(f,[xmin xmax])` uses the specified *x*-domain instead of the default `[-2*pi, 2*pi]`.

`ezplot(f,[xmin xmax],fig)` uses the specified figure number instead of the current figure. It also omits the title of the graph.

Examples Either of the following commands:

```
ezplot('erf(x)')
ezplot erf(x)
```

plots a graph of the error function:

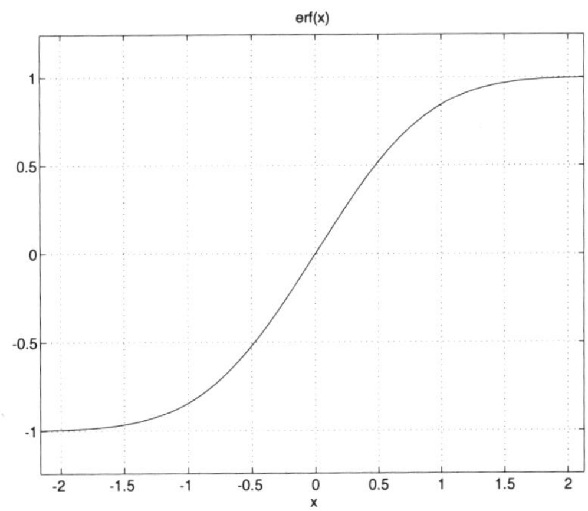

Algorithm `ezplot` determines the interval of the *x*-axis by sampling the function between `-2*pi` and `2*pi` and then selecting a subinterval where the variation is significant. For the range of the *y*-axis, `ezplot` omits extreme values associated with singularities.

See Also `symvar`

factor

Purpose Symbolic factorization.

Synopsis

```
R = factor(N)
R = factor(S)
```

Description `factor(N)` returns the prime factorization of each element of `N`, where `N` is a matrix of integers.

`factor(S)` factors each element of the symbolic matrix `S`.

Examples `factor('x^3 + x – 3*x^2 – 3')` returns

```
(x–3)*(x^2+1)
```

`factor(10:15)` returns

```
[(2)*(5), (11), (2)^2*(3), (13), (2)*(7), (3)*(5)]
```

The statements

```
X = sym('x^3–1;x^4–y^4')
R = factor(X)
```

return

```
X =
[  x^3–1]
[x^4–y^4]

R =
[       (x–1)*(x^2+x+1)]
[(x–y)*(x+y)*(x^2+y^2)]
```

See Also `horner`, `simplify`, `expand`, `simple`, `collect`

finverse

Purpose Functional inverse.

Synopsis

```
g = finverse(f)
g = finverse(f,'u')
```

Description `g = finverse(f)` returns the functional inverse of `f`. If `f` is a symbolic expression representing a function of exactly one symbolic variable, say `x`, then `g` is a symbolic expression that satisfies $g(f(x)) = x$.

`g = finverse(f,'u')` uses the symbolic variable `u` as the independent variable. The result `g` is a symbolic expression that satisfies $g(f(u)) = u$. Use this form when `f` contains more than one symbolic variable.

Examples `finverse('1/tan(x)')` returns

`arctan(1/x)`

`finverse('exp(u–2*v)','u')` returns

`log(u)+2*v`

See Also `compose`, `symvar`

fourier

Purpose Fourier integral transform.

Synopsis

```
F = fourier(f)
F = fourier(f,'v')
F = fourier(f,'v','x')
F = fourier
```

Description `F= fourier(f)` is the Fourier transform of the symbolic expression `f`, `F(w) = int(f(t)*exp(-i*w*t),'t',-inf,inf)`.

`F = fourier(f,'v')` is a function of `'v'` instead of `'w'`.

`F = fourier(f,'v','x')` assumes `f` is a function of `'x'` instead of `'t'`.

`F = fourier`, with no input or output arguments, transforms the previous result.

Example `fourier exp(-t)*Heaviside(t)` is `1/(1+i*w)`.

See Also `invfourier`, `laplace`, `ztrans`

funtool

Purpose Function calculator.

Synopsis

```
funtool
```

Description `funtool` is an interactive graphing calculator that manipulates functions of a single variable. At any time, there are two functions displayed, `f1` and `f2`, which are symbolic expressions representing *f1(x)* and *f2(x)*. The result of any operation replaces the active function, which is indicated by a yellow button on color monitors and a white button on monochrome monitors. Clicking the corresponding button for the inactive function makes it active. The title of each graph is editable text that can be changed at any time to install a new function.

The control buttons on the top row are unary function operators that involve only the active function. These operators are:

`D f`	Symbolically differentiate $f(x)$.
`I f`	Symbolically integrate $f(x)$.
`Simp f`	Simplify $f(x)$, if possible.
`Num f`	Extract the numerator of $f(x)$.
`Den f`	Extract the denominator of $f(x)$.
`1/f`	Replace $f(x)$ by $1/f(x)$.
`finv`	Replace $f(x)$ by its inverse function.

The operators `If` and `finv` may fail if the corresponding symbolic expressions do not exist in closed form.

The second row of buttons translates and scales the active function by a quantity a, which is specified by editable text. The operations are:

`f + a`	Replace $f(x)$ by $af(x)$.
`f * a`	Replace $f(x)$ by $f(x)^a$.
`f ^ a`	Replace $f(x)$ by $f(x)$ ^ a.
`x + a`	Replace $f(x)$ by $f(x + a)$.
`x * a`	Replace $f(x)$ by $f(ax)$.

The buttons on the third row are binary function operators that combine both functions and install the result in place of the active one. The operations are:

`f1 + f2`	Replace $f(x)$ by $f1(x) + f2(x)$.
`f1 - f2`	Replace $f(x)$ by $f1(x) - f2(x)$.
`f1 * f2`	Replace $f(x)$ by $f1(x)\ 2(x)$.
`f1 / f2`	Replace $f(x)$ by $f1(x) / f2(x)$.
`f1 @ f2`	Replace $f(x)$ by $f1(f2(x))$.
`f1 = f2`	Replace $f1(x)$ by $f2(x)$.
`f2 = f1`	Replace $f2(x)$ by $f1(x)$.

The fourth row of buttons begins with editable text specifying the range of `x`. Initially, the range is `[-2*pi, 2*pi]`. The other buttons are as follows:

`help`	Prints the online help for `funtool`.
demo	Demonstrates funtool.
`reset`	Sets `f1 = 1` and `f2 = 0`.
`close`	Closes all three windows.

See Also `ezplot`, `symvar`

---

horner

Purpose Horner polynomial representation.

Synopsis

```
R = horner(P)
```

Description `horner(P)` transforms each element of `P` into its Horner, or nested, representation. `P` is a matrix of symbolic polynomials.

Examples `horner('x^3-6*x^2+11*x-6')` returns

```
-6+(11+(-6+x)*x)*x
```

The statements

```
P = sym('x^2+x;y^3-2*y')
R = horner(P)
```

return

```
P =
[   x^2+x]
[y^3-2*y]

R =
[    (1+x)*x]
[(-2+y^2)*y]
```

See Also `simplify`, `simple`, `expand`, `factor`

---

int

Purpose Integrate.

Synopsis

```
R = int(S)
R = int(S,'v')
```

```
R = int(S,a,b)
R = int(S,'v',a,b)
int
```

Description `int(S)` is the indefinite integral of each element of `S` with respect to the symbolic variable of `S`. See `symvar` for the definition of symbolic variable.

`int(S,'v')` is the indefinite integral of each element of `S` with respect to the symbolic variable *v*.

`int(S,a,b)` is the definite integral from `a` to `b` of each element of `S` with respect to the symbolic variable of `S`. `a` and `b` are symbolic or numeric expressions.

`int(S,'v',a,b)` is the definite integral of each element of `S` with respect to *v* from `a` to `b`. `a` and `b` are symbolic or numeric expressions.

`int`, by itself, is the indefinite integral of the previous symbolic matrix with respect to its symbolic variable.

Examples The statements

```
s = int('-2*x/(1+x^2)^2')
int
```

return

```
s =
1/(1+x^2)

ans =
atan(x)
```

`int('sin(u*w)','w')` returns

```
-cos(u*w)/u
```

`int('x*log(1+x)',0,1)` returns

```
1/4
```

`int('4*x','sin(t)',2)` returns

```
8-2*sin(t)^2
```

`int(sym('exp(t),exp(a*t)'))` returns

```
[exp(t), 1/a*exp(a*t)]
```

See Also `symvar`, `diff`, `symsum`

---

inverse

Purpose Symbolic matrix inverse.

Synopsis

```
R = inverse(A)
R = inverse(vpa(A))
```

Description `inverse(A)` returns the symbolic inverse of `A`, where `A` is a symbolic or numeric matrix.

`inverse(vpa(A))` computes the inverse of the matrix `A` using variable precision arithmetic.

Examples The statements

```
B = inverse(hilb(3))
digits(10);
C = inverse(vpa(hilb(3)))
```

return

```
B =
[  9,  -36,   30]
[-36,  192, -180]
[ 30, -180,  180]

C =
[ 9.000000082, -36.00000039,  30.00000035]
[-36.00000039,  192.0000021, -180.0000019]
[ 30.00000035, -180.0000019,  180.0000019]
```

`inverse(sym(2,2,'1/(i+j-t)'))` returns

```
[      -(-3+t)^2*(-2+t), (-3+t)*(-2+t)*(-4+t)]
[(-3+t)*(-2+t)*(-4+t),       -(-3+t)^2*(-4+t)]
```

See Also `symdiv,vpa`

---

invfourier

Purpose Inverse Fourier integral transform.

Synopsis

```
f = invfourier(F)
f = invfourier(F,'x')
f = invfourier(F,'x','v')
f = invfourier
```

Description `f = invfourier(F)` is the inverse Fourier transform of the expression `F`, `f(t) = 1/(2*pi)*int(F(w)*exp(i*w*t),'w',-inf,inf)`

`f = invfourier(F,'x')` is a function of `'x'` instead of `'t'`.

`f = invfourier(F,'x','v')` assumes F is a function of `'v'` instead of `'w'`.

`f = invfourier`, with no input or output arguments, transforms the previous result.

Examples `invfourier exp(-w^2)` is `1/2/pi^(1/2)*exp(-1/4*t^2)`

`invfourier 1/(w-i)` is `i*exp(-t)*Heaviside(t)`

See Also `fourier`, `invlaplace`, `invztrans`

---

invlaplace

Purpose Inverse Laplace transform.

Synopsis

```
f = invlaplace(F)
f = invlaplace(F,'x')
f = invlaplace(F,'x','v')
f = invlaplace
```

Description `f = invlaplace(F)` is the inverse Laplace transform of the expression `F`, `f(t) = int(F(s)*exp(s*t),'s',0,inf)`

`f = invlaplace(F,'x')` is a function of 'x' instead of 't'.

`f = invlaplace(F,'x','v')` assumes F is a function of 'v' instead 's'.

`f = invlaplace`, with no input or output arguments, transforms the previous result.

Examples `invlaplace 1/(s-1)` is `exp(t)`

`invlaplace('(2*s^2+2+s^3)/s^3/(s^2+1)')` is `t^2+sin(t)`

`invlaplace('t^(-5/2)','x')` is `4/3/pi^(1/2)*x^(3/2)`

`invlaplace('laplace(f(t))')` is `f(t)`

See Also `laplace`, `invfourier`, `invztrans`

---

invztrans

Purpose Inverse *z*-transform.

Synopsis

```
f = invztrans(F)
f = invztrans(F,'x')
f = invztrans(F,'x','v')
f = invztrans
```

Description `f = invztrans(F)` is the inverse z-transform of the expression `F`, `f(n) = 1/(2*pi*i)*(a complex contour integral of F(z)*z^(n-1) dz).`

`f = invztrans(F,'x')` is a function of `'x'` instead of `'n'`.

`f = invztrans(F,'x','v')` assumes `F` is a function of `'v'` instead of `'z'`.

`f = invztrans`, with no input arguments, transforms the previous result.

Examples `invztrans z/(z-1)` is `1`

`invztrans z/(z-a)` is `a^n`

`invztrans('exp(x/z)','k','z')` is `x^k/k!`

`invztrans(ztrans('f(n)'))` is `f(n)`

See Also `ztrans`, `invlaplace`, `invfourier`

---

jacobian

Purpose Jacobian matrix.

Synopsis

```
R = jacobian(w,v)
```

Description `jacobian(w,v)` computes the Jacobian of `w` with respect to `v`. `w` is a symbolic scalar expression or a symbolic vector. `v` is a symbolic vector. The `(i,j)`-th entry of the result is ∂`w(i)`/∂`v(j)`.

Examples The statements

```
w = sym('x*y*z; y; x+z');
v = sym('x,y,z');
R = jacobian(w,v)
b = jacobian(sym(w,3,1),v)
```

return

```
R =
[y*z, x*z, x*y]
[  0,   1,   0]
[  1,   0,   1]

b =
[1, 0, 1]
```

See Also `diff`

jordan

Purpose Jordan canonical form.

Synopsis

```
J = jordan(A)
[V,J] = jordan(A)
```

Description `jordan(A)` computes the Jordan canonical (normal) form of `A`, where `A` is a symbolic or numeric matrix. Because any errors in `A` may completely change its Jordan form, `A` must be known exactly. Therefore, its elements must be integers or ratios of small integers.

`[V,J] = jordan(A)` computes both `J`, the Jordan canonical form, and `V`, the matrix whose columns are the generalized eigenvectors.

Examples The statements

```
A = [1 -3 -2; -1  1 -1; 2 4 5]
[V,J] = jordan(A)
```

return

```
A =
      1    -3    -2
     -1     1    -1
      2     4     5

V =
[-1, -2, 2]
[ 0, -2, 0]
[ 0,  4, 0]

J =
[3, 0, 0]
[0, 2, 1]
[0, 0, 2]
```

Then the statements

```
V = numeric(V)
V\A*V
```

return

```
V =
     -1    -2    2
      0    -2    0
      1     4    0

ans =
      3    0    0
      0    2    1
      0    0    2
```

See Also `charpoly`, `eigensys`

---

laplace

Purpose Laplace transform

Synopsis

```
F = laplace(f)
F = laplace(f,'v')
F = laplace(f,'v','x')
F = laplace
```

Description `F = laplace(f)` is the Laplace transform of the symbolic expression `f`, `F(s) = int(f(t)*exp(-s*t),'t',0,inf)`

`F = laplace(f,'v')` is a function of `'v'` instead of `'s'`.

`F = laplace(f,'v','x')` assumes `f` is a function of `'x'` instead of `'t'`.

`F = laplace`, with no input arguments, transforms the previous result.

Examples `laplace exp(t)` is `1/(s-1)`

`laplace t^2+sin(t)` is `(2*s^2+2+s^3)/s^3/(s^2+1)`

`laplace('y^(3/2)','z')` is `3/4*pi^(1/2)/z^(5/2)`

`laplace(diff('F(t)'))` is `laplace(F(t),t,s)*s-F(0)`

See Also `invlaplace`, `fourier`, `ztrans`

---

latex

Purpose LaTeX represention of symbolic output.

Synopsis

```
latex(S)
latex(S,'filename')
```

Description `latex(S)` prints the LaTeX representation of `S`.

`latex(S,'filename')` also prints it to the specified file.

Examples The example is

```
r = '(1+2*x+3*x^2)/(4+5*x+6*x^2)'
latex(r)
{\frac {1+2\,x+3\,x^{2}}{4+5\,x+6\,x^{2}}}
H = hilb(3);
latex(H,'hilb.tex')
\left [\begin {array}{ccc}
1&1/2&1/3\\\noalign{\medskip}1/2&1/3&1/4
\\\noalign{\medskip}1/3&1/4&1/5\end {array}\right ]
```

See Also `pretty`

---

linsolve

Purpose Solution of simultaneous linear equations.

Synopsis

```
X = linsolve(A,b)
[X,Z] = linsolve(A,b)
```

Description `X = linsolve(A,b)` solves the system `A*X = b`. `A` is an `n`-by-`n` symbolic or numeric matrix. `b` is symbolic or numeric column vector of length `n`. A warning message is printed if the matrix is singular.

`[X,Z] = linsolve(A,b)` also computes `Z`, a basis for the nullspace of `A`. The general solution to the linear system is

```
X + Z*p
```

where `p` is a vector (or matrix) of free parameters.

Example The statements

```
A = sym('1,2,t;0,1,0;t,1,0')
b = sym('2;1;3')
x = linsolve(A,b)
```

return

```
A =
[1,2,t]
[0,1,0]
[t,1,0]

b =
[2]
[1]
[3]

x =
[   2/t]
[     1]
[-2/t^2]
```

See Also `nullspace`, `solve`
`\ (backslash)` in *MATLAB Reference*

---

nullspace

Purpose Basis for nullspace.

Synopsis

```
Z = nullspace(A)
```

Description `nullspace(A)` returns a matrix whose columns form a basis for the nullspace of `A`. `A` is a symbolic or numeric matrix. If `A` has full rank, `nullspace(A)` is the empty matrix.

If `Z = nullspace(A)`, `symsize(Z,2)` is the nullity of `A`, and `symmul(A,Z)` is a zero matrix.

Example The statements

```
A = sym('1,1,1,0;t,1,0,1')
Z = nullspace(A)
```

return

```
A =
[1,1,1,0]
[t,1,0,1]

Z =
[ -1, -1]
[  1,  0]
[  0,  1]
[t-1,  t]
```

See Also `colspace`, `linsolve`
`null` in *MATLAB Reference*

numden

Purpose Numerator and denominator.

Synopsis

```
[N,D] = numden(A)
```

Description `[N,D] = numden(A)` converts each element of `A` to a rational form where the numerator and denominator are relatively prime polynomials with integer coefficients. `A` is a symbolic or a numeric matrix. `N` is the symbolic matrix of numerators, and `D` is the symbolic matrix of denominators.

Examples `[n,d] = numden(4/5)` returns `n = 4` and `d = 5`.

`[n,d] = numden('x/y + y/x')` returns

```
n =
x^2+y^2

d =
y*x
```

The statements

```
A = sym('a, 1/b')
[n,d] = numden(A)
```

return

```
A =
[a, 1/b]

n =
[a, 1]

d =
[1, b]
```

numeric

Purpose Convert symbolic matrix to MATLAB numeric form.

Synopsis

```
R = numeric(S)
R = numeric
```

Description `numeric(S)` converts the symbolic matrix `S` to numeric form. `S` must not contain any symbolic variables.

`numeric` converts the previous symbolic matrix.

Examples `numeric('(1+sqrt(5))/2')` returns `1.6180`.

The following statements

```
S = sym('cos(4*t),sin(4*t)')
T = subs(S,pi/6)
numeric
```

return

```
S =
[cos(4*t),sin(4*t)]

T =
[cos(2/3*pi), sin(2/3*pi)]

ans =
   -0.5000    0.8660
```

See Also vpa

---

poly2sym

Purpose Convert polynomial coefficient vector to symbolic polynomial.

Synopsis

```
r = poly2sym(c)
r = poly2sym(c,'v')
```

Description `r = poly2sym(c)` returns a symbolic representation of the polynomial whose coefficents are in the numeric vector `c`. The symbolic variable is `x`. If `c = [c1 c2 ... cn]`, then `r` has the form

$$c_1 x^{n-1} + c_2 x^{n-2} + ... + c_n$$

`poly2sym(c,'v')` generates the polynomial in *v*, rather than the default `x`.

The coefficients are approximated, if necessary, by the rational values obtained from `symrat`. If `x` has a numeric value and the elements of `c` are reproduced exactly by `symrat`, `eval(poly2sym(c))` returns the same value as `polyval(c,x)`.

Examples `poly2sym([1 3 2])` returns

```
x^2 + 3*x + 2
```

`poly2sym([1 0 2],'y')` returns

```
y^2 + 2
```

See Also `sym2poly`, `symrat`
`polyval` in *MATLAB Reference*

pretty

Purpose Pretty print symbolic output.

Synopsis

```
pretty(S)
pretty(S,n)
pretty
```

Description The `pretty` function prints symbolic output in a format that resembles typeset mathematics.

`pretty(S)` pretty prints the symbolic matrix `S` using the default line width of `79`.

`pretty(S,n)` pretty prints `S` using line width `n` instead of `79`.

`pretty`, with no arguments, pretty prints the most recent result returned from Maple. Note that this is not necessarily `ans`.

Example The following statements:

```
A = sym(pascal(2))
B = eigensys(A)
pretty
```

return

```
A =
[1, 1]
[1, 2]

B =
[3/2+1/2*5^(1/2)]
[3/2-1/2*5^(1/2)]

[              1/2 ]
[ 3/2 + 1/2 5      ]
[                  ]
[              1/2 ]
[ 3/2 - 1/2 5      ]
```

rsums

Purpose Interactive evaluation of Riemann sums.

Synopsis

```
rsums(f)
```

Description `rsums(f)` interactively approximates the integral of *f(x)* by Riemann sums. `rsums(f)` displays a graph of *f(x)*. You can then adjust the number of terms taken in the Riemann sum by using the slider below the graph. The number of terms available ranges from 2 to 256.

Example

```
rsums exp(-5*x^2)
```

creates the following plot

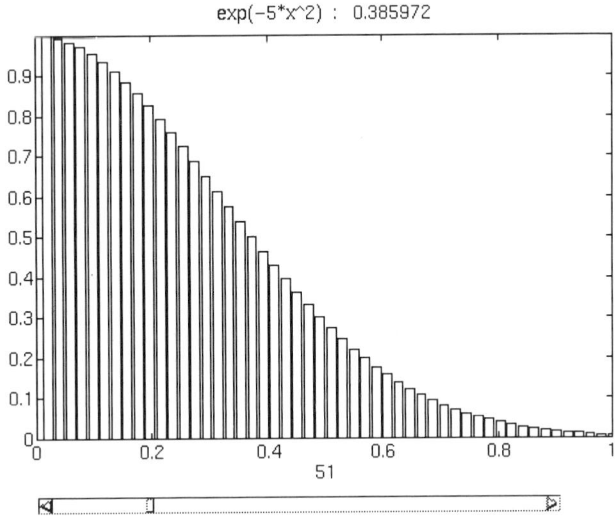

simple

Purpose Search for simplest form.

Synopsis

```
r = simple('expr')
[r,how] = simple('expr')
simple
```

Description `simple('expr')` performs several different algebraic simplifications of the symbolic expression *expr*. `simple('expr')` displays the simplifications that shorten the length of *expr* and returns the shortest one.

`[r,how] = simple('expr')` does not display intermediate simplifications. `r` is the shortest simplification found, and `how` is a string describing this simplification.

`simple` uses the previous expression.

Examples

expr	r	how
cos(x)^2+sin(x)^2	1	simplify
2*cos(x)^2–sin(x)^2	3*cos(x)^2–1	simplify
cos(x)^2–sin(x)^2	cos(2*x)	combine(trig)
cos(x)+(–sin(x)^2)^(1/2)	cos(x)+i*sin(x)	radsimp
cos(x)+i*sin(x)	exp(i*x)	convert(exp)
(x+1)*x*(x–1)	x^3–x	collect(x)
x^3+3*x^2+3*x+1	(x+1)^3	factor
cos(3*acos(x))	4*x^3–3*x	expand

See Also `simplify`, `factor`, `expand`, `collect`, `horner`

---

simplify

Purpose Symbolic simplification.

Synopsis

```
R = simplify(S)
```

Description `simplify(S)` simplifies each element of the symbolic matrix `S` using Maple simplification rules.

Examples `simplify('sin(x)^2 + cos(x)^2')` returns

```
1
```

`simplify(exp(c*ln(sqrt(a+b))))` returns

```
(a+b)^(1/2*c)
```

The statements

```
S = sym('(x^2+5*x+6)/(x+2),sqrt(16)');
R = simplify(S)
```

return

```
R = [x+3,4]
```

See Also `simple`, `factor`, `expand`, `collect`, `horner`

singvals

Purpose Symbolic matrix singular values and singular vectors.

Synopsis

```
singvals(A)
singvals(VPA(A))
[U,S,V] = singvals(VPA(A))
```

Description `singvals(A)` computes symbolic singular values of a matrix A.

`singvals(VPA(A))` computes numerical singular values using variable precision arithmetic.

`[U,S,V] = singvals(VPA(A))` produces two variable precision orthogonal matrices, `U` and `V`, and a diagonal `vpa` matrix, `S`, so that

```
symop(U,'*',S,'*',transpose(V))
```

equals `A`.

Symbolic singular vectors are not directly available.

Examples The example is

```
A = sym('[a, b, c; 0, a, b; 0, 0, a]');
s = singvals(A)
```

The example is

```
A = magic(8);
s = singvals(A)
[U,S,V] = singvals(vpa(A))
```

See Also `eigensys, vpa`

solve

Purpose Symbolic solution of algebraic equations.

Synopsis

```
g = solve(f)
g = solve(f,'v')
g = solve(f1,f2,...,fn)
g = solve(f1,f2,...,fn,'v1,v2,...,vn')
[g1,g2,...,gn] = solve(...)
```

Description

Single Equation

`solve(f)` solves the symbolic equation `f` for its symbolic variable as determined by `symvar`. The result is a symbolic column vector. `f` may also be a symbolic expression. In this case, the symbolic equation is implicitly `f = 0`.

`solve(f,'v')` solves the equation `f` for the variable `v`.

System of Equations

`solve(f1,f2,...,fn)` solves the `n` symbolic equations specified by `f1, f2,..., fn` for the `n` unknowns as determined by `symvar`. The result is a string that lists the solutions.

`solve(f1,f2,...,fn,'v1,v2,...,vn')` solves the `n` symbolic equations specified by `f1, f2,..., fn` using `v1,v2,...,vn` as the unknowns.

`[g1,g2,...,gn] = solve(...)` returns the solution in the form of symbolic column vectors `g1,g2,...,gn`.

For both single equations and systems of equations, numeric solutions are returned if symbolic solutions cannot be found.

Examples

`solve('a*x^2 + b*x + c')` returns

```
[1/2/a*(-b+(b^2-4*a*c)^(1/2))]
[1/2/a*(-b-(b^2-4*a*c)^(1/2))]
```

`solve('a*x^2 + b*x + c','b')` returns

```
-(a*x^2+c)/x
```

`solve('x + y  = 1','x - 11*y = 5')` returns

```
 y = -1/3, x = 4/3
```

The statement

```
[a,u,v] = solve('a*u^2 + v^2', 'u - v = 1', 'a^2 - 5*a + 6')
```

returns

```
a =
[2]
[3]

u =
[RootOf(3*_Z^2+4*_Z+2)+1]
[RootOf(4*_Z^2+6*_Z+3)+1]

v =
[RootOf(3*_Z^2+4*_Z+2)]
[RootOf(4*_Z^2+6*_Z+3)]
```

See Also `allvalues`, `symvar`, `dsolve`, `linsolve`

subs

Purpose Symbolic substitution in a symbolic expression or matrix.

Synopsis

```
R = subs(S,new)
R = subs(S,new,old)
```

Description `subs(S,`*`new`*`)` replaces all occurrences of the symbolic variable in `S` by *new*. `S` is a symbolic expression or matrix. *new* is a symbolic variable or a symbolic expression.

`subs(S,`*`new`*`,`*`old`*`)` replaces all occurrences of *`old`* in `S` by *new*. `S` is a symbolic expression or matrix. *new* and *`old`* can be symbolic variables or symbolic expressions.

Examples `subs sin(z) x+i*y` returns

```
sin(x+i*y)
```

`subs(sym('sin(u),cos(u)'),'a+w*t')` returns

```
[sin(a+w*t), cos(a+w*t)]
```

`subs('F(pi*r^2)','sqrt(x^2+y^2)','r')` returns

```
F(pi*(x^2+y^2))
```

See Also `simplify`

svdvpa

Purpose Variable precision singular value decomposition.

Synopsis

```
R = svdvpa(A)
[U,S,V] = svdvpa(A)
```

Description

`svdvpa(A)` uses variable precision arithmetic, with an accuracy determined by `digits`, to compute the singular values of `A`, a numeric real or a variable precision real matrix. (Note that computation of singular values of symbolic matrices is not available.)

`[U,S,V] = svdvpa(A)` returns two orthogonal variable precision matrices, `U` and `V`, and a diagonal variable precision matrix `S` such that

$$A = USV'$$

Examples

The statements

```
digits(25)
svdvpa(vpa(magic(3)))
```

return

```
[15.00000000000000000000002]
[6.928203230275509174109786]
[3.464101615137754587054892]
```

The statements

```
digits(20)
X = [1 1;1 1;0 0]
[u,s,v] = svdvpa(X)
```

return

```
X =
      1     1
      1     1
      0     0

u =
[ -.70710678118654752440, -.70710678118654752440,  0]
[-.70710678118654752440,   .70710678118654752440,  0]
[                      0,                      0, 1.]

s =
[2.0000000000000000000, 0]
[                    0, 0]
[                    0, 0]

v =
[-.70710678118654752440, -.70710678118654752440]
[-.70710678118654752440,  .70710678118654752440]
```

See Also

`vpa,digits,eigensys`

sym

Purpose Create, access, or modify a symbolic matrix.

Synopsis

```
R = sym(X)
R = sym(m,n,'expr')
R = sym(m,n,'p','q','expr')
R = sym('[s11,s12,...,s1n;s21,s22,...;...,smn]')
r = sym(S,i,j)
R = sym(S,i,j,s)
```

Description `sym(X)` converts the numeric matrix `X` to its symbolic form with exact rational representation (obtained using the `symrat` function) of the elements.

`sym(m,n,'expr')` creates an `m`-by-`n` symbolic matrix by expanding the symbolic expression *expr*. *expr* typically contains the characters `i` and `j` to denote row and column indices. Each element of the result is *expr* evaluated at the corresponding row and column. *expr* can contain other symbolic variables as well.

`sym(m,n,'p','q','expr')` uses the symbolic variables *p* and *q* as the row and column indices instead of `i` and `j`, respectively.

`sym('[s11,s12,...,s1n;s21,s22,...; ...,smn]')` creates an `m`-by-`n` symbolic matrix using the symbolic elements *s11*, *s12*, ..., *smn*. The brackets are optional. This form of `sym` mimics the generation of numeric matrices in MATLAB. Individual elements are separated by commas or blanks. Semicolons terminate rows.

`sym(S,i,j)` returns the element in the `i`th row and `j`th column of the symbolic matrix `S`.

`sym(S,i,j,s)` replaces the element in the `i`th row and `j`th column of the symbolic matrix `S` with the symbolic or numeric expression `s`.

Examples `sym(hilb(3))` creates the rational symbolic representation of the 3-by-3 Hilbert matrix:

```
[  1, 1/2, 1/3]
[1/2, 1/3, 1/4]
[1/3, 1/4, 1/5]
```

`sym(3,3,'1/(i+j-t)')` generalizes the Hilbert matrix:

```
[1/(2-t), 1/(3-t), 1/(4-t)]
[1/(3-t), 1/(4-t), 1/(5-t)]
[1/(4-t), 1/(5-t), 1/(6-t)]
```

As another example, `S = sym('a,2*b,3*c;0,5*b,6*c;0,0,7*c')` creates the following upper triangular symbolic matrix:

```
[a,2*b,3*c]
[0,5*b,6*c]
[0,  0,7*c]
```

`sym(S,1,2,'4*c')` modifies `S` to be:

```
[a,4*c,3*c]
[0,5*b,6*c]
[0,  0,7*c]
```

Finally, `sym(S,1,3)` returns `3*c`.

See Also `vpa`, `symop`

sym2poly

Purpose Convert symbolic polynomial to polynomial coefficient vector.

Synopsis

```
c = sym2poly(s)
```

Description `c = sym2poly(s)` returns the vector of coefficients of the symbolic polynomial `s`. `s` has the form `a1*x^(n-1) + a2*x^(n-2) + ... + an`, where `a1,...,an` are numbers or numeric expressions and `x` is the symbolic variable of `s`. The order of the terms may vary. The result is a numeric vector whose elements are the coefficients of `s` ordered in descending powers of `x`, that is, `c = numeric([a1 a2 ... an])`.

Examples `sym2poly('x^3 - 2*x - 5')` returns

```
1     0    -2    -5
```

`sym2poly('u^4 - 3 + 5*u^2')` returns

```
1     0     5     0    -3
```

`sym2poly('sin(pi/6)*v + exp(1)*v^2')` returns

```
2.7183    0.5000         0
```

See Also `poly2sym`
`polyval` in *MATLAB Reference*

symadd

Purpose Symbolic addition.

Synopsis

```
R = symadd(A,B)
```

Description

`symadd(A,B)` symbolically computes `A+B`. `A` and `B` must have the same dimensions. Each argument can be symbolic or numeric.

Examples

`symadd('cos(t)','t')` returns

```
cos(t)+t
```

`symadd('sqrt(x)',1)` returns

```
x^(1/2)+1
```

The following statements

```
A = sym('x^2, 1, 2')
B = sym('y, 0, y')
R = symadd(A,B)
```

return

```
A =
[x^2, 1, 2]

B =
[y, 0, y]

R =
[x^2+y, 1, 2+y]
```

See Also

`symsub`, `symop`, `symsum`

---

symdiv

Purpose

Symbolic division.

Synopsis

```
R = symdiv(A,B)
```

Description

`symdiv(A,B)` is the symbolic computation `A/B`, which is `A*inverse(B)`. Each input argument can be symbolic or numeric. `B` must be a square matrix whose dimensions are the same as the number of columns of `A`.

Examples

`symdiv('t^2-4','t-2')` returns `(t^2-4)/(t-2)`.
`simplify(symdiv('t^2-4','t-2'))` returns `t+2`.

`symdiv('2*cos(t)+6',3)` returns `2/3*cos(t)+2`.

Suppose

```
A =
[   2, a + 3/2]
[ 7/6, a/2 + 1]

B =
[   1, 1/2]
[ 1/2, 1/3]
```

Then `symdiv(A,B)` returns

```
[   -1-6*a, 6+12*a]
[-4/3-3*a,   5+6*a]
```

`symdiv(B,5)` returns

```
[ 1/5, 1/10]
[1/10, 1/15]
```

See Also `symmul`, `sympow`, `symop`, `inverse`

---

symmul

Purpose Symbolic multiplication.

Synopsis

```
R = symmul(A,B)
```

Description `symmul(A,B)` is the symbolic linear algebraic product of the matrices `A` and `B`. The number of columns of `A` must equal the number of rows of `B`. Each input argument can be symbolic or numeric.

Examples `symmul('x','exp(x)')` returns

```
x*exp(x)
```

`symmul('y+1',3)` returns

```
3*y+3
```

The statements

```
A = sym('a,b;c,d')
v = sym('x;y')
R = symmul(A,v)
```

return

```
A =
[ a, b]
[ c, d]

v =
[ x]
[ y]

R =
[a*x+b*y]
[c*x+d*y]
```

See Also symdiv, sympow, symop

symop

Purpose Symbolic operations.

Synopsis

```
R = symop(arg1,arg2,arg3,...)
```

Description symop(arg1,arg2,arg3,...) concatenates the input arguments and returns the resulting symbolic matrix. Each argument can be a symbolic matrix, a numeric matrix, or one of the following operators:

```
+ - * / ^ ( )
```

Input operand arguments must have the appropriate dimensions for the corresponding operator.

Scalar-matrix multiplications must be done in one of the following orders:

$$\text{scalar}_1*\ldots*\text{scalar}_m * \text{matrix}_1*\ldots*\text{matrix}_n$$
$$\text{matrix}_1*\ldots*\text{matrix}_n * \text{scalar}_1*\ldots*\text{scalar}_m$$

That is, scalar multiplications and matrix multiplications must be grouped together, not interspersed.

symop allows up to 16 arguments.

Examples To minimize typing when using symop, each argument should contain the largest contiguous text string possible. For example, suppose s is sin(x). symop('4*x*',s) and symop('4','*','x','*',s) both produce 4*x*sin(x). However, the first form requires fewer keystrokes.

The statements

```
x = 'x';
f = symop(1,'+',x,'+',x,'^',2,'/',2)
g = symop(f,'-',x)
```

return

```
f =
   1+x+1/2*x^2
g =
   1+1/2*x^2
```

The statements

```
G = sym('[c, s; -s, c]');
H = symop(G,'*',transpose(G));
pretty(H)
```

return

```
[  2    2           ]
[ c  + s      0     ]
[                   ]
[            2    2 ]
[    0      c  + s  ]
```

See Also `symadd`, `symsub`, `symmul`, `symdiv`, `sympow`

---

sympow

Purpose Power of a symbolic expression or matrix.

Synopsis

```
R = sympow(S,p)
```

Description `sympow(S,p)` computes `S^p`. If `S` is a scalar symbolic expression, `p` can be a scalar symbolic or scalar numeric expression. If `S` is a symbolic matrix, `S` must be square, and `p` must be an integer.

Examples `sympow('exp(t)',2)` returns `exp(t)^2`.

`sympow(10,'y')` returns `10^y`.

The statements

```
S = sym('a, 1; 0, d')
R = sympow(S,3)
```

return

```
S =
[ a, 1]
[ 0, d]

R =
[a^3, a^2+d^2+a*d]
[  0,         d^3]
```

See Also `symmul`, `symop`

symrat

Purpose Symbolic rational approximation.

Synopsis

```
symrat(x)
```

Description `symrat(x)`, for scalar `x`, is a string representation of an integer, the ratio of two integers, the ratio of two integers times `pi`, or an integer times a power of 2. When the string is evaluated with MATLAB floating point arithmetic, the result reproduces `x` exactly.

Examples

x	symrat(x)
22/7	22/7
2*pi/3	2*pi/3
1.e12	1000000000000
eps	2^(-52)

See Also `poly2sym`, `sym`

symsize

Purpose Symbolic matrix dimensions.

Synopsis

```
d = symsize(A)
[m,n] = symsize(A)
m = symsize(A,1)
n = symsize(A,2)
```

Description Suppose `A` is an `m`-by-`n` symbolic or numeric matrix. The statement `d = symsize(A)` returns a numeric vector with two integer components, `d = [m,n]`.

The multiple assignment statement `[m,n] = symsize(A)` returns the two integers in two separate variables.

The statements `m = symsize(A,1)` and `n = symsize(A,2)` return the number of rows and columns of `A`, respectively.

Example Suppose `S` is

```
[atan(u+v), exp(t), t]
[      u*v,      2, t]
```

Then the statement `symsize(S)` returns `[2 3]` and the statement `symsize(S,2)` returns 3.

symsub

Purpose Symbolic subtraction.

Synopsis

```
R = symsub(A,B)
```

Description `symsub(A,B)` symbolically computes `A–B`. `A` and `B` must have the same dimensions. Each argument may be symbolic or numeric.

Example The following statements

```
A = sym('sin(t), 1, 2')
B = sym('t, 1, t')
R = symsub(A,B)
```

return

```
A =
[sin(t), 1, 2]

B =
[t, 1, t]

R =
[sin(t)-t, 0, 2-t]
```

See Also `symadd`, `symop`

symsum

Purpose Symbolic summation.

Synopsis

```
r = symsum(s)
r = symsum(s,'v')
symsum
r = symsum(s,a,b)
r = symsum(s,'v',a,b)
```

Description `symsum(s)` is the indefinite summation of the symbolic expression `s` with respect to its symbolic variable as determined by `symvar`.

`symsum(s,'v')` is the indefinite summation of the symbolic expression `s` with respect to the symbolic variable `v`.

`symsum` is the indefinite summation of the previous expression with repect to its symbolic variable.

`symsum(s,a,b)` is the definite summation of the symbolic expressions with respect to its symbolic variable from `a` to `b`.

`symsum(s,'v',a,b)` is the definite summation of the symbolic expressions with respect to `v` from `a` to `b`.

Examples `symsum k^2` returns

```
1/3*k^3-1/2*k^2+1/6*k
```

`symsum('k^2',0,'n-1')` returns

```
1/3*n^3-1/2*n^2+1/6*n
```

`symsum('k^2',0,10)` returns

```
385
```

`symsum('x^k/k!','k',0,Inf)` returns

```
exp(x)
```

See Also `symvar,int,symadd`

---

symvar

Purpose Find the symbolic variables in a symbolic matrix.

Synopsis

```
r = symvar(S)
r = symvar(S,n)
r = symvar(S,N)
```

Description A symbolic variable in a symbolic matrix is a single lower-case alphabetic character, other than `i` or `j`, that is not part of a word formed from several alphabetic characters.

`symvar(S)` searches the symbolic matrix `S` for its symbolic variable. The result is returned according to the following table:

Symbolic variable of S	**symvar(S)**
Exists and is unique.	Symbolic variable in `S`.
Does not exist.	`x`
Not unique.	Alphabetically closest variable to `x`.

`symvar(S,n)`, where `n` is a scalar integer, searches `S` for `n` different symbolic variables. If exactly `n` symbolic variables exist, `symvar(S,n)` returns a list of them. If not, an error results.

`symvar(S,N)` searches `S` for several different symbolic variables. `N` is an integer vector with at least two components. With this form, `symvar` never returns an error. If the number found is between `min(N)` and `max(N)`, a list of these symbolic variables is returned. If the number found is less than `min(N)`, an empty matrix is returned. If the number found is greater than `max(N)`, `NaN` is returned.

Examples

`symvar('sin(pi*t)')` returns `t`

`symvar('3*i+4*j')` returns `x`

`symvar('a+y')` returns `y`

If you have the following statements:

```
f = '3*x+4*y'
g = 'x+y+t'
```

then

`symvar(f)` returns `x`

`symvar(f,2)` returns `x,y`

`symvar(g,1:2)` returns `NaN`

`symvar(g,1:3)` returns `t,x,y`

`symvar(g,2)` is an error.

See Also

`diff`, `int`

---

taylor

Purpose

Taylor series expansion.

Synopsis

```
r = taylor('f')
r = taylor('f','v')
r = taylor('f',n)
```

Description

`taylor('f')` returns the Taylor series expansion of *f*. *f* is a symbolic expression representing a function in one variable. `taylor('f')` returns the first six terms of *f*'s Taylor series expansion about zero and the order of the resulting truncation error. The expansion variable is determined by `symvar`.

`taylor('f','v')` uses the variable *v* in the Taylor series expansion.

`taylor('f',n)` returns n terms in the series rather than the default six terms.

Examples `taylor('exp(-x)')` with pretty-printing return

```
                 2         3          4           5      6
[1 - x + 1/2 x  - 1/6 x  + 1/24 x - 1/120 x +  O(x )]
```

This is the first six terms of the Taylor expansion and the resulting truncation error, which is of order x^6.

To get two additional terms in the series, you can use

```
taylor('exp(-x)',8)
```

which returns

```
                  2         3          4           5
[ 1 - x + 1/2 x  - 1/6 x  + 1/24 x  - 1/120 x

                 6            7    8
      + 1/720 x  - 1/5040 x  + O) ]
```

`taylor('cos(u+v)','u',3)` returns

```
cos(v)+(-sin(v))*u+(-1/2*cos(v))*u^2+O(u^3)
```

See Also `symvar`

---

transpose

Purpose Symbolic matrix transpose.

Synopsis

```
R = transpose(A)
```

Description `transpose(A)` returns the transpose of `A`, which is a symbolic or numeric matrix.

Example The statements

```
S = sym('a, b; 0, c')
R = transpose(S)
```

return

```
S =
[ a, b]
[ 0, c]

R =
[a, 0]
[b, c]
```

vpa

Purpose Variable precision arithmetic.

Synopsis

```
R = vpa(A)
R = vpa(A,d)
vpa
```

Description `vpa(A)` numerically evaluates each element of `A` using variable precision floating-point arithmetic with `d` decimal digit accuracy. `d` is the current setting of `digits`. Each element of the result is a symbolic expression.

`vpa(A,d)` uses `d` digits, instead of the current setting of `digits`.

`vpa` evaluates the previous symbolic matrix.

Examples The statements

```
digits(25)
T = vpa([pi,sin(pi/6)])
w = vpa('(1+sqrt(5))/2')
```

return

```
T =
[3.141592653589793238462643, .5000000000000000000000000]

w =
1.618033988749894848204587
```

`vpa pi 75` computes π to 75 digits.

The statements

```
A = vpa(hilb(2),25)
B = vpa(hilb(2),5)
```

return

```
A =
[                        1., .5000000000000000000000000]
[.5000000000000000000000000, .3333333333333333333333333]

B =
[    1., .50000]
[.50000, .33333]
```

See Also `digits`, `numeric`

ztrans

Purpose z-transform.

Synopsis

```
F = ztrans(f)
F = ztrans(f,'v')
F = ztrans(f,'v','x')
F = ztrans
```

Description F = ztrans(f) is the z-transform of the symbolic expression `f`,
`F(z) = symsum(f(n)/z^n,'n',0,inf)`

`F = ztrans(f,'v')` is a function of `'v'` instead of 'z'.

`F = ztrans(f,'v','x')` assumes `f` is a function of `'x'` instead `'n'`.

`F = ztrans`, with no input arguments, transforms the previous result.

Examples `ztrans 1` is `z/(z-1)`

`ztrans a^n` is `z/(z-a)`

`ztrans sin(n*pi/2)` is `z/(1+z^2)`

`ztrans('x^k/k!','z','k')` is `exp(1/z*x)`

`ztrans('f(n+1)')` is `z*ztrans(f(n),n,z)-f(0)*z`

See Also `invztrans`, `laplace`, `fourier`

10 Signals and Systems Toolbox Reference

10.1 Reference Tables

Model Building	
`append`	Append system dynamics.
`cloop`	Close loops of system.
`feedback`	Feedback system connection.
`rmodel`	Generate random continuous model.
`series`	Series system connection.
`ssselect`	Select subsystem from larger system.

Model Conversions	
`c2d`	Continuous to discrete-time conversion.
`c2dm`	Continuous to discrete-time conversion with method.
`ss2tf`	State-space to transfer function conversion.
`ss2zp`	State-space to zero-pole conversion.
`tf2ss`	Transfer function to state-space conversion.
`tf2zp`	Transfer function to zero-pole conversion.
`zp2tf`	Zero-pole to transfer function conversion.
`zp2ss`	Zero-pole to state-space conversion.

Model Realizations	
`canon`	Conversion of system to canonical form.

Model Properties	
`ctrb`	Controllability matrix.
`damp`	Damping factors and natural frequencies.
`ddamp`	Discrete damping factors and natural frequencies.
`dsort`	Sort discrete eigenvalues by magnitude.
`esort`	Sort continuous eigenvalues by real part.
`obsv`	Observability matrix.
`tzero`	Transmission zeros.

Time Response	
`dimpulse`	Discrete unit sample response.
`dlsim`	Discrete simulation to arbitrary inputs.
`dstep`	Discrete step response.
`impulse`	Impulse response.
`lsim`	Continuous simulation to arbitrary inputs.
`step`	Step response.

Frequency Response	
`bode`	Bode plots.
`freqs`	Laplace-transform.
`freqz`	z- transform.
`nyquist`	Nyquist plots.

Root Locus	
`rlocus`	Evans root-locus.

Gain Selection	
`acker`	SISO pole placement.
`lqe`	Linear-quadratic estimator design.
`lqr`	Linear-quadratic regulator design.
`place`	Pole placement.

Waveform Generation	
`sinc`	Sinc or $\sin(\pi x)/\pi x$ function.

Filter Analysis/Implementation	
`freqs`	Laplace transform frequency response.
`freqz`	z-transform frequency response.
`grpdelay`	Group delay.
`zplane`	Discrete pole-zero plot.

Linear System Transformations	
`residuez`	Partial fraction expansion.
`sos2ss`	Second-order sections to state-space conversion.
`sos2tf`	Second-order sections to transfer function conversion.
`sos2zp`	Second-order sections to zero-pole conversion.
`ss2sos`	State-space to second-order sections conversion.
`ss2tf`	State-space to transfer function conversion.
`ss2zp`	State-space to zero-pole conversion.
`tf2ss`	Transfer function to state-space conversion.
`tf2zp`	Transfer function to zero-pole conversion.
`zp2sos`	Zero-pole to second-order sections conversion.
`zp2ss`	Zero-pole to state-space conversion.
`zp2tf`	Zero-pole to transfer function conversion.

IIR Filter Design - Classical and Direct	
`butter`	Butterworth filter design.
`cheby1`	Chebyshev type I filter design.
`cheby2`	Chebyshev type II filter design.
`ellip`	Elliptic filter design.
`yulewalk`	Yule-Walker filter design.

FIR Filter Design	
`remez`	Parks-McClellan optimal FIR filter design.

Statistical Signal Processing	
`psd`	Power spectrum estimation.

Windows	
`boxcar`	Rectangular window.
`hamming`	Hamming window.
`hanning`	Hanning window.
`kaiser`	Kaiser window.
`triang`	Triangular window.

Specialized Operations	
`specgram`	Spectrogram.

Analog Prototype Design	
`buttap`	Butterworth filter prototype.
`cheb1ap`	Chebyshev type I filter (passband ripple) prototype.
`cheb2ap`	Chebyshev type II filter (stopband ripple) prototype.
`ellipap`	Elliptic filter prototype.

Frequency Translation	
lp2bp	Lowpass to bandpass transformation.
lp2bs	Lowpass to bandstop transformation.
lp2hp	Lowpass to highpass transformation
lp2lp	Lowpass to lowpass transformation.

Filter Discretization	
bilinear	Bilinear transformation.
impinvar	Impulse invariance method of A/D filter conversion.

Other	
detrend	Linear trend removal.
polystab	Polynomial stabilization.
strips	Strip plot.

10.2 Signals and Systems Toolbox Reference

append

Purpose Combine dynamics of two state-space systems.

Synopsis `[a,b,c,d] = append(a1,b1,c1,d1,a2,b2,c2,d2)`

Description append appends the dynamics of two state-space systems, forming an augmented model:

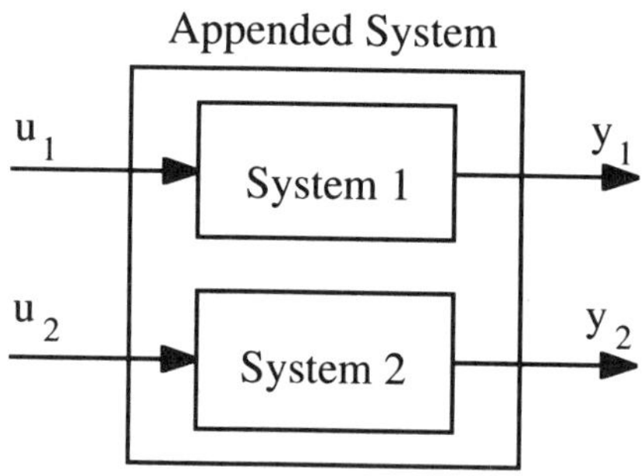

`[a,b,c,d] = append(a1,b1,c1,d1,a2,b2,c2,d2)` produces an aggregate state-space system consisting of the appended dynamics of systems 1 and 2. The resulting system is

$$\begin{bmatrix} \dot{x}_1 \\ \dot{x}_2 \end{bmatrix} = \begin{bmatrix} A_1 & 0 \\ 0 & A_2 \end{bmatrix} \begin{bmatrix} x_1 \\ x_2 \end{bmatrix} + \begin{bmatrix} B_1 & 0 \\ 0 & B_2 \end{bmatrix} \begin{bmatrix} u_1 \\ u_2 \end{bmatrix}$$

$$\begin{bmatrix} y_1 \\ y_2 \end{bmatrix} = \begin{bmatrix} C_1 & 0 \\ 0 & C_2 \end{bmatrix} \begin{bmatrix} x_1 \\ x_2 \end{bmatrix} + \begin{bmatrix} D_1 & 0 \\ 0 & D_2 \end{bmatrix} \begin{bmatrix} u_1 \\ u_2 \end{bmatrix}$$

See Also `augstate`, `parallel`, `series`, `cloop`, `feedback`, `are`

---

bilinear

Purpose Map variables using bilinear transformation.

Synopsis

```
[zd,pd,kd] = bilinear(z,p,k,Fs)
[zd,pd,kd] = bilinear(z,p,k,Fs,Fp)
[numd,dend] = bilinear(num,den,Fs)
[numd,dend] = bilinear(num,den,Fs,Fp)
[Ad,Bd,Cd,Dd] = bilinear(A,B,C,D,Fs)
[Ad,Bd,Cd,Dd] = bilinear(A,B,C,D,Fs,Fp)
```

Description The *bilinear transformation* is a mathematical mapping of variables. In digital filtering, it is a standard method of mapping the s or analog plane into the z or digital plane. It transforms analog filters, designed using classical filter design techniques, into their discrete equivalents.

The bilinear transformation maps the s-plane into the z-plane by

$$H(z) = H(s)\Big|_{s=2f_s\frac{z-1}{z+1}}$$

This transformation maps the $j\Omega$ axis (from $\Omega = -\infty$ to $+\infty$) repeatedly around the unit circle ($\exp(j\omega)$, from $\omega = -\pi$ to π) by

$$\omega = 2\tan^{-1}\left(\frac{\Omega}{2f_s}\right)$$

`bilinear` can accept an optional parameter `Fp` that specifies prewarping. `Fp`, in Hertz, indicates a "match" frequency, that is, a frequency for which the frequency responses before and after mapping match exactly. In prewarped mode, the bilinear transformation maps the s-plane into the z-plane with

$$H(z) = H(s)\Big|_{s=\frac{2\pi f_p}{\tan\left(\pi\frac{f_p}{f_s}\right)}\frac{(z-1)}{(z+1)}}$$

With the prewarping option, `bilinear` maps the $j\Omega$ axis (from $\Omega = -\infty$ to $+\infty$) repeatedly around the unit circle ($\exp(j\omega)$, from $\omega = -\pi$ to π) by

$$\omega = 2\tan^{-1}\left(\frac{\Omega\tan\left(\pi\frac{f_p}{f_s}\right)}{2\pi f_p}\right)$$

In prewarped mode, `bilinear` matches the frequency $2\pi f_p$ (in radians/sec) in the s-plane to the normalized frequency $2\pi f_p/f_s$ (in radians/sec) in the z-plane.

The `bilinear` function works with three different linear system representations: zero-pole-gain, transfer function, and state-space form.

Zero-pole-gain

`[zd,pd,kd] = bilinear(z,p,k,Fs)` and
`[zd,pd,kd] = bilinear(z,p,k,Fs,Fp)` convert the s-domain transfer function specified by `z`, `p`, and `k` to a discrete equivalent. Inputs `z` and `p` are column vectors containing the zeros and poles, while `k` is a scalar gain. `Fs` is the sampling frequency in Hertz. `bilinear` returns the discrete equivalent in column vectors `zd`, `pd`, and scalar `kd`. `Fp` is the optional match frequency, in Hertz, for prewarping.

Transfer Function

`[numd,dend] = bilinear(num,den,Fs)` and
`[numd,dend] = bilinear(num,den,Fs,Fp)` convert an s-domain transfer function given by `num` and `den` to a discrete equivalent. Row vectors `num` and `den` spec-

ify the coefficients of the numerator and denominator, respectively, in descending powers of s

$$\frac{num(s)}{den(s)} = \frac{num(1)s^{nn} + \cdots + num(nn)s + num(nn+1)}{den(1)s^{nd} + \cdots + den(nd)s + den(nd+1)}$$

and Fs is the sampling frequency in Hertz. bilinear returns the discrete equivalent in row vectors numd and dend in descending powers of z (ascending powers of z^{-1}). Fp is the optional match frequency, in Hertz, for prewarping.

State-space

[Ad,Bd,Cd,Dd] = bilinear(A,B,C,D,Fs) and
[Ad,Bd,Cd,Dd] = bilinear(A,B,C,D,Fs,Fp) convert the continuous-time state-space system in matrices A, B, C, D,

$$\begin{aligned} \dot{x} &= Ax + Bu \\ y &= Cx + Du \end{aligned}$$

to the discrete-time system

$$\begin{aligned} x[n+1] &= A_d x[n] + B_d u[n] \\ y[n] &+ C_d x[n] + D_d u[n] \end{aligned}$$

As before, Fs is the sampling frequency in Hertz. bilinear returns the discrete equivalent in matrices Ad, Bd, Cd, Dd. Fp is the optional match frequency, in Hertz, for prewarping.

Algorithm

bilinear uses one of two algorithms depending on the format of the input linear system you supply. One algorithm works on the zero-pole-gain format, and the other on the state-space format. For transfer function representations, bilinear converts to state-space form, performs the transformation, and converts the resulting state-space system back to transfer function form.

Zero-pole-gain Algorithm

For a system in zero-pole-gain form, bilinear performs four steps:

1. If Fp is present, k = 2*pi*Fp/tan(pi*Fp/Fs); else k = 2*Fs.
2. It strips any zeros at plus or minus infinity using

```
z = z(find(finite(z)));
```

3. It transforms the zeros, poles and gain using

```
pd = (1+p/k)./(1-p/k);
zd = (1+z/k)./(1-z/k);
kd = real(k*prod(fs-z)./prod(fs-p));
```

4. It adds extra zeros at −1 so the resulting system has equivalent numerator and denominator order.

State-space Algorithm

For a system in state-space form, `bilinear` performs two steps:

1. If `Fp` is present, `k = 2*pi*Fp/tan(pi*Fp/Fs)`; else `k = 2*Fs`.
2. It computes `Ad`, `Bd`, `Cd`, and `Dd` in terms of `A`, `B`, `C` and `D` using

$$A_d = \left(I + \left(\tfrac{1}{k}\right)A\right)\left(I - \left(\tfrac{1}{k}\right)A\right)^{-1}$$

$$B_d = \tfrac{2k}{r}\left(I - \left(\tfrac{1}{k}\right)A\right)^{-1}B$$

$$C_d = rC\left(I - \left(\tfrac{1}{k}\right)A\right)^{-1}$$

$$D_d = \left(\tfrac{1}{k}\right)C\left(I - \left(\tfrac{1}{k}\right)A\right)^{-1}B + D$$

`bilinear` implements these relations using conventional MATLAB statements. The scalar `r` is arbitrary; `bilinear` uses `sqrt(2/k)` to ensure good quantization noise properties in the resulting system.

Diagnostics `bilinear` requires that the numerator order is no greater than the denominator order. If this is not the case, `bilinear` displays:

```
Numerator cannot be higher order than denominator.
```

For `bilinear` to distinguish between the zero-pole-gain and transfer function linear system formats, the first two input parameters must be vectors with the same orientation in these cases.If this is not the case, `bilinear` displays:

```
First two arguments must have the same orientation.
```

See Also `impinvar`, `lp2bp`, `lp2bs`, `lp2hp`, `lp2lp`

References

[1] Parks, T.W, and C. S. Burrus. *Digital Filter Design*, pp. 209–213. New York: John Wiley & Sons, 1987.

[2] Oppenheim, A. V., and R.W. Schafer. *Discrete-Time Signal Processing*, pp. 415-430. Englewood Cliffs, NJ: Prentice-Hall, 1989.

---

bode

Purpose Bode frequency response plots.

Synopsis

```
[mag,phase,w] = bode(a,b,c,d);
[mag,phase,w] = bode(a,b,c,d,iu)
[mag,phase,w] = bode(a,b,c,d,iu,w)
[mag,phase,w] = bode(num,den)
[mag,phase,w] = bode(num,den,w)
```

Description `bode` computes the magnitude and phase frequency response of continuous-time LTI systems. Bode plots are used to analyze system properties including gain margin, phase margin, D.C. gain, bandwidth, disturbance rejection, and stability. When invoked without left-hand arguments, `bode` produces a Bode plot on the screen.

`bode(a,b,c,d)` produces a series of Bode plots, one for each input of the continuous state-space system:

$$\dot{x} = Ax + Bu$$
$$y = Cx + Du$$

with the frequency range automatically determined. More points are used where the response is changing rapidly.

`bode(a,b,c,d,iu)` produces the Bode plot from the single input `iu` to all the outputs of the system with the frequency range automatically determined. The scalar `iu` is an index into the inputs of the system and specifies which input to use for the Bode response.

`bode(num,den)` draws the Bode plot of the continuous polynomial transfer function `G(s) = num(s)/den(s)` where `num` and `den` contain the polynomial coefficients in descending powers of s.

`bode(a,b,c,d,iu,w)` or `bode(num,den,w)` uses the user-supplied frequency vector `w`. The vector `w` specifies the frequencies in radians/sec at which the Bode response is calculated. See `logspace` to generate frequency vectors that are equally spaced logarithmically in frequency.

When invoked with lefthand arguments:

```
[mag,phase,w] = bode(a,b,c,d,iu)
[mag,phase,w] = bode(a,b,c,d,iu,w)
[mag,phase,w] = bode(num,den)
[mag,phase,w] = bode(num,den,w)
```

returns the frequency response of the system in the matrices `mag`, `phase`, and `w`. No plot is drawn on the screen. The matrices `mag` and `phase` contain the magnitude and phase response of the system evaluated at the frequency values `w`. `mag` and `phase` have as many columns as outputs and one row for each element in `w`:

$$G(s) = C(sI - A)^{-1}B + D$$
$$mag(\omega) = |G(j\omega)|$$
$$phase(\omega) = \angle G(j\omega)$$

The phase is returned in degrees. The magnitude can be converted to decibels with

```
magdb = 20*log10(mag).
```

You can use `fbode` instead of `bode` for diagonalizable systems. It uses a faster algorithm based on diagonalization of the system A matrix.

Examples Plot the magnitude and phase responses of a second-order system with a natural frequency $\omega_n = 1$ and a damping factor of $\zeta = 0.2$:

```
[a,b,c,d] = ord2(1,.2);
bode(a,b,c,d);
title('Bode plot')
```

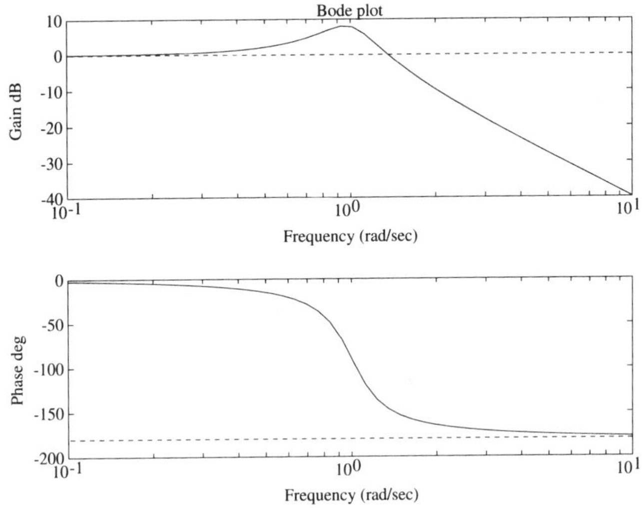

Plot the magnitude and phase of the system in Hertz instead of radians per second:

```
[mag,phase,w]=bode(a,b,c,d);
subplot(211), semilogx(w/2*pi,20*log10(mag))
xlabel('Frequency (Hz)'), ylabel('Gain dB')
title('Bode plot')
subplot(212), semilogx(w/2*pi,phase)
xlabel('Frequency (Hz)'), ylabel('Phase deg')
```

Algorithm `bode` uses the Hessenberg algorithm from [1]. The A, B, C, D matrices are balanced and A is reduced to upper Hessenberg form. Next, the linear equation $C(j\omega - A)^{-1}B$ is solved directly at each frequency point, taking advantage of the Hessenberg form. The reduction to Hessenberg form provides a good compromise between efficiency and reliability. You can comment out the last line of `bode.m` to get the magnitude returned in decibels.

The auto-selection of frequency points is performed in the M-file `freqint.m` which uses the poles and zeros of the system to estimate regions of rapid change.

`bode` adds ±360 degrees to the phase, as appropriate, when the phase goes through ±180 to make the phase continuous above or below ±180, instead of jumping and staying within ±180. You can fool this adjustment algorithm if you have only a few frequency points and the phase is changing rapidly. The M-file `unwrap.m` in the main MATLAB Toolbox performs this phase unwrapping. You can comment it out in `bode.m` if you prefer the phase to always be within ±180 degrees.

Diagnostics If there is a system pole on the *jw* axis and the `w` vector happens to contain this frequency point, the gain is infinite, $(j\omega I\text{-}A)$ is singular, and `bode` produces the warning message:

```
Matrix is singular to working precision.
```

You may also see this message during the calculation of the frequency interval.

See Also `logspace`, `dbode`, `margin`, `nyquist`

References [1] Laub, A.J., Efficient Multivariable Frequency Response Computations, *IEEE Transactions on Automatic Control*, Vol. AC-26, No. 2, April 1981, pp. 407-408.

---

boxcar

Purpose Rectangular window.

Synopsis `w = boxcar(n)`

Description `boxcar(n)` returns a rectangular window of length `n`. This function is provided for completeness; a rectangular window is equivalent to no window at all.

Algorithm

```
w = ones(n,1);
```

See Also `hamming`, `hanning`, `kaiser`

---

buttap

Purpose Butterworth analog lowpass filter prototype.

Synopsis `[z,p,k] = buttap(n)`

Description `[z,p,k] = buttap(n)` returns the zeros, poles, and gain of an order `n` Butterworth analog lowpass filter prototype. It returns the poles in the length `n` column vector `p`, and the gain in scalar `k`. `z` is an empty matrix, because there are no zeros. The transfer function is

$$H(s) = \frac{z(s)}{p(s)} = \frac{k}{(s - p(1)(s - p(2))\cdots(s - p(n))}$$

Butterworth filters are characterized by a magnitude response that is maximally flat in the passband and monotonic overall. In the lowpass case, the first 2n–1 derivatives of the squared magnitude response are zero at $\omega = 0$. The squared magnitude response function is

$$|H(\omega)|^2 = \frac{1}{1+(\omega/\omega_0)^{2n}}$$

corresponding to a transfer function with poles equally spaced around a circle in the left half plane. The magnitude response at the cutoff frequency ω_0 is always `1/sqrt(2)`, regardless of the filter order. `buttap` sets ω_0 to 1 for a normalized result.

Algorithm

```
z = [];
p = exp(sqrt(-1)*(pi*(1:2:2*n-1)/(2*n)+pi/2)).';
k = real(prod(-p));
```

References

[1] Parks, T.W., and C. S. Burrus. *Digital Filter Design*, Chapter 7. New York: John Wiley & Sons, 1987.

See Also `besselap`, `butter`, `cheb1ap`, `cheb2ap`, `ellipap`

---

butter

Purpose Butterworth analog and digital filter design.

Synopsis

```
[b,a] = butter(n,Wn)
[b,a] = butter(n,Wn,'ftype')
[b,a] = butter(n,Wn,'s')
[b,a] = butter(n,Wn,'ftype','s')
[z,p,k] = butter(...)
[A,B,C,D] = butter(...)
```

Description `butter` designs lowpass, bandpass, highpass, and bandstop digital and analog Butterworth filters. Butterworth filters are characterized by a magnitude response that is maximally flat in the passband and monotonic overall.

Butterworth filters sacrifice rolloff steepness for monotonicity in the pass- and stopbands. Unless the Butterworth filter's smoothness is needed, an elliptic or Chebyshev filter can generally provide steeper rolloff characteristics with a lower filter order.

Digital Domain

`[b,a] = butter(n,Wn)` designs an order `n` lowpass digital Butterworth filter with cutoff frequency `Wn`. It returns the filter coefficients in length `n + 1` row vectors `b` and `a`, with coefficients in descending powers of z,

$$H(z) = \frac{B(z)}{A(z)} = \frac{b(1) + b(2)z^{-1} + \cdots + b(n+1)z^{-n}}{1 + a(2)z^{-1} + \cdots + a(n+1)z^{-n}}$$

Cutoff frequency is that frequency where the filter's magnitude response is `sqrt(1/2)`. For `butter`, the cutoff frequency `Wn` must be a number between 0 and 1, where 1 corresponds to half the sampling frequency (the Nyquist frequency).

If `Wn` is a two element vector, `Wn = [w1 w2]`, `butter` returns an order `2*n` digital bandpass filter with passband `w1 < ω < w2`.

`[b,a] = butter(n,Wn,'ftype')` designs a highpass or bandstop filter, where *`ftype`* is

- `high for a highpass digital filter with cutoff frequency Wn.`
- `stop` for an order `2*n` bandstop digital filter if `Wn` is a 2-element vector, `Wn = [w1 w2]`. The stopband is `w1` < ω < `w2`.

With different numbers of output arguments, `butter` directly obtains other realizations of the filter. To obtain zero-pole-gain form, use three output arguments:

```
[z,p,k] = butter(n,Wn)
```

or

```
[z,p,k] = butter(n,Wn,'ftype')
```

`butter` returns the zeros and poles in length `n` column vectors `z` and `p`, and the gain in the scalar `k`.

To obtain state-space form, use four output arguments:

```
[A,B,C,D] = butter(n,Wn)
```

or

```
[A,B,C,D] = butter(n,Wn,'ftype')
```

where `A`, `B`, `C`, and `D` are

$$x[n+1] = Ax[n] + Bu[n]$$
$$y[n] = Cx[n] + Du[n]$$

and u is the input, x is the state vector, and y is the output.

Analog Domain

`[b,a] = butter(n,Wn,'s')` designs an order `n` lowpass analog Butterworth filter with cutoff frequency `Wn`. It returns the filter coefficients in the length `n + 1` row vectors `b` and `a`, in descending powers of s,

$$H(s) = \frac{B(s)}{A(s)} = \frac{b(1)s^n + b(2)s^{n-1} + \cdots + b(n+1)}{s^n + a(2)s^{n-1} + \cdots + a(n+1)}$$

`butter`'s cutoff frequency `Wn` must be greater than 0.

If `Wn` is a two-element vector with `w1 < w2`, `butter(n,Wn,'s')` returns an order `2*n` bandpass analog filter with passband `w1` < ω < `w2`.

`[b,a] = butter(n,Wn,'ftype','s')` designs a highpass or bandstop filter, where *`ftype`* is

- `high for a highpass analog filter with cutoff frequency Wn.`
- `stop` for an order `2*n` bandstop analog filter if `Wn` is a two-element vector, `Wn = [w1 w2]`. The stopband is `w1` < ω < `w2`.

With different numbers of output arguments, `butter` directly obtains other realizations of the analog filter. To obtain zero-pole-gain form, use three output arguments:

```
[z,p,k] = butter(n,Wn,'s')
```

or

```
[z,p,k] = butter(n,Wn,'ftype','s')
```

`butter` returns the zeros and poles in length `n` or `2*n` column vectors `z` and `p`, and the gain in the scalar `k`.

To obtain state-space form, use four output arguments:

```
[A,B,C,D] = butter(n,Wn,'s')
```

or

```
[A,B,C,D] = butter(n,Wn,'ftype','s')
```

where `A`, `B`, `C`, and `D` are

$$\dot{x} = Ax + Bu$$
$$y = Cx + Du$$

and u is the input, x is the state vector, and y is the output.

Examples For data sampled at 1000 Hz, design a 9th-order highpass Butterworth filter with cutoff frequency of 300 Hz:

```
[b,a] = butter(9,300/500,'high')
```

The filter's frequency response is

```
freqz(b,a,128,1000)
```

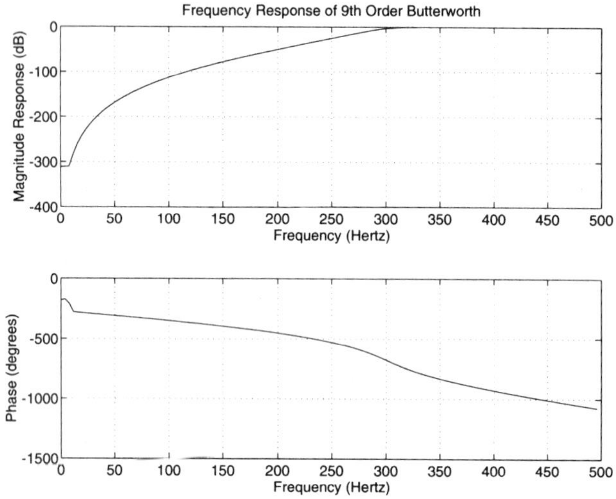

Design a 10th-order bandpass Butterworth filter with a passband from 100 to 200 Hz and plot its impulse response, or *unit sample response*:

```
n = 5; Wn = [100 200]/500;
[b,a] = butter(n,Wn);
[y,t] = impz(b,a,101);
stem(t,y)
```

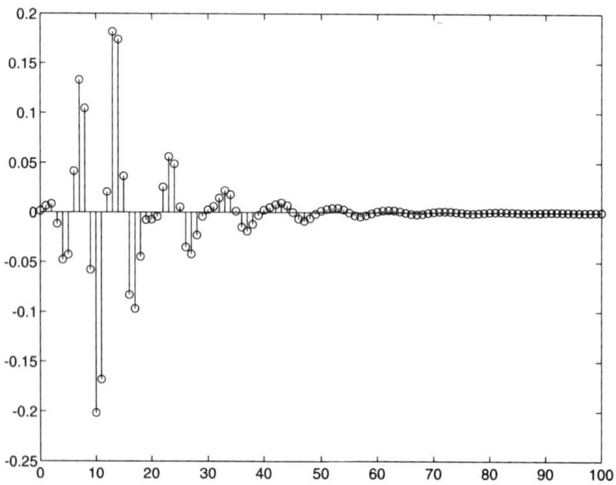

Limitations For high order filters, the state-space form is the most numerically accurate, followed by the zero-pole-gain form. The transfer function coefficient form is the least accurate, and numerical problems can arise for filter orders as low as fifteen.

Algorithm `butter` uses a five-step algorithm:

1. It finds the lowpass analog prototype poles, zeros, and gain using the `buttap` function.
2. It converts the poles, zeros, and gain into state-space form.
3. It transforms the lowpass filter into a bandpass, highpass, or bandstop filter with desired cutoff frequencies using a state-space transformation.
4. For digital filter design, `butter` uses `bilinear` to convert the analog filter into a digital filter through a bilinear transformation with frequency prewarping. Careful frequency adjustment guarantees that the analog filters and the digital filters will have the same frequency response magnitude at `Wn` or `w1` and `w2`.
5. It converts the state-space filter back to transfer function or zero-pole-gain form, as required.

See Also `besself`, `buttap`, `buttord`, `cheby1`, `cheby2`, `ellip`

---

canon

Purpose State-space to canonical form transformation.

Synopsis `[ab,bb,cb,db,T] = canon(a,b,c,d,'type')`

Description `canon` transforms the continuous state-space system:

$$\dot{x} = Ax + Bu$$
$$y = Cx + Du$$

into either modal canonical form (*type*=`'modal'`) or companion canonical form (*type*=`'companion'`). If no `'type'` is specified, the modal canonical form is assumed.

The transformed system has the same input-output relationships (i.e., the transfer functions are the same) but the states are different. The companion canonical form is badly conditioned; avoid using it if possible.

`[ab,bb,cb,db] = canon(a,b,c,d,'modal')` transforms the state-space system into modal form where the real eigenvalues appear on the diagonal of the `A` matrix and the complex eigenvalues appear in a 2×2 block on the diagonal of the A matrix. For a system with eigenvalues $(-\lambda_1, \sigma \pm j\omega, \lambda_2)$ the modal A matrix is

$$A = \begin{bmatrix} \lambda_1 & 0 & 0 & 0 \\ 0 & \sigma & \omega & 0 \\ 0 & -\omega & \sigma & 0 \\ 0 & 0 & 0 & \lambda_2 \end{bmatrix}$$

`[ab,bb,cb,db] = canon(a,b,c,d,'companion')` transforms the state-space system to companion canonical form where the characteristic polynomial of the system appears explicitly in the right column of the A matrix. For a system with characteristic polynomial:

$$s^n + a_1 s^{n-1} + \ \dots \ + a_{n-1} s + a_n$$

the corresponding companion A matrix is

$$A = \begin{bmatrix} 0 & 0 & 0 & \cdots & -a_n \\ 1 & 0 & 0 & \cdots & \vdots \\ 0 & 1 & 0 & \cdots & -a_3 \\ \vdots & \vdots & \vdots & \ddots & -a_2 \\ 0 & \cdots & \cdots & 1 & -a_1 \end{bmatrix}$$

With an additional output argument:

```
[ab,bb,cb,db,T] = canon(a,b,c,d,'type')
```

also returns the transformation vector T where

$$z = Tx$$

Algorithm The transformation to modal form uses a transformation matrix formed from the eigenvectors of the system:

$$P = \begin{bmatrix} \vdots & \vdots & & \vdots \\ v_1 & v_2 & \cdots & v_n \\ \vdots & \vdots & & \vdots \end{bmatrix}$$

The state-space system after the similarity transform is

$$\dot{x} = P^{-1}APx + P^{-1}Bu$$
$$y = CPx + Du$$

The transformation to companion canonical form uses a transformation based on the controllability matrix. The transformation matrix returned is the inverse of P:

$$T = P^{-1}$$

Limitations The modal transformation requires that the A matrix be diagonalizable. A sufficient condition for diagonalizability is that the A matrix contain no repeated roots.

The companion transformation requires that the system be controllable.

See Also `ss2ss`, `ctrb`, `ctrbf`

References [1] Kailath, T., *Linear Systems*, Prentice-Hall, 1980.

---

c2d

Purpose Conversion from continuous to discrete time.

Synopsis `[ad,bd] = c2d(a,b,Ts)`

Description `c2d` converts state space models from continuous time to discrete time assuming a zero-order hold on the inputs.

`[ad,bd] = c2d(a,b,Ts)` converts the continuous-time state-space system:

$$\dot{x} = Ax + Bu$$

to the discrete-time system:

$$x[n+1] = A_d x[n] + B_d u[n]$$

assuming the control inputs are piecewise constant over the sample time `Ts`.

`c2d` and `d2c`, like `expm` and `logm`, are inverse operations.

Algorithm `c2d` uses a matrix exponential, which is calculated via Pade's approximation in the function `expm`.

See Also `expm`, `d2c`, `logm`, `funm`

References [1] Franklin, G.F. and Powell, J.D., *Digital Control of Dynamic Systems*, Addison-Wesley, 1980.

c2dm

Purpose Converts state-space models from continuous to discrete time.

Synopsis

```
[ad,bd,cd,dd] = c2dm(a,b,c,d,Ts,'method')
[numd,dend] = c2dm(num,den,Ts,'method')
```

Description `c2dm` converts state-space models from continuous time to discrete time using one of several conversion methods:

`'zoh'`	converts to discrete time assuming a zero-order hold on the inputs – the control inputs are assumed piecewise constant over the sample time `Ts`. (This is the same as `c2d`.)
`'foh'`	converts to discrete time assuming a first-order hold on the inputs – the control inputs are assumed piecewise linear over the sample time `Ts`. This conversion is not invertible; i.e., no `d2cm(...,'foh')` exists. (This is called the triangle-hold approximation in `[,J]`.
`'tustin'`	converts to discrete time using the bilinear (Tustin) approximation to the derivative.
`'prewarp'`	converts to discrete time using the bilinear (Tustin) approximation with frequency prewarping.
`'matched'`	converts the SISO system to discrete time using the matched pole-zero method of [1]. A warning is produced if the system is MIMO.

If `method` is unspecified then `'zoh'` is assumed:

`[ad,bd,cd,dd] = c2dm(a,b,c,d,Ts,'method')` converts the continuous-time state-space system:

$$\dot{x} = Ax + Bu$$
$$y = Cx + Du$$

to the discrete-time system:

$$x[n+1] = A_d x[n] + B_d u[n]$$
$$y[n] = C_d x[n] + D_d u[n]$$

using the method specified by `'method'`.

c2dm can be used with polynomial transfer function models if (a,b,c,d) is replaced by (num,den):

```
[numd,dend] = c2dm(num,den,Ts,'method')
```

When invoked without left-hand arguments

```
c2dm(a,b,c,d,Ts,'method')
c2dm(num,den,Ts,'method')
```

a comparison singular value or Bode gain plot is drawn on the screen. The response of the continuous system is drawn with a solid line and the response of the discrete system is drawn with a dashed line.

The bilinear approximation (or Tustin's method) can transform discrete systems from the ω-domain to the z-domain. In this case d2cm(...,'tustin')transforms a discrete system to the ω-domain and c2dm(...,'tustin')transforms it back. Control systems in the ω-domain can be analyzed and designed using continuous techniques and then transformed back for implementation. For more information, see [1].

c2dm and d2cm, like expm and logm, are inverse operations so long as the Nyquist rate is not exceeded.

Examples Discretize a state-space system with two inputs and three outputs using Tustin's method. Compare the continuous singular value response with the discrete singular value response:

```
[ad,bd,cd,dd] = c2dm(a,b,c,d,Ts,'tustin');
c2dm(a,b,c,d,Ts,'tustin')   % Plot comparison graph
title('Continuous/Discrete Singular Value Comparison')
```

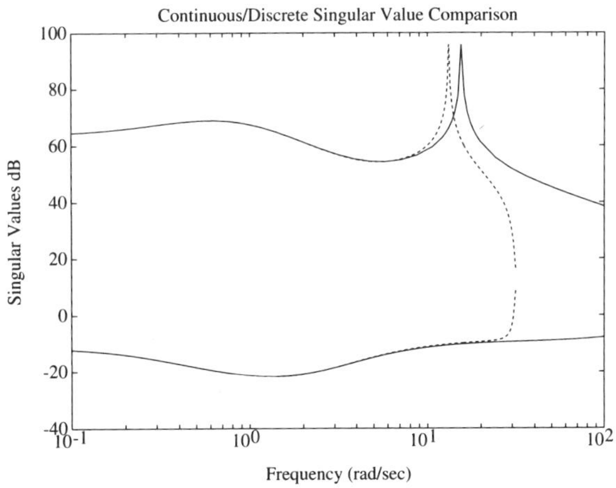

Algorithm c2dm is implemented in an M-file. The zero-order-hold and first-order-hold calculations use a matrix exponential, which is calculated via Padé approximation in the function expm. The Tustin and prewarped Tustin calculations use the trapezoid approximation to the derivative. The matched pole-zero method uses the algorithm in [1].

See Also expm, c2d, d2cm, logm, funm

References
[1] Franklin, G.F. and Powell, J.D., *Digital Control of Dynamic Systems*, Addison-Wesley, 1980.

---

cheb1ap

Purpose Chebyshev type I analog lowpass filter prototype.

Synopsis `[z,p,k] = cheb1ap(n,Rp)`

Description `[z,p,k] = cheb1ap(n,Rp)` returns the zeros, poles, and gain of an order n Chebyshev type I analog lowpass filter prototype with Rp decibels of ripple in the passband. It returns the poles in the length n column vector p and the gain in scalar k. z is an empty matrix, because there are no zeros. The transfer function is

$$H(s) = \frac{z(s)}{p(s)} = \frac{k}{(s - p(1)(s - p(2))\cdots(s - p(n))}$$

Chebyshev type I filters are equiripple in the passband and monotonic in the stopband. The poles are evenly spaced about an ellipse in the left half plane. The Chebyshev type I cutoff frequency ω_0 is set to 1.0 for a normalized result. This is the frequency at which the passband ends and the filter has magnitude response of $10^{-Rp/20}$.

See Also buttap, besselap, cheb2ap, cheby1, ellipap

References
[1] Parks, T. W., and C. S. Burrus. *Digital Filter Design*, Chapter 7. New York: John Wiley & Sons, 1987.

---

cheb2ap

Purpose Chebyshev type II analog lowpass filter prototype.

Synopsis `[z,p,k] = cheb2ap(n,Rs)`

Description `[z,p,k] = cheb2ap(n,Rs)` finds the zeros, poles, and gain of an order n Chebyshev type II analog lowpass filter prototype with stopband ripple Rs decibels down from the passband peak value. cheb2ap returns the zeros and poles in length n column vectors z and p, and the gain in scalar k. If n is odd, z is length n–1. The transfer function is

$$H(s) = \frac{z(s)}{p(s)} = k\frac{(s - z(1))(s - z(2))\cdots(s - z(n))}{(s - p(1))(s - p(2))\cdots(s - p(n))}$$

Chebyshev type II filters are monotonic in the passband and equiripple in the stopband. The pole locations are the inverse of the pole locations of cheb1ap, whose poles are evenly spaced about an ellipse in the left half plane. The Chebyshev type II cutoff frequency ω_0 is set to 1 for a normalized result. This is the frequency at which the stopband begins and the filter has magnitude response of $10^{-Rs/20}$.

Algorithm Chebyshev type II filters are sometimes called *inverse Chebyshev* filters because of their relationship to Chebyshev type I filters. The cheb2ap function is a modification of the Chebyshev type I prototype algorithm:

1. cheb2ap replaces the frequency variable ω with $1/\omega$, turning the lowpass filter into a highpass filter while preserving the performance at $\omega = 1$.
2. cheb2ap subtracts the filter transfer function from unity.

See Also besselap, buttap, cheb1ap, cheby2, ellipap

References
[1] Parks, T.W., and C. S. Burrus. *Digital Filter Design*, Chapter 7. New York: John Wiley & Sons, 1987.

---

cheby1

Purpose Chebyshev type I filter design (passband ripple).

Synopsis

```
[b,a] = cheby1(n,Rp,Wn)
[b,a] = cheby1(n,Rp,Wn,'ftype')
[b,a] = cheby1(n,Rp,Wn,'s')
[b,a] = cheby1(n,Rp,Wn,'ftype','s')
[z,p,k] = cheby1(...)
[A,B,C,D] = cheby1(...)
```

Description cheby1 designs lowpass, bandpass, highpass, and bandstop digital and analog Chebyshev type I filters. Chebyshev type I filters are equiripple in the passband and monotonic in the stopband. Type I filters roll off faster than type II filters, but at the expense of greater deviation from unity in the passband.

Digital Domain

[b,a] = cheby1(n,Rp,Wn) designs an order n lowpass digital Chebyshev filter with cutoff frequency Wn and Rp decibels of ripple in the passband. It returns the

filter coefficients in the length `n+1` row vectors `b` and `a`, with coefficients in descending powers of z:

$$H(z) = \frac{B(z)}{A(z)} = \frac{b(1) + b(2)z^{-1} + \cdots + b(n+1)z^{-n}}{1 + a(2)z^{-1} + \cdots + a(n+1)z^{-n}}$$

Cutoff frequency is the frequency at which the filter's magnitude response is equal to –`Rp` decibels. For `cheby1`, the cutoff frequency `Wn` is a number between 0 and 1, where 1 corresponds to half the sampling frequency (the Nyquist frequency). Smaller values of passband ripple `Rp` lead to wider transition widths (shallower rolloff characteristics).

If `Wn` is a two element vector, `Wn = [w1 w2]`, `cheby1` returns an order `2*n` bandpass filter with passband `w1 < ω < w2`.

`[b,a] = cheby1(n,Rp,Wn,'ftype')` designs a highpass or bandstop filter design, where *`ftype`* is

- `high for a highpass digital filter with cutoff frequency Wn.`
- `stop` for an order `2*n` bandstop digital filter if `Wn` is a 2-element vector, `Wn = [w1 w2]`. The stopband is `w1` $< \omega <$ `w2`.

With different numbers of output arguments, `cheby1` directly obtains other realizations of the filter. To obtain zero-pole-gain form, use three output arguments:

```
[z,p,k] = cheby1(n,Rp,Wn)
```

or

```
[z,p,k] = cheby1(n,Rp,Wn,'ftype')
```

`cheby1` returns the zeros and poles in length `n` column vectors `z` and `p`, and the gain in the scalar `k`.

To obtain state-space form, use four output arguments:

```
[A,B,C,D] = cheby1(n,Rp,Wn)
```

or

```
[A,B,C,D] = cheby1(n,Rp,Wn,'ftype')
```

where `A`, `B`, `C`, and `D` are

$$x[n+1] = Ax[n] + Bu[n]$$
$$y[n] = Cx[n] + Du[n]$$

and u is the input, x is the state vector, and y is the output.

Analog Domain

`[b,a] = cheby1(n,Rp,Wn,'s')` designs an order `n` lowpass analog Chebyshev type I filter with cutoff frequency `Wn`. It returns the filter coefficients in length `n + 1` row vectors `b` and `a`, in descending powers of *s*,

$$H(s) = \frac{B(s)}{A(s)} = \frac{b(1)s^n + b(2)s^{n-1} + \cdots + b(n+1)}{s^n + a(2)s^{n-1} + \cdots + a(n+1)}$$

Cutoff frequency is the frequency at which the filter's magnitude response is –`Rp` dB. `cheby1`'s cutoff frequency `Wn` is greater than 0.

If `Wn` is a two element vector with `w1 < w2`, `cheby1(n,Rp,Wn,'s')` returns an order `2*n` bandpass analog filter with passband `w1` < ω < `w2`.

`[b,a] = cheby1(n,Rp,Wn,'ftype','s')` designs a highpass or bandstop filter design, where *`ftype`* is

- `high for a highpass analog filter with cutoff frequency Wn.`
- `stop` for an order `2*n` bandstop analog filter if `Wn` is a 2-element vector, `Wn = [w1 w2]`. The stopband is `w1` < ω < `w2`.

You can supply different numbers of output arguments for `cheby1` to directly obtain other realizations of the analog filter. To obtain zero-pole-gain form, use three output arguments:

```
[z,p,k] = cheby1(n,Rp,Wn,'s')
```

or

```
[z,p,k] = cheby1(n,Rp,Wn,'ftype','s')
```

`cheby1` returns the zeros and poles in length `n` or `2*n` column vectors `z` and `p`, and the gain in the scalar `k`.

To obtain state-space form, use four output arguments:

```
[A,B,C,D] = cheby1(n,Rp,Wn,'s')
```

or

```
[A,B,C,D] = cheby1(n,Rp,Wn,'ftype','s')
```

where `A`, `B`, `C`, and `D` are defined as

$$\dot{x} = Ax + Bu$$
$$y = Cx + Du$$

and *u* is the input, *x* is the state vector, and *y* is the output.

Examples For data sampled at 1000 Hz, design a 9th-order lowpass Chebyshev type I filter with 0.5 dB of ripple in the passband and a cutoff frequency of 300 Hz:

```
[b,a] = cheby1(9,0.5,300/500);
```

The filter's frequency response is

```
freqz(b,a,512,1000);
```

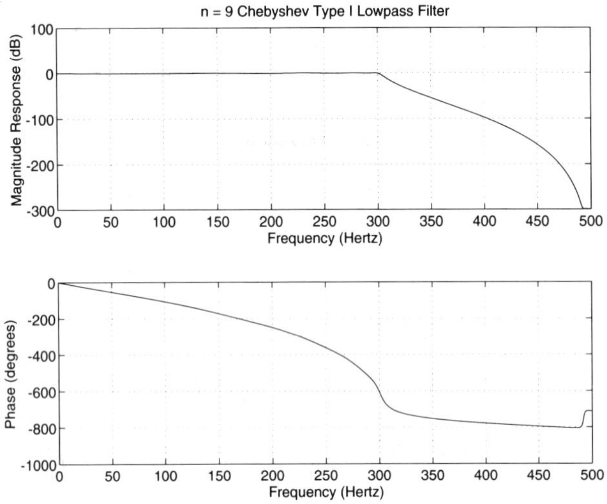

Design a 10th-order bandpass Chebyshev type I filter with a passband from 100 to 200 Hz and plot its impulse response:

```
n = 10; Rp = 0.5;
Wn = [100 200]/500;
[b,a] = cheby1(n,Rp,Wn);
[y,t] = impz(b,a,101); stem(t,y)
```

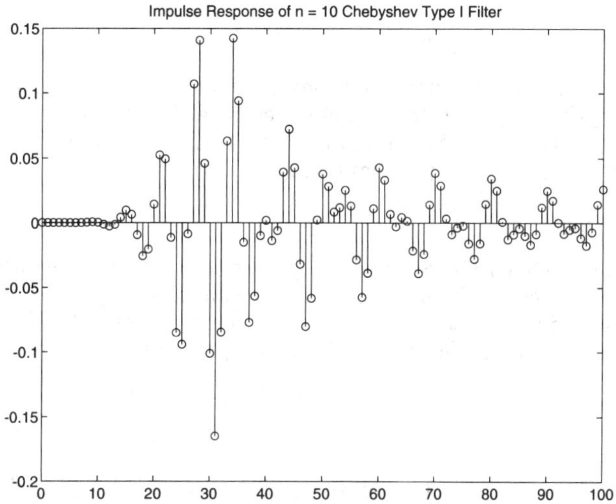

Limitations For high order filters, the state-space form is the most numerically accurate, followed by the zero-pole-gain form. The transfer function form is the least accurate, and numerical problems can arise for filter orders as low as fifteen.

Algorithm `cheby1` uses a five-step algorithm:

1. It finds the lowpass analog prototype poles, zeros, and gain using the `cheb1ap` function.
2. It converts the poles, zeros, and gain into state-space form.
3. It transforms the lowpass filter into a bandpass, highpass, or bandstop filter with desired cutoff frequencies using a state-space transformation.
4. For digital filter design, `cheby1` uses `bilinear` to convert the analog filter into a digital filter through a bilinear transformation with frequency prewarping. Careful frequency adjustment guarantees that the analog filters and the digital filters will have the same frequency response magnitude at `Wn` or `w1` and `w2`.
5. It converts the state-space filter back to transfer function, or zero-pole-gain form, as required.

See Also `besself`, `butter`, `cheb1ap`, `cheb1ord`, `cheby2`, `ellip`

cheby2

Purpose Chebyshev type II filter design (stopband ripple).

Synopsis

```
[b,a] = cheby2(n,Rs,Wn)
[b,a] = cheby2(n,Rs,Wn,'ftype')
[b,a] = cheby2(n,Rs,Wn,'s')
[b,a] = cheby2(n,Rs,Wn,'ftype','s')
[z,p,k] = cheby2(...)
[A,B,C,D] = cheby2(...)
```

Description `cheby2` designs lowpass, highpass, bandpass, and bandstop digital and analog Chebyshev type II filters. Chebyshev type II filters are monotonic in the passband and equiripple in the stopband. Type II filters do not roll off as fast as type I filters, but are free of passband ripple.

Digital Domain

`[b,a] = cheby2(n,Rs,Wn)` designs an order `n` lowpass digital Chebyshev type II filter with cutoff frequency `Wn` and stopband ripple `Rs` decibels down from the peak passband value. It returns the filter coeffi-cients in the length `n + 1` row vectors `b` and `a`, with coefficients in descending powers of z,

$$H(z) = \frac{B(z)}{A(z)} = \frac{b(1) + b(2)z^{-1} + \cdots + b(n+1)z^{-n}}{1 + a(2)z^{-1} + \cdots + a(n+1)z^{-n}}$$

Cutoff frequency is the beginning of the stopband, where the filter's magnitude response is equal to –`Rs` decibels. For `cheby2`, the cutoff frequency `Wn` is a number between 0 and 1, where 1 corresponds to half the sampling frequency (the Nyquist frequency). Larger values of stopband attenuation `Rs` lead to wider transition widths (shallower rolloff characteristics).

If `Wn` is a two-element vector, `Wn = [w1 w2]`, `cheby2` returns an order `2*n` bandpass filter with passband `w1` $< \omega <$ `w2`.

`[b,a] = cheby2(n,Rs,Wn,'ftype')` designs a highpass or bandstop filter design, where `ftype` is

- `high for a highpass digital filter with cutoff frequency Wn.`
- `stop` for an order `2*n` bandstop digital filter if `Wn` is a two-element vector, `Wn = [w1 w2]`. The stopband is `w1` $< \omega <$ `w2`.

With different numbers of output arguments, `cheby2` directly obtains other realizations of the filter. To obtain zero-pole-gain form, use three output arguments:

```
[z,p,k] = cheby2(n,Rs,Wn)
```

or

```
[z,p,k] = cheby2(n,Rs,Wn,'ftype')
```

`cheby2` returns the zeros and poles in length `n` column vectors `z` and `p`, and the gain in the scalar `k`.

To obtain state-space form, use four output arguments:

```
[A,B,C,D] = cheby2(n,Rs,Wn)
```

or

```
[A,B,C,D] = cheby2(n,Rs,Wn,'ftype')
```

where `A`, `B`, `C`, and `D` ar

$$x[n+1] = Ax[n] + Bu[n]$$
$$y[n] = Cx[n] + Du[n]$$

and u is the input, x is the state vector, and y is the output.

Analog Domain

`[b,a] = cheby2(n,Rs,Wn,'s')` designs an order `n` lowpass analog Chebyshev type II filter with cutoff frequency `Wn`. It returns the filter coefficients in the length `n + 1` row vectors `b` and `a`, with coefficients in descending powers of s,

$$H(s) = \frac{B(s)}{A(s)} = \frac{b(1)s^n + b(2)s^{n-1} + \cdots + b(n+1)}{s^n + a(2)s^{n-1} + \cdots + a(n+1)}$$

Cutoff frequency is the frequency at which the filter's magnitude response is equal to –`Rs` decibels. For `cheby2`, the cutoff frequency `Wn` must be greater than 0.

If `Wn` is a two element vector, `Wn = [w1 w2]` with `w1 < w2`, `cheby2(n,Rs,Wn,'s')` returns an order `2*n` bandpass analog filter with passband `w1` < ω < `w2`.

`[b,a] = cheby2(n,Rs,Wn,'ftype','s')` lets you select a highpass or bandstop filter design, where *`ftype`* can be:

- `high for a highpass analog filter with cutoff frequency Wn.`
- `stop` for an order `2*n` bandstop analog filter if `Wn` is a 2-element vector, `Wn = [w1 w2]`. The stopband is `w1` < ω < `w2`.

With different numbers of output arguments,`cheby2` directly obtains other realizations of the analog filter. To obtain zero-pole-gain form, use three output arguments:

```
[z,p,k] = cheby2(n,Rs,Wn,'s')
```

or

```
[z,p,k] = cheby2(n,Rs,Wn,'ftype','s')
```

`cheby2` returns the zeros and poles in length `n` or `2*n` column vectors `z` and `p`, and the gain in the scalar `k`.

To obtain state-space form, use four output arguments:

```
[A,B,C,D] = cheby2(n,Rs,Wn,'s')
```

or

```
[A,B,C,D] = cheby2(n,Rs,Wn,'ftype','s')
```

where A, B, C, and D are

$$\dot{x} = Ax + Bu$$
$$y = Cx + Du$$

and u is the input, x is the state vector, and y is the output.

Examples For data sampled at 1000 Hz, design a ninth-order lowpass Chebyshev type II filter with stopband attenuation 20 dB down from the passband, and cutoff frequency of 300 Hz:

```
[b,a] = cheby2(9,20,300/500);
```

The filter's frequency response is

```
freqz(b,a,512,1000)
```

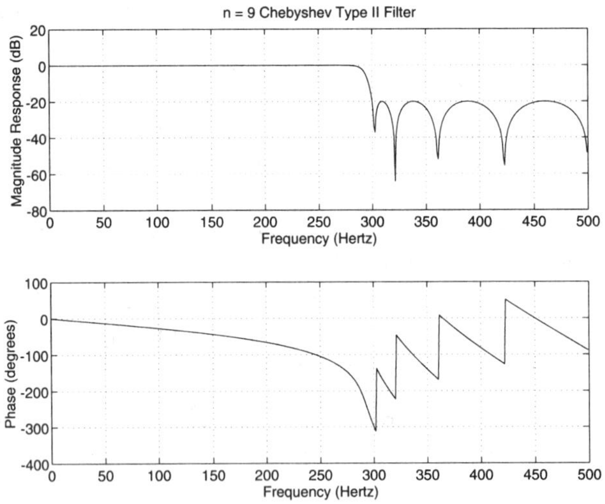

Design a fifth-order bandpass Chebyshev type II filter with passband from 100 to 200 Hz, and plot the filter's impulse response:

```
n = 5; r = 20;
Wn = [100 200]/500;
[b,a] = cheby2(n,r,Wn);
[y,t] = impz(b,a,101); stem(t,y)
```

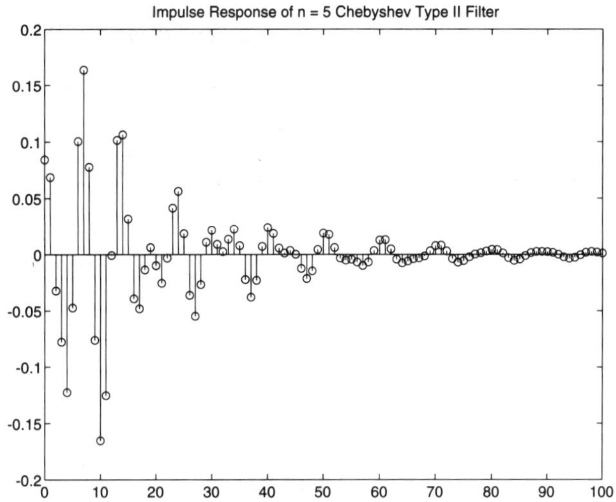

Limitations For high order filters, the state-space form is the most numerically accurate, followed by the zero-pole-gain form. The transfer function coefficient form is the least accurate, and numerical problems can arise for filter orders as low as fifteen.

Algorithm `cheby2` uses a five-step algorithm:

1. It finds the lowpass analog prototype poles, zeros, and gain using the `cheb2ap` function.
2. It converts poles, zeros, and gain into state-space form.
3. It transforms the lowpass filter into a bandpass, highpass, or bandstop filter with desired cutoff frequencies using a state-space transformation.
4. For digital filter design, `cheby2` uses `bilinear` to convert the analog filter into a digital filter through a bilinear transformation with frequency prewarping. Careful frequency adjustment guarantees that the analog filters and the digital filters will have the same frequency response magnitude at `Wn` or `w1` and `w2`.
5. It converts the state-space filter back to transfer function, or zero-pole-gain form, as required.

See Also `besself`, `butter`, `cheb2ap`, `cheb2ord`, `cheby1`, `ellip`

cloop

Purpose Close loops of state-space system.

Synopsis

```
[ac,bc,cc,dc] = cloop(a,b,c,d,sign)
[ac,bc,cc,dc] = cloop(a,b,c,d,outputs,inputs)
[numc,denc] = cloop(num,den,sign)
```

Description `cloop` forms the closed-loop system obtained by feeding the outputs into the inputs of the system. All the inputs and outputs from the open-loop system remain in the closed-loop system. `cloop` works with both continuous and discrete systems.

`[ac,bc,cc,dc] = cloop(a,b,c,d,sign)` produces a state-space model of the closed-loop system obtained by feeding back all the outputs to all the inputs:

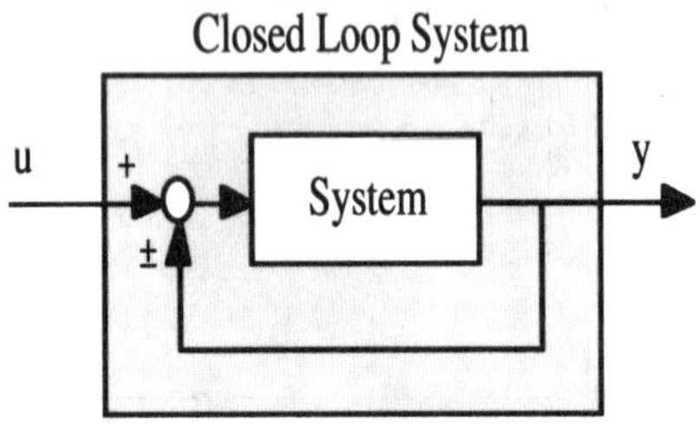

`sign = 1` specifies positive feedback and `sign = -1` specifies negative feedback. If no sign parameter is specified, negative feedback is assumed. The resulting closed-loop system is

$$\dot{x} = \left[A \pm B(I \mp D)^{-1}C\right]x + \left[B(I \mp D)^{-1}\right]u$$
$$y = \left[C \pm D(I \mp D)^{-1}C\right]x + \left[D(I \mp D)^{-1}\right]u$$

where the top sign corresponds to positive feedback and the bottom sign corresponds to negative feedback.

`[numc,denc] = cloop(num,den,sign)` produces a polynomial transfer function of the closed-loop system obtained by unity feedback with sign `sign`. `num` and `den` contain the polynomial coefficients in descending powers of s or z. If no sign parameter is specified, negative feedback is assumed. The resulting closed-loop system is

$$\frac{num_c(s)}{den_c(s)} = \frac{G(s)}{1 \mp G(s)} = \frac{num(s)}{den(s) \mp num(s)}$$

`[ac,bc,cc,dc] = cloop(a,b,c,d,outputs,inputs)` produces a state-space model of the closed-loop system obtained by feeding back the specified outputs to the specified inputs:

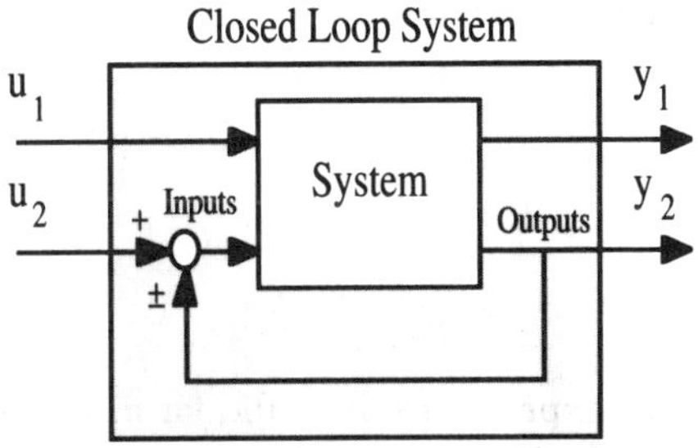

The vector `outputs` contains indices into the output vector of the system and specifies which outputs are to be fed back into the inputs specified by the vector `inputs` (see example below). Positive feedback is used. To close with negative feedback, use negative values in the vector `inputs`. (See Algorithm below for more details)

Examples Consider a state-space system `(a,b,c,d)` with five outputs and eight inputs. To feed back outputs 1, 3 and 5 to the inputs 2, 8 and 7 with negative feedback:

```
outputs = [1 3 5];
inputs = [2 8 7]
[ac,bc,cc,dc] = cloop(a,b,c,d,outputs,-inputs)
```

Given a state-space system:

$$\dot{x} = Ax + \begin{bmatrix} B_1 & B_2 \end{bmatrix} \begin{bmatrix} u_1 \\ u_2 \end{bmatrix}$$

$$\begin{bmatrix} y_1 \\ y_2 \end{bmatrix} = \begin{bmatrix} C_1 \\ C_2 \end{bmatrix} x + \begin{bmatrix} D_{11} & D_{12} \\ D_{21} & D_{22} \end{bmatrix} \begin{bmatrix} u_1 \\ u_2 \end{bmatrix}$$

suppose that the closed-loop system is to be formed by feeding back the outputs y_2 to the inputs u_2 as shown in the block diagram above.

The resulting state-space system is

$$\dot{x} = \begin{bmatrix} A \pm B_2 E C_2 \end{bmatrix} x + \begin{bmatrix} B_1 \pm B_2 E D_{21} & B_2 E \end{bmatrix} \begin{bmatrix} u_1 \\ u_2 \end{bmatrix}$$

$$\begin{bmatrix} y_1 \\ y_2 \end{bmatrix} = \begin{bmatrix} C_1 \pm D_{12} E C_2 \\ C_2 \pm D_{22} E C_2 \end{bmatrix} x + \begin{bmatrix} D_{11} \pm D_{12} E D_{21} & D_{12} E \\ D_{21} \pm D_{22} E D_{21} & D_{22} E \end{bmatrix} \begin{bmatrix} u_1 \\ u_2 \end{bmatrix}$$

where

$$E = (I \mp D_{22})^{-1}$$

If D_{22} is zero this simplifies to

$$\dot{x} = [A \pm B_2 C_2]x + [B_1 \pm B_2 D_{21} \quad B_2]\begin{bmatrix} u_1 \\ u_2 \end{bmatrix}$$

$$\begin{bmatrix} y_1 \\ y_2 \end{bmatrix} = \begin{bmatrix} C_1 \pm D_{12}C_2 \\ C_2 \end{bmatrix} x + \begin{bmatrix} D_{11} \pm D_{12}D_{21} & D_{12} \\ D_{21} & 0 \end{bmatrix} \begin{bmatrix} u_1 \\ u_2 \end{bmatrix}$$

These same expressions are valid for discrete models with differentiation replaced by differences and s-plane transfer functions replaced by z-plane transfer functions.

Limitations The matrix $(I \pm D_{22})$ must be invertible.

See Also `append`, `connect`, `feedback`, `parallel`, `series`

---

ctrb, obsv

Purpose Controllability and observability matrices.

Synopsis

```
co = ctrb(a,b)
ob = obsv(a,c)
```

Description `ctrb` and `obsv` return the controllability and observability matrices for state-space systems.

For an n-by-n matrix A, an n-by-m matrix B, and an p-by-n matrix C, `ctrb(a,b)` returns the n-by-n times m controllability matrix:

$$C_o = [B \quad AB \quad A^2B \quad \cdots \quad A^{n-1}B]$$

and `obsv(a,c)` returns the n times m-by-n observability matrix:

$$O_b = \begin{bmatrix} C \\ CA \\ CA^2 \\ \vdots \\ CA^{n-1} \end{bmatrix}$$

The system is controllable if C_o has rank n and observable if O_b has rank n.

Examples `ctrb` and `obsv` check if the system `(a,b,c,d)` is controllable and observable using

```
co = ctrb(a,b)
% Number of uncontrollable states
unco = length(a)–rank(co)
ob = obsv(a,c)
% Number of unobservable states
unob = length(a)–rank(ob)
```

but the calculation may be ill-conditioned (see note below). It is better to determine the controllability of a system using either `bode`, or the functions `ctrbf` and `obsvf`.

Algorithm Use these functions with considerable caution. The matrices C_o and O_b have a tendency to become highly ill-conditioned with respect to inversion. An indication of the kind of trouble that can arise can be seen from the simple example.

$$A = \begin{bmatrix} 1 & \delta \\ 0 & 1 \end{bmatrix}, \quad B = \begin{bmatrix} 1 \\ \delta \end{bmatrix}$$

This pair is controllable if $\delta \neq 0$ but if $\delta < \sqrt{eps}$, where *eps* is the relative machine precision, then it is easily seen that `ctrb` returns

$$[B \quad AB] = \begin{bmatrix} 1 & 1 \\ \delta & \delta \end{bmatrix}$$

from which an erroneous conclusion would be drawn.

See Also `ctrbf`, `obsvf`, `bode`

---

damp, ddamp

Purpose Damping factors and natural frequencies.

Synopsis

```
[Wn,Z] = damp(a)
mag = ddamp(a)
[mag,Wn,Z] = ddamp(a,Ts)
```

Description `damp` and `ddamp` calculate natural frequencies and damping factors. When invoked without left-hand arguments a table of the eigenvalues, damping ratios and natural frequencies are displayed on the screen.

`[Wn,Z] = damp(a)` returns column vectors `Wn` and `Z` containing the natural frequencies, ω_n, and damping factors, ζ, of the continuous eigenvalues computed from a. The variable a can be in one of several formats:

- `If a is square, it is treated as the state-space A matrix.`

- `If a is a row vector, it is assumed to be a vector containing the polynomial coefficients of a transfer function.`
- When a is a column vector, it is assumed to contain root locations.

`mag = ddamp(a)` returns the column vector `mag` containing the magnitude of the discrete eigenvalues computed from a. a can be in the formats described above.

`[mag,Wn,Z] = ddamp(a,Ts)` returns the vectors, `mag`, `Wn` and `Z` containing the magnitude, equivalent s-plane natural frequencies and equivalent s-plane damping ratios of the eigenvalues of a. `Ts` is the sample time. The equivalent s-plane damping ratio and natural frequency for a discrete eigenvalue, λ, are:

$$\omega_n = \left|\frac{\log \lambda}{T_s}\right| \qquad \zeta = -\cos(\angle \log \lambda)$$

Examples Compute and display the eigenvalues, natural frequencies and damping ratios of the continuous transfer function:

$$H(s) = \frac{2s^2 + 5s + 1}{s^2 + 2s + 3}$$

```
num = [2 5 1];
den = [1 2 3];
damp(den);
   Eigenvalue          Damping     Freq.(rad/sec)]
-1.0000 + 1.4142i      0.5774          1.7321
-1.0000 - 1.4142i      0.5774          1.7321
```

Compute and display the eigenvalues, magnitude, and equivalent s-plane frequency and damping ratios of the discrete transfer function with sample time T_s = 0.1:

$$H(z) = \frac{2z^2 - 3.4z + 1.5}{z^2 - 1.6z + 0.8}$$

```
num = [2 -3.4 1.5];
den = [1 -1.6 0.8];
ddamp(den,0.1);
  Eigenvalue      Magnitude  Equiv.Damping    Equiv.Freq
0.8000+0.4000i     0.8944       0.2340          4.7688
0.8000-0.4000i     0.8944       0.2340          4.7688
```

detrend

Purpose Remove linear trends.

Synopsis

```
y = detrend(x)
y = detrend(x,0)
```

Description `detrend` removes the mean value or linear trend from a vector or matrix, usually for FFT processing.

`y = detrend(x)` removes the best straight line fit from vector `x` and returns it in `y`. If `x` is a matrix, `detrend` removes the trend from each column of the matrix.

`y = detrend(x,0)` removes just the mean value from vector `x`, or if `x` is a matrix, the mean value from each column.

Algorithm `detrend` computes the least squares fit of a straight line to the data and subtracts the resulting function from the data. The main part of the algorithm is

```
m = length(x);
a = [(1:m)'/m ones(m,1)];
y = x - a*(a\x);
```

To obtain the equation of the straight line fit, use `polyfit`.

See Also `polyfit` in the *MATLAB Reference*

dimpulse

Purpose Discrete unit impulse response.

Synopsis

```
[y,x] = dimpulse(a,b,c,d)
[y,x] = dimpulse(a,b,c,d,iu)
[y,x] = dimpulse(a,b,c,d,iu,n)
[y,x] = dimpulse(num,den)
[y,x] = dimpulse(num,den,n)
```

Description `dimpulse` calculates the unit impulse response of a discrete-time linear system. Invoked without left-hand arguments, `dimpulse` plots the impulse response on the screen.

`dimpulse(a,b,c,d)` produces a series of impulse response plots, one for each input and output combination of the discrete LTI system:

$$x[n+1] = Ax[n] + Bu[n]$$
$$y[n] = Cx[n] + Du[n]$$

with the number of sample points automatically determined.

`dimpulse(a,b,c,d,iu)` produces an impulse response plot from the single input `iu` to all the outputs of the system with the number of sample points automatically determined. The scalar `iu` is an index into the inputs of the system and specifies which input to be used for the impulse response.

`dimpulse(num,den)` produces the impulse response plot of the polynomial transfer function `G(z) = num(z)/den(z)` where `num` and `den` contain the polynomial coefficients in descending powers of z.

`dimpulse(a,b,c,d,iu,n)` or `dimpulse(num,den,n)` uses the user-supplied number of sample points `n`.

Invoked with left-hand arguments:

```
[y,x] = dimpulse(a,b,c,d,iu)
[y,x] = dimpulse(a,b,c,d,iu,n)
[y,x] = dimpulse(num,den)
[y,x] = dimpulse(num,den,n)
```

returns the output and state responses of the system. No plot is drawn on the screen. `y` has as many columns as outputs and `x` as many columns as states.

Examples Plot the impulse response of the system:

$$H(z) = \frac{2z^2 - 3.4z + 1.5}{z^2 - 1.6z + 0.8}$$

```
num = [2 -3.4 1.5];
den = [1 -1.6 0.8];
dimpulse(num,den);
title('Discrete Impulse Response')
```

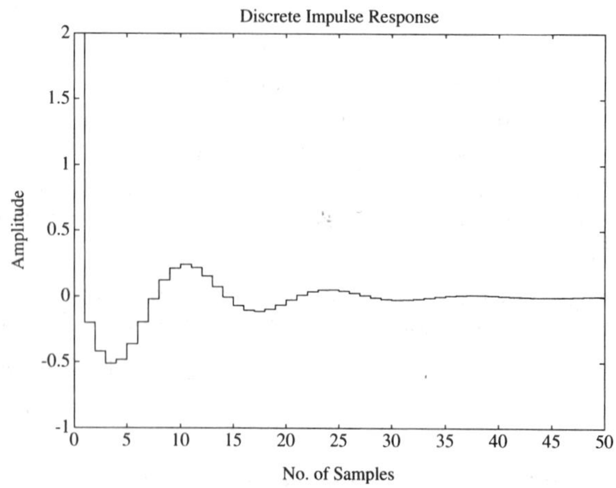

See Also `impulse`, `dlsim`, `dstep`, `dinitial`

dlsim

Purpose Discrete system simulation to arbitrary inputs.

Synopsis

```
[y,x] = dlsim(a,b,c,d,u)
[y,x] = dlsim(a,b,c,d,u,x0)
[y,x] = dlsim(num,den,u)
```

Description `dlsim` simulates discrete-time linear systems with arbitrary inputs. When invoked without left-hand arguments, `dlsim` produces a plot on the screen.

Given the LTI system:

$$x[n+1] = Ax[n] + Bu[n]$$
$$y[n] = Cx[n] + Du[n]$$

`dlsim(a,b,c,d,u)` produces a plot of the time response of the system to the input sequence in the matrix `u`. The matrix `u` must have as many columns as there are inputs u. Each row of `u` corresponds to a new time point. When used with an extra righthand argument, initial conditions on the states are specified:

```
dlsim(a,b,c,d,u,x0)
```

`dlsim(num,den,u)` plots the time response of the polynomial transfer function `G(z) = num(z)/den(z)` where `num` and `den` contain the polynomial coefficients in descending powers of z..

`dlsim(num,den,u)` is equivalent to `filter(num,den,u)` for SISO systems if the sizes of the numerator and denominator polynomials are the same.

Invoked with left-hand arguments

```
[y,x] = dlsim(a,b,c,d,u)
[y,x] = dlsim(a,b,c,d,u,x0)
][y,x] = dlsim(num,den,u)
```

returns the matrices `x` and `y`, where `y` is the output response of the system and `x` is the state response of the system. No plot is drawn on the screen. `x` has as many columns as there are outputs y, and one row for each row of `u`. `y` has as many columns as there are states, x, and has one row for each row of `u`.

Examples Simulate the response of the system:

$$H(z) = \frac{2z^2 - 3.4z + 1.5}{z^2 - 1.6z + 0.8}$$

to 100 samples of random noise:

```
num=[2 -3.4 1.5];
den = [1 -1.6 0.8];
rand('normal');
u = rand(100,1);
dlsim(num,den,u);
title('Noise Response')
```

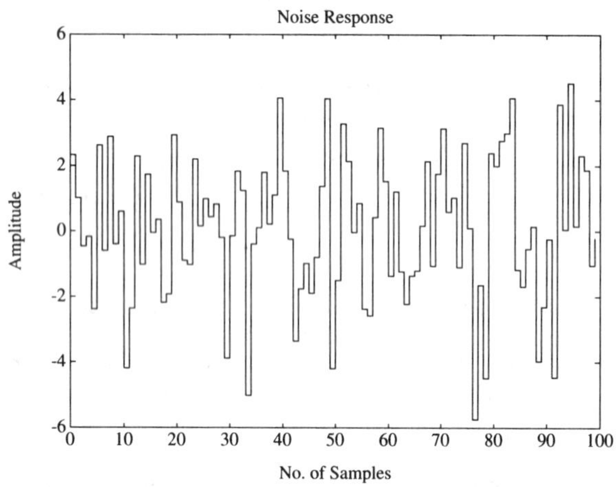

Algorithm `dlsim` uses the built-in function `ltitr` to propagate the states. When `dlsim` is given a SISO *z*-transform, `filter` is used.

See Also `dimpulse`, `dstep`, `impulse`, `step`, `lsim`, `filter`

---

dstep

Purpose Discrete unit step response.

Synopsis

```
[y,x] = dstep(a,b,c,d)
[y,x] = dstep(a,b,c,d,iu)
[y,x] = dstep(a,b,c,d,iu,n)
[y,x] = dstep(num,den)
[y,x] = dstep(num,den,n)
```

Description `dstep` calculates the unit step response of a discrete-time linear system. Invoked without left-hand arguments, `dstep` plots the step response on the screen.

`dstep(a,b,c,d)` produces a series of step response plots, one for each input and output combination of the discrete LTI system:

$$H(z) = \frac{2z^2 - 3.4z + 1.5}{z^2 - 1.6z + 0.8}$$

with the number of sample points automatically determined.

`dstep(a,b,c,d,iu)` produces a step response plot from the single input `iu` to all the outputs of the system with the number of sample points automatically determined. The scalar `iu` is an index into the inputs of the system and specifies which input to be used for the impulse response.

`dstep(num,den)` produces the step response plot of the polynomial transfer function `G(z) = num(z)/den(z)` where `num` and `den` contain the polynomial coefficients in descending powers of `z` .

`dstep(a,b,c,d,iu,n)` or `dstep(num,den,n)` uses the user-supplied number of sample points `n`.

Invoked with left-hand arguments

```
[y,x] = dstep(a,b,c,d,iu)
[y,x] = dstep(a,b,c,d,iu,n)
[y,x] = dstep(num,den)
[y,x] = dstep(num,den,n)
```

`dstep` returns the output and state responses of the system. No plot is drawn on the screen. `y` has as many columns as outputs and `x` has as many columns as states.

Examples Plot the step response of the system:

$$H(z) = \frac{2z^2 - 3.4z + 1.5}{z^2 - 1.6z + 0.8}$$

```
num=[2 -3.4 1.5];
den = [1 -1.6 0.8];
dstep(num,den);
title('Discrete Step Response')
```

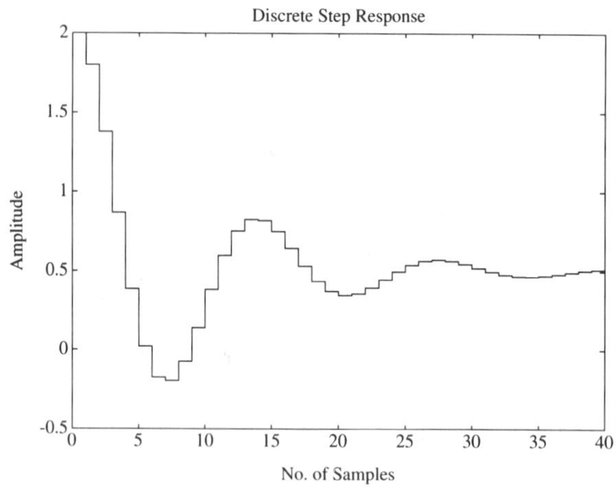

See Also `step`, `dlsim`, `dimpulse`

---

ellip

Purpose Elliptic or Cauer filter design.

Synopsis

```
[b,a] = ellip(n,Rp,Rs,Wn)
[b,a] = ellip(n,Rp,Rs,Wn,'ftype')
[b,a] = ellip(n,Rp,Rs,Wn,'s')
[b,a] = ellip(n,Rp,Rs,Wn,'ftype','s')
[z,p,k] = ellip(...)
[A,B,C,D] = ellip(...)
```

Description `ellip` designs lowpass, bandpass, highpass, and bandstop digital and analog elliptic filters. Elliptic filters offer steeper rolloff characteristics than Butterworth or Chebyshev filters, but are equiripple in both the pass- and stopbands. In general, elliptic filters meet given performance specifications with the lowest order of any filter type.

Digital Domain

`[b,a] = ellip(n,Rp,Rs,Wn)` designs an order `n` lowpass digital elliptic filter with cutoff frequency `Wn`, `Rp` decibels of ripple in the passband, and a stopband `Rs` decibels down from the peak value in the passband. It returns the filter coefficients in the length `n + 1` row vectors `b` and `a`, with coefficients in descending powers of z,

$$H(z) = \frac{B(z)}{A(z)} = \frac{b(1) + b(2)z^{-1} + \cdots + b(n+1)z^{-n}}{1 + a(2)z^{-1} + \cdots + a(n+1)z^{-n}}$$

Cutoff frequency is the edge of the passband, at which the filter's magnitude response is –`Rp` decibels. `ellip`'s cutoff frequency `Wn` is a number between 0 and 1, where 1 corresponds to half the sample frequency (Nyquist frequency). Smaller values of passband ripple `Rp` and larger values of stopband attenuation `Rs` both lead to wider transition widths (shallower rolloff characteristics).

If `Wn` is a two element vector, `Wn = [w1 w2]`, `ellip` returns an order `2*n` bandpass filter with passband `w1 < ω < w2`.

`[b,a] = ellip(n,Rp,Rs,Wn,'ftype')` designs a highpass or bandstop filter design, where *`ftype`* is

- `high for a highpass digital filter with cutoff frequency Wn.`
- `stop` for an order `2*n` bandstop digital filter if `Wn` is a 2-element vector, `Wn = [w1 w2]`. The stopband is `w1` < ω < `w2`.

With different numbers of output arguments, `ellip` directly obtains other realizations of the filter. To obtain zero-pole-gain form, use three output arguments:

```
[z,p,k] = ellip(n,Rp,Rs,Wn)
```

or

```
[z,p,k] = ellip(n,Rp,Rs,Wn,'ftype')
```

`ellip` returns the zeros and poles in length `n` column vectors `z` and `p`, and the gain in the scalar `k`.

To obtain state-space form, use four output arguments:

```
[A,B,C,D] = ellip(n,Rp,Rs,Wn)
```

or

```
[A,B,C,D] = ellip(n,Rp,Rs,Wn,'ftype')
```

where `A`, `B`, `C`, and `D` are

$$\begin{aligned} x[n+1] &= \mathrm{a}x[n] + \mathrm{b}u[n] \\ y[n] &= \mathrm{c}x[n] + \mathrm{d}\,u[n] \end{aligned}$$

and u is the input, x is the state vector, and y is the output.

Analog Domain

`[b,a] = ellip(n,Rp,Rs,Wn,'s')` designs an order `n` lowpass analog elliptic filter with cutoff frequency `Wn` and returns the filter coefficients in the length `n + 1` row vectors `b` and `a`, in descending powers of *s*,

$$H(s) = \frac{B(s)}{A(s)} = \frac{b(1)s^n + b(2)s^{n-1} + \cdots + b(n+1)}{s^n + a(2)s^{n-1} + \cdots + a(n+1)}$$

Cutoff frequency is the edge of the passband, at which the filter's magnitude response is –`Rp` decibels. `ellip`'s cutoff frequency `Wn` is greater than 0.

If `Wn` is a two-element vector with `w1 < w2`, `ellip(n,Rp,Rs,Wn,'s')` returns an order `2*n` bandpass analog filter with passband `w1` < ω < `w2`.

`[b,a] = ellip(n,Rp,Rs,Wn,'ftype','s')` designs a highpass or bandstop filter design, where `ftype` is

- `high for a highpass analog filter with cutoff frequency Wn.`
- `stop` for an order `2*n` bandstop analog filter if `Wn` is a two-element vector, `Wn = [w1 w2]`. The stopband is `w1` < ω < `w2`.

With different numbers of output arguments, `ellip` directly obtains other realizations of the analog filter. To obtain zero-pole-gain form, use three output arguments:

```
[z,p,k] = ellip(n,Rp,Rs,Wn,'s')
```

or

```
[z,p,k] = ellip(n,Rp,Rs,Wn,'ftype','s')
```

`ellip` returns the zeros and poles in length `n` or `2*n` column vectors `z` and `p`, and the gain in the scalar `k`.

To obtain state-space form, use four output arguments:

```
[A,B,C,D] = ellip(n,Rp,Rs,Wn,'s')
```

or

```
[A,B,C,D] = ellip(n,Rp,Rs,Wn,'ftype','s')
```

where `A`, `B`, `C`, and `D` are

$$\dot{x} = Ax + Bu$$
$$y = Cx + Du$$

and *u* is the input, *x* is the state vector, and *y* is the output.

Example For data sampled at 1000 Hz, design a sixth-order lowpass elliptic filter with a cutoff frequency of 300 Hz, 3 dB of ripple in the passband, and 50 dB of ripple on the stopband;

```
[b,a] = ellip(6,3,50,300/500);
```

The filter's frequency response is

```
freqz(b,a,512,1000)
```

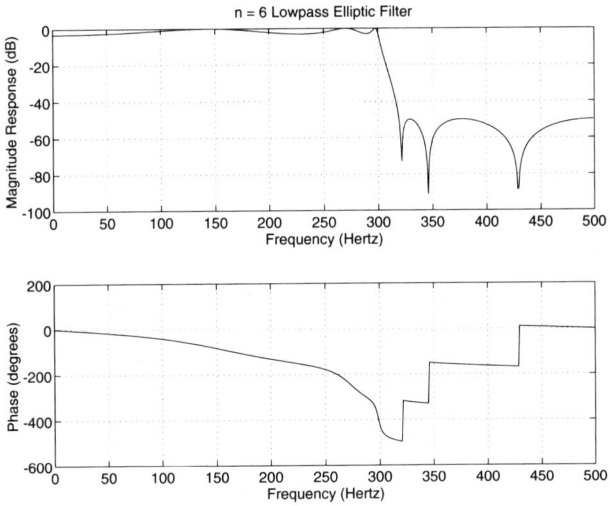

Design a 20th-order bandpass elliptic filter with a passband from 100 to 200Hz and plot its impulse response:

```
n = 10; Rp = 0.5; Rs = 20;
Wn = [100 200]/500;
[b,a] = ellip(n,Rp,Rs,Wn);
[y,t] = impz(b,a,101); stem(t,y)
```

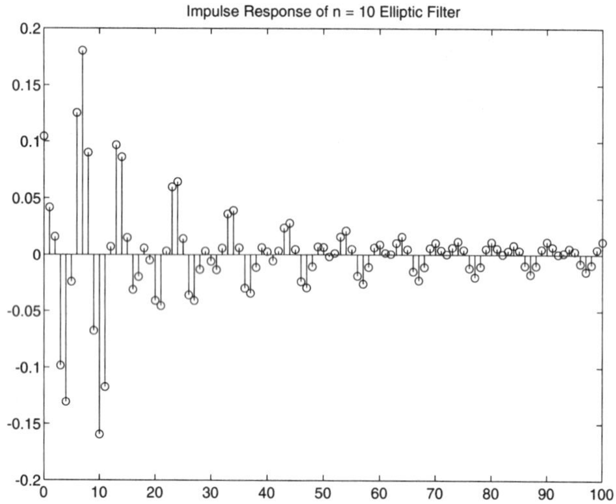

Limitations For high order filters, the state-space form is the most numerically accurate, followed by the zero-pole-gain form. The transfer function form is the least accurate, and numerical problems can arise for filter orders as low as fifteen.

Algorithm The design of elliptic filters is the most difficult and computationally intensive of the Butterworth, Chebyshev type I and II, and elliptic designs. `ellip` uses a five-step algorithm:

1. It finds the lowpass analog prototype poles, zeros, and gain using the `ellipap` function.
2. It converts the poles, zeros, and gain into state-space form.
3. It transforms the lowpass filter to a bandpass, highpass, or bandstop filter with the desired cutoff frequencies using a state-space transformation.
4. For digital filter design, `ellip` uses `bilinear` to convert the analog filter into a digital filter through a bilinear transformation with frequency prewarping. Careful frequency adjustment guarantees that the analog filters and the digital filters will have the same frequency response magnitude at `Wn` or `w1` and `w2`.
5. It converts the state-space filter back to transfer function, or zero-pole-gain form, as required.

See Also `besself`, `butter`, `cheby1`, `cheby2`, `ellipord`, `ellipap`

ellipap

Purpose Elliptic analog lowpass filter prototype.

Synopsis

```
[z,p,k] = ellipap(n,Rp,Rs)
```

Description `[z,p,k] = ellipap(n,Rp,Rs)` returns the zeros, poles, and gain of an order `n` elliptic analog lowpass filter prototype, with `Rp` decibels of ripple in the passband, and a stopband `Rs` decibels down from the peak value in the passband. The zeros and poles are returned in length `n` column vectors `z` and `p`, and the gain in scalar `k`. If `n` is odd, `z` is length `n – 1`. The transfer function is

$$H(s) = \frac{z(s)}{p(s)} = k\frac{(s - z(1))(s - z(2))\cdots(s - z(n))}{(s - p(1))(s - p(2))\cdots(s - p(n))}$$

Elliptic filters are equiripple in both the pass- and stopbands. They offer steeper rolloff characteristics than Butterworth and Chebyshev filters, but at the expense of pass- and stopband ripple. Of the four classical filter types, elliptic filters usually meet a given set of filter performance specifications with the lowest filter order.

`ellip` sets the cutoff frequency ω_0 of the elliptic filter to 1 for a normalized result. The *cutoff frequency* is the frequency at which the passband ends and the filter has magnitude response of $10^{-Rp/20}$.

Algorithm `ellipap` uses the algorithm outlined in [1]. It employs the M-file `ellipk` to calculate the complete elliptic integral of the first kind and the M-file `ellipj` to calculate Jacobi elliptic functions.

See Also `buttap`, `cheb1ap`, `cheb2ap`, `ellip`

References

[1] Parks, T.W, and C. S. Burrus. *Digital Filter Design*, Chapter 7. New York: John Wiley & Sons, 1987.

esort, dsort

Purpose Sort eigenvalues by real part or complex magnitude.

Synopsis

```
[s,ndx] = esort(p)
[s,ndx] = dsort(p)
```

Description `s = esort(p)` sorts the complex eigenvalues in the vector `p` in descending order by real part. For continuous eigenvalues, unstable eigenvalues appear first.

`s = dsort(p)` sorts the complex eigenvalues in the vector `p` in descending order by magnitude. For discrete eigenvalues, unstable eigenvalues appear first.

`[s,ndx] = esort(p)` or `[s,ndx] = dsort(p)` also returns the vector `ndx` containing the indices used in the sort.

Limitations The eigenvalues in the vector p must appear in complex conjugate pairs.

See Also sort

feedback

Purpose Feedback connection of two systems.

Synopsis

```
[a,b,c,d] = feedback(a1,b1,c1,d1,a2,b2,c2,d2)
[a,b,c,d] = feedback(a1,b1,c1,d1,a2,b2,c2,d2,sign)
[a,b,c,d] = feedback(a1,b1,c1,d1,a2,b2,c2,d2,inp1,out1)
[num,den] = feedback(num1,den1,num2,den2)
[num,den] = feedback(num1,den1,num2,den2,sign)
```

Description `feedback` connects two systems in feedback. Typically, system 1 is a plant and system 2 is a feedback controller. `feedback` works with both continuous and discrete systems.

`[a,b,c,d] = feedback(a1,b1,c1,d1,a2,b2,c2,d2,sign)` produces an aggregate state-space system consisting of the feedback connection of systems 1 and 2:

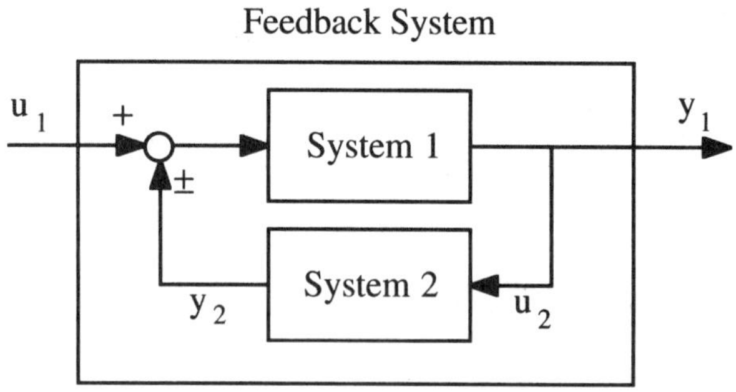

The feedback system is formed by connecting all the outputs of system 1 to all the inputs of system 2 and all the outputs of system 2 to all the inputs of system 1. `sign = 1` specifies that the connection of y_2 to u_1 be with positive sign and `sign = -1` specifies connection with negative sign. If no `sign` parameter is specified, negative feedback is assumed. The feedback system has the same inputs and outputs as system 1.

The resulting system is

$$\begin{bmatrix}\dot{x}_1\\ \dot{x}_2\end{bmatrix}=\begin{bmatrix}A_1 \pm B_1ED_2C_1 & \pm B_1EC_2\\ B_2C_1 \pm B_2D_1ED_2C_1 & A_2 \pm B_2D_1EC_2\end{bmatrix}\begin{bmatrix}x_1\\ x_2\end{bmatrix}+\begin{bmatrix}B_1(I \pm ED_2D_1)\\ B_2D_1(I \pm ED_2D_1)\end{bmatrix}u_1$$

$$y_1=\begin{bmatrix}C_1 \pm D_1ED_2C_1 & \pm D_1EC_2\end{bmatrix}\begin{bmatrix}x_1\\ x_2\end{bmatrix}+\begin{bmatrix}D_1(I \pm ED_2D_1)\end{bmatrix}u_1$$

where

$$E=(I \mp D_2D_1)^{-1}$$

and the upper sign corresponds to postive feedback while the lower sign corresponds to negative feedback.

`[num,den] = feedback(num1,den1,num2,den2,sign)` produces a polynomial transfer function of the feedback system obtained with feedback with sign `sign`. The polynomial transfer function `G(s) = num(s)/den(s)` where `num` and `den` contain polynomial coefficients in descending powers of s or z. If no `sign` parameter is specified, negative feedback is assumed. The resulting feedback system is

$$\frac{num(s)}{den(s)}=\frac{G_1(s)}{1 \mp G_1(s)G_2(s)}=\frac{num_1(s)\,den_2(s)}{den_1(s)\,den_2(s) \mp num_1(s)\,num_2(s)}$$

To connect two transfer function systems together in feedback where both systems appear in the forward path, use a combination of `series` and `cloop`.

`[a,b,c,d] = feedback(a1,b1,c1,d1,a2,b2,c2,d2,inp1,out1)` produces the feedback system formed by feeding the specified outputs of system 1 into all the inputs of system 2 and by feeding all the outputs of system 2 to the specified inputs of system 1.

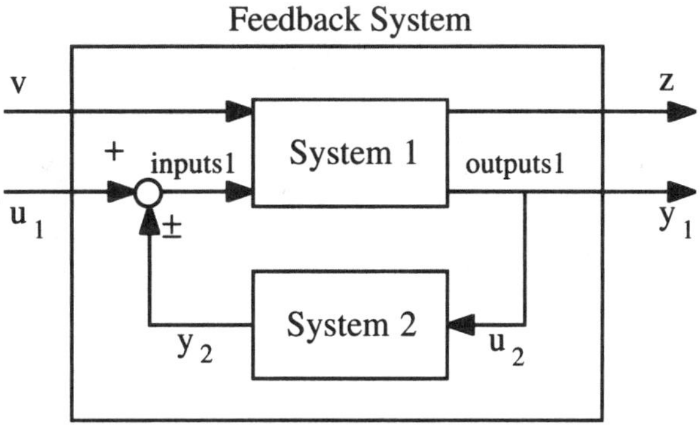

The vector `inp1` contains indices into the input vector of system 1 and specifies which inputs of system 1 are used for the feedback connection. Similarly, the vector `out1` contains indices into the outputs of system 1 and specifies which outputs of system 1 are fed into the inputs of system 2. Positive feedback is assumed. To connect with negative feedback, use negative values in the vector `inp1`. The resulting feedback system has the same inputs and outputs as system 1. (See Algorithm below for more details)

For more complicated feedback connections use a combination of `append` and `cloop`.

Examples

Example 1

Connect the plant

$$G(s) = \frac{2s^2 + 5s + 1}{s^2 + 2s + 3}$$

with the controller

$$H(s) = \frac{5(s+2)}{s+10}$$

in negative feedback, use

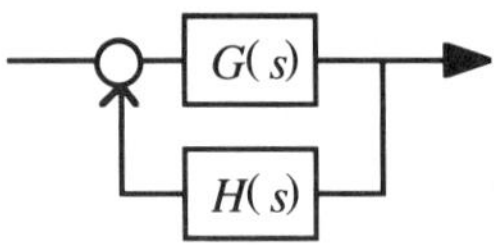


```
numg = [2 5 1];
deng = [1 2 3];
numh = [5 10];
denh = [1 10];
[num,den] = feedback(numg,deng,numh,denh);
```

Example 2

Consider a state-space plant `(a1,b1,c1,d1)` with five inputs and four outputs and a state-space feedback controller `(a2,b2,c2,d2)` with two inputs and three outputs. To connect plant outputs 1, 3, and 4 to the inputs of the controller and the controller outputs to inputs 4 and 2 of the plant, use

```
inp1 = [4 2];
out1 = [1 3 4];
[a,b,c,d] = feedback(a1,b1,c1,d1,a2,b2,c2,d2,inp1,out1);
```

Example 3

Form the negative feedback closed-loop systems for the SISO systems

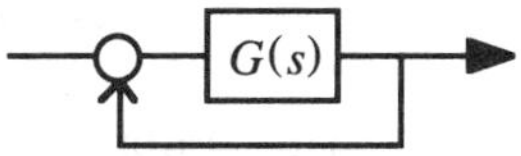


```
[numc,denc] = feedback(num,den,1,1)
```

and

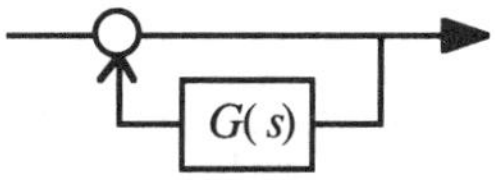


```
[numc,denc] = feedback(1,1,num,den)
```

Given two state-space systems

$$\dot{x}_1 = A_1 x_1 + \begin{bmatrix} B_1 & G \end{bmatrix} \begin{bmatrix} u_1 \\ v \end{bmatrix}$$

$$\begin{bmatrix} y_1 \\ z \end{bmatrix} = \begin{bmatrix} C_1 \\ H \end{bmatrix} x_1 + \begin{bmatrix} D_{11} & D_{12} \\ J_{11} & J_{12} \end{bmatrix} \begin{bmatrix} u_1 \\ v \end{bmatrix}$$

and

$$\begin{aligned} \dot{x}_2 &= A_2 x_2 + B_2 u_2 \\ y_2 &= C_2 x_2 + D_2 u_2 \end{aligned}$$

suppose the feedback system is to be formed by connecting the outputs y_1 to the inputs u_2 and the outputs y_2 to the inputs u_1 as shown in an earlier diagram.

Retaining the inputs and outputs of system 1, the resulting state-space system is

$$\begin{bmatrix} \dot{x}_1 \\ \dot{x}_2 \end{bmatrix} = \begin{bmatrix} A_1 \pm B_1 E D_2 C_1 & \pm B_1 E C_2 \\ B_2 C_1 \pm B_2 D_{11} E D_2 C_1 & A_2 \pm B_2 D_{11} E C_2 \end{bmatrix} \begin{bmatrix} x_1 \\ x_2 \end{bmatrix}$$

$$+ \begin{bmatrix} B_1 F & G \pm B_1 E D_2 D_{12} \\ B_2 D_1 F & B_2 D_{12} \pm B_2 D_{11} E D_2 D_{12} \end{bmatrix} \begin{bmatrix} u_1 \\ v \end{bmatrix}$$

$$\begin{bmatrix} y_1 \\ z \end{bmatrix} = \begin{bmatrix} C_1 \pm D_{11} E D_2 C_1 & \pm D_{11} E C_2 \\ H \pm J_{11} E D_2 C_1 & \pm J_{11} E C_2 \end{bmatrix} \begin{bmatrix} x_1 \\ x_2 \end{bmatrix}$$

$$+ \begin{bmatrix} D_{11} F & D_{12} \pm D_{11} E D_2 D_{12} \\ J_{11} F & J_{12} \pm J_{11} E D_2 D_{12} \end{bmatrix} \begin{bmatrix} u_1 \\ v \end{bmatrix}$$

where

$$E = (I \mp D_2 D_{11})^{-1}$$
$$F = (I \pm E D_2 D_{11})$$

If D_2 and D_{11} are zero this simplifies to

$$\begin{bmatrix} \dot{x}_1 \\ \dot{x}_2 \end{bmatrix} = \begin{bmatrix} A_1 & \pm B_1 C_2 \\ \pm B_2 C_1 & A_2 \end{bmatrix} \begin{bmatrix} x_1 \\ x_2 \end{bmatrix} + \begin{bmatrix} B_1 & G \\ B_2 D_1 & B_2 D_{12} \end{bmatrix} \begin{bmatrix} u_1 \\ v \end{bmatrix}$$

$$\begin{bmatrix} y_1 \\ z \end{bmatrix} = \begin{bmatrix} C_1 & 0 \\ H & \pm J_{11} C_2 \end{bmatrix} \begin{bmatrix} x_1 \\ x_2 \end{bmatrix} + \begin{bmatrix} 0 & D_{12} \\ J_{11} & J_{12} \end{bmatrix} \begin{bmatrix} u_1 \\ v \end{bmatrix}$$

These same expressions are valid for discrete models with differentiation replaced by differences and s-plane transfer functions replaced by z-plane transfer functions.

Limitations The matrix $(I \pm D_2 D_{11})$ must be invertible.

See Also `append`, `cloop`, `connect`, `parallel`, `series`

---

freqs

Purpose Frequency response of analog filters.

Synopsis

```
h = freqs(b,a,w)
[h,w] = freqs(b,a)
[h,w] = freqs(b,a,n)
freqs(b,a)
```

Description `freqs` computes the complex frequency response $H(jw)$ (Laplace transform) of an analog filter,

$$H(s) = \frac{B(s)}{A(s)} = \frac{b(1)s^{nb} + b(2)s^{(nb-1)} + \cdots + b(nb+1)}{s^{na} + a(2)s^{(na-1)} + \cdots + a(na+1)}$$

given the numerator and denominator coefficients in vectors `b` and `a`.

`h = freqs(b,a,w)` returns the complex frequency response of the analog filter specified by coefficient vectors `b` and `a`. `freqs` evaluates the frequency response along the imaginary axis in the complex plane at the frequencies specified in real vector `w`.

`[h,w] = freqs(b,a)` automatically picks a set of 200 frequency points `w` on which to compute the frequency response `h`.

`[h,w] = freqs(b,a,n)` picks `n` frequencies on which to compute the frequency response `h`.

`freqs` with no output arguments plots the magnitude and phase response versus frequency in the current figure window.

`freqs` works only for real input systems and positive frequencies.

Example Find and graph the frequency response of the transfer function given by

$$H(s) = \frac{0.2s^2 + 0.3s + 1}{s^2 + 0.4s + 1}$$

```
a = [1 0.4 1]; b = [0.2 0.3 1];
w = logspace(-1,1);
freqs(b,a,w)
```

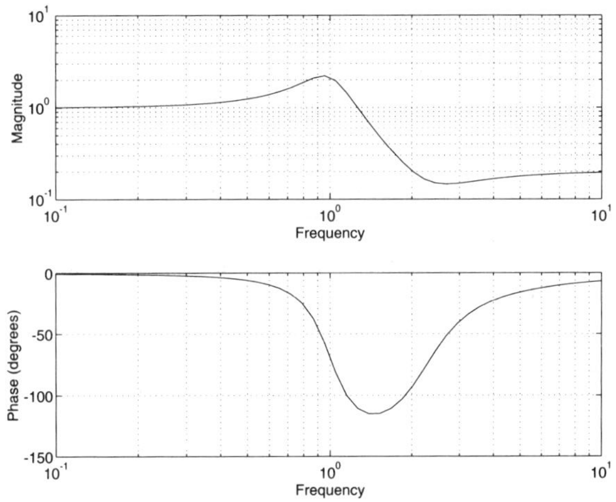

You can also create the plot with

```
h = freqs(b,a,w);
mag = abs(h); phase = angle(h);
subplot(2,1,1), loglog(w,mag)
subplot(2,1,2), semilogx(w,phase)
```

To convert to Hertz, degrees, and decibels, use

```
f = w/(2*pi);
mag = 20*log10(mag); phase = phase*180/pi;
```

Algorithm `freqs` evaluates the polynomials at each frequency point, then divides the numerator response by the denominator response:

```
s = i*w;
h = polyval(b,s)./polyval(a,s);
```

See Also `abs`, `angle`, `freqz`, `invfreqs`

`logspace`, `polyval` in the *MATLAB Reference*

---

freqz

Purpose Frequency response of digital filters.

Synopsis

```
[h,w] = freqz(b,a,n)
[h,f] = freqz(b,a,n,Fs)
[h,w] = freqz(b,a,n,'whole')
[h,f] = freqz(b,a,n,'whole',Fs)
h = freqz(b,a,w)
h = freqz(b,a,f,Fs)
freqz(b,a)
```

Description `freqz` returns the complex frequency response $H(e^{jw})$ of a digital filter given the numerator and denominator coefficients in vectors `b` and `a`.

`[h,w] = freqz(b,a,n)` returns the `n`-point complex frequency response of the digital filter,

$$H(z)=\frac{B(z)}{A(z)}=\frac{b(1)+b(2)z^{-1}+\cdots+b(nb+1)z^{-nb}}{1+a(2)z^{-1}+\cdots+a(na+1)z^{-na}}$$

given the coefficient vectors `b` and `a`. `freqz` returns both `h`, the complex frequency response, and `w`, a vector containing the `n` frequency points. `freqz` evaluates the frequency response at `n` points equally spaced around the upper half of the unit circle, so `w` contains `n` points between 0 and π.

It is best, although not necessary, to choose a value for `n` that is an exact power of two, because this allows fast computation using an FFT algorithm. If you do not specify a value for `n`, it defaults to 512.

`[h,f] = freqz(b,a,n,Fs)` allows you to specify a positive sampling frequency `Fs`, where `Fs` is in Hertz. It returns a vector `f` containing the actual frequency points between 0 and `Fs/2` at which it calculated the frequency response. `f` is of length `n`.

`[h,w] = freqz(b,a,n,'whole')` and

`[h,f] = freqz(b,a,n,'whole',Fs)` use `n` points around the whole unit circle (from 0 to 2π, or from 0 to `Fs`).

`h = freqz(b,a,w)` returns the frequency response at the frequencies in vector `w`. These frequencies must be between 0 and 2π.

`h = freqz(b,a,f,Fs)` returns the frequency response at the frequencies in vector `f`, where the elements of `f` are between 0 and `Fs`.

`freqz` with no output arguments plots the magnitude and phase response versus frequency in the current figure window.

`freqz` works for both real and complex input systems.

Example Plot the magnitude and phase response of a Butterworth filter:

```
[b,a] = butter(5,.2);
freqz(b,a,128)
```

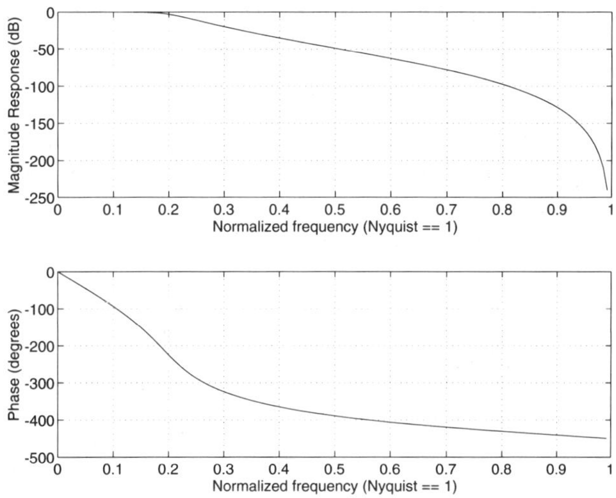

Algorithm `freqz` uses an FFT algorithm when argument `n` is present. It computes the frequency response as the ratio of the transformed numerator and denominator coefficients, padded with zeros to the desired length:

```
h = fft(b,n)./fft(a,n)
```

If `n` is not a power of two, the FFT algorithm is not as efficient and may cause long computation times.

When a frequency vector `w` or `f` is present, or if `n` is less than `max(length(b),length(a)`, `freqz` evaluates the polynomials at each frequency point using Horner's method of polynomial evaluation, and then divides the numerator response by the denominator response.

See Also `abs`, `angle`, `fft`, `filter`, `freqs`, `impz`, `invfreqz`

`logspace` in the *MATLAB Reference*

grpdelay

Purpose Average filter delay (group delay).

Synopsis

```
[gd,w] = grpdelay(b,a,n)
[gd,f] = grpdelay(b,a,n,Fs)
[gd,w] = grpdelay(b,a,n,'whole')
[gd,f] = grpdelay(b,a,n,'whole',Fs)
gd = grpdelay(b,a,w)
gd = grpdelay(b,a,f,Fs)
grpdelay(b,a)
```

Description The *group delay* of a filter is a measure of the average delay of the filter as a function of frequency. It is the negative first derivative of a filter's phase response. If the complex frequency response of a filter is $H(e^{jw})$, then the group delay is

$$\tau_g(\omega) = -\frac{d\theta(\omega)}{d\omega}$$

where ω is frequency and θ is the phase angle of $H(e^{jw})$.

`[gd,w] = grpdelay(b,a,n)` returns the `n`-point group delay, $\tau_g(\omega)$, of the digital filter

$$H(z) = \frac{B(z)}{A(z)} = \frac{b(1) + b(2)z^{-1} + \cdots + b(nb+1)z^{-nb}}{1 + a(2)z^{-1} + \cdots + a(na+1)z^{-na}}$$

given the numerator and denominator coefficients in vectors `b` and `a`. `grpdelay` returns both `gd`, the group delay, and `w`, a vector containing the `n` frequency points in radians. `grpdelay` evaluates the group delay at `n` points equally spaced around the upper half of the unit circle, so `w` contains `n` points between 0 and π. A value for `n` that is an exact power of two allows fast computation using an FFT algorithm.

`[gd,f] = grpdelay(b,a,n,Fs)` allows you to specify a positive sampling frequency `Fs` in Hertz. It returns a length `n` vector `f` containing the actual frequency points at which the group delay is calculated, also in Hertz. `f` contains `n` points between 0 and `Fs/2`.

`[gd,w] = grpdelay(b,a,n,'whole')` and `[gd,f] = grpdelay(b,a,n,'whole',Fs)` use `n` points around the whole unit circle (from 0 to 2π, or from 0 to `Fs`).

`gd = grpdelay(b,a,w)` and `gd = grpdelay(b,a,f,Fs)` return the group delay evaluated at the points in `w` (in radians) or `f` (in Hertz) respectively, where `Fs` is the sampling frequency in Hertz.

`grpdelay` with no output arguments plots the group delay versus frequency in the current figure window.

`grpdelay` works for both real and complex input systems.

Examples Plot the group delay of Butterworth filter *b(z)/a(z)*:

```
[b,a]=butter(6,.2);
grpdelay(b,a,128)
```

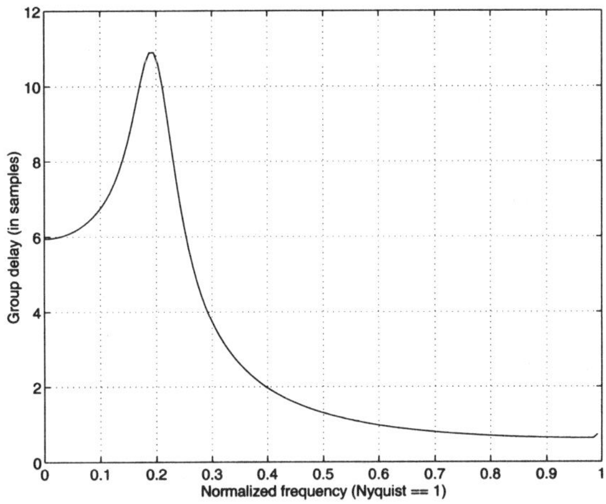

Plot both the group and phase delays of a system on the same graph:

```
gd = grpdelay(b,a,512);
gd(1) = []; % avoid NaNs
[h,w] = freqz(b,a,512); h(1) = []; w(1) = [];
pd = -unwrap(angle(h))./w;
plot(w,gd,w,pd)
```

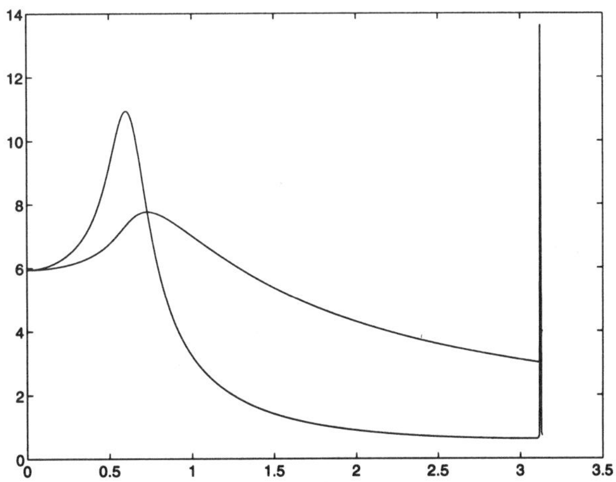

Algorithm `grpdelay` multiplies the filter coefficients by a unit ramp. After Fourier transformation, this process corresponds to differentiation.

See Also `cceps`, `fft`, `freqz`, `hilbert`, `rceps`

---

hamming

Purpose Hamming window.

Synopsis `w = hamming(n)`

Description `hamming(n)` returns an `n`-point Hamming window in a column vector. The coefficients of a Hamming window are

$$w[k+1] = 0.54 - 0.46\cos\left(2\pi\frac{k}{n-1}\right), \quad k = 0,\ldots,n-1$$

See Also `boxcar`, `hanning`, `kaiser`

---

hanning

Purpose Hanning window.

Synopsis `w = hanning(n)`

Description `hanning(n)` returns an `n`-point Hanning window in a column vector. The coefficients of a Hanning window are

$$w[k] = 0.5\left(1-\cos\left(2\pi\frac{k}{n+1}\right)\right), \quad k = 1,\ldots,n$$

See Also `boxcar`, `hamming`, `kaiser`

---

impinvar

Purpose Impulse invariance method of analog-to-digital filter conversion.

Synopsis
```
[bz,az] = impinvar(b,a,Fs)
[bz,az] = impinvar(b,a)
```

Description `[bz,az] = impinvar(b,a,Fs)` creates a digital filter with numerator and denominator coefficients `bz` and `az`, respectively, whose impulse response is equal to the impulse response of the analog filter with coefficients `b` and `a` sampled at frequency `Fs` Hz.

`[bz,az] = impinvar(b,a)` uses the default value for `Fs`, 1 Hz.

Example Convert an analog lowpass filter to a digital filter using `impinvar` with a sampling frequency of 10 Hertz:

```
[b,a] = butter(4,.3,'s');
[bz,az] = impinvar(b,a,10)
real(bz)
ans =
        1.0e-05 *
        0.0000    0.1324    0.5192    0.1273
real(az)
ans =
        Columns 1 through 4
        1.0000   –3.9216    5.7679   –3.7709    0.9246
```

Algorithm `impinvar` performs the impulse-invariant method of analog-to-digital transfer function conversion discussed in [1]:

1. It finds the partial fraction expansion of the system represented by `b` and `a` using `residue`.
2. It replaces the poles `p` by the poles `exp(p/Fs)`.
3. It finds the transfer function coefficients of the system using the function `residuez` with the new poles.

References
[1] T. W. Parks and C. S. Burrus, *Digital Filter Design*, pp. 206–209, John Wiley & Sons, 1987.

See Also `bilinear`, `residuez`

`residue` in *MATLAB Reference*

---

impulse

Purpose Unit impulse response.

Synopsis

```
[y,x,t] = impulse(a,b,c,d)
[y,x,t] = impulse(a,b,c,d,iu)
[y,x,t] = impulse(a,b,c,d,iu,t)
[y,x,t] = impulse(num,den)
[y,x,t] = impulse(num,den,t)
```

Description `impulse` calculates the unit impulse response of a linear system. Invoked without left-hand arguments, `impulse` plots the impulse response on the screen.

`impulse(a,b,c,d)` produces a series of impulse response plots, one for each input and output combination of the continuous LTI system:

$$\dot{x} = Ax + Bu$$
$$y = Cx + Du$$

with the time vector automatically determined.

`impulse(a,b,c,d,iu)` produces an impulse response plot from the single input `iu` to all the outputs of the system with the time vector automatically determined. The scalar, `iu`, is an index into the inputs of the system and specifies which input to be used for the impulse response.

`impulse(num,den)` produces the impulse response plot of the polynomial transfer function `G(s) = num(s)/den(s)` where `num` and `den` contain the polynomial coefficients in descending powers of s.

`impulse(a,b,c,d,iu,t)` or `impulse(num,den,t)` uses the user-supplied time vector `t`. The vector `t` specifies the times at which the impulse response is computed and must be regularly spaced.

Invoked with left-hand arguments:

```
[y,x,t] = impulse(a,b,c,d,iu)
[y,x,t] = impulse(a,b,c,d,iu,t)
[y,x,t] = impulse(num,den)
[y,x,t] = impulse(num,den,t)
```

returns the output and state responses of the system and the time vector `t`. No plot is drawn on the screen. The matrices `y` and `x` contain the output and state response of the system evaluated at the time points `t`. `y` has as many columns as outputs and one row for each element in `t`. `x` has as many columns as states and one row for each element in `t`.

Examples Plot the impulse response of the second-order state-space system:

$$\begin{bmatrix}\dot{x}_1\\ \dot{x}_2\end{bmatrix}=\begin{bmatrix}-0.5572 & -0.7814\\ 0.7814 & 0\end{bmatrix}\begin{bmatrix}x_1\\ x_2\end{bmatrix}+\begin{bmatrix}1\\ 0\end{bmatrix}u$$

$$y=\begin{bmatrix}1.9691 & 6.4493\end{bmatrix}\begin{bmatrix}x_1\\ x_2\end{bmatrix}+\begin{bmatrix}0\end{bmatrix}u$$

```
a = [-0.5572   -0.7814;0.7814  0];
b = [1;0];
c = [1.9691  6.4493];
d = [0];
impulse(a,b,c,d);
title('Impulse Response')
```

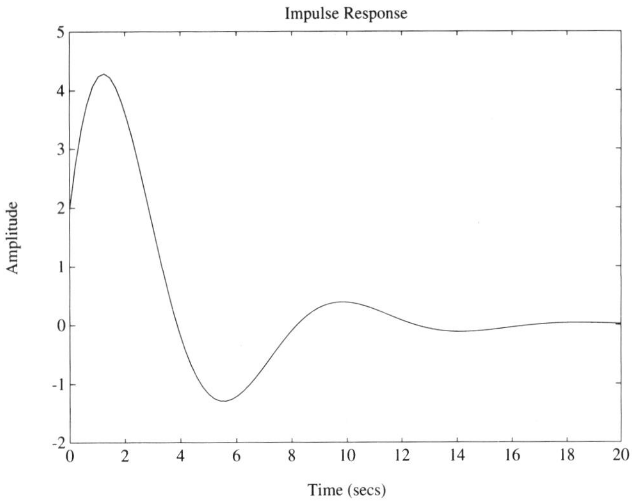

Algorithm `impulse` converts the continuous system to a discrete one using matrix exponentials and the interval between the time points in vector `t`. The discrete system is then propagated over the number of points in `t`. The time vector is assumed to start at `t` = 0.

See Also `dimpulse`, `lsim`, `step`

kaiser

Purpose Kaiser window.

Synopsis `kaiser(n,beta)`

Description `kaiser(n,beta)` returns an `n`-point Kaiser (I_0 – sinh) window in a column vector. `beta` is the Kaiser window β parameter that affects the sidelobe attenuation of the Fourier transform of the window.

The minimum sidelobe attenuation α, in decibels, is heuristically related to `beta` by

$$\beta = \begin{cases} 0.1102(\alpha - 8.7), & \alpha > 50 \\ 0.5842(\alpha - 21)^{0.4} + 0.07886(\alpha - 21), & 50 \geq \alpha \geq 21 \\ 0, & \alpha < 21 \end{cases}$$

Increasing `beta` widens the mainlobe and decreases the amplitude of the sidelobes (increases the attenuation).

See Also `boxcar`, `hamming`, `hanning`

References

[1] Kaiser, J. F. “Nonrecursive Digital Filter Design Using the I_0 – sinh Window Function.” *Proc. 1974 IEEE Symp. Circuits and Syst.* April 1974: 20–23.

[2] IEEE. *Digital Signal Processing II. IEEE Press*, New York: John Wiley & Sons, 1975.

lp2bp

Purpose Lowpass to bandpass analog filter transformation.

Synopsis

```
[bt,at] = lp2bp(b,a,Wo,Bw)
[At,Bt,Ct,Dt] = lp2bp(A,B,C,D,Wo,Bw)
```

Description `lp2bp` transforms analog lowpass filter prototypes with cutoff frequency 1 radian/sec into bandpass filters with desired bandwidth and center frequency. The transformation is one step in the digital filter design process for the `butter`, `cheby1`, `cheby2`, and `ellip` functions.

`lp2bp` can perform the transformation on two different linear system representations: transfer function form and state-space form. In both cases, the input system must be an analog filter prototype.

Transfer Function Form (Polynomial)

`[bt,at] = lp2bp(b,a,Wo,Bw)` transforms an analog lowpass filter prototype given by polynomial coefficients into a bandpass filter with center frequency `Wo` and bandwidth Bw. Row vectors b and a specify the coefficients of the numerator and denominator of the prototype in descending powers of s,

$$\frac{b(s)}{a(s)} = \frac{b(1)s^{nn} + \cdots + b(nn)s + b(nn+1)}{a(1)s^{nd} + \cdots + a(nd)s + a(nd+1)}$$

Scalars Wo and Bw specify the center frequency and bandwidth in units of radians/sec. For a filter with lower band edge w1 and upper band edge w2, use `Wo = sqrt(w1*w2)` and `Bw = w2 - w1`.

`lp2bp` returns the frequency transformed filter in row vectors `bt` and `at`.

State-space Form

`[At,Bt,Ct,Dt] = lp2bp(A,B,C,D,Wo,Bw)` converts the continuous-time state-space lowpass filter prototype in matrices A, B, C, D

$$\dot{x} = Ax + Bu$$
$$y = Cx + Du$$

into a bandpass filter with center frequency Wo and bandwidth Bw. For a filter with lower band edge w1 and upper band edge w2, use `Wo = sqrt(w1*w2)` and `Bw = w2 - w1`.

The bandpass filter is returned in matrices `At`, `Bt`, `Ct`, `Dt`.

Algorithm `lp2bp` is a highly accurate state-space formulation of the classic analog filter frequency transformation. Consider the state-space system

$$\dot{x} = Ax + Bu$$
$$y = Cx + Du$$

where u is the input, x is the state vector, and y is the output. The Laplace transform of the first equation is

$$sx = Ax + Bu$$

Now if a bandpass filter is to have center frequency ω_0 and bandwidth B_w, the standard s-domain transformation is

$$s = Q(p^2 + 1)/p$$

where $Q = \omega_0/B_w$ and $p = s/\omega_0$. Substituting this for s in the Laplace transformed state-space equation, and considering the operator p as d/dt,

$$Q\ddot{x} + Qx = A\dot{x} + B\dot{u} \qquad \text{or} \qquad Q\ddot{x} - A\dot{x} - B\dot{u} = -Qx$$

Now define

$$Q\dot{\omega} = -Qx$$

which when substituted, leads to

$$Q\dot{x} = Ax + Q\omega + Bu$$

The last two equations give equations of state. Write them in standard form and multiply the differential equations by ω_0 to recover the time/frequency scaling represented by p, and find state matrices for the bandpass filter:

```
Q = Wo/Bw; [ma,na] = size(A);
At = Wo*[A/Q eye(ma,na);-eye(ma,na) zeros(ma,na)];
Bt = Wo*[B/Q; zeros(ma,nb)];
Ct = [C zeros(mc,ma)];
Dt = d;
```

If the input to `lp2bp` is in transfer function form, the function transforms it into state-space form before applying this algorithm.

See Also `bilinear`, `lp2bs`, `lp2hp`, `lp2lp`

---

lp2bs

Purpose Lowpass to bandstop analog filter transformation.

Synopsis

```
[bt,at] = lp2bs(b,a,Wo,Bw)
[At,Bt,Ct,Dt] = lp2bs(A,B,C,D,Wo,Bw)
```

Description `lp2bs` transforms analog lowpass filter prototypes with cutoff frequency 1 radian/sec into bandstop filters with desired bandwidth and center frequency. The transformation is one step in the digital filter design process for the `butter`, `cheby1`, `cheby2`, and `ellip` functions.

`lp2bs` can perform the transformation on two different linear system representations: transfer function form and state-space form. In both cases, the input system must be an analog filter prototype.

Transfer Function Form (Polynomial)

`[bt,at] = lp2bs(b,a,Wo,Bw)` transforms an analog lowpass filter prototype given by polynomial coefficients into a bandstop filter with center frequency `Wo` and bandwidth `Bw`. Row vectors `b` and `a` specify the coefficients of the numerator and denominator of the prototype in descending powers of s,

$$\frac{b(s)}{a(s)} = \frac{b(1)s^{nn} + \cdots + b(nn)s + b(nn+1)}{a(1)s^{nd} + \cdots + a(nd)s + a(nd+1)}$$

Scalars `Wo` and `Bw` specify the center frequency and bandwidth in units of radians/sec. For a filter with lower band edge `w1` and upper band edge `w2`, use `Wo = sqrt(w1*w2)` and `Bw = w2 – w1`.

lp2bs returns the frequency transformed filter in row vectors bt and at.

State-space Form

[At,Bt,Ct,Dt] = lp2bs(A,B,C,D,Wo,Bw) converts the continuous-time state-space lowpass filter prototype in matrices A, B, C, D,

$$\dot{x} = Ax + Bu$$
$$y = Cx + Du$$

into a bandstop filter with center frequency Wo and bandwidth Bw. For a filter with lower band edge w1 and upper band edge w2, use Wo = sqrt(w1*w2) and Bw = w2 – w1.

The bandstop filter is returned in matrices At, Bt, Ct, Dt.

Algorithm lp2bs is an M-file with a highly accurate state-space formulation of the classic analog filter frequency transformation. If a bandstop filter is to have center frequency ω_0 and bandwidth B_w, the standard s-domain transformation is

$$s = \frac{p}{Q(p^2 + 1)}$$

where $Q = \omega_0/B_w$ and $p = s/\omega_0$. The state-space version of this transformation is

```
Q = Wo/Bw;
At = [Wo/Q*inv(A) Wo*eye(ma); ...
-Wo*eye(ma) zeros(ma)];
Bt = -[Wo/Q*(A B); zeros(ma,nb)];
Ct = [C/A zeros(mc,ma)];
Dt = D - C/A*B;
```

See lp2bp for a derivation of the bandpass version of this transformation.

See Also bilinear, lp2bp, lp2hp, lp2lp

---

lp2hp

Purpose Lowpass to highpass analog filter transformation.

Synopsis

```
[bt,at] = lp2hp(b,a,Wo)
[At,Bt,Ct,Dt] = lp2hp(A,B,C,D,Wo)
```

Description lp2hp transforms analog lowpass filter prototypes with cutoff frequency 1 radian/sec into highpass filters with desired cutoff frequency. The transformation is one step in the digital filter design process for the butter, cheby1, cheby2, and ellip functions.

The `lp2hp` function can perform the transformation on two different linear system representations: transfer function form and state-space form. In both cases, the input system must be an analog filter prototype.

Transfer Function Form (Polynomial)

`[bt,at] = lp2hp(b,a,Wo)` transforms an analog lowpass filter prototype given by polynomial coefficients into a highpass filter with cutoff frequency `Wo`. Row vectors `b` and `a` specify the coefficients of the numerator and denominator of the prototype in descending powers of s,

$$\frac{b(s)}{a(s)} = \frac{b(1)s^{nn} + \cdots + b(nn)s + b(nn+1)}{a(1)s^{nd} + \cdots + a(nd)s + a(nd+1)}$$

Scalar `Wo` specifies the cutoff frequency in units of radians/sec. The frequency transformed filter is returned in row vectors `bt` and `at`.

State-space Form

`[At,Bt,Ct,Dt] = lp2hp(A,B,C,D,Wo)` converts the continuous-time state-space lowpass filter prototype in matrices `A`, `B`, `C`, `D`,

$$\dot{x} = Ax + Bu$$
$$y = Cx + Du$$

into a highpass filter with cutoff frequency `Wo`. The highpass filter is returned in matrices `At`, `Bt`, `Ct`, `Dt`.

Algorithm `lp2hp` is an M-file with a highly accurate state-space formulation of the classic analog filter frequency transformation. If a highpass filter is to have cutoff frequency ω_0, the standard s-domain transformation is

$$s = \frac{\omega_o}{p}$$

The state-space version of this transformation is

```
At = Wo*inv(A);
Bt = –Wo*(A\B);
Ct = C/A;
Dt = D – C/A*B;
```

See `lp2bp` for a derivation of the bandpass version of this transformation.

See Also `bilinear`, `lp2bp`, `lp2bs`, `lp2lp`

lp2lp

Purpose Lowpass to lowpass analog filter transformation.

Synopsis

```
[bt,at] = lp2lp(b,a,Wo)
[At,Bt,Ct,Dt] = lp2lp(A,B,C,D,Wo)
```

Description `lp2lp` transforms an analog lowpass filter prototype with cutoff frequency 1 radian/sec into a lowpass filter with any specified cutoff frequency. The transforma tion is one step in the digital filter design process for the `butter`, `cheby1`, `cheby2`, and `ellip` functions.

The `lp2lp` function can perform the transformation on two different linear system representations: transfer function form and state-space form. In both cases, the input system must be an analog filter prototype.

Transfer Function Form (Polynomial)

`[bt,at] = lp2lp(b,a,Wo)` transforms an analog lowpass filter prototype given by polynomial coefficients into a lowpass filter with cutoff frequency `Wo`. Row vectors `b` and `a` specify the coefficients of the numerator and denominator of the prototype in descending powers of s,

$$\frac{b(s)}{a(s)} = \frac{b(1)s^{nn} + \cdots + b(nn)s + b(nn+1)}{a(1)s^{nd} + \cdots + a(nd)s + a(nd+1)}$$

Scalar `Wo` specifies the cutoff frequency in units of radians/sec. `lp2lp` returns the frequency transformed filter in row vectors `bt` and `at`.

State-space Form

`[At,Bt,Ct,Dt] = lp2lp(A,B,C,D,Wo)` converts the continuous-time state-space lowpass filter prototype in matrices `A`, `B`, `C`, `D`,

$$\dot{x} = Ax + Bu$$
$$y = Cx + Du$$

into a lowpass filter with cutoff frequency `Wo`. `lp2lp` returns the lowpass filter in matrices `At`, `Bt`, `Ct`, `Dt`.

Algorithm `lp2lp` is an M-file with a highly accurate state-space formulation of the classic analog filter frequency transformation. If a lowpass filter is to have cutoff frequency ω_0, the standard s-domain transformation is

$$s = p / \omega_o$$

The state-space version of this transformation is

```
At = Wo*A;
Bt = Wo*B;
Ct = C;
Dt = D;
```

See `lp2bp` for a derivation of the bandpass version of this transformation.

See Also `bilinear`, `lp2bp`, `lp2bs`, `lp2hp`

---

lqe

Purpose Linear-quadratic estimator design.

Synopsis

```
[L,P,E] = lqe(a,g,c,Q,R)
[L,P,E][L,P,E] = lqe(a,g,c,Q,R,N)
```

Description `lqe` solves the continuous-time linear-quadratic estimator problem and the associated Riccati equation.

For the continuous-time system with state and measurement equations

$$\dot{x} = Ax + Bu + Gw$$
$$y = Cx + Du + v$$

and process and measurement noise covariances

$$E[w] = E[v] = 0, \quad E[ww'] = Q, \quad E[vv'] = R, \quad E[wv'] = 0$$

`L = lqe(a,g,c,Q,R)` returns the gain matrix `L` such that the continuous, stationary Kalman filter:

$$\dot{x} = Ax + Bu + L(y - Cx - Du)$$

produces an LQG optimal estimate of x.

`[L,P,E] = lqe(a,g,c,Q,R)` also returns the Riccati equation solution `p` which is the estimation error covariance and the closed-loop eigenvalues of the estimator `E = eig(A–L*C)`.

`[L,P,E] = lqe(a,g,c,Q,R,N)` calculates the Kalman gain matrix when the process and sensor noise are correlated.

$$E[w] = E[v] = 0, \quad E[ww'] = Q, \quad E[vv'] = R, \quad E[wv'] = N$$

Limitations Certain assumptions must be met in order for a unique positive definite solution to the LQE problem to exist:

- `Matrix Q must be symmetric and positive semi-definite.`

- Matrix R must be symmetric and positive definite.
- The (A,C) pair must be observable.

Algorithm lqe uses lqr to calculate the estimator gains using duality.

See Also lqr, dlqe, dlqew

---

lqr

Purpose Linear-quadratic regulator design.

Synopsis

```
[K,S,E] = lqr(a,b,Q,R)
[K,S,E] = lqr(a,b,Q,R,N)
```

Description lqr solves the continuous-time linear-quadratic regulator problem and the associated Riccati equation.

K = lqr(a,b,Q,R) calculates the optimal feedback gain matrix K such that the feedback law

$$u = -Kx$$

minimizes the cost function

$$J = \int (x'Qx + u'Ru)dt$$

subject to the constraint equation

$$\dot{x} = Ax + Bu$$

[K,S,E] = lqr(a,b,Q,R) also returns S, the unique positive definite solution to the associated matrix Riccati equation:

$$0 = SA + A'S - SBR^{-1}B'S + Q$$

and E, the closed-loop eigenvalues, E=eig(A–B*K):

[K,S,E] = lqr(a,b,Q,R,N) calculates the optimal feedback gain matrix that minimizes the cost function with the cross weighting matrix N:

$$J = \int (x'Qx + 2u'Nx + u'Ru)dt$$

Limitations Certain assumptions must be met in order for a unique positive definite solution to the LQR problem to exist:

- Matrix Q must be symmetric and positive semi-definite.
- Matrix R must be symmetric and positive definite.

- The (A,B) pair must be controllable.

Algorithm lqr uses eigenvector decomposition of an associated Hamiltonian matrix.

See Also lqe, dlqr, dlqry

---

lsim

Purpose Continuous system simulation to arbitrary inputs.

Synopsis

```
[y,x] = lsim(a,b,c,d,u,t)
[y,x] = lsim(a,b,c,d,u,t,x0)
[y,x] = lsim(num,den,u,t)
```

Description lsim simulates continuous-time linear systems with arbitrary inputs. When invoked without left-hand arguments, lsim produces a plot on the screen.

Given the LTI system

$$\dot{x} = Ax + Bu$$
$$y = Cx + Du$$

lsim(a,b,c,d,u,t) produces a plot of the time response of the system to the input time history in matrix U. Matrix U must have as many columns as there are inputs u. Each row of U corresponds to a new time point and U must have length(t) rows. Vector t specifies the time axis for the simulation and must be regularly spaced. When used with an extra right-hand argument, initial conditions on the states can be specified:

```
lsim(a,b,c,d,u,t,x0)
```

lsim(num,den,u,t) plots the time response of the polynomial transfer function G(s) = num(s)/den(s) where num and den contain the polynomial coefficients in descending powers of s.

Invoked with left-hand arguments:

```
[y,x] = lsim(a,b,c,d,u,t)
[y,x] = lsim(a,b,c,d,u,t,x0)
[y,x] = lsim(num,den,u,t)
```

returns the matrices x and y, where y is the output response and x is the state response of the system. No plot is drawn on the screen. y has as many columns as there are outputs y and one row for each row of u. x has as many columns as there are states x and has one row for each row of u.

Examples Simulate and plot the response of the system:

$$H(s) = \frac{2s^2 + 5s + 1}{s^2 + 2s + 3}$$

to 10 seconds of a square wave

```
num = [2 5 1];
den = [1 2 3];
t = (0:.1:10)';% define time axis
```

Generate the square wave with a period of four seconds:

```
period = 4;
u = (rem(t,period) >= period./2);
lsim(num,den,u,t);
title('Square wave response')
```

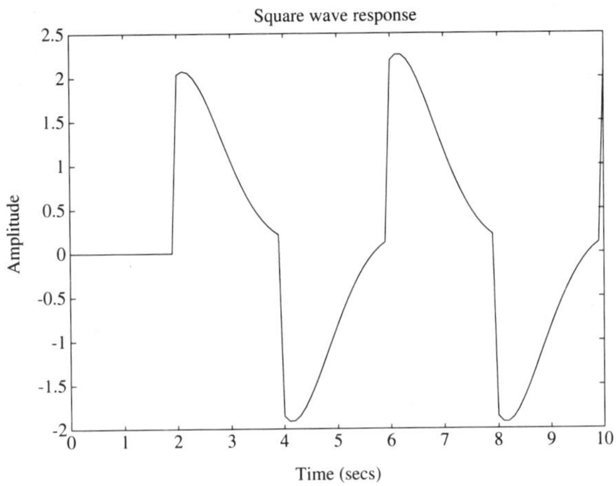

Algorithm `lsim` converts the continuous system to a discrete system using matrix exponentials and the interval between the time points in vector `t`. The time vector is assumed to start at t=0. The discrete system is then propagated over the number of points in `t` using `ltitr`.

See Also `impulse`, `step`, `dimpulse`, `dstep`, `dlsim`, `filter`, `c2d`, `ltitr`

---

nyquist

Purpose Nyquist frequency response plots.

Synopsis

```
[re,im,w] = nyquist(a,b,c,d)
[re,im,w] = nyquist(a,b,c,d,iu)
[re,im,w] = nyquist(a,b,c,d,iu,w)
[re,im,w] = nyquist(num,den)
[re,im,w] = nyquist(num,den,w)
```

Description `nyquist` calculates the Nyquist frequency response of continuous-time LTI systems. Nyquist plots are used to analyze system properties including gain margin, phase margin, and stability. When invoked without left-hand arguments, `nyquist` produces a Nyquist plot on the screen.

`nyquist` can determine the stability of a unity feedback system. Given the Nyquist plot of the open-loop transfer function *G(s)*, the closed-loop transfer function:

$$G_{cl}(s) = \frac{G(s)}{1 + G(s)}$$

is stable if the Nyquist plot encircles the $-1 + j0$ point exactly P times in the counterclockwise direction, where P is the number of unstable open-loop poles.

`nyquist(a,b,c,d)` produces a series of Nyquist plots, one for each input and output combination of the continuous state-space system:

$$\dot{x} = Ax + Bu$$
$$y = Cx + Du$$

with the frequency range automatically determined. More points are used where the response is changing rapidly.

`nyquist(a,b,c,d,iu)` produces a Nyquist plot from the single input `iu` to all the outputs of the system with the frequency range determined automatically. The scalar `iu` is an index into the inputs of the system and specifies which input to be used for the Nyquist response.

`nyquist(num,den)` draws the Nyquist plot of the continuous polynomial transfer function `G(s) = num(s)/den(s)` where `num` and `den` contain the polynomial coefficients in descending powers of *s*.

`nyquist(a,b,c,d,iu,w)` or `nyquist(num,den,w)` uses the user-supplied frequency vector `w`. The vector `w` specifies the frequencies in radians/sec at which the Nyquist response is calculated. See `logspace` to generate frequency vectors that are equally spaced logarithmically in frequency.

When invoked with left-hand arguments:

```
[re,im,w] = nyquist(a,b,c,d,iu)
[re,im,w] = nyquist(a,b,c,d,iu,w)
[re,im,w] = nyquist(num,den)
[re,im,w] = nyquist(num,den,w)
```

returns the frequency response of the system in the matrices re, im, and w. No plot is drawn on the screen. The matrices re and im contain the real and imaginary parts of the frequency response of the system evaluated at the frequency values w. re and im have as many columns as outputs and one row for each element in w.

Examples Plot the Nyquist response of the system:

$$H(s) = \frac{2s^2 + 5s + 1}{s^2 + 2s + 3}$$

```
num = [2 5 1];
den = [1 2 3];
nyquist(num,den);
title('Nyquist Plot')
```

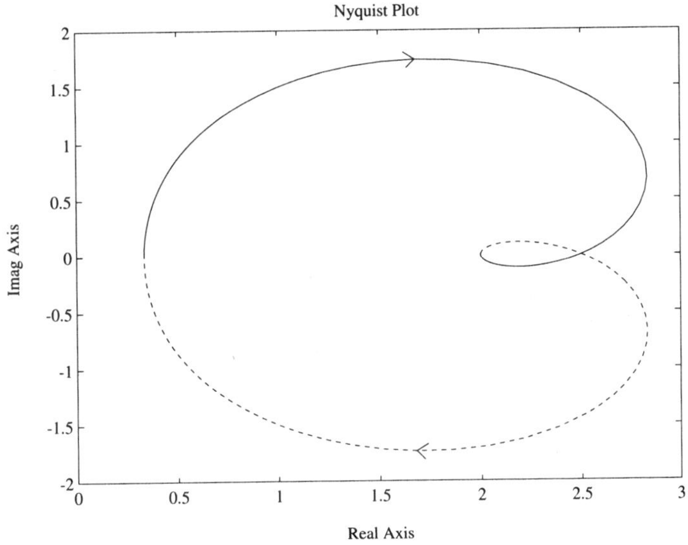

Since there are no encirclements of the $-1 + j0$ point and $P = 0$, the closed-loop system associated with $H(s)$ is stable.

Algorithm See bode for a description of the algorithm used.

Diagnostics If there is a system pole on the $j\omega$ axis and the w vector happens to contain this frequency point, the gain is infinite, $(j\omega I–A$ is singular, and nyquist produces the message:

```
Matrix is singular to working precision.
```

See Also dnyquist, bode, logspace

References [1] [1] Laub, A.J., Efficient Multivariable Frequency Response Computations, *IEEE Proceedings*, Vol. AC-26, No. 2, April 1981, pp. 407-408.

---

place, acker

Purpose Pole placement gain selection.

Synopsis

```
K = place(a,b,p)
K = acker(a,b,p)
```

Description `place` and `acker` are used for pole placement gain selection.

`K = acker(a,b,p)` uses Ackermann's formula [1] to calculate the feedback gain vector `k` such that the single-input system:

$$\dot{x} = Ax + Bu$$

with a full-state feedback law of $u = -kx$, has closed-loop poles at the values specified in vector `p`; that is

```
p = eig(a–b*k)
```

`K = place(a,b,p)` computes the feedback gain matrix to achieve the desired closed-loop pole locations assuming all the inputs of the plant are control inputs. `place` works for multi-input systems and is based on the algorithm from [2]. This algorithm uses the extra degrees of freedom to find a solution that minimizes the sensitivity of the poles to perturbations in the closed-loop system *A-BK*.

`place` and `acker` can also be used for estimator gain selection by transposing matrices and substituting `c'` for `b`:

```
l = place(a',c',p).'
```

Examples Consider a state-space system `(a,b,c,d)` with two inputs, three outputs and three states. Compute the feedback gain matrix needed to place the closed-loop poles at `p = [1.1 23 5.0]`:

```
p = [1 1.23 5.0];
K = place(a,b,p)
```

Algorithm `acker` uses Ackermann's formula [1]. Note that this method is not numerically reliable and starts to break down rapidly for problems of order greater than 5 or for weakly controllable systems.

`place` uses the reliable algorithm of [2] which, for multi-input systems, optimizes the choice of eigenvectors for a robust solution. We recommend `place` rather than `acker` even for single input systems.

With full-order pole-placement designs in general, you can demand pole locations that result in very large gains. The sensitivity problems which accompany large gains suggest caution in the use of pole placement techniques. See [3] for results from extensive numerical testing.

Diagnostics For both these functions, a warning message is printed if the nonzero closed-loop poles are greater than 10% from the desired locations specified in `p`.

See Also `lqr`, `lqe`, `rlocus`

References

[1] Kailath, T., *Linear Systems*, Prentice-Hall, 1980, p 201.

[2] Kautsky, J. and Nichols N.K, Robust Eigenstructure Assignment in State Feedback Control, *Numerical Analysis Report NA/2/83*, School of Mathematical Sciences, Flinders U., Bedford Park, S.A. 5042, Australia.

[3] Laub, A.J. and Wette, M., *Algorithms and Software for Pole Assignment and Observers*, UCRL-15646 Rev. 1, EE Dept, Univ. of Calif., Santa Barbara, CA, Sept. 1984.

---

polystab

Purpose Stabilize polynomial.

Synopsis

```
b = polystab(a)
```

Description `polystab` stabilizes a polynomial with respect to the unit circle; it reflects roots with magnitudes greater than 1 inside the unit circle.

`b = polystab(a)`, where a is a vector of polynomial coefficients, normally in the z-domain,

$$a(z) = a(1) + a(2)z^{-1} + \cdots + a(na+1)z^{-na}$$

returns a row vector containing the stabilized polynomial.

Example `polystab` can convert a linear phase filter into a minimum phase filter with the same magnitude response:

```
h = fir1(25,0.4);
hmin = polystab(h)*norm(h)/norm(polystab(h));
```

Algorithm `polystab` finds the roots of the polynomial and maps those roots found outside the unit circle to the inside of the unit circle:

```
v = roots(a);
vs = 0.5*(sign(abs(v)-1)+1);
v = (1-vs).*v + vs./conj(v);
b = a(1)*poly(v);
```

See Also `roots`

psd

Purpose Estimate the power spectral density (PSD) of a signal.

Synopsis

```
Pxx = psd(x)
Pxx = psd(x,nfft)
[Pxx,f] = psd(x,nfft,Fs)
Pxx = psd(x,nfft,Fs,window)
Pxx = psd(x,nfft,Fs,window,noverlap)
Pxx = psd(x,...,'dflag')
[Pxx,Pxxc,f] = psd(x,nfft,Fs,window,noverlap,p)
psd(x)
```

Description `Pxx = psd(x)` estimates the power spectrum of the sequence `x` using the Welch method of spectral estimation. `Pxx = psd(x)` uses the following default values:

- `nfft = min(256,length(x))`
- `Fs = 2`
- `window = hanning(nfft)`
- `noverlap = 0`

`nfft` specifies the FFT length that `psd` uses. This value determines the frequencies at which the power spectrum is estimated. `Fs` is a scalar that specifies the sampling frequency. `window` specifies a windowing function and the number of samples `psd` uses in its sectioning of the `x` vector. `noverlap` is the number of samples by which the sections overlap. Any arguments that you omit from the end of the input parameter list use the default values shown above.

If `x` is real, `psd` estimates the spectrum at positive frequencies only; in this case, the output `Pxx` is a column vector of length `nfft/2+1` for `nfft` even, and `(nfft+1)/2` for `nfft` odd. If `x` is complex, `psd` estimates the spectrum at both positive and negative frequencies, and `Pxx` has length `nfft`.

`Pxx = psd(x,nfft)` uses the specified FFT length `nfft` in estimating the power spectrum for `x`. Use a power of 2 for `nfft` for fastest execution.

`[Pxx,f] = psd(x,nfft,Fs)` returns a vector `f` of frequencies at which the function evaluates the PSD. `Fs` is the sampling frequency. `f` is the same size as `Pxx`, so `plot(f,Pxx)` plots the power spectrum versus properly scaled frequency. `Fs` has no effect on the output `Pxx`; it is only a frequency scaling multiplier.

`Pxx = psd(x,nfft,Fs,window)` specifies a windowing function and the number of samples per section of the `x` vector. If you supply a scalar for `window`, `psd` uses a Hanning window of that length. The length of the window must be less than or equal to `nfft`; `psd` zero pads the sections if `nfft` exceeds the length of the window.

`Pxx = psd(x,nfft,Fs,window,noverlap)` overlaps the sections of `x` by `noverlap` samples.

You can use the empty matrix `[]` to specify the default value for any input argument except `x`. For example,

```
psd(x,[],10000)
```

is equivalent to

```
psd(x)
```

but with a sampling frequency of 10000 Hertz instead of the default of 2 Hertz.

`Pxx = psd(x,...,'dflag')` lets you specify a detrend option, where *`dflag`* can be:

- `linear` to remove the best straight-line fit from the prewindowed sections of `x` (default).
- `mean` to remove the mean from the prewindowed sections of `x`.
- `none` for no detrending.

The *`dflag`* parameter must appear last in the list of input arguments. `psd` recognizes a *`dflag`* string no matter how many intermediate arguments are omitted.

`[Pxx,Pxxc,f] = psd(x,nfft,Fs,window,noverlap,p)`, where `p` is a positive scalar between 0 and 1, returns a vector `Pxxc` that contains an estimate of the `p*100` percent confidence interval for `Pxx`. `Pxxc` is the same size as `Pxx`. The interval `[Pxx–Pxxc, Pxx+Pxxc]` covers the true PSD with probability `p.plot(f,[Pxx Pxx–Pxxc Pxx+Pxxc])` plots the power spectrum inside the `p*100` percent confidence interval. If unspecified, `p` defaults to 0.95.

`psd` with no output arguments plots the PSD versus frequency in the current figure window. If the `p` parameter is specified, the plot includes the confidence interval.

Example Generate a colored noise signal and plot its PSD with a confidence interval of 95%. This example uses a length 1024 FFT, a 512 point Kaiser window with no overlap, and a sampling frequency of 10 kHz:

```
h = fir1(30,.2,boxcar(31));   % design a lowpass filter
r = randn(16384,1);            % white noise
x = filter(h,1,r);             % color the noise
psd(x,1024,0,kaiser(512,5),10000,.95)
```

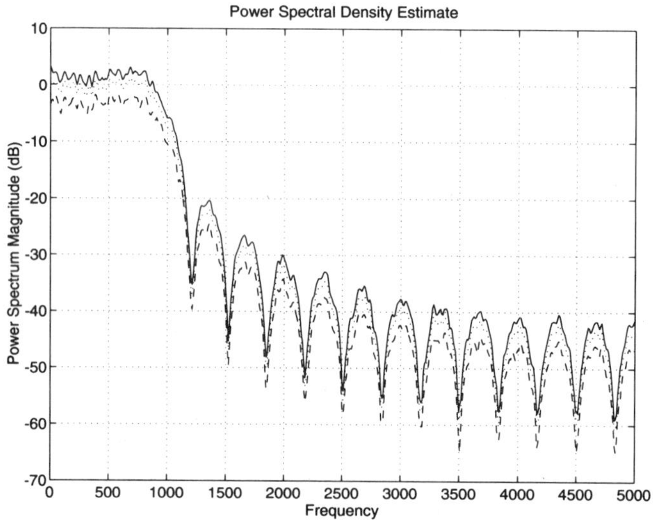

Algorithm psd calculates the power spectral density using Welch's method [1], [2]:

1. It applies the window specified by the `window` vector to successive detrended sections of `x`.
2. It transforms each section with an `nfft`-point FFT.
3. It forms the periodogram of each section by scaling the magnitude squared of each transform.
4. It averages the periodograms of the overlapping sections to form `Pxx`, the power spectrum of `x`.

The number of sections that `psd` averages is

```
k = fix((length(x)-noverlap)/(length(window)-noverlap))
```

Diagnostics The following diagnostic messages arise from incorrect use of psd:

```
Requires window's length to be no greater than FFT length.
Requires NOVERLAP to be strictly less than the window length.
Requires positive integer values for NFFT and NOVERLAP.
Requires confidence parameter to be a scalar between 0 and 1.
Requires vector input.
```

References

[1] Rabiner, L.R., and B. Gold. *Theory and Application of Digital Signal Processing*, 399–419. Englewood Cliffs, NJ: Prentice-Hall, 1975.

[2] Peter D. Welch, "The Use of Fast Fourier Transform for the Estimation of Power Spectra: A Method Based on Time Averaging Over Short, Modified Periodograms," *IEEE Trans. Audio Electroacoust.* AU-15 (June 1967): pp. 70–73.

[3] Oppenheim, A. V., and R.W. Schafer. *Discrete-Time Signal Processing*, pp. 311–312. Englewood Cliffs, NJ: Prentice-Hall, 1989.

---

remez

Purpose Parks-McClellan optimal FIR filter design.

Synopsis

```
b = remez(n,f,m)
b = remez(n,f,m,w)
b = remez(n,f,m,'ftype')
b = remez(n,f,m,w,'ftype')
```

Description remez designs linear-phase FIR filters using the Parks-McClellan algorithm. The Parks-McClellan algorithm [1] uses the Remez exchange algorithm and Chebyshev approximation theory to design filters with an optimal fit between the desired and actual frequency responses. The filters are optimal in the sense that the maximum error between the desired frequency response and the actual frequency response is minimized. Filters designed this way exhibit an equiripple behavior in their frequency responses, and hence are sometimes called *equiripple* filters.

b = remez(n,f,m) returns row vector b containing the n+1 coefficients of the order n FIR filter whose frequency-magnitude characteristics match those given by vectors f and m.

The output filter coefficients (taps) in b obey the symmetry relation

$$b(k) = b(n+2-k), \quad k = 1,\ldots,n+1$$

Vectors f and m specify the filter's frequency-magnitude characteristics:

- f is a vector of pairs of frequency points, specified in the range between 0 and 1, where 1 corresponds to half the sampling frequency (the Nyquist frequency). The frequencies must be in increasing order.

- m is a vector containing the desired magnitude response at the points specified in f.

  The desired magnitude function at frequencies between pairs of points (f(k),f(k+1)) for k odd is the line segment connecting the points (f(k),m(k)) and (f(k+1),m(k+1)).

  The desired magnitude function at frequencies between pairs of points (f(k),f(k+1)) for k even is unspecified. The areas between such points are transition or "don't care" regions.

- f and m must be the same length. The length must be an even number.

  The relationship between the f and m vectors in defining a desired frequency response is

```
f = [0 .3 .4 .6 .7 .9]
m = [0  1  0  0 .5 .5]
```

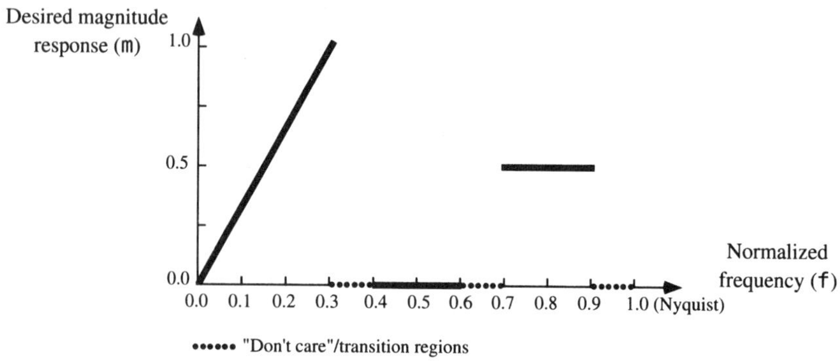

remez(n,f,m,w) uses the weights in vector w to weight the fit in each frequency band. The length of w is half the length of f and m, so there is exactly one weight per band.

b = remez(n,f,m,'*ftype*') and b = remez(n,f,m,w,'ftype') specify a filter type, where *ftype* is

- hilbert for linear-phase filters with odd symmetry (type III and type IV): the output coefficients in b obey the relation $b(k) = -b(n + 2 - k)$,
  $k = 1,\ldots,n + 1$. This class of filters includes the Hilbert transformer, which has a desired magnitude of one across the entire band.

  For example,

```
h = remez(30,[.1 .9],[1 1],'Hilbert');
```

  designs an approximate FIR Hilbert transformer of length 31.

- `differentiator` also designs type III and IV filters, but uses a special weighting technique. For nonzero magnitude bands, it weights the error by a factor of (1/*f*) so the error at low frequencies is much smaller than at high frequencies. For FIR differentiators, which have a magnitude characteristic proportional to frequency, these filters minimize the maximum relative error (the maximum of the ratio of the error to the desired magnitude).

Example Graph the desired and actual frequency responses of a 17th-order Parks-McClellan bandpass filter:

```
f = [0 0.3 0.4 0.6 0.7 1]; m = [0 0 1 1 0 0];
b = remez(17,f,m);
[h,w] = freqz(b,1,512);
plot(f,m,w/pi,abs(h))
```

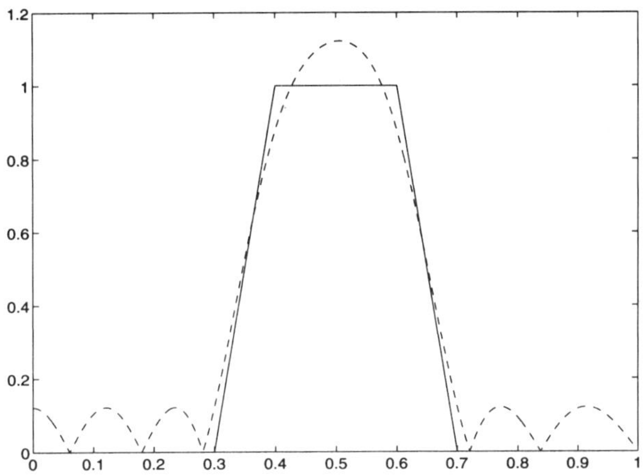

Algorithm `remez` is an M-file version of the algorithm from [1]. It can take some time to execute, particularly for higher order filters. On most machines, a MEX-file version of the original FORTRAN code, altered to design arbitrarily long filters with arbitrary many linear bands, provides fast execution.

`remez` designs type I, II, III, and IV linear-phase filters. Type I and II are the defaults for `n` even and odd respectively, while Type III (`n` even) and IV (`n` odd) are obtained with the `'hilbert'` and `'differentiator'` flags. The different types of filters have different symmetries and certain constraints on their frequency responses. (See [5] for more details).

Linear Phase Filter Type	Filter Order n	Symmetry of Coefficients	Response H(f), f = 0	Response H(f), f = 1 (Nyquist)
Type I	Even	even:	No restriction	No restriction
Type II	Odd	$b(k) = b(n+2-k),\quad k = 1,\dots,n+1$	No restriction	$H(1) = 0$
Type III	Even	odd:	$H(0) = 0$	$H(1) = 0$
Type IV	Odd	$b(k) = -b(n+2-k),\quad k = 1,\dots,n+1$	$H(0) = 0$	No restriction

Diagnostics The following diagnostic messages arise from incorrect usage of `remez`:

```
Filter order must be 3 or more.
There should be one weight per band.
Frequency and amplitude vectors must be the same length.
The number of frequency points must be even.
Frequencies must lie between 0 and 1.
Frequencies must be specified in bands.
Frequencies must be nondecreasing.
Adjacent bands not allowed.
```

A more serious message that could occur is

```
— Failure to Converge —
Probable cause is machine rounding error.
```

In the rare event that you see this message, it is possible that the filter design may still be correct. Verify the design by checking its frequency response.

See Also `butter`, `cheby1`, `cheby2`, `ellip`, `fir1`, `fir2`, `firls`, `remezord`, `yulewalk`

References

[1] IEEE. *Programs for Digital Signal Processing, II*, Algorithm 5.1. *IEEE Press*, New York: John Wiley & Sons, 1979.

[2] IEEE. *Selected Papers in Digital Signal Processing, II. IEEE Press*, New York: John Wiley & Sons, 1979.

[3] Parks, T. W., and C. S. Burrus. *Digital Filter Design*, 83, New York: John Wiley and Sons, 1987.

[4] Rabiner, L. R., J. H. McClellan, and T. W. Parks. "FIR Digital Filter Design Techniques Using Weighted Chebyshev Approximations." *Proc. IEEE 63*, 1975.

[5] Oppenheim, A. V., and R.W. Schafer. *Discrete-Time Signal Processing*, pp. 256–266. Englewood Cliffs, NJ: Prentice-Hall, 1989.

residuez

Purpose z-transform partial-fraction expansion.

Synopsis

```
[r,p,k] = residuez(b,a)
[b,a] = residuez(r,p,k)
```

Description `residuez` converts a discrete time system expressed as the ratio of two polynomials to partial fraction expansion, or residue, form. It also converts the partial fraction expansion back to the original polynomial coefficients.

`[r,p,k] = residuez(b,a)` finds the residues, poles and direct terms of a partial fraction expansion of the ratio of two polynomials, *b(z)* and *a(z)*. Vectors `b` and `a` specify the coefficients of the polynomials of the discrete-time system *b(z)/a(z)* in descending powers of z,

$$b(z) = b_0 + b_1 z^{-1} + b_2 z^{-2} + \ldots + b_m z^{-m}$$
$$a(z) = a_0 + a_1 z^{-1} + a_2 z^{-2} + \ldots + a_n z^{-n}$$

If there are no multiple roots and `m` > `n–1`,

$$\frac{b(z)}{a(z)} = \frac{r(1)}{1-p(1)z^{-1}} + \cdots + \frac{r(n)}{1-p(n)z^{-1}} + k(1) + k(2)z^{-1} + \cdots + k(m-n+1)z^{-(m-n)}$$

The returned column vector `r` contains the residues, column vector `p` contains the pole locations, and row vector `k` contains the direct terms. The number of poles is

```
n = length(a)–1 = length(r) = length(p)
```

The direct term coefficient vector `k` is empty if `length(b) < length(a)`; otherwise

```
length(k) = length(b) – length(a) + 1
```

If `p(j) = ... = p(j+s–1)` is a pole of multiplicity `s`, then the expansion includes terms of the form

$$\frac{r(j)}{1-p(j)z^{-1}} + \frac{r(j+1)}{(1-p(j)z^{-1})^2} + \cdots + \frac{r(j+s-1)}{(1-p(j)z^{-1})^s}$$

`[b,a] = residuez(r,p,k)`, with three input arguments and two output arguments, converts the partial fraction expansion back to polynomials with coefficients in row vectors `b` and `a`.

The `residue` function in the MATLAB environment is very similar to `residuez`. It computes the partial fraction expansion of continuous-time systems in the Laplace domain [see 1], rather than discrete-time systems in the z-domain as does `residuez`.

Algorithm `residuez` applies standard MATLAB functions and partial fraction techniques to find `r`, `p`, and `k` from `b` and `a`:

1. It finds the direct terms a using `deconv` (polynomial long division) when `length(b)>length(a)-1`.
2. It finds the poles using `p = roots(a)`. `mpoles` finds repeated poles and reorders the poles according to their multiplicities.
3. It finds the residue for each nonrepeating pole p_i by multiplying $b(z)/a(z)$ by $1/(1-p_iz^{-1})$, and evaluating the resulting rational function at $z = p_i$. It finds the residues for the repeated poles by solving

```
S2*r2 = h - S1*r1
```

for `r2` using `\`. `h` is the impulse response of the reduced $b(z)/a(z)$, `S1` is a matrix whose columns are impulse responses of the first-order systems made up of the nonrepeating roots, and `r1` is a column containing the residues for the nonrepeating roots. Each column of matrix `S2` is an impulse response. For each root p_j of multiplicity s_j, `S2` contains s_j columns representing the impulse responses of each of the following systems:

$$\frac{1}{1-p_jz^{-1}}, \frac{1}{(1-p_jz^{-1})^2}, \ldots, \frac{1}{(1-p_jz^{-1})^{s_j}}$$

The vector `h` and matrices `S1` and `S2` have `n + xtra` rows, where `n` is the total number of roots, and the internal parameter `xtra`, set to one by default, determines the degree of overdetermination of the system of equations.

Diagnostics If `a(1) == 0` while computing the partial fraction decomposition using `[r,p,k] = residuez(b,a)`, `residuez` produces the following error message:

```
First coefficient in A vector must be nonzero.
```

If the number of residues `r` and poles `p` is not the same, `residuez` produces the following error message:

```
R and P vectors must be the same size.
```

See Also `ss2zp`, `tf2ss`, `zp2ss`

`deconv`, `poly`, `residue`, `roots` in the *MATLAB Reference*

References

[1] Oppenheim, A. V., and R.W. Schafer. *Digital Signal Processing*, pp. 166–170. Englewood Cliffs, NJ: Prentice-Hall, 1975.

rlocus

Purpose Evans root-locus.

Synopsis

```
r = rlocus(num,den)
r = rlocus(num,den,k)
r = rlocus(a,b,c,d)
r = rlocus(a,b,c,d,k)
```

Description `rlocus` calculates the Evans root-locus of a SISO system. Root-loci are used to study the effects of varying feedback gain on system pole locations, indirectly providing information on time and frequency responses. For a plant with transfer function *g(s)* and feedback compensator *k*f(s)*, the closed-loop transfer function is

$$h(s) = \frac{g(s)}{1 + k\,g(s)f(s)} = \frac{g(s)}{q(s)}$$

When invoked without left-hand arguments, the root locus plot is drawn on the screen. `rlocus` works for both continuous- and discrete-time systems.

`rlocus(num,den)` plots the locus of the roots of

$$q(s) = 1 + k\frac{num(s)}{den(s)} = 0$$

with the gain vector k automatically determined. Vectors `num` and `den` specify the numerator and denominator coefficients in descending powers of s or z:

$$\frac{num(s)}{den(s)} = \frac{num(1)s^{nn-1} + num(2)s^{nn-2} + \ldots + num(nn)}{den(1)s^{nd-1} + den(2)s^{nd-2} + \ldots + den(nd)}$$

`rlocus(a,b,c,d)` plots the root locus of the continuous-time or discrete-time SISO state-space system `(a,b,c,d)` with the gain vector automatically determined.

`rlocus(num,den,k)` or `rlocus(a,b,c,d,k)` uses the user supplied gain vector `k`. The vector `k` contains the gain for which the closed-loop roots are to be computed.

Invoked with left-hand arguments:

```
[r,k] = rlocus(num,den)
[r,k] = rlocus(num,den,k)
[r,k] = rlocus(a,b,c,d)
[r,k] = rlocus(a,b,c,d,k)
```

returns the matrix `r` and the gain vector `k`. `r` has `length(k)` rows and `(length(den)-1)` columns containing the complex root locations. Each row of the matrix corresponds to a gain from vector `k`. The root-locus can be plotted with `plot(r,'x')`.

Examples Find and plot the root-locus of the system:

$$H(s) = \frac{2s^2 + 5s + 1}{s^2 + 2s + 3}$$

```
num = [2 5 1];
den = [1 2 3];
rlocus(num,den);
title('Root Locus')
```

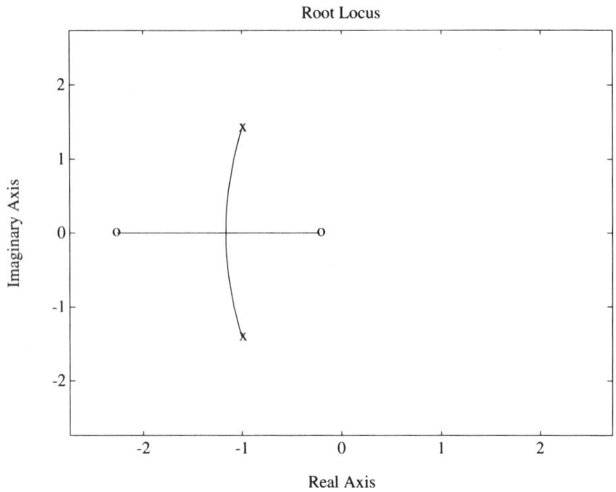

The `rlocus` function solves the most commonly posed special case of the general root-locus problem. MATLAB can also find the root-loci as a function of any parameter in a system. For example, here is a symmetric root-locus, found with repeated applications of `lqr`:

```
i = 1;
for r = 0:1:10
[K,S,E] = lqr(a,b,q,r);
R(i,:) = E.';
i = i + 1;
end
plot(r,'x')
```

Algorithm `rlocus` uses repetitive application of the `roots` function.

See Also `rlocfind`, `pzmap`, `place`, `lqr`

rmodel, drmodel

Purpose Random stable nth order model.

Synopsis

```
[a,b,c,d] = rmodel(n)
[a,b,c,d] = rmodel(n,p,m)
[num,den] = rmodel(n)
[num,den] = rmodel(n,p)
[a,b,c,d] = drmodel(n)
[a,b,c,d] = drmodel(n,p,m)
[num,den] = drmodel(n)
[num,den] = drmodel(n,p)
```

Description `[a,b,c,d] = rmodel(n)` produces a random nth order stable state-space model `(a,b,c,d)`. The resulting model has one input and one output.

`[a,b,c,d] = rmodel(n,p,m)` produces a random nth order stable model with `m` inputs and `p` outputs.

`[num,den] = rmodel(n)` produces a random nth order stable transfer function model. `num` and `den` contain the transfer function polynomial coefficients in descending powers of s.

`[num,den] = rmodel(n,p)` produces a random nth order stable SIMO model with one input and `m` outputs.

`drmodel` produces stable discrete-time random models.

series

Purpose Series system connection of two state-space systems.

Synopsis

```
[a,b,c,d] = series(a1,b1,c1,d1,a2,b2,c2,d2)
[a,b,c,d] = series(a1,b1,c1,d1,a2,b2,c2,d2,outputs1,inputs2)
[num,den] = series(num1,den1,num2,den2)
```

Description `series` forms the series connection of two systems. `series` works for both continuous-time and discrete-time systems.

`[a,b,c,d] = series(a1,b1,c1,d1,a2,b2,c2,d2)` produces an aggregate state-space system with all the outputs of system 1 connected to all the inputs of system 2, $u_2 = y_1$

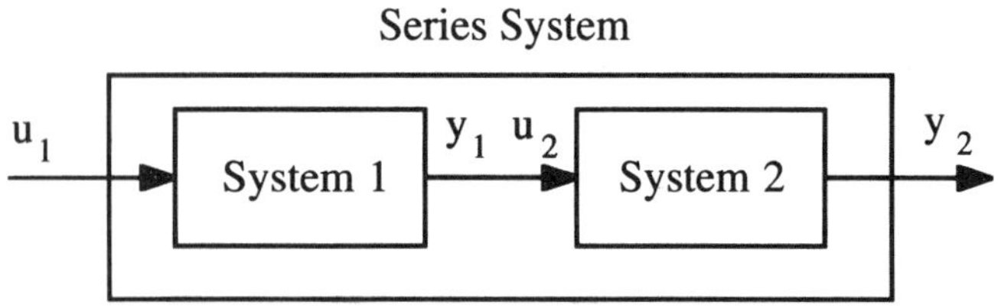

The resulting system is

$$\begin{bmatrix} \dot{x}_1 \\ \dot{x}_2 \end{bmatrix} = \begin{bmatrix} A_1 & 0 \\ B_2 C_1 & A_2 \end{bmatrix} \begin{bmatrix} x_1 \\ x_2 \end{bmatrix} + \begin{bmatrix} B_1 \\ B_2 D_1 \end{bmatrix} u_1$$

$$y_2 = \begin{bmatrix} D_2 C_1 & C_2 \end{bmatrix} \begin{bmatrix} x_1 \\ x_2 \end{bmatrix} + \begin{bmatrix} D_2 D_1 \end{bmatrix} u_1$$

`[num,den] = series(num1,den1,num2,den2)` produces a polynomial transfer function of the series connected system. `num` and `den` contain the polynomial coefficients in descending powers of s or z. The resulting series system is

$$\frac{num(s)}{den(s)} = G_1(s)G_2(s) = \frac{num_1(s)num_2(s)}{den_1(s)den_2(s)}$$

```
[a,b,c,d] = ...
series(a1,b1,c1,d1,a2,b2,c2,d2,outputs1,inputs2)
```

produces an aggregate state-space system from the series connection of systems 1 and 2. The specified outputs of system 1 are connected to the specified inputs of system 2:

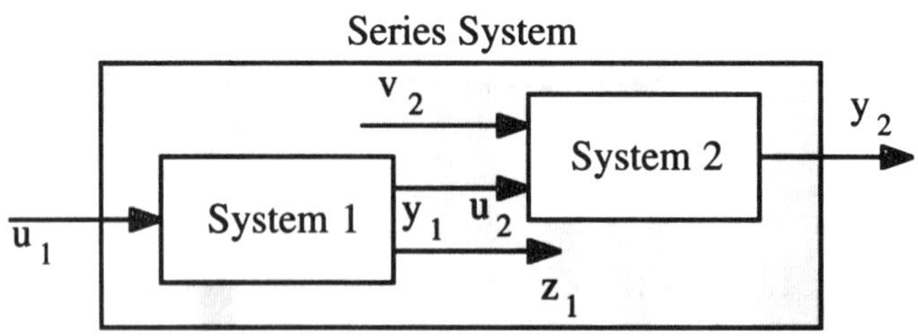

The vector `outputs1` contains indices into the outputs of system 1 and specifies which outputs of system 1 to connect to the inputs of system 2 specified by `inputs2`. The outputs of system 1 are paired with the inputs of system 2; i.e., the output of system 1 specified by `outputs1(1)` is paired with the input of system 2 specified by `inputs2(1)` and so on.

The resulting system has the same inputs as system 1 and the same outputs as system 2 (for more details see Algorithm below). For a more general series connection use a combination of `append` and `cloop`.

Examples Connect the two transfer functions in series:

$$G(s) = \frac{3}{s+4} \qquad H(s) = \frac{2s+4}{s^2+2s+3}$$

```
num1=3;
den1=[1 4];
num2=[2 4];
den2=[1 2 3];
[num,den] = series(num1,den1,num2,den2)
```

Consider a state-space system `(a1,b1,c1,d1)` with five inputs and four outputs and another system `(a2,b2,c2,d2)` with two inputs and three outputs. Connect the two systems in series by connecting outputs 2 and 4 of system 1 with the inputs 1 and 2 of system 2:

```
outputs1 = [2 4];
inputs2 = [1 2];
[a,b,c,d] = series(a1,b1,c1,d1,a2,b2,c2,d2,...
  outputs2,inputs1)
```

Algorithm Given two state-space systems

$$\dot{x}_1 = A_1 x_1 + B_1 u_1$$

$$\begin{bmatrix} y_1 \\ z_1 \end{bmatrix} = \begin{bmatrix} C_1 \\ H_1 \end{bmatrix} x + \begin{bmatrix} D_{11} \\ J_{11} \end{bmatrix} u_1$$

and

$$\dot{x}_2 = A_2 x_2 + \begin{bmatrix} B_2 & G_2 \end{bmatrix} \begin{bmatrix} u_2 \\ v_2 \end{bmatrix}$$

$$y_2 = C_2 x_2 + \begin{bmatrix} D_{21} & D_{22} \end{bmatrix} \begin{bmatrix} u_2 \\ v_2 \end{bmatrix}$$

suppose the series system is to be formed by connecting output y_1 with input u_2.

The resulting state-space system is

$$\begin{bmatrix} \dot{x}_1 \\ \dot{x}_2 \end{bmatrix} = \begin{bmatrix} A_1 & 0 \\ B_2C_1 & A_2 \end{bmatrix} \begin{bmatrix} x_1 \\ x_2 \end{bmatrix} + \begin{bmatrix} B_1 \\ B_2D_{11} \end{bmatrix} u_1$$

$$y_2 = \begin{bmatrix} D_{21}C_1 & C_2 \end{bmatrix} \begin{bmatrix} x_1 \\ x_2 \end{bmatrix} + \begin{bmatrix} D_{21}D_{11} \end{bmatrix} u_1$$

These same expressions are valid for discrete models with differentiation replaced by differences and s-plane transfer functions replaced by z-plane transfer functions.

See Also `append`, `cloop`, `feedback`, `connect`

---

sinc

Purpose Sinc or sin(πt)/πt function.

Synopsis `y = sinc(x)`

Description `sinc` computes the `sinc` function of an input vector or matrix, where the `sinc` function is

$$\text{sinc}(t) = \begin{cases} 1, & t = 0 \\ \dfrac{\sin(\pi t)}{\pi t}, & t \neq 0 \end{cases}$$

This function is the continuous inverse Fourier transform of the rectangular pulse of width 2π and height 1:

$$\text{sinc}(t) = \frac{1}{2\pi} \int_{-\pi}^{\pi} e^{j\omega t} d\omega$$

`y = sinc(x)` returns a matrix `y` the same size as `x` whose elements are the `sinc` function of the elements of `x`.

The space of functions bandlimited in the frequency band $\omega \in [-\pi, \pi]$ is spanned by the infinite (yet countable) set of sinc functions shifted by integers. Thus any such bandlimited function $g(t)$ can be reconstructed from its samples at integer spacings:

$$g(t) = \sum_{n=-\infty}^{\infty} g(n)\,\text{sinc}(t-n)$$

Example Perform ideal bandlimited interpolation by assuming the signal to be interpolated is zero outside of the given time interval and that it has been sampled at exactly the Nyquist frequency:

```
t = (1:10)'; % a column vector of time samples
x = randn(size(t)); % a column vector of data
% ts is times at which to interpolate data
ts = linspace(-5,15,600)';
y = sinc(ts(:,ones(size(t))) - t(:,ones(size(ts)))')*x;
plot(t,x,'o',ts,y);
```

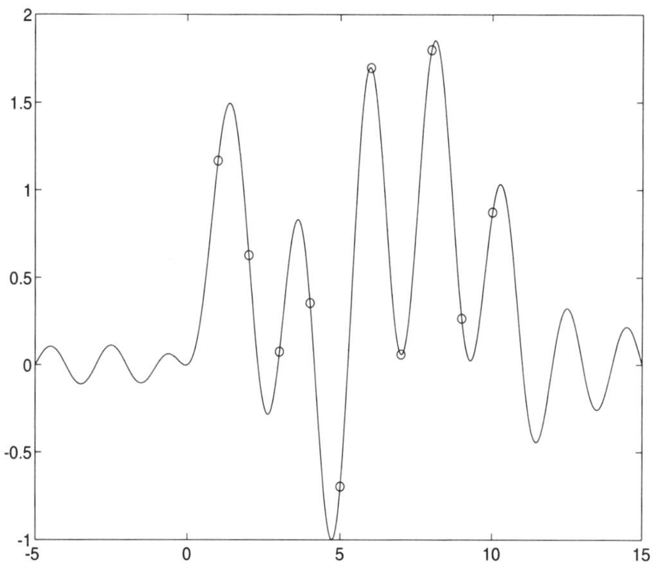

---

sos2ss

Purpose Second-order section to state-space conversion.

Synopsis `[A,B,C,D] = sos2ss(sos)`

Description `sos2ss` converts a second-order section representation of a given system to an equivalent state-space representation.

`[A,B,C,D] = sos2ss(sos)` converts the system `sos`, in second-order section form, to a single-input, single-output state-space representation

$$\begin{aligned} x[n+1] &= \mathrm{A}x[n] + \mathrm{B}u[n] \\ y[n] &= \mathrm{C}x[n] + \mathrm{D}u[n] \end{aligned}$$

The discrete transfer function in second-order section form is given by

$$H(z) = \prod_{k=1}^{L} H_k(z) = \prod_{k=1}^{L} \frac{b_{0k} + b_{1k}z^{-1} + b_{2k}z^{-2}}{a_{0k} + a_{1k}z^{-1} + a_{2k}z^{-2}}$$

where L is the number of rows in `sos`. `sos` is a L-by-6 matrix organized as

$$sos = \begin{bmatrix} b_{01} & b_{11} & b_{21} & a_{01} & a_{11} & a_{21} \\ b_{02} & b_{12} & b_{22} & a_{02} & a_{12} & a_{22} \\ \vdots & \vdots & \vdots & \vdots & \vdots & \vdots \\ b_{0L} & b_{1L} & b_{2L} & a_{0L} & a_{1L} & a_{2L} \end{bmatrix}$$

The entries of `sos` must be real for proper conversion to state space. The returned matrix `A` is size $N \times N$ where $N = 2L–1$, `B` is a length $N–1$ column vector, `C` is a length $N–1$ row vector, and `D` is a scalar.

Example Compute the state-space representation of a simple second-order section form system:

```
sos = [1  1  1  1  0 –1; –2  3  1  1 10  1];
[A,B,C,D] = sos2ss(sos)
A =
    –10.8990   –3.1463         0         0
      3.1463         0         0         0
     –9.8990   –2.8284    0.8990    0.3178
           0         0    0.3178         0
B =
     1
     0
     1
     0
C =
     19.7980    5.6569    1.2020    2.5106
D =
    –2
```

Algorithm `sos2ss` first finds the zeros and poles of the second-order sections using `roots`, then uses `zp2ss` to find a state-space representation of the system.

See Also sos2tf, sos2zp, ss2sos, zp2sos

---

sos2tf

Purpose Second-order section to transfer function conversion.

Synopsis `[b,a] = sos2tf(sos)`

Description `sos2tf` converts a second-order section representation of a given system to an equivalent transfer function representation.

`[b,a] = sos2tf(sos)` returns the numerator coefficients `b` and denominator coefficients a of the transfer function that describes a discrete-time system given by `sos` in second-order section form. The second-order section format of $H(z)$ is given by

$$H(z) = \prod_{k=1}^{L} H_k(z) = \prod_{k=1}^{L} \frac{b_{0k} + b_{1k}z^{-1} + b_{2k}z^{-2}}{a_{0k} + a_{1k}z^{-1} + a_{2k}z^{-2}}$$

where L is the number of rows of `sos`. `sos` is an L-by-6 matrix which contains the coefficients of each second-order section stored in its rows:

$$sos = \begin{bmatrix} b_{01} & b_{11} & b_{21} & a_{01} & a_{11} & a_{21} \\ b_{02} & b_{12} & b_{22} & a_{02} & a_{12} & a_{22} \\ \vdots & \vdots & \vdots & \vdots & \vdots & \vdots \\ b_{0L} & b_{1L} & b_{2L} & a_{0L} & a_{1L} & a_{2L} \end{bmatrix}$$

Row vectors `b` and `a` contain the numerator and denominator coefficients of $H(z)$ stored in descending powers of z

$$H(z) = \frac{B(z)}{A(z)} = \frac{b(1) + b(2)z^{-1} + \ldots + b(n+1)z^{-n}}{a(1) + a(2)z^{-1} + \ldots + a(m+1)z^{-m}}$$

Algorithm `sos2tf` uses the `conv` function to multiply all of the numerator and denominator second-order polynomials together.

Example Compute the transfer function representation of a simple second-order section form system:

```
sos = [1  1  1  1  0 –1; –2  3  1  1 10  1];
[b,a] = sos2tf(sos)
b =
    –2     1     2     4     1
a =
     1    10     0   –10    –1
```

See Also sos2ss, sos2zp, ss2sos,zp2sos

---

sos2zp

Purpose Second-order section to zero-pole-gain conversion.

Synopsis `[z,p,k] = sos2zp(sos)`

Description sos2zp converts a second-order section representation of a given system to an equivalent zero-pole-gain representation.

`[z,p,k] = sos2zp(sos)` returns the zeros z, poles p, and gain k of the system given by sos in second-order section form. The second-order section format of $H(z)$ is given by

$$H(z) = \prod_{k=1}^{L} H_k(z) = \prod_{k=1}^{L} \frac{b_{0k} + b_{1k}z^{-1} + b_{2k}z^{-2}}{a_{0k} + a_{1k}z^{-1} + a_{2k}z^{-2}}$$

where L is the number of rows of sos. sos is an L-by-6 matrix which contains the coefficients of each second-order section stored in its rows:

$$sos = \begin{bmatrix} b_{01} & b_{11} & b_{21} & a_{01} & a_{11} & a_{21} \\ b_{02} & b_{12} & b_{22} & a_{02} & a_{12} & a_{22} \\ \vdots & \vdots & \vdots & \vdots & \vdots & \vdots \\ b_{0L} & b_{1L} & b_{2L} & a_{0L} & a_{1L} & a_{2L} \end{bmatrix}$$

Column vectors z and p contain the zeros and poles of the transfer function $H(z)$

$$H(z) = k\frac{(z - z(1))(z - z(2))\cdots(z - z(N))}{(p - p(1))(p - p(2))\cdots(p - p(M))}$$

where the orders N and M are determined by the matrix sos.

Example Compute the poles, zeros and gain of a simple system in second-order section form:

```
sos = [1  1  1  1  0 -1; -2  3  1  1 10  1];
[z,p,k] = sos2zp(sos)
z =
  -0.5000 + 0.8660i
  -0.5000 - 0.8660i
   1.7808
  -0.2808
p =
   -1.0000
    1.0000
   -9.8990
   -0.1010
k =
    -2
```

Algorithm sos2zp finds the roots and poles of each second-order section using the roots command. sos2zp returns the roots and poles with conjugate pairs in consecutive locations, with the order of the pairs determined by their row in the sos matrix. The gain k is the product of the gains of the sections:

$$k = \prod_{k=1}^{L} \frac{b_{0k}}{a_{0k}}$$

See Also sos2ss, sos2tf, ss2sos, zp2sos

---

specgram

Purpose Time-dependent frequency analysis (spectrogram).

Synopsis

```
B = specgram(a)
B = specgram(a,nfft)
[B,f] = specgram(a,nfft,Fs)
[B,f,t] = specgram(a,nfft,Fs)
B = specgram(a,nfft,Fs,window)
B = specgram(a,nfft,Fs,window,noverlap)
B = specgram(a,...,'dflag')
specgram(a)
```

Description specgram computes the windowed discrete-time Fourier transform of a signal using a sliding window. The spectrogram is the magnitude of this function.

B = specgram(a) calculates the spectrogram for the signal in vector a.
B = specgram(a) uses the default values:

- nfft = min(256, length(a))
- Fs = 2
- window = hanning(nfft)
- noverlap = length(window)/2

nfft specifies the FFT length that specgram uses. This value determines the frequencies at which the discrete-time Fourier transform is computed. Fs is a scalar that specifies the sampling frequency. window specifies a windowing function and the number of samples specgram uses in its sectioning of vector a. noverlap is the number of samples by which the segments overlap. Any arguments that you omit from the end of the input parameter list use the default values shown above.

If a is real, specgram computes the discrete-time Fourier transform at positive frequencies only. If n is even, specgram returns nfft/2+1 rows (including the zero and Nyquist frequency terms). If n is odd, specgram returns nfft/2 rows. The number of columns in B is

```
k = fix((n–noverlap)/(length(window)–noverlap))
```

If a is complex, specgram computes the discrete-time Fourier transform at both positive and negative frequencies. In this case, B is a complex matrix with nfft rows. Time increases linearly across the columns of B, starting with sample 1 in column 1. Frequency increases linearly down the rows, starting at 0.

B = specgram(a,nfft) uses the specified FFT length nfft in its calculations. Use a power of 2 for nfft for fastest execution.

[B,f] = specgram(a,nfft,Fs) returns a vector f of frequencies at which the function computes the discrete-time Fourier transform. Fs has no effect on the output B; it is only a frequency scaling multiplier.

[B,f,t] = specgram(a,nfft,Fs) returns frequency and time vectors f and t respectively. t is a column vector of scaled times with length equal to the number of columns of B. t(j) is the earliest time at which the jth window intersects a. t(1) is always equal to zero.

B = specgram(a,nfft,Fs,window) specifies a windowing function and the number of samples per section of the x vector. If you supply a scalar for window, specgram uses a Hanning window of that length. The length of the window must be less than or equal to nfft; specgram zero pads the sections if nfft exceeds the length of the window.

B = specgram(a,nfft,Fs,window,noverlap) overlaps the sections of x by noverlap samples.

You can use the empty matrix [] to specify the default value for any input argument. For example,

```
B = specgram(x,[],[],[],10000)
```

is equivalent to

```
B = specgram(x)
```

but with a sampling frequency of 10000 Hz instead of the default 2 Hz.

`B = specgram(a,...,'dflag')` lets you specify a detrend option, where *`dflag`* can be:

- `linear to remove the best straight-line fit from the prewindowed sections of x (default).`
- `mean to remove the mean from the prewindowed sections of x.`
- none for no detrending.

The *`dflag`* parameter must appear last in the list of input arguments. `specgram` recognizes a *`dflag`* string no matter how many intermediate arguments are omitted.

`specgram` with no output arguments displays the scaled logarithm of the spectrogram in the current figure window using

```
imagesc(t,f,20*log10(abs(b))), axis xy, colormap(jet)
```

The `axis xy` mode displays the low frequency content of the first portion of the signal in the lower-left corner of the axes. `specgram` uses `Fs` to label the axes according to true time and frequency.

Algorithm

`specgram` calculates the spectrogram for a given signal as follows:

1. It splits the signal into overlapping segments, detrends them and applies the window specified by the `window` parameter to each segment.
2. It computes the discrete-time Fourier transform of each segment with a length `nfft` FFT to produce an estimate of the short-term frequency content of the signal; these transforms make up the columns of `B`. `specgram` zero pads the windowed segments if `nfft > length(window)`, so the quantity
3. `(length(window) - noverlap)`
4. specifies how many samples `specgram` shifts the window.
5. For real input, `specgram` truncates the spectrogram to the first `nfft/2 + 1` points for `nfft` even, and `(nfft + 1)/2` for `nfft` odd.

Example Plot the spectrogram of a digitized speech signal:

```
load mtlb
specgram(mtlb,512,475,kaiser(500,5),Fs)
```

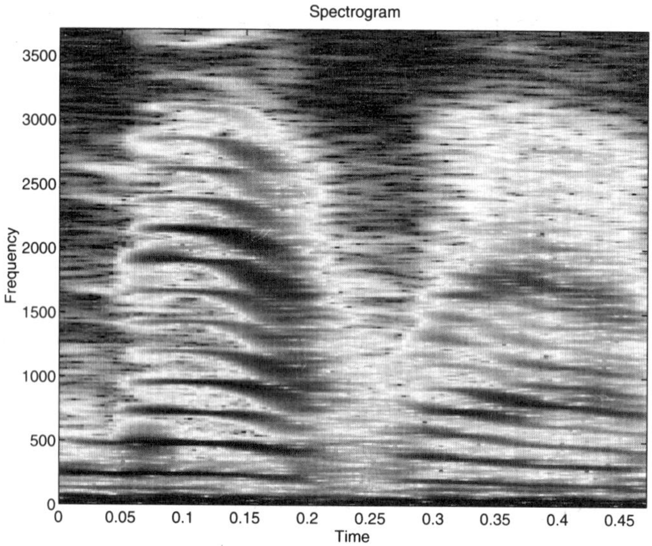

Diagnostics The following diagnostic messages arise from incorrect use of `specgram`:

```
Requires window's length to be no greater than the FFT length.
Requires NOVERLAP to be strictly less than the window length.
Requires positive integer values for NFFT and NOVERLAP.
Requires vector input.
```

References

[1] Oppenheim, A. V., and R.W. Schafer. *Discrete-Time Signal Processing*, pp. 713–718. Englewood Cliffs, NJ: Prentice-Hall, 1989.

[2] Rabiner, L. R., and Schafer, R. W. *Digital Processing of Speech Signals*. Englewood Cliffs, NJ: Prentice-Hall, 1978.

---

ss2sos

Purpose State-space to second-order section conversion.

Synopsis

```
sos = ss2sos(A,B,C,D)
sos = ss2sos(A,B,C,D,iu)
sos = ss2sos(A,B,C,D,'order')
```

```
sos = ss2sos(A,B,C,D,iu,'order')
```

Description ss2sos converts a state-space representation of a given system to an equivalent second-order section representation.

sos = ss2sos(A,B,C,D) finds a matrix sos in second-order section form that is equivalent to the state-space system represented by input arguments A, B, C and D. The input system must be single input and real. sos is an L by 6 matrix,

$$sos = \begin{bmatrix} b_{01} & b_{11} & b_{21} & a_{01} & a_{11} & a_{21} \\ b_{02} & b_{12} & b_{22} & a_{02} & a_{12} & a_{22} \\ \vdots & \vdots & \vdots & \vdots & \vdots & \vdots \\ b_{0L} & b_{1L} & b_{2L} & a_{0L} & a_{1L} & a_{2L} \end{bmatrix}$$

whose rows contain the numerator and denominator coefficients b_{ik} and a_{ik} of the second-order sections of $H(z)$:

$$H(z) = \prod_{k=1}^{L} H_k(z) = \prod_{k=1}^{L} \frac{b_{0k} + b_{1k}z^{-1} + b_{2k}z^{-2}}{a_{0k} + a_{1k}z^{-1} + a_{2k}z^{-2}}$$

sos = ss2sos(A,B,C,D,iu) specifies a scalar iu that determines which output of the state-space system A, B, C, D is used in the conversion. If unspecified, iu defaults to 1.

sos = ss2sos(A,B,C,D,'down') orders the sections so the first row of sos contains the poles closest to the unit circle. Without the 'down' flag, ss2sos orders the sections so the last row of sos contains the poles closest to the unit circle.

sos = ss2sos(A,B,C,D,'*order*') and sos = ss2sos(A,B,C,D,iu,'*order*') specifies the order of the rows in sos, where *order* is

- down to order the sections so the first row of sos contains the poles closest to the unit circle.
- up to order the sections so the first row of sos contains the poles farthest from the unit circle (default).

Example Find a second-order section form of a Butterworth lowpass filter:

```
[A,B,C,D] = butter(5,.2);
sos = ss2sos(A,B,C,D)
sos =
      0.2330    0.2329         0    1.0000   -0.5095         0
      0.0647    0.1294    0.0647    1.0000   -1.0966    0.3554
      0.0851    0.1701    0.0851    1.0000   -1.3693    0.6926
```

Algorithm ss2sos uses a four-step algorithm to determine the second-order section representation for an input state-space system

1. It finds the poles and zeros of the system given by A, B, C and D.

2. It uses the function `zp2sos`, which first groups the zeros and poles into complex conjugate pairs using the `cplxpair` function. `zp2sos` then forms the second-order sections by matching the pole and zero pairs according to the following rules:

   a. Match the poles closest to the unit circle with the zeros closest to those poles.

   b. Match the poles next closest to the unit circle with the zeros closest to those poles.

   c. Continue until all of the poles and zeros are matched.

   `ss2sos` groups real poles into sections with the real poles closest to them in absolute value. The same rule holds for real zeros.

3. It orders the sections according to the proximity of the pole pairs to the unit circle. `ss2sos` normally orders the sections with poles closest to the unit circle last in the cascade. You can tell `ss2sos` to order the sections in the reverse order by specifying the `'down'` flag.

   Putting "high Q" sections at the beginning of the cascade (the `'down'` flag operation) reduces the filter response's sensitivity to quantization noise near those poles. Putting "high Q" sections at the end of the cascade (the default operation) prevents reduction in signal power level early in the cascade. `ss2sos` orders all-zero sections according to the minimum of $|z_i|$ and $|z_i^{-1}|$, where z_i, $i = 1, 2$, are the zeros in the section. [1] and [2] provide a detailed discussion of section ordering.

4. `ss2sos` scales the sections so the maximum of the magnitude of the transfer function of the first N sections in cascade is less than 1,

$$\max_{|\omega| \le \pi} \left| \prod_{i=1}^{N} H_i(e^{j\omega}) \right| < 1, \quad N = 1, \ldots, L-1$$

subject to the constraint that the overall gain, `k`, stays the same

$$k = \prod_{k=1}^{L} \frac{b_{0k}}{a_{0k}}$$

This scaling is an attempt to minimize overflow in some standard fixed point implementations of filtering.

Diagnostics `ss2sos` generates an error message if there is more than one input to the system:

```
State-space system must have only one input.
```

See Also `sos2ss`, `sos2tf`, `sos2zp`, `zp2sos`

References [1] Oppenheim, A. V., and R.W. Schafer. *Discrete-Time Signal Processing*, pp. 363–370. Englewood Cliffs, NJ: Prentice-Hall, 1989.

[2] Jackson, Leland B. *Digital Filters and Signal Processing*, pp. 319–324. Boston: Kluwer Academic Publishers, 1989.

---

ss2tf

Purpose State-space to transfer function conversion.

Synopsis

```
[num,den] = ss2tf(a,b,c,d,iu)
```

Description `ss2tf` converts state-space systems to transfer function form.

`[num,den] = ss2tf(a,b,c,d,iu)` returns the transfer function

$$H(s) = \frac{NUM(s)}{den(s)} = C(sI - A)^{-1}B + D$$

of the system

$$\dot{x} = Ax + Bu$$
$$y = Cx + Du$$

from the `iu`th input. Vector `den` contains the coefficients of the denominator in descending powers of s. The numerator coefficients are returned in matrix `NUM` with as many rows as there are outputs y. The function `tf2ss` is the inverse of `ss2tf`. `ss2tf` also works with systems in discrete-time, in which case it returns the z-transform representation.

Algorithm `ss2tf` uses `poly` to find the characteristic polynomial *det(sI–A)*, and the equality

$$H(s) = c(sI - A)^{-1}b = \frac{\det(sI - A + bc) - \det(sI - A)}{\det(sI - A)}$$

See Also `ss2zp`, `tf2ss`, `tf2zp`, `zp2ss`, `zp2tf`

---

ss2zp

Purpose State-space to zero-pole-gain conversion.

Synopsis

```
[Z,p,k] = ss2zp(A,B,C,D,iu)
```

Description `ss2zp` converts a state-space representation of a given system to an equivalent zero-pole-gain representation. The zeros, poles, and gains of state-space systems represent the transfer function in factored form.

`[Z,p,k] = ss2tf(A,B,C,D,iu)` calculates the transfer function in factored form

$$H(s) = \frac{z(s)}{p(s)} = k\frac{(s-z(1))(s-z(2))\cdots(s-z(n))}{(s-p(1))(s-p(2))\cdots(s-p(n))}$$

of the system

$$\dot{x} = Ax + Bu$$
$$y = Cx + Du$$

from the `iu`th input. Returned column vector `p` contains the pole locations of the denominator coefficients of the transfer function. Matrix `Z` contains the numerator zeros in its columns, with as many columns as there are outputs y. Column vector `k` contains the gains for each numerator transfer function.

The function `zp2ss` is the inverse of `ss2zp`. `ss2zp` also works with systems in discrete-time, in which case it returns the z-transform representation. The input state-space system must be real.

Example Here are two ways of finding the zeros, poles, and gains of a system:

```
num = [2 3];
den = [1 0.4 1];
[z,p,k] = tf2zp(num,den)
z =
-1.5000
p =
 -0.2000 + 0.9798i
-0.2000 - 0.9798i
k =
2
[A,B,C,D] = tf2ss(num,den);
[z,p,k] = ss2zp(A,B,C,D,1)
z =
-1.5000
p =
-0.2000 + 0.9798i
-0.2000 - 0.9798i
k =
2
```

Algorithm `ss2zp` finds the poles from the eigenvalues of the `A` matrix. The zeros are the finite solutions to a generalized eigenvalue problem:

```
z = eig([A B;C D], diag([ones(1,n) 0]);
```

In many situations this algorithm produces spurious large, but finite zeros. `ss2zp` interprets these large zeros as infinite.

ss2zp finds the gains by solving for the first nonzero Markov parameters.

See Also ss2tf, tf2ss, zp2ss

References [1] Laub, A. J., and B.C. Moore. "Calculation of Transmission Zeros Using QZ Techniques." *Automatica* 14 (1978): p. 557.

---

ssselect

Purpose Select subsystem from larger state-space system.

Synopsis

```
[ae,be,ce,de] = ssselect(a,b,c,d,inputs,outputs)
[ae,be,ce,de] = ssselect(a,b,c,d,inputs,outputs,states)
```

Description Given a state-space system:

$$\dot{x} = Ax + \begin{bmatrix} B_1 & B_2 \end{bmatrix} \begin{bmatrix} u_1 \\ u_2 \end{bmatrix}$$

$$\begin{bmatrix} y_1 \\ y_2 \end{bmatrix} = \begin{bmatrix} C_1 \\ C_2 \end{bmatrix} x + \begin{bmatrix} D_{11} & D_{12} \\ D_{21} & D_{22} \end{bmatrix} \begin{bmatrix} u_1 \\ u_2 \end{bmatrix}$$

[ae,be,ce,de] = ssselect(a,b,c,d,inputs,outputs) produces the subsystem with the inputs and outputs specified. The vector inputs contains indices into the input vector of the system and specifies which inputs are retained in the selected subsystem. Similarly, the vector outputs contains indices into the output vector of the system and specifies which outputs are retained in the selected subsystem. The system is returned with the inputs and outputs ordered as in the vectors inputs and outputs. Because of this property, ssselect can be used to reorder the inputs, outputs and states of a model.

Suppose u_1 are the desired inputs and y_1 are the desired outputs of the selected system, then the state-space system produced is

$$\dot{x} = Ax + B_1 u_1$$
$$y_1 = C_1 x + D_{11} u_1$$

[ae,Be,Ce,De] = ssselect(a,b,c,d,inputs,outputs,

states) produces the subsystem with the specified inputs, outputs and states. The inputs, outputs, and states are reordered as in the vectors inputs, outputs, and states.

ssselect works with both continuous- and discrete-time models.

Examples Consider a state-space system (a,b,c,d) with five outputs and four inputs. To select the subsystem with inputs 1 and 2 and outputs 2, 3 and 4:

```
inputs = [1 2];
outputs = [2 3 4];
[ae,be,ce,de] = ssselect(a,b,c,d,inputs,outputs);
```

step

Purpose Unit step response.

Synopsis

```
[y,x,t] = step(a,b,c,d)
[y,x,t] = step(a,b,c,d,iu)
[y,x,t] = step(a,b,c,d,iu,t)
[y,x,t] = step(num,den)
[y,x,t] = step(num,den,t)
```

Description step calculates the unit step response of a linear system. Invoked without left-hand arguments, step plots the impulse response on the screen.

step(a,b,c,d) produces a series of step response plots, one for each input and output combination of the continuous LTI system:

$$\dot{x} = Ax + Bu$$
$$y = Cx + Du$$

with the time vector automatically determined.

step(a,b,c,d,iu) produces a step response plot from the single input iu to all the outputs of the system with the time vector automatically determined. The scalar iu is an index into the inputs of the system and specifies which input to be used for the impulse response.

step(num,den) produces the step response plot of the polynomial transfer function G(s) = num(s)/den(s) where num and den contain the polynomial coefficients in descending powers of s.

step(a,b,c,d,iu,t) or step(num,den,t) uses the user-supplied time vector t. The vector t specifies the times at which the step response is computed and must be regularly spaced.

Invoked with lefthand arguments:

```
[y,x,t] = step(a,b,c,d,iu)
[y,x,t] = step(a,b,c,d,iu,t)
[y,x,t] = step(num,den)
[y,x,t] = step(num,den,t)
```

returns the output and state responses of the system and the time vector `t`. No plot is drawn on the screen. The matrices `y` and `x` contain the output and state response of the system respectively evaluated at the time points `t`. `y` has as many columns as outputs and one row for each element in `t`. `x` as many columns as states and one row for each element in `t`.

Examples Plot the step response of the second-order state-space system:

$$\begin{bmatrix} \dot{x}_1 \\ \dot{x}_2 \end{bmatrix} = \begin{bmatrix} -0.5572 & -0.7814 \\ 0.7814 & 0 \end{bmatrix} \begin{bmatrix} x_1 \\ x_2 \end{bmatrix} + \begin{bmatrix} 1 \\ 0 \end{bmatrix} u$$

$$y = \begin{bmatrix} 1.9691 & 6.4493 \end{bmatrix} \begin{bmatrix} x_1 \\ x_2 \end{bmatrix} + \begin{bmatrix} 0 \end{bmatrix} u$$

```
a = [-0.5572   -0.7814;  0.7814    0];
b = [1;0];
c = [1.9691    6.4493];
d = [0];
step(a,b,c,d);
title('Step Response')
```

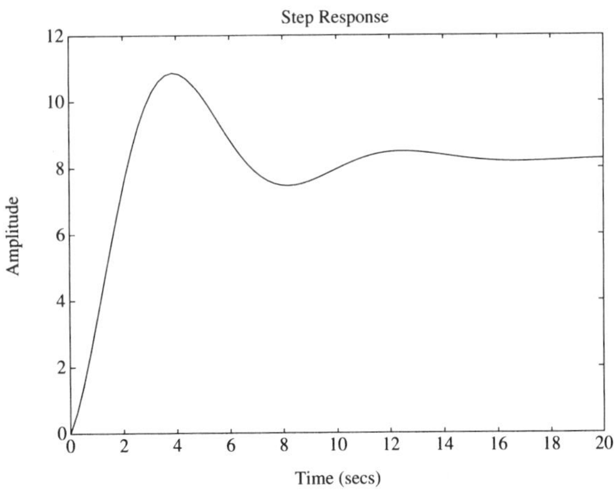

Algorithm `step` converts the continuous system to a discrete one using matrix exponentials and the interval between the time points in vector `t`. The discrete system is then propagated over the number of points in `t`. The time vector is assumed to start at `t` = 0.

See Also `lsim`, `impulse`, `dstep`

strips

Purpose Strip plot.

Synopsis

```
strips(x)
strips(x,n)
strips(x,sd,Fs)
```

Description `strips(x)` plots vector `x` in horizontal strips of length 250. If `x` is a matrix, `strips(x)` plots each column of `x`.

`strips(x,n)` plots vector `x` in strips that are each `n` samples long. If `x` is a matrix, `strips(x,n)`plots the different columns of `x` on the same strip plot.

`strips(x,sd,Fs)` plots vector `x` in strips of duration `sd` seconds given a sampling frequency of `Fs` samples per second. If `x` is a matrix, `strips(x,sd,Fs)`plots the different columns of `x` on the same strip plot.

`strips` ignores the imaginary part of `x` if it is complex.

Example Plot two seconds of a frequency modulated sinusoid in .25 second strips.

```
Fs = 1000; % sampling frequency
t = 0:1/Fs:2; % time vector
x = vco(sin(2*pi*t),[10 490],Fs); % FM waveform
strips(x,.25,Fs)
```

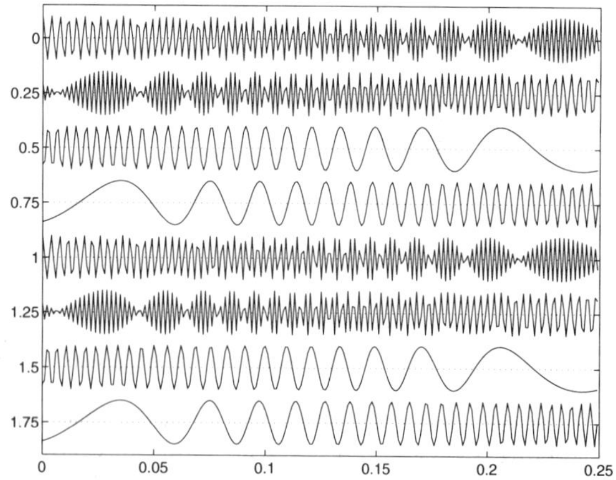

Diagnostics If `x` is a matrix and you have supplied more than one input, `strips` displays:

```
X must be a vector.
```

See Also `plot`, `stem`

---

tf2ss

Purpose Transfer function to state-space conversion.

Synopsis `[A,B,C,D] = tf2ss(num,den)`

Description `tf2ssp` converts a transfer function representation of a given system to an equivalent state-space representation.

`[A,B,C,D] = tf2ss(num,den)` finds a state-space representation

$$\dot{x} = Ax + Bu$$
$$y = Cx + Du$$

given a system in transfer function form

$$H(s) = \frac{num(s)}{den(s)} = C(sI - A)^{-1}B + D$$

from a single input. Input vector `den` contains the denominator coefficients in descending powers of s. Matrix `num` contains the numerator coefficients with as many rows as there are outputs y. `tf2ss` returns the `A`, `B`, `C`, and `D` matrices in controller canonical form.

`tf2ss` also works for discrete systems, but you must pad the numerator with trailing zeros to make it the same length as the denominator.

The function `ss2tf` is the inverse of `tf2ss`.

Example Consider the system

$$H(s) = \frac{\begin{bmatrix} 2s + 3 \\ s^2 + 2s + 1 \end{bmatrix}}{s^2 + 0.4s + 1}$$

To convert this system to state-space:

```
num = [0 2 3; 1 2 1];
den = [1 0 .4 1];
[A,B,C,D] = tf2ss(num,den)
A =
    -0.4000   -1.0000
     1.0000         0
B =
    1
    0
C =
    2.0000    3.0000
    1.6000         0
D =
    0
    1
```

There is disagreement in the literature on naming conventions for the canonical forms. It is easy, however, to generate similarity transformations that convert to other forms. For example:

```
T = fliplr(eye(n));
A = T\A*T;
```

Algorithm `tf2ss` writes the output in controller canonical form by inspection.

See Also `ss2tf`, `ss2zp`, `zp2ss`

tf2zp

Purpose Transfer function to zero-pole-gain conversion.

Synopsis `[z,p,k] = tf2zp(num,den)`

Description `tf2zp` finds the zeros, poles, and gains of systems in polynomial transfer function form.

`[z,p,k] = tf2zp(NUM,den)` finds the single-input, multi-output (SIMO) factored transfer function form,

$$H(s) = \frac{Z(s)}{p(s)} = k\frac{(s - Z(1))(s - Z(2))...(s - Z(m))}{(s - p(1))(s - p(2))...(s - p(n))}$$

given a SIMO system in polynomial transfer function form,

$$\frac{NUM(s)}{den(s)} = \frac{NUM(1)s^{nn-1} + \ldots + NUM(nn-1)s + NUM(nn)}{den(1)s^{nd-1} + \ldots + den(nd-1)s + den(nd)}$$

Vector `den` specifies the coefficients of the denominator in descending powers of s. Matrix `NUM` indicates the numerator coefficients with as many rows as there are outputs. The zero locations are returned in the columns of matrix `z`, with as many columns as there are rows in `NUM`. The pole locations are returned in column vector `p`, and the gains for each numerator transfer function in vector `k`.

`tf2zp` also works for discrete systems. The function `zp2tf` does the inverse of `tf2zp`.

Example Find the zeros, poles, and gains of the system

$$H(s) = \frac{2s + 3}{s^2 + 0.4s + 1}$$

```
num = [2 3];
den = [1 0.4 1];
[z,p,k] = tf2zp(num,den)
z =
   -1.5000
p =
   -0.2000 + 0.9798i
   -0.2000 - 0.9798i
k =  2
```

Algorithm The system is converted to state-space using `tf2ss` and then to zeros, poles, and gains using `ss2zp`.

See Also `ss2tf`, `ss2zp`, `tf2ss`, `zp2ss`, `zp2tf`

---

triang

Purpose Triangular window.

Synopsis `w = triang(n)`

Description `triang(n)` returns an `n`-point triangular window in a column vector. The coefficients of a triangular window are

For n *odd*

$$w[k] = \begin{cases} \dfrac{2k}{n+1}, & 1 \le k \le \dfrac{n+1}{2} \\[2ex] \dfrac{2(n-k+1)}{n+1}, & \dfrac{n+1}{2} \le k \le n \end{cases}$$

For n *even*

$$w[k] = \begin{cases} \dfrac{2k-1}{n}, & 1 \le k \le \dfrac{n}{2} \\[2ex] \dfrac{2(n-k+1)}{n}, & \dfrac{n}{2}+1 \le k \le n \end{cases}$$

The triangular window is very similar to a Bartlett window. The Bartlett window always ends with zeroes at samples 1 and n, whereas the triangular window is nonzero at those points. For n odd, the center n–2 points of `triang(n–2)` are equivalent to `bartlett(n)`.

See Also `boxcar`, `hamming`, `hanning`, `kaiser`

---

tzero

Purpose Transmission zeros.

Synopsis `z = tzero(a,b,c,d)`

Description `tzero` finds the invariant zeros of state-space systems. For minimal systems, the invariant zeros are the transmission zeros. Transmission zeros are a multivariable generalization of the classical SISO zeros of individual transfer functions. Physically, they relate to the situation when the output of a multivariable system is zero, while the states and inputs of the system are non-zero. A complex number λ_z is an invariant zero if the following equation is true for all time for some input vector u_0:

$$\begin{bmatrix} \dot{x}(t) \\ 0 \end{bmatrix} = \begin{bmatrix} A & B \\ C & D \end{bmatrix} \begin{bmatrix} x(t) \\ u_o e^{\lambda_z t} \end{bmatrix}$$

`tzero(a,b,c,d)` returns the transmission zeros of the state-space system:

$$\dot{x} = Ax + Bu$$
$$y = Cx + Du$$

`tzero` can also compute the input and output decoupling zeros of a state-space system. To compute the input decoupling zeros use

```
zid = tzero(a,b,[],[])
```

To compute the output decoupling zeros use

```
zod = tzero(a,[],c,[])
```

Algorithm The transmission zeros are computed using the algorithm in [1].

See Also `ss2zp`

References
[1] Emami-Naeini, A and Van Dooren, P., Computation of Zeros of Linear Multivariable Systems, *Automatica*, Vol 18, No.4, 1982, pp. 415-430.

yulewalk

Purpose Recursive digital filter design.

Synopsis `[b,a] = yulewalk(n,f,m)`

Description `yulewalk` designs recursive IIR digital filters using a least squares fit to a specified frequency response.

`[b,a] = yulewalk(n,f,m)` returns row vectors `b` and `a` containing the `n + 1` coefficients of the order `n` IIR filter whose frequency-magnitude characteristics approximately match those given in vectors `f` and `m`:

- `f` is a vector of frequency points, specified in the range between `0` and `1`, where `1` corresponds to half the sample frequency (the Nyquist frequency).The first point of `f` must be `0` and the last point `1`, with all intermediate points in increasing order. Duplicate frequency points are allowed, corresponding to steps in the frequency response.
- `m` is a vector containing the desired magnitude response at the points specified in `f`.
- `f` and `m` must be the same length.
- `plot(f,m)` displays the filter shape.

The output filter coefficients are ordered in descending powers of z,

$$\frac{B(z)}{A(z)} = \frac{b(1) + b(2)z^{-1} + \cdots + b(n+1)z^{-n}}{1 + a(2)z^{-1} + \cdots + a(n+1)z^{-n}}$$

When specifying the frequency response, avoid excessively sharp transitions from passband to stopband. You may need to experiment with the slope of the transition region to get the best filter design.

Example Design an 8th-order lowpass filter and overplot the desired frequency response with the actual frequency response:

```
f = [0 .6 .6 1];
m = [1 1 0 0];
[b,a] = yulewalk(8,f,m);
[h,w] = freqz(b,a,128);
plot(f,m,w/pi,abs(h),'– –')
```

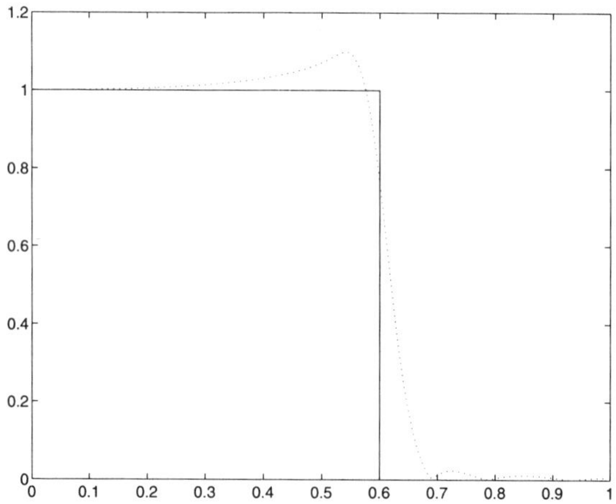

Algorithm `yulewalk` performs a least squares fit in the time domain. It computes the denominator coefficients using modified Yule-Walker equations, using correlation coefficients computed by inverse Fourier transformation of the specified frequency response. To compute the numerator:

1. It computes a numerator polynomial corresponding to an additive decomposition of the power frequency response.
2. It evaluates the complete frequency response corresponding to the numerator and denominator polynomials.
3. It uses a spectral factorization technique to obtain the impulse response of the filter.
4. It obtains the numerator polynomial by a least squares fit to this impulse response.

See Also `butter`, `cheby1`, `cheby2`, `ellip`, `fir2`, `remez`, `yulewalk`

References
[1] [1] Friedlander, B. and B. Porat. "The Modified Yule-Walker Method of ARMA Spectral Estimation." *IEEE Transactions on Aerospace Electronic Systems*, AES-20, No. 2, (March 1984): pp. 158-173.

zp2sos

Purpose Zero-pole-gain to second-order section conversion.

Synopsis

```
sos = zp2sos(z,p,k)
sos = zp2sos(z,p,k,'order')
```

Description zp2sos converts a zero-pole-gain representation of a given system to an equivalent second-order section representation.

sos = zp2sos(z,p,k) finds a matrix sos in second-order section form equivalent to the zero-pole-gain system represented by input arguments z, p and k. Vectors z and p contain the zeros and poles of the system *H(z)*, not necessarily in any order:

$$H(z) = k\frac{(z-z(1))(z-z(2))\cdots(z-z(N))}{(p-p(1))(p-p(2))\cdots(p-p(M))}$$

The zeros and poles must be real or come in complex conjugate pairs. sos is an *L*-by-6 matrix,

$$sos = \begin{bmatrix} b_{01} & b_{11} & b_{21} & a_{01} & a_{11} & a_{21} \\ b_{02} & b_{12} & b_{22} & a_{02} & a_{12} & a_{22} \\ \vdots & \vdots & \vdots & \vdots & \vdots & \vdots \\ b_{0L} & b_{1L} & b_{2L} & a_{0L} & a_{1L} & a_{2L} \end{bmatrix}$$

whose rows contain the numerator and denominator coefficients b_{ik} and a_{ik} of the second-order sections of *H(z)*:

$$H(z) = \prod_{k=1}^{L} H_k(z) = \prod_{k=1}^{L} \frac{b_{0k} + b_{1k}z^{-1} + b_{2k}z^{-2}}{a_{0k} + a_{1k}z^{-1} + a_{2k}z^{-2}}$$

The number of rows *L* of matrix sos is the maximum of the ceiling of *N*/2 and the ceiling of *M*/2, where *N* and *M* are the lengths of z and p respectively.

sos = zp2sos(z,p,k,'*order*') specifies the order of the rows in sos, where *order* is

- down to order the sections so the first row of sos contains the poles closest to the unit circle.
- up to order the sections so the first row of sos contains the poles farthest from the unit circle (default).

Example Find a second-order section form of a Butterworth lowpass filter:

```
[z,p,k] = butter(5,.2);
sos = zp2sos(z,p,k);
```

Algorithm zp2sos uses a four-step algorithm to determine the second-order section representation for an input zero-pole-gain system:

1. It groups the zeros and poles into complex conjugate pairs using the cplxpair function.
2. It forms the second-order section by matching the pole and zero pairs according to the following rules:
   a. Match the poles closest to the unit circle with the zeros closest to those poles.
   b. Match the poles next closest to the unit circle with the zeros closest to those poles.
   c. Continue until all of the poles and zeros are matched.

   zp2sos groups real poles into sections with the real poles closest to them in absolute value. The same rule holds for real zeros.
3. It orders the sections according to the proximity of the pole pairs to the unit circle. zp2sos normally orders the sections with poles closest to the unit circle last in the cascade. You can tell zp2sos to order the sections in the reverse order by specifying the down flag.

   Putting "high Q" sections at the beginning of the cascade (the down flag operation) reduces the filter response's sensitivity to quantization noise near those poles. On the other hand, putting "high Q" sections at the end of the cascade (the default operation) prevents reduction in signal power level early in the cascade. [1] and [2] provide detailed discussions of section ordering. zp2sos orders all zero sections according to the minimum of $|z_i|$ and $|z_i^{-1}|$ where z_i (for $i = 1, 2$) are the zeros in the section.
4. zp2sos scales the sections so the maximum of the magnitude of the transfer function of the first N sections in cascade is less than 1, that is subject to the

$$\max_{|\omega| \le \pi} \left| \prod_{i=1}^{N} H_i(e^{j\omega}) \right| < 1, \quad N = 1, \ldots, L-1$$

   constraint that the overall gain, k, stays the same. This scaling is an attempt

$$\prod_{k=1}^{L} \frac{b_{0k}}{a_{0k}} = k$$

   to minimize overflow in some standard fixed point implementations of filtering.

See Also sos2ss, sos2tf, sos2zp, ss2sos

References [1] [1] Oppenheim, A. V. and R. W. Schafer, *Discrete-Time Signal Processing*, pp. 363–370. Englewood Cliffs, NJ: Prentice-Hall, 1989.

[2] [2] Jackson, Leland B. *Digital Filters and Signal Processing*, pp. 319–324. Boston: Kluwer Academic Publishers, 1989.

---

zp2ss

Purpose Zero-pole-gain to state-space conversion.

Synopsis `[a,b,c,d] = zp2ss(z,p,k)`

Description `zp2ss` forms state-space models from the zeros, poles, and gains of systems in transfer function form.

`[a,b,c,d] = zp2ss(z,p,k)` finds a SIMO state-space representation

$$\dot{x} = Ax + Bu$$
$$y = Cx + Du$$

given a system in factored transfer function form:

$$H(s) = \frac{Z(s)}{p(s)} = k\frac{(s - Z(1))(s - Z(2))...(s - Z(m))}{(s - p(1))(s - p(2))...(s - p(n))}$$

Column vector `p` specifies the pole locations and matrix `z` the zero locations with as many columns as there are outputs. The gains for each numerator transfer function are in vector `k`. The `A,B,C,D` matrices are returned in controller canonical form.

`Inf`'s can be used as place holders in `z` if some columns have fewer zeros. See the *Tutorial* section.

Algorithm `zp2ss`, for single-output systems, groups complex pairs together into two-by-two blocks down the diagonal of the `A` matrix. For multi-output systems, a less reliable algorithm is used. The system is converted to transfer function form using `zp2tf` and then to state-space using `tf2ss`.

See Also ss2tf, ss2zp, tf2ss, tf2zp, zp2tf

---

zp2tf

Purpose Zero-pole-gain to transfer function conversion.

Synopsis `[num,den] = zp2tf(z,p,k)`

Description `zp2tf` forms transfer function polynomials from the zeros, poles, and gains of systems in factored form.

`[num,den] = zp2tf(z,p,k)` finds a rational transfer function:

$$\frac{NUM(s)}{den(s)} = \frac{NUM(1)s^{nn-1} + \ldots + NUM(nn-1)s + NUM(nn)}{den(1)s^{nd-1} + \ldots + den(nd-1)s + den(nd)}$$

given a system in factored transfer function form

$$H(s) = \frac{Z(s)}{p(s)} = k\frac{(s-Z(1))(s-Z(2))\ldots(s-Z(m))}{(s-p(1))(s-p(2))\ldots(s-p(n))}$$

Column vector `p` specifies the pole locations and matrix `z` the zero locations with as many columns as there are outputs. The gains for each numerator transfer function are in vector `k`. The polynomial coefficients are returned in vectors, the denominator coefficients in row vector `den`, and the numerator coefficients in matrix `num` with as many rows as there are columns of `z`.

`Infs` can be used as place holders in `z` if some columns have fewer zeros.

Algorithm The system is converted to transfer function form using `poly` with `p` and the columns of `z`.

See Also `ss2tf`, `ss2zp`, `tf2ss`, `tf2zp`, `zp2ss`

---

zplane

Purpose Zero-pole plot.

Synopsis

```
zplane(z,p)
zplane(b,a)
[hz,hp,ht] = zplane(z,p)
```

Description This function displays the poles and zeros of discrete-time systems.

`zplane(z,p)` plots the zeros in column vector `z` and the poles in column vector `p` in the current figure window. The symbol `'o'` represents a zero and the symbol `'x'` represents a pole. The plot includes the unit circle for reference. If `z` and `p` are matrices, `zplane` plots the poles and zeros in the columns of `z` and `p` respectively in different colors.

You can override the automatic scaling of `zplane` using

`axis([xmin xmax ymin ymax])` or
`set(gca,'ylim',[ymin ymax])` or
`set(gca,'xlim',[xmin xmax])`

after calling zplane. This is useful in the case where one or a few of the zeros or poles have such a large magnitude that the others are grouped around the origin and are thus hard to distinguish.

zplane(b,a), where b and a are row vectors, first uses roots to find the zeros and poles of the transfer function represented by numerator coefficients b and denominator coefficients a.

[hz,hp,ht] = zplane(z,p) returns vectors of handles to the zero lines, hz, and the pole lines, hp. ht is a vector of handles to the axes /unit circle line and to text objects, which are present when there are multiple zeros or poles. If there are no zeros or no poles, hz or hp is set to the empty matrix [].

Examples Plot the poles and zeros of a 5th-order Butterworth lowpass digital filter with cutoff frequency of 0.2:

```
[z,p,k] = butter(5,.2);
zplane(z,p)
```

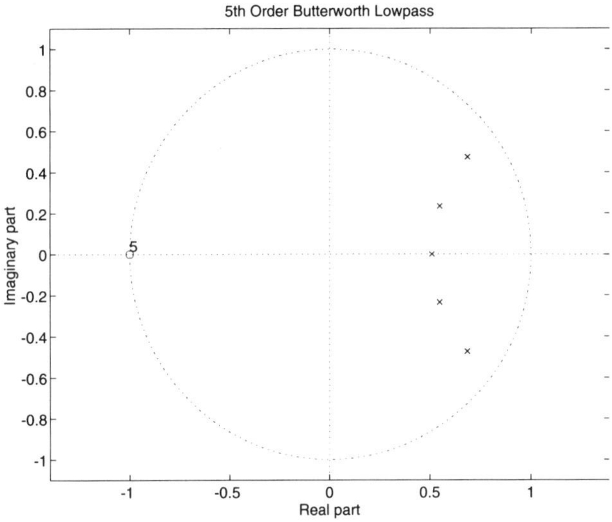

To generate the same plot with a transfer function representation of the filter:

```
[b,a] = butter(5,.2);    % transfer function
zplane(b,a);
```

Index

Symbols

Numerics

A

B

C

D

E

G

H

I

J

K

L

N

O

P